Green Chemistry and Sustainable Technology

Aims and Scope

The series *Green Chemistry and Sustainable Technology* aims to present cutting-edge research and important advances in green chemistry, green chemical engineering and sustainable industrial technology. The scope of coverage includes (but is not limited to):

- Environmentally benign chemical synthesis and processes (green catalysis, green solvents and reagents, atom-economy synthetic methods etc.)
- Green chemicals and energy produced from renewable resources (biomass, carbon dioxide etc.)
- Novel materials and technologies for energy production and storage (bio-fuels and bioenergies, hydrogen, fuel cells, solar cells, lithium-ion batteries etc.)
- Green chemical engineering processes (process integration, materials diversity, energy saving, waste minimization, efficient separation processes etc.)
- Green technologies for environmental sustainability (carbon dioxide capture, waste and harmful chemicals treatment, pollution prevention, environmental redemption etc.)

The series *Green Chemistry and Sustainable Technology* is intended to provide an accessible reference resource for postgraduate students, academic researchers and industrial professionals who are interested in green chemistry and technologies for sustainable development.

More information about this series at http://www.springer.com/series/11661

Xuefeng Zhu • Weishen Yang

Mixed Conducting Ceramic Membranes

Fundamentals, Materials and Applications

Springer

Xuefeng Zhu
Dalian Institute of Chemical Physics
Dalian, China

Weishen Yang
Dalian Institute of Chemical Physics
Dalian, China

ISSN 2196-6982 ISSN 2196-6990 (electronic)
Green Chemistry and Sustainable Technology
ISBN 978-3-662-53532-5 ISBN 978-3-662-53534-9 (eBook)
DOI 10.1007/978-3-662-53534-9

Library of Congress Control Number: 2016953191

Printed on acid-free paper

This Springer imprint is published by Springer Nature
The registered company is Springer-Verlag GmbH Germany
The registered company address is: Heidelberger Platz 3, 14197 Berlin, Germany

Preface

Air separation for pure oxygen production is an important process for many industries, such as steel production, glassmaking, ammonia synthesis, coal gasification, natural gas conversion, environmental protection, etc. The mixed ionic–electronic conducting (MIEC) membrane is deemed as the next-generation technology for oxygen separation since it was proposed in the 1980s, because it exhibits high energy efficiency and low cost for pure oxygen production as well as facile integration with other processes with improved efficiency. Much effort has been made by researchers from both academia and industry in the past 30 years. In the academic community, the research was focused on the development of membrane materials, preparation techniques of ceramic powders and membranes, basic principles related to oxygen permeation, and catalytic reactions in MIEC membrane reactors. The research conducted by companies, for instance, Air Products & Chemicals Inc., was focused on the fabrication of MIEC membrane modules in large scale, operation of membrane modules, and design of the membrane process.

Our group started the MIEC program in the 1990s and published more than 100 papers on the MIEC membranes. Our research was focused on the development of MIEC membranes, such as $Ba_{0.5}Sr_{0.5}Co_{0.8}Fe_{0.2}O_{3-\delta}$, dual-phase membranes, catalytic reactions in membrane reactors, and development of permeation model. Recently, we extended the operation temperature of MIEC membranes from the typical high temperatures (700–1000 °C) to low temperatures (350–600 °C) by solving the problems related to permeation stability.

In the past 30 years, thousands of research papers and dozens of reviews and book chapters have been published, but there is no work giving a comprehensive introduction to the knowledge related to MIEC membranes. Most of the reviews and book chapters are focused on the development of MIEC materials and catalytic reactions in membrane reactors. Therefore, we came to the idea to write a book to describe the critical aspects of MIEC membrane to readers, including the theoretical foundation, experimental methods, membrane materials, potential applications, and development of the MIEC technology. Another purpose of this book is to summarize the progress of MIEC membranes made in the past 30 years. This book contains ten chapters. The background is introduced in Chap. 1, and then the basic principles, ionic conductors, experimental methods, model development, membrane materials, and reactions in MIEC membrane reactors are introduced in Chaps. 2, 3, 4, 5, 6, 7, 8, and 9, and finally, the commercialization of the MIEC membranes is introduced in Chap. 10. This book cannot cover all the aspects of MIEC membrane technology. Comments and queries from readers will be gratefully received and can be sent to zhuxf@dicp.ac.cn or yangws@dicp.ac.cn.

Dalian, China

Xuefeng Zhu
Weishen Yang

Contents

Chapter 1
Introduction to Mixed Ionic–Electronic Conducting Membranes

Abstract Mixed ionic–electronic conducting (MIEC) membranes have attracted considerable attention from the industrial and academic communities over the past 30 years. This is deemed as a next-generation technology for pure oxygen production and powerful catalytic membrane reactors with high process intensification and large energy savings. This chapter briefly presents the history of MIEC membrane and provides an introduction to membrane reactors, the permeation principle, and the types of MIEC membranes, as well as outlines the scope of this book.

Keywords Mixed ionic–electronic conducting membranes • Dense ceramic membranes • Oxygen permeation • Perovskite • Membrane reactors • Bulk diffusion • Interfacial oxygen exchange

1.1 Introduction

Oxygen is an important industrial gas that is widely used in chemical industries, ferrous metallurgy, glassmaking, clean power generation, environmental protection, etc. At present, cryogenic distillation is the only commercialization technology for the large-scale production of pure oxygen. Over the past 100 years, the cryogenic technology has matured gradually, and there is only marginal room to reduce the energy consumption in the production of pure oxygen. However, the current production cost of pure oxygen is still too high to be accepted by many industrial processes, such as the clean power generation process, i.e., the oxyfuel process with CO_2 capture, in which air is replaced by pure oxygen for the combustion of fossil fuels. Another matured technology for oxygen separation from air is pressure swing adsorption (PSA) using zeolites as the absorbents. However, the purity of oxygen produced via the PSA process is lower than 95 %, and it is economical only in mid- or small-scale oxygen separation. Therefore, the PSA technology is unsuitable for the abovementioned large-scale industrial processes.

Mixed ionic–electronic conducting (MIEC) membranes are a type of dense ceramic membranes, in which oxygen ions can diffuse from one side to the other side via the lattice vacancies or interstitial sites under the driving force of the oxygen chemical potential gradient across the membranes. The permeation selectivity of such membranes is 100 % to oxygen, according to the permeation

X. Zhu, W. Yang, *Mixed Conducting Ceramic Membranes*, Green Chemistry and Sustainable Technology, DOI 10.1007/978-3-662-53534-9_1

mechanism. These membranes are operated at elevated temperatures (typically in the range of 700–1000 °C) with elevated air pressures (1–2 MPa) fed to the retention side and vacuum in the permeation side, according to the Air Products & Chemicals Inc.'s reports, which have resulted in great progress in the commercialization of the MIEC membrane technology for pure oxygen productions. An industrial demonstration factory for oxygen separation based on the MIEC membrane technology with a production capability of 100 metric tons oxygen per day is under construction. Compared with the cryogenic distillation technology, the MIEC membrane technology for pure oxygen production has low energy consumption and less capital investment and is easily integrated with the abovementioned large-scale industrial processes.

Although the solid-state oxygen ionic conductors were invented by Nernst in 1899 [1], the concept for oxygen permeation using a mixed ionic–electronic conductor was inadvertently introduced by Cales and Baumard in 1982 [2]. They used calcia-stabilized zirconia tubes to determine the oxygen permeation flux at 1800 °C with steam fed to the retention side and equimolar CO–CO_2 fed to the permeation side. However, their target was to determine the partial electronic conductivity of the solid electrolyte at elevated temperatures. The concept of separating pure oxygen from air using an MIEC membrane was explicitly proposed by Teraoka and his coworkers in 1985 [3]. They reported an investigation on the oxygen permeation through perovskite-type oxides, i.e., $La_{1-x}Sr_xCo_{1-y}Fe_yO_{3-\delta}$ ($0 \le x \le 1$, $0 \le y \le 1$), in the temperature range of 450–877 °C, and achieved high oxygen permeation flux up to 3.1 mL cm^{-2} min^{-1} through a 1-mm-thick $SrCo_{0.8}Fe_{0.2}O_{3-\delta}$ membrane at 850 °C. Subsequently, perovskite-type oxides (ABO_3) attracted considerable attention, and the field of MIEC membranes was gradually formed. In the initial stage, researchers focused on the $La_{1-x}Sr_xCo_{1-y}Fe_yO_{3-\delta}$ system, in particular $SrCo_{0.8}Fe_{0.2}O_{3-\delta}$ because of its high oxygen permeability among the materials in this series. However, subsequent research showed that the cubic $SrCo_{0.8}Fe_{0.2}O_{3-\delta}$ phase transfers to an oxygen vacancy-ordered brownmillerite phase as the oxygen partial pressure decreases below 10 kPa and temperature decreases below 790 °C [4]. The use of La to partially replace Sr and the reduction of Co content are effective methods for inhibiting the phase transformation, but they result in a significant reduction in oxygen permeability. Yang's group used Ba to partially replace Sr and prepared the optimal membrane material in this series, $Ba_{0.5}Sr_{0.5}Co_{0.8}Fe_{0.2}O_{3-\delta}$, which shows high oxygen permeability and good stability at temperatures higher than 850 °C [5]. The development of new materials has not stopped over the past 30 years, and the target has gradually become clear, i.e., the stability of membrane materials is more important than its permeability for the commercialization of the MIEC membrane technology. Therefore, to improve the stability, many kinds of membrane materials have been developed by improving the cobalt-based perovskite materials through doping of cations with stable valence state in the B site and by preparing cobalt-free perovskite materials and dual-phase materials.

Besides air separation for pure oxygen production, another important application of MIEC membranes is as membrane reactors in which catalytic reactions are integrated with the oxygen separation process [6, 7]. The typical catalytic reaction in MIEC membranes reactors is the partial oxidation of natural gas to produce syngas ($H_2 + CO$). In the membrane reactor, air is fed to the retention side and natural gas is fed to the permeation side. Ni-based catalysts are usually filled on the permeation side. The permeated oxygen from the retention side reacts with natural gas to produce syngas on the permeation side, and thus, this membrane reactor couples the oxygen separation and catalytic reaction together with enhanced energy efficiency. The catalytic reaction consumes the permeated oxygen, thus increasing the oxygen chemical potential gradient across the membrane, and the oxygen flux is thereby enhanced by about ten times compared with the flux for pure oxygen production. Oxidative dehydrogenation of light alkanes to olefins and oxidation coupling of methane to ethane and ethylene were also considered in MIEC membrane reactors [8, 9]. The main advantage of the catalytic membrane reactors is that the selectivity of these reactions can be improved compared with the traditional fixed-bed reactors, because lattice oxygen and not gaseous oxygen is the active oxygen species for the oxidation reactions. In the abovementioned MIEC membrane reactors, oxygen is selectively introduced into the reaction system; of course, oxygen can be selectively removed from the reaction system as long as there is an oxygen chemical potential difference across the membrane. For example, as steam mixed with a proper ratio of air is fed to the retention side and natural gas is fed to the permeation side, partial oxidation reaction to syngas takes place on the permeation side, and the water-splitting reaction to hydrogen takes place on the retention side, accompanied by the oxygen permeation from the retention side to the permeation side [10]. For the retention side, the oxygen in water and air is moved to the permeation side and reacts with natural gas to produce syngas. After cooling and drying, ammonia syngas ($H_2 + N_2$) and liquid fuel syngas ($H_2 + CO$) are produced, respectively, from the retention side and the permeation side. The energy consumption of this MIEC membrane reactor process can be saved by more than 60 % compared to the traditional industrial processes. Therefore, the MIEC membrane reactors will attract considerable attention from the industrial and academic communities.

1.2 Principle of Oxygen Permeation

Oxygen permeates through MIEC membranes via three main steps, i.e., oxygen exchange at the gas–solid interface of retention side, oxygen–electron (or hole) ambipolar diffusion through the membrane bulk, and oxygen exchange at the gas–solid interface of permeation side, as shown in Fig. 1.1. The reactions related to oxygen exchange at the gas–solid interfaces are also called surface exchange reactions. The surface exchange reactions are complicated, because there are four electrons related to the reactions. If electrons and oxygen ions are considered as the

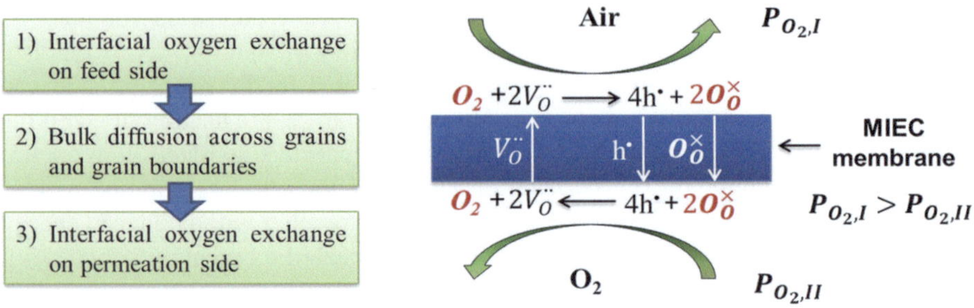

Fig. 1.1 Mechanism of oxygen permeation through an MIEC membrane

charge carriers and the diffusion path of oxygen ions is not addressed, the total reactions can be simply written as

$$\text{For the retention side}: \quad O_2 + 4e^- \leftrightarrow 2O^{2-} \tag{1.1}$$

$$\text{For the permeation side}: \quad 2O^{2-} \leftrightarrow O_2 + 4e^- \tag{1.2}$$

If holes and oxygen vacancies are considered as the charge carriers, the surface exchange reactions can be written by using the Kröger–Vink notations as follows:

$$\text{For the retention side}: \quad O_2 + 2V_O^{\cdot\cdot} \leftrightarrow 2O_O^x + 4h^{\cdot} \tag{1.3}$$

$$\text{For the permeation side}: \quad 2O_O^{\times} + 4h^{\cdot} \leftrightarrow O_2 + 2V_O^{\cdot\cdot} \tag{1.4}$$

where the Kröger–Vink notations, $O_O^{\times}$, $V_O^{\cdot\cdot}$, and $h^{\cdot}$, denote the oxygen ion occupying the oxygen lattice site, the oxygen vacancy, and the hole, respectively. Although the overall reactions look very simple, they contain many steps. For the oxygen exchange reaction on the retention side, which has been extensively investigated in the field of solid oxide fuel cells, steps like adsorption, catching electrons, dissociation, catching another electron, and incorporation into the crystal lattice would occur. The detailed mechanism is still not clear, and it may change depending on the material. The following is a widely accepted mechanism for the surface oxygen exchange on the retention side though it is still not well verified by experiments.

$$\text{Adsorption}: \quad O_{2(g)} \rightarrow O_{2(ad)} \tag{1.5a}$$

$$\text{Combining with electron}: \quad O_{2(ad)} + e^- \rightarrow O^-_{2(ad)} \tag{1.5b}$$

$$\textit{Changing}\text{ the adsorption state}: \quad O^-_{2(ad)} \rightarrow O^-_{2(ad,bi)} \tag{1.5c}$$

$$\text{Combining with electron}: \quad O^-_{2(ad,bi)} + e^- \rightarrow O^{2-}_{2(ad,bi)} \tag{1.5d}$$

$$\text{Dissociation}: \quad O^-_{2(ad,bi)} \rightarrow 2O^-_{(ad)} \tag{1.5e}$$

$$\text{Combining with electron}: \quad O^{-}_{(ad)} + e^{-} \rightarrow O^{2-}_{(ad)} \tag{1.5f}$$

$$\text{Incorporating into lattice}: \quad O^{2-}_{(ad)} + V_{O}^{\cdot\cdot} \leftrightarrow O_{O}^{\times} \tag{1.5g}$$

Every step shown above can become the rate-determining step for the surface exchange reaction on the retention side. The surface exchange reaction on the permeation side is also not well understood because there is no wide application related to the reaction to drive the investigation. The elementary steps in the surface exchange reaction on the retention side are not simple as the reverse reactions numbered from Eqs. 1.5a to 1.5g.

There are two diffusion paths for oxygen ions through the membrane bulk: one is via the oxygen vacancies and the other is via the oxygen interstitial sites. For most ABO_3 perovskite oxides and fluorite oxides, the transport of oxygen ions follows the vacancy diffusion mechanism, while for the K_2NiF_4-type oxides, the transport of oxygen ions follows the interstitial diffusion mechanism or a mixed interstitial and vacancy diffusion mechanism. In the oxygen permeation process, oxygen ions transport from the retention side to the permeation side, while the electrons transfer in an inverse direction to keep the electric neutrality throughout the MIEC membranes. Unlike an oxygen pump, which is based on pure oxygen ionic conductors, such as yttria-stabilized zirconia (YSZ), and needs an outside electric circle, the MIEC membranes can transfer electrons internally, and thus, the complexity of the setup is significantly reduced and facile to the large-scale applications.

1.3 Types of Membranes

To understand the MIEC membranes from different angles, five types of classifications are frequently mentioned in the literature. Figure 1.2 shows the classification of MIEC membranes based on the crystal structure, phase composition, chemical composition, geometrical shape, and configuration of the dense layer.

Most MIEC membranes have a crystal structure of perovskite (ABO_3), in which A is the larger cation and B is the smaller cation. Perovskite is a framework structure constructed from BO_6 octahedra with A ions located in 12-coordinated interstices. Some MIEC membranes have a crystal structure similar to that of perovskite, such as the Ruddlesden–Popper (R–P) series with a formula of $A_{n+1}B_nO_{3n+1}$ ($n = 1, 2, 3......\infty$). The crystal structure of the R–P phase is similar to that of perovskite, in which a number of perovskite blocks (n is the number of perovskite blocks) having corner-shared BO_6 octahedra alter with AO rock salt layers along the c-axis. K_2NiF_4-type oxides and perovskite oxides are two extremes for this series of materials ($n = 1$ for K_2NiF_4 type and $n = \infty$ for perovskite). Some MIEC membranes have a fluorite structure, and the typical example is CeO_2-based materials. Thus, according to the crystal structure, MIEC membranes are classified as perovskite type, perovskite-related type, and fluorite type.

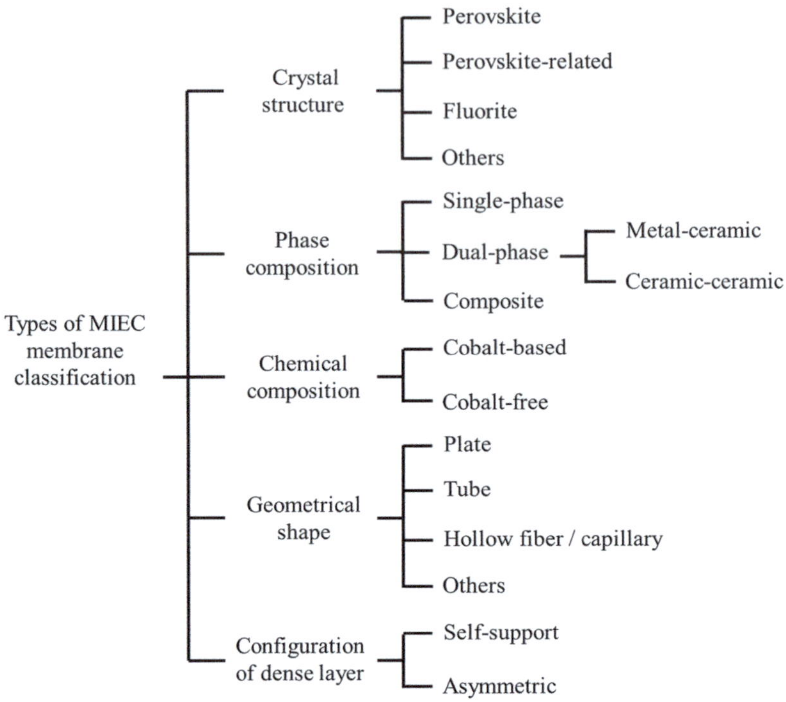

Fig. 1.2 Classification of MIEC membranes based on the crystal structure, phase composition, chemical composition, geometrical shape, and configuration of the dense layer

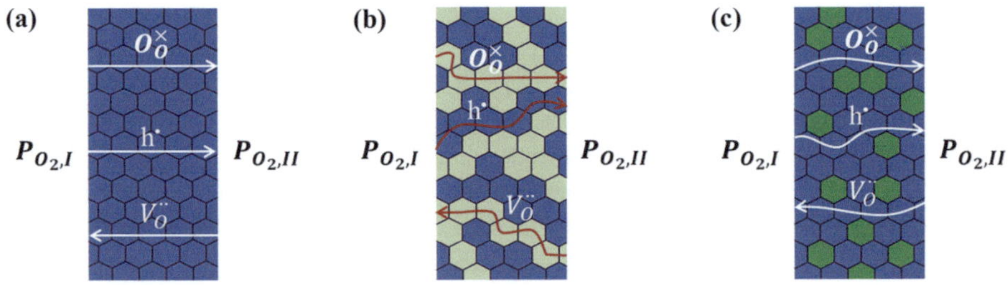

Fig. 1.3 Types of membranes classified according to phase composition

If a membrane has only one crystalline phase, it is called a single-phase membrane. Most of perovskite membranes are single-phase membrane, such as $La_{1-x}Sr_xCo_{1-y}Fe_yO_{3-\delta}$ ($0 \le x \le 1$, $0 \le y \le 1$). If a membrane has two phases and both contribute to oxygen permeation, it is called a dual-phase membrane. The first dual-phase membrane is YSZ–Pd, which contains a fluorite-type oxide phase, YSZ, for the transport of oxygen ions and a metal phase (face-centered cubic structure), Pd, for the transfer of electrons. Figure 1.3 shows the types of membranes classified according to phase composition. If a membrane has two or more phases and only one phase contributes to oxygen permeation, it is called a composite membrane. The inert phase is added in the membrane to improve the mechanical strength or other properties of the materials. For example, the $SrCo_{0.8}Fe_{0.2}O_{3-\delta}$–$SrSnO_3$ composite membrane comprises two perovskite phases, in which $SrSnO_3$ is inert to oxygen permeation, but it can improve the mechanical strength of the membrane [11].

In the early stage of the development of perovskite membranes, research was focused on perovskite with a cobalt element in the B site because cobalt-based perovskite membranes exhibit high oxygen permeability, such as $Ba_{0.5}Sr_{0.5}Co_{0.8}Fe_{0.2}O_{3-\delta}$ [5]. However, cobalt cations are easily reduced to a lower valence state owing to the weak Co–O bond, which results in the membranes being unstable in strong reducing environments. Therefore, cobalt-free perovskite membranes were proposed. For example, $BaCe_{0.05}Fe_{0.95}O_{3-\delta}$ shows not only high oxygen permeability even comparable to most of the cobalt-based perovskite membranes but also high structural stability even in H_2 atmosphere at elevated temperatures [12].

Similar to other types of membranes, the most common geometric shapes of MIEC membranes are plate, tubular, and hollow fibers. Figure 1.4 shows typical images of the plate, tubular, and hollow fiber MIEC membranes. As shown in Fig. 1.5, if a membrane has a thin dense layer for the selective permeation of

Fig. 1.4 Typical images of the plate, tubular, and hollow fiber MIEC membranes

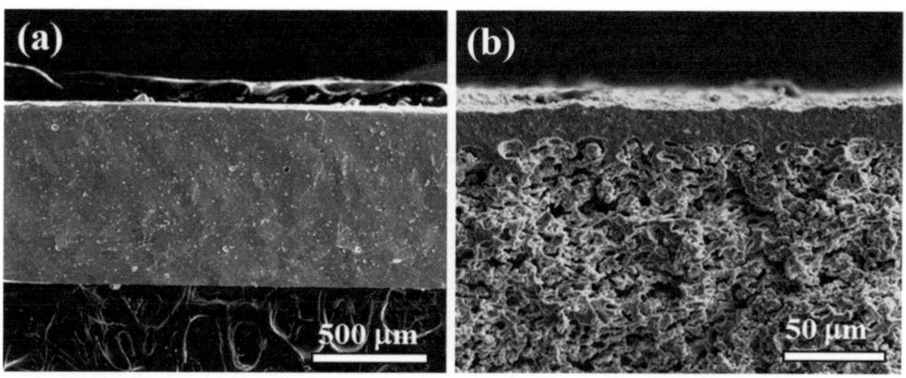

Fig. 1.5 SEM cross-views of self-support membrane (**a**) and asymmetric membrane (**b**)

oxygen and a thick porous layer to support the dense layer, the membrane is frequently called an asymmetric membrane; while if the dense layer has enough thickness to support the integrity of the membrane, it is called a self-support membrane.

1.4 Scope of This Book

Although the earliest research on MIEC membranes for oxygen permeation can be traced back to 30 years ago, no book has yet given a systematic description of the MIEC membranes. In this book, we present the scientific and technological progresses as well as the basic knowledge related to MIEC membranes. Solid defect chemistry and solid electrochemistry are part of the basic knowledge required in the field. The contents, such as point defects, defect equilibrium, grain boundary, and oxygen ionic diffusion of oxygen ionic conductors, are introduced in Chap. 2. This knowledge is important to help readers to understand the performance of materials and design new materials. For MIEC membrane materials, the ambipolar diffusion of oxygen ions and electrons through the membrane bulk is the critical step in the process of oxygen permeation, while the ionic conductivity of MIEC materials is several orders lower than their electronic conductivity for most perovskite membranes, and thus, the improvement of oxygen ionic conductivity is the key to enhancing the permeability of MIEC membrane. In Chap. 3, the oxygen ionic conductors, such as fluorite type, perovskite type, and other types, are introduced, after they are properly doped by elements with variable valence states or mixed with a secondary phase with electronic conduction; they can be made into MIEC membranes for oxygen permeation. In addition, the factors influencing the ionic conductivity of perovskite-type ionic conductors or mixed conductors are introduced in this chapter, and thus, researchers can design their MIEC materials according to their application targets. MIEC membranes are all operated at high temperatures, and thus, the special properties related to high temperature are briefly introduced.

Except for the inherent properties, the preparation technology of a membrane, as shown in Chap. 4, has a large influence on its permeation performance. The preparation technology includes the preparation of a ceramic powder, green body shaping, and sintering to a dense form. All these steps are critical to successfully fabricate a membrane, especially for the larger-scale preparation of membranes for the commercialization of the MIEC membrane technology. A brief introduction to the characterization technologies and testing methods of MIEC membranes is also presented in this chapter to help the beginners to know how a MIEC membrane is investigated in laboratories. Although the oxygen permeation process can be simply depicted in Fig. 1.1, the mechanism related to oxygen exchange on the gas–solid interfaces is not clear so far, and the transport of oxygen ions across or along the grain boundaries is difficult to be determined in situ. Thus, the roles of membrane surfaces and grain boundary are still under investigation. Several permeation models were proposed to describe the oxygen permeation behavior of MIEC membranes in the past 30 years, and these are introduced in Chap. 5. Although all the equations were derived based on many assumptions, these permeation models can help readers to understand the mechanism of oxygen permeation through MIEC membranes.

As mentioned in the "Introduction" section, most MIEC membranes reported in the past 30 years have the perovskite structure (ABO_3), in which the cations in the A and B sites can be substituted by many cations with the valence state from +1 to +3 for the A site and +2 to +6 for the B site. Therefore, the properties of perovskite membranes can be adjusted over a wide range, and the substitution effects are introduced in Chap. 6. In Chap. 7, the dual-phase type membranes are addressed owing to the rapid development of this type of membranes in recent years. Oxygen permeability of dual-phase membranes is greatly improved as mixed conducting perovskite oxides are used to replace the traditional pure electronic conducting compounds (noble metals and metallic oxides). The design principle of high-performance dual-phase membranes is introduced, and the factors having great influence on the permeability of dual-phase membranes are discussed in this chapter. One of the important applications of MIEC membranes is to fabricate them into membrane reactors for the process intensification, i.e., coupling separation and reactions into one membrane reactor. The most widely investigated reaction is partial oxidation of methane (POM) to syngas ($H_2 + CO$), while coupling the reaction of water splitting and POM in one membrane reactor, the use of MIEC oxygen permeable membranes to separate hydrogen has attracted increasing attention recently. The requirement of membrane materials for the membrane reactor depends on the reactions, and the stability under reducing environment is the primary problem that needs to be considered during the selection of membrane materials to construct membrane reactors. All the above points are discussed in Chap. 8 in detail.

Operating MIEC membranes at low temperatures (400–600 °C) can significantly decrease the energy consumption for oxygen separation, while few membranes exhibit stable permeation flux at that temperature range. Recently, Yang's group made good progress on the degradation mechanism and stabilization of MIEC membrane at low temperatures, and thus, this progress is introduced in Chap. 9. The US Department of Energy and Air Products & Chemicals Inc. have made great efforts to develop MIEC membrane technology for oxygen separation, and they recently achieved remarkable progress in commercialization. In addition, combining MIEC membrane with oxyfuel process is deemed as a promising route to capture CO_2 with less energy efficiency loss, and the simulation results of many researchers have revealed that this route is feasible and has a bright future. Therefore, in Chap. 10 we introduce some techniques developed by Air Products & Chemicals Inc. and its collaborators for the commercialization of MIEC membranes, as well as other institutes' efforts in the applications of the membrane technology.

Although all the above contents cannot cover the whole area of MIEC membranes, and some points in this book are not well addressed, we wish this book can help the researchers who are working in the field of MIEC membranes.

References

1. Nernst W (1899) On the electrolytic conduction of solid bodies at high temperatures. Z Electrochem 6:41–43
2. Cales B, Baumard JF (1982) Oxygen semipermeability and electronic conductivity in calcia-stabilized zirconia. J Mater Sci 17:3243–3248
3. Teraoka Y, Zhang HM, Furukawa S, Yamazoe N (1985) Oxygen permeation through perovskite-type oxides. Chem Lett 11:1743–1746
4. Kruidhof H, Bouwmeester HJM, van Doorn RHE, Burggraaf AJ (1993) Influence of order–disorder transitions on oxygen permeability through selected nonstoichiometric perovskite-type oxides. Solid State Ionics 63–65:816–822
5. Shao ZP, Yang WS, Cong Y, Dong H, Tong JH, Xiong GX (2000) Investigation of the permeation behavior and stability of a $Ba_{0.5}Sr_{0.5}Co_{0.8}Fe_{0.2}O_{3-\delta}$ oxygen membrane. J Membr Sci 172:177–188
6. Yang WS, Wang HH, Zhu XF, Lin LW (2005) Development and application of oxygen permeable membrane in selective oxidation of light alkanes. Top Catal 35:155–167
7. Balachandran U, Dusek JT, Maiya PS, Ma B, Mieville RL, Kleefisch MS, Udovich CA (1997) Ceramic membrane reactor for converting methane to syngas. Catal Today 36:265–272
8. Nozaki T, Fujimoto K (1994) Oxide ion transport for selective oxidative coupling of methane with new membrane reactor. AIChE J 40:870–877
9. Wang HH, Cong Y, Yang WS (2002) High selectivity of oxidative dehydrogenation of ethane to ethylene in an oxygen permeable membrane reactor. Chem Commun 14:1468–1469
10. Li WP, Zhu XF, Chen SG, Yang WS (2016) Integration of nine steps for producing ammonia and liquid-fuel synthesis gases in one membrane reactor. Angew Chem Int Ed 55:8566–8570
11. Fan CG, Deng ZQ, Zuo YB, Liu W, Chen CS (2004) Preparation and characterization of $SrCo_{0.8}Fe_{0.2}O_{3-\delta}$-$SrSnO_3$ oxygen-permeable composite membrane. Solid State Ionics 166:339–342
12. Zhu XF, Wang HH, Cong Y, Yang WS (2006) Structure and oxygen permeation of cerium light doped $BaFeO_{3-\delta}$ ceramic membranes. Solid State Ionics 117:2917–2921

Chapter 2
Defects and Diffusion

Abstract Point defects are the main defect type concerned in the field of MIEC membranes, because they are closely related to the motion of oxygen vacancies and holes in crystalline solids. In the first part of this chapter, the concept of point defects is introduced briefly, and then, the knowledge about point defects in MIEC oxides, defect association, and defect equilibrium are introduced. Grain boundary is mentioned as an important two-dimensional defect because of its potential influences on the mechanical, electric, magnetic, catalytic, and thermal properties of materials. In the second part, the concept of diffusion, the diffusion mechanism, and diffusion paths of oxygen ions in perovskite- and fluorite-type ionic conductors are presented. There are several types of diffusion coefficients derived from different measuring methods, and the difference and relationships between these diffusion coefficients are provided.

Keywords Point defects • Electron and holes • Defect association • Grain boundary • Ionic diffusion • Ionic conductivity • Diffusion path • Grain boundary diffusion

2.1 Defects Concerned in MIEC Materials

An ideal (defect-free) crystal should be a solid with a periodic arrangement of atoms (ions, molecules) in the three-dimensional (3-D) space, i.e., in which the atoms are located at specific positions that can be repeated by moving the arrangement toward a certain direction on a constant distance. However, there is no ideal crystal in nature, and it is also impossible to obtain an ideal crystal via artificial synthesis. Defects widely exist in real crystals, and they play an important role in both the chemical and physical behavior of solids. Most of the new upcoming and future technologies are based on the development and application of the science on solid defects, such as batteries, fuel cells, solar cells, and ionic transport membranes. There are many types of defects in solids, and their sizes range from electronic, atomic sizes to micrometers. A common classification of the defects is, thus, based on their 3-D dimensions, i.e.:

- Zero-dimensional defects or point defects
- One-dimensional defects or linear defects

X. Zhu, W. Yang, *Mixed Conducting Ceramic Membranes*, Green Chemistry and Sustainable Technology, DOI 10.1007/978-3-662-53534-9_2

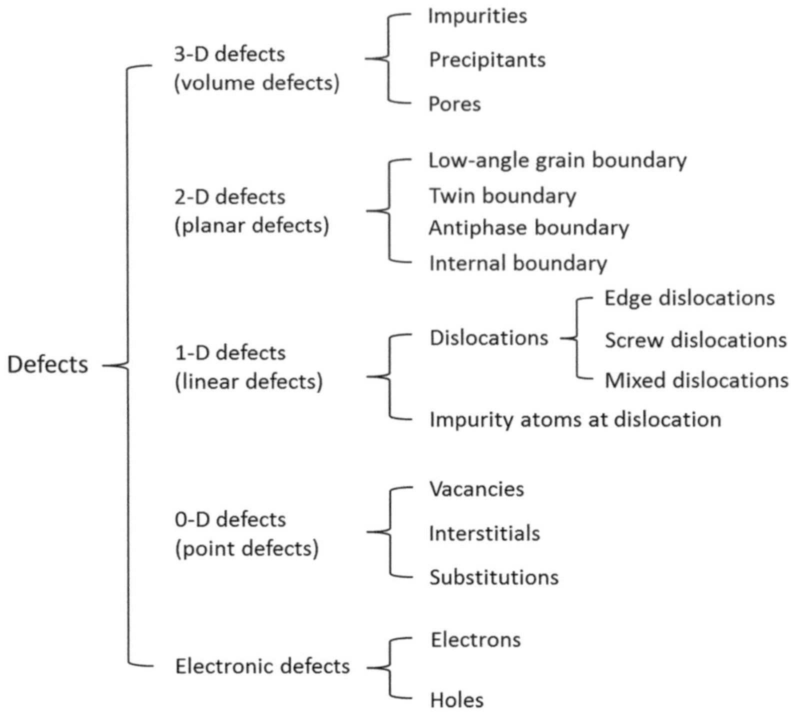

Fig. 2.1 Classification of the defects based on their structural characteristics

- Two-dimensional defects or planar defects
- Three-dimensional defects or volume defects

Apart from the abovementioned defects, ones smaller than an atom in size are called electronic defects (electrons and holes). Figure 2.1 shows a more detailed classification of the defects based on their structural characteristics. Point defects are one of the major aspects in the field of solid-state chemistry, especially in the field of MIEC membranes related to the motions of oxygen vacancies (a kind of point defect) and holes in crystalline solids. Thus, the relevant aspects of point defects will be specially introduced in this chapter. If readers are interested in the other types of defects in crystals, they may refer to the literature for a comprehensive understanding [1].

2.1.1 Point Defects

Point defects are where an atom (or ion) is missing (vacancy defect), or in an irregular place in the lattice structure (interstitial defect) where no atom would ordinarily appear, or is replaced by another atom (substitutional defect). Vacancy

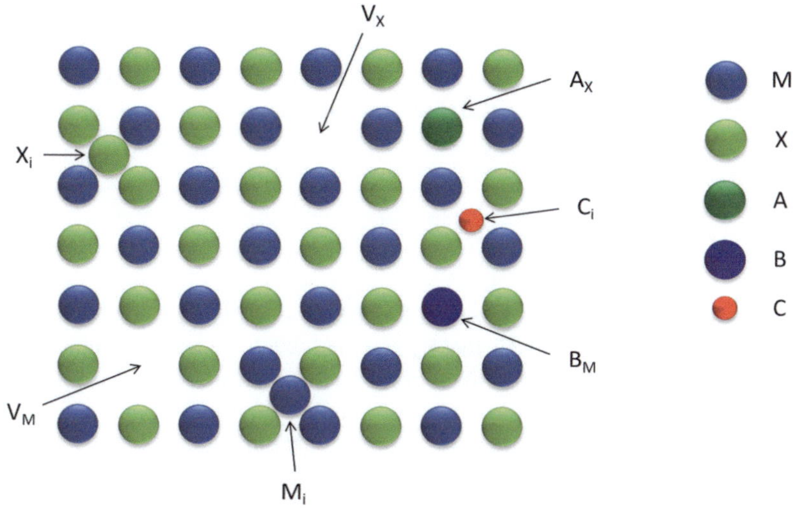

Fig. 2.2 Point defect types in a solid

defects are common in solid materials, as shown in Fig. 2.2; in particular, at high temperatures, atoms frequently depart from their normal lattice sites and leave behind empty lattice sites. In most cases, vacancies are the most important diffusion path for atoms in solids. For example, oxygen ionic conducting solid-state electrolytes can transport oxygen ions at elevated temperatures via the motion of oxygen vacancies in lattices. Interstitial defects can be self-interstitial atoms or interstitial impurity atoms, as shown in Fig. 2.2. A self-interstitial atom is an extra atom that is located at an interstitial void in the lattice structure. Interstitial impurity atoms are usually much smaller than the atoms in the bulk matrix. A classic example of interstitial impurity atoms is the carbon atoms located at the interstitial sites of the iron atomic lattices, which makes the flexible iron become hard steel because the carbon atoms distort and highly stress the tightly packed lattice structure. A substitutional defect is one in which the bulk atoms are replaced by a different type of atom in the lattice. The size of the substitutional atoms is usually similar to that of the bulk atoms. For example, MgO has a NaCl-type structure, in which Mg^{2+} ions with a radius of 0.086 nm can be substituted by Ni^{2+} ions with a radius of 0.083 nm, and the resultant material is a good catalyst for methane reforming to syngas.

The point defects also can be divided into intrinsic defects, which appear in a pure material, and extrinsic defects, which are caused by substitutional or impurity atoms. An intrinsic defect forms as a bulk atom leaves from its normal lattice site and creates a vacancy or the atom crowds into the interstitial site. The well-known Schottky defect, involving paired vacancies on the cation and anion sublattices, and the Frenkel defect, which is made up of a paired vacancy and interstitial, are both intrinsic defects. Extrinsic defects are referred to as foreign atoms, which are called substitution (or doping) if they are intentionally added in a material and called impurity if they are not. Lattice sites and interstitial sites can both be occupied by the foreign atoms. For MIEC oxides, the substitutional extrinsic defects are significantly concerned because the properties of the materials can be manipulated by altering the substitution elements and the doping dosage.

2.1.2 Point Defect Notations

Various notation systems have been used in the development of defect chemistry to describe point defects, but only the Kröger and Vink notation system is widely accepted now. This system describes crystals in terms of structural elements, in which the major symbol is used to describe its chemical content, a subscript indicates the site that it occupies, and a superscript indicates the effective charges that it holds, as shown in Fig. 2.3. In this notation system, vacancy is denoted by "V"; the substitutional, impurity, and interstitial defects are denoted by their elemental symbols; electron and hole defects are denoted by "e" and "h," respectively. The character in the subscript position indicates the position of the defect in the lattice. If it is an elemental symbol, it means that the substitutional or impurity defect locates at the position where the element should be; if it is "i," it means that it is an interstitial defect. The symbol in the superscript position indicates the effective charges that the defect holds, i.e., "×," "•," and "/" denote that the defect is electroneutral, holds a positive effective charge, or holds a negative effective charge, respectively. If the defect holds more than one effective charge, one can just put the same number of "•" or "/" in the superscript position. For example, the Mg vacancy and the substitution of Mg by Ni in the MgO lattice can be denoted, respectively, as V_{Mg}'' and $Ni_{Mg}^{\times}$, and the notations indicate that V_{Mg}'' holds two negative effective charges and $Ni_{Mg}^{\times}$ is electroneutral.

The effective charge is different from the real charge, and the difference is calculated by using the total charge around the defect to subtract the charge of an

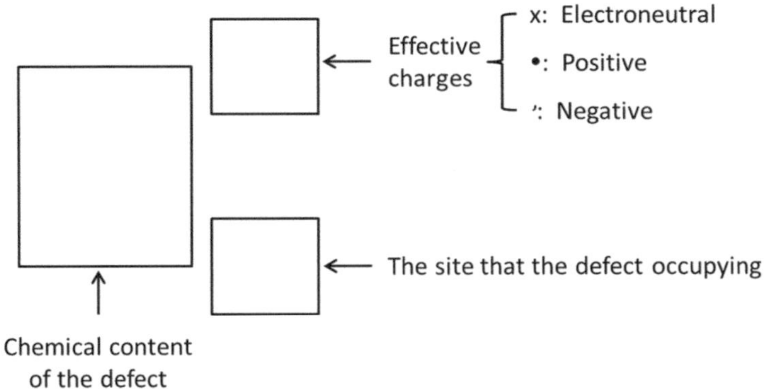

Fig. 2.3 Kröger and Vink notation system

ideal crystal in the same position. For both electrons and holes, their effective charges are equal to their real charges. In complex compounds, the effective charge of a point defect is usually not equal to the charge of the substitutional ion. For example, in yttria-stabilized zirconia (YSZ), the Y^{3+} ions occupy about half of the sites belonging to Zr^{4+} ions; meanwhile oxygen vacancies are created as a result of the doping. The substitutional defect and the oxygen vacancy can then be described as $Y_{Zr}^{'}$ and $V_{O}^{\cdot\cdot}$ with one negative effective charge and two positive effective charges, respectively.

2.1.3 Electrons and Holes

The appearance of point defects destroys the periodicity of the crystal lattice, and thus, the electronic energy levels around the point defects are different from those of atoms on the normal crystal lattice; as a result, there are many local energy levels produced with different energies in the forbidden band of the crystal. These point defects can exchange charges with the energy bands (conduction band and valence band) and be dissociated at a certain condition, so the electric properties of the real crystals containing point defects are much different from that of ideal crystals. The charge carriers can be electrons or holes, but for a special crystal, the charge carrier (whether electrons or holes) depends on the defects types. A classic example is the introduction of impurity atoms into the silicon atomic crystal. Each silicon atom has four outer electrons available for chemical bonding, and these electrons all participate in the formation of tetrahedral-network-type sp^3-hybrid bonds. Since phosphorus (P) atoms with five outer electrons are incorporated into the silicon lattice, one electron is left per phosphorus atom after four electrons are used to form the four sp^3-hybrid bonds. These electrons are easily excited to the conduction band by the thermal energy, and they then become quasifree to move through the crystal. The excitation process can be described by the following equation:

$$P_{Si}^{\times} \overset{E_D}{\rightarrow} P_{Si}^{\cdot} + e' \tag{2.1}$$

where E_D is the excitation energy to make $P_{Si}^{\times}$ release an electron to the conduction band. It is only 0.04 eV, which means the defect energy level is very close to the bottom of the conduction band. Because the charge carriers in the phosphorus-doped silicon are electrons with negative charge, the material is called an n-type semiconductor. In terms of energy band theory, point defects that can release electrons under excitation are called donors, and the corresponding energy level is called the donor level.

Similarly, as gallium (Ga) atoms with three outer electrons are doped into the silicon lattice, there is one electron short to form the four bonds. The missing electron can be regarded as a hole, denoted as $h^{\cdot}$. The superscript reveals that the hole carries a positive charge relative to the surroundings. These holes are easily excited to leave from the impurity atom by thermal energy, and they then become quasifree to move through the crystal. As the hole carries positive charge, the material is called a p-type semiconductor. The defects are called acceptors since they can be regarded as accepting electrons from the valence band. The excitation process can be described by the following equation:

$$Ga_{Si}^{\times} \xrightarrow{E_A} Ga_{Si}' + h^{\cdot} \tag{2.2}$$

where E_A is the excitation energy to make $Ga_{Si}^{\times}$ release a hole to the valence band. It is only 0.065 eV, which means the defect energy level is very close to the top of the valence band.

It can be seen from the above introduction that the electrons or holes are bound to the defects, so they are not free as in the case of metal crystals. The conduction of electrons or holes is realized through the hopping from one defect to another, so this type of conduction mechanism is called hopping of small polarons. In principle, the dissociation of a bound electron or hole from a defect is similar to the ionization of hydrogen atoms, and therefore, a hydrogen-like atom model can be used to describe the ionization of defects. The ionization energy of a hydrogen atom is

$$\frac{me^4}{8h^2\varepsilon_0^2} = 13.6\,\text{eV} \tag{2.3}$$

where m, h, and ε_0 are the electronic mass, Planck constant, and vacuum permittivity, respectively. If we consider the ionization of $P_{\text{Si}}^{\times}$ defect in a silicon crystal, the electronic mass and vacuum permittivity can be replaced by the effective electronic mass (m^*, $m^*/m = 0.3$) and the relative permittivity of the silicon crystal ($\varepsilon_{\text{Si}} = 12$) in a hydrogen-like atom model. Thus, the ionization energy is

$$E_{\text{D}} = \frac{m^*}{m}\frac{13.6}{\varepsilon_{\text{Si}}^2} = 0.03\,\text{eV} \tag{2.4}$$

This ionization energy of $P_{\text{Si}}^{\times}$ is very close to the experimental result (0.04 eV), which reveals that the hydrogen-like atom model is reliable to estimate the ionization energy of point defects.

As the aliovalent substitution occurs on an ionic crystal, the electrons or holes may be introduced into the crystal lattice through the ionization of the substitutional defects. For example, as Al^{3+} is doped into ZnO to form a transparent conductive oxide, electrons produced by the doping are loosely bound to the defects. The ionization of the defects can be described as

$$Al_{\text{Zn}}^{\times} \xrightarrow{E_{\text{D}}} Al_{\text{Zn}}^{\cdot} + e' \tag{2.5}$$

The transparent conductive oxide film has high electronic conductivity up to the order of 10^5 S cm^{-1}. There have been rapid developments in such films in recent years, and they are promising candidates to replace traditional In–Sn–O (ITO) transparent conductive oxides.

In the above substitution, the number of $Al_{\mathrm{Zn}}^{\cdot}$ is equal to the number of produced electrons. However, for metallic ions, their valence states are variable; the aliovalent substitution becomes complex, for instance, Li^{+}-doped NiO. Li^{+} has a similar ionic radius to that of Ni^{2+}, and it can occupy the Ni^{2+} site in the NiO lattice. The ionization of the defects can be described as

$$Li_{\mathrm{Ni}}^{\times} \stackrel{E_{\mathrm{A}}}{\rightarrow} Li_{\mathrm{Ni}}^{\prime} + h^{\cdot} \tag{2.6}$$

However, the holes are easy to be captured by the Ni^{2+} ions by increasing the valence state to +3. This process can be denoted as

$$Ni_{\mathrm{Ni}}^{\times} + h^{\cdot} \rightarrow Ni_{\mathrm{Ni}}^{\cdot} \tag{2.7}$$

Therefore, in the Li^{+}-doped NiO crystal, the number of holes is much lower than that of $Li_{\mathrm{Ni}}^{\prime}$. According to the electroneutral principle, the number of various defects in the Li^{+}-doped NiO crystal should fulfill the following relationship:

$$[Li_{\mathrm{Ni}}^{\prime}] = [Ni_{\mathrm{Ni}}^{\cdot}] + [h^{\cdot}] \tag{2.8}$$

where the square-bracketed defects denote the volume concentration of the defects.

2.1.4 Defects in MIEC Oxides

2.1.4.1 Fluorite-Type Ceria-Based Materials

Ceria is an important material for researchers in the field of solid oxide fuel cells and oxygen transport membranes; however, pure ceria exhibits extremely lower oxygen ionic conductivity owing to the limited oxygen vacancy concentration in crystals. Substitution of Ce^{4+} with other rare earth ions is effective to improve the ionic conductivity by creating more oxygen vacancies in the materials. Sm^{3+} and Gd^{3+} ions are good choices to dope ceria, and the doped ceria exhibits high oxygen vacancies and ionic conductivity that is one order of magnitude higher than that of YSZ. The doping can be described as

$$\frac{1}{2}Sm_2O_3 \stackrel{CeO_2}{\rightarrow} Sm_{\mathrm{Ce}}^{\prime} + \frac{1}{2}V_{\mathrm{O}}^{\cdot\cdot} + \frac{3}{2}O_{\mathrm{O}}^{\times} \tag{2.9}$$

However, unlike YSZ, ceria is not a pure oxygen ionic conductor and shows considerable electronic conduction in a strong reducing atmosphere. The loss of oxygen and the generation of oxygen vacancies occur at a low oxygen partial pressure. The formation of oxygen vacancies is balanced by the introduction of electrons:

$$O_O^{\times} \rightleftharpoons V_O^{\cdot\cdot} + \frac{1}{2}O_2 + 2e' \tag{2.10}$$

The electrons are easily captured by Ce^{4+} ions to form Ce^{3+}:

$$Ce_{Ce}^{\times} + e' \rightleftharpoons Ce'_{Ce} \tag{2.11}$$

In an oxidizing atmosphere, for instance, in air, the volume concentration of Ce'_{Ce} in the crystal is extremely low, and thus, there is little electron transfer between Ce^{4+} and Ce^{3+} ions, and the material shows predominantly oxygen ionic conductivity. As the reducing atmosphere becomes stronger, more Ce^{4+} ions are reduced to Ce^{3+} ions, leading to mixed ionic and electronic n-type conductivity. Therefore, in fuel cells with ceria-based materials as the electrolyte, the open cell voltages are lower than those of fuel cells with YSZ as the electrolyte due to the partial inner short of electronic current for ceria-based electrolyte. If the ceria-based materials are designed as membranes to separate oxygen gas from air, their electronic conductivities are insufficient. A second phase is needed to improve the electronic conductivity; the resultant dual-phase membranes are discussed in detail in Chap. 7.

2.1.4.2 Perovskite-Type Materials

For MIEC perovskite-type oxides (ABO_3), the defect chemistry becomes more complex. A- and/or B-site-doped $LaCoO_3$ perovskite has been widely investigated as the cathodes of solid oxide fuel cells and as membranes for oxygen separation. The parent material $LaCoO_3$ is virtually stoichiometric when prepared in air although the loss of oxygen can occur under the changes of oxygen partial pressure and temperature. Sr^{2+} ion with the similar ionic radius to that of La^{3+} is the most popular doping cation in the La^{3+} sites. The ionization of the substitutional defect can be written as

$$Sr_{La}^{\times} \rightarrow Sr'_{La} + h^{\cdot} \tag{2.12}$$

This acceptor substitution can be compensated through the formation of oxygen vacancies and oxidizing Co^{3+} to Co^{4+}, i.e.,

$$O_O^{\times} + 2h^{\cdot} \rightleftharpoons V_O^{\cdot\cdot} + \frac{1}{2}O_2 \tag{2.13}$$

$$Co_{Co}^{\times} + h^{\cdot} \rightleftharpoons Co_{Co}^{\cdot} \tag{2.14}$$

If it is assumed that the defects are single and unassociated, then the dilute solution model involving randomly distributed defects in the descriptions of the defect

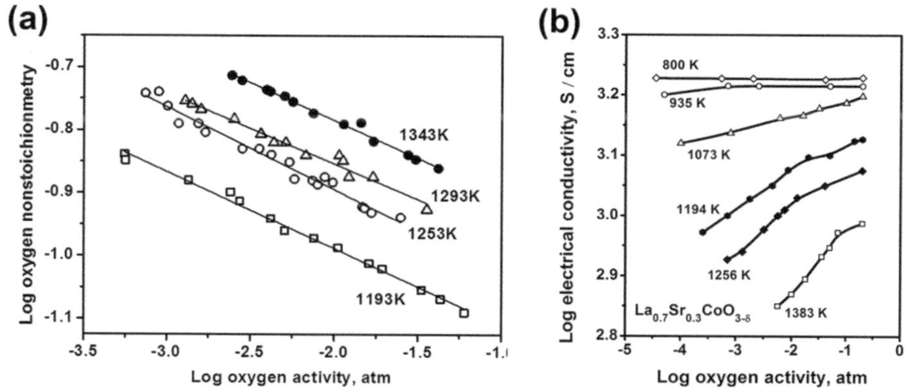

Fig. 2.4 Dependences of oxygen nonstoichiometry (**a**) and electronic conductivity (**b**) of $La_{0.7}Sr_{0.3}CoO_{3-y}$ as functions of the oxygen activity and temperature (Reproduced from Refs. [2, 14] with permission of Elsevier)

equilibrium may be used, and thus, the overall electroneutrality condition is given by

$$[Sr'_{La}] = [Co^{\cdot}_{Co}] + 2[V^{\cdot\cdot}_{O}] \tag{2.15}$$

Oxygen partial pressure and temperature have a significant effect on the compensation mechanisms. Figure 2.4a shows the oxygen deficiency, i.e., oxygen vacancies, in $La_{0.7}Sr_{0.3}CoO_{3-\delta}$ as a function of the oxygen partial pressure in the range of $1–10^{-3}$ atm [2]. As shown in the figure, the oxygen deficiency increases with temperature and decreases with oxygen partial pressure, and thus, it can be inferred that $[Co^{\cdot}_{Co}]$ decreases with temperature and increases with oxygen partial pressure according to Eq. (2.15). As shown in Fig. 2.4b, the electronic conductivity of $La_{0.7}Sr_{0.3}CoO_{3-\delta}$ decreases with temperature and increases with oxygen partial pressure.

Iron ions are the most common cations to substitute the cobalt ions in $LaCoO_3$-based perovskite oxides since iron ions have the similar ionic radii to those of cobalt ions. As Fe^{3+} is incorporated into the B site, it is partially oxidized to Fe^{4+} by Co^{4+}. The reaction can be written as

$$Co^{\cdot}_{Co} + Fe^{\times}_{Co} \rightleftharpoons Fe^{\cdot}_{Co} + Co^{\times}_{Co} \tag{2.16}$$

The acceptor substitution is competitively compensated through the formation of oxygen vacancies and $M^{\cdot}_{Co}$ (M = Co, Fe) defects [3], and the iron doping increases the $M^{\cdot}_{Co}$ concentration and decreases the oxygen vacancy concentration, as shown in Fig. 2.5.

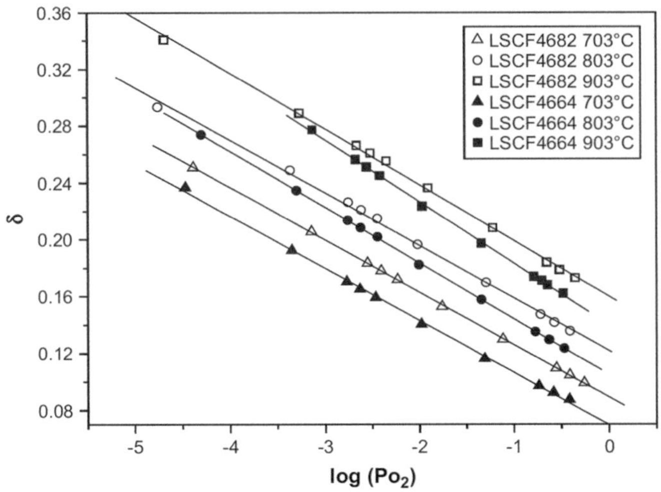

Fig. 2.5 Oxygen nonstoichiometry as a function of oxygen partial pressure (Reproduced from Ref. [3] with permission of Elsevier)

2.1.5 Association of Defects

In the above sections, isolated point defects are introduced, i.e., substitutional and impurity defects, intrinsic and extrinsic defects, electrons and holes, and electroneutral and charged defects. The behaviors of defects are predicted by the ideal, dilute solution model, which is generally only applicable for defect concentrations smaller than about 1 %. For crystals containing high defect concentrations, it is conceivable that there are high probabilities for different defects occupying the adjacent sublattice positions if they are randomly distributed in the crystal. The adjacent defects, thus, can associate with each other to form double or triple defect associates. The terms of "defect clusters" and "ordering" are frequently used to describe the defect associates and association, respectively. Since the concentration of defects is low, there is a small probability for different defects occupying the adjacent sublattice positions, and thus, the concentration of the defect associates is low. Owing to the association of such defects, the concentration of free or quasifree defects does not increase linearly with the increase in defect concentration. Electrostatic attracting force is the most important dynamic force for the formation of defect associates. The energy that is needed to disassociate the defect clusters is called the association energy. Therefore, defect associates are easier to form at low temperatures, and they are inclined to disassociate at high temperatures. As the temperature is increased from low to high, an ordering-disordering transition can be observed for many perovskite materials, in which a decrease in the conducting activation energy is always found.

For example, in $La_{0.3}Sr_{0.7}CoO_{3-\delta}$ perovskite, the defect clusters probably form through the association between substitutional defect of strontium ion (with one

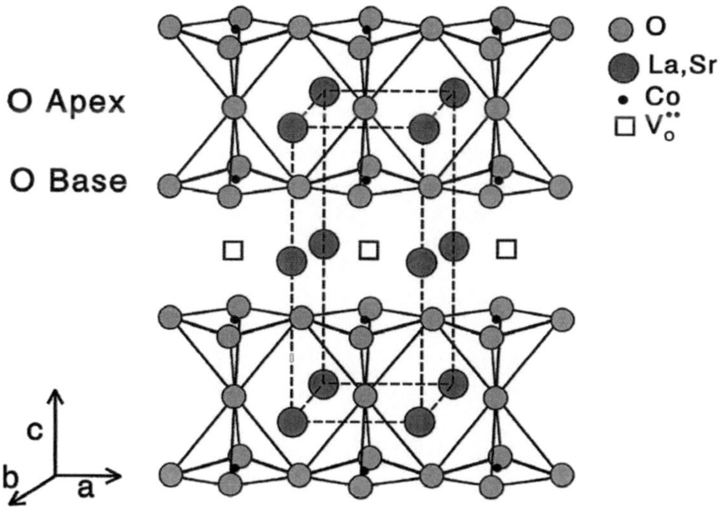

Fig. 2.6 Superstructure of oxygen vacancy-ordered $La_{0.3}Sr_{0.7}CoO_{3\text{-}\delta}$ perovskite (Reproduced from Ref. [4] with permission of Elsevier)

negative charge) and oxygen vacancies (with two positive charges) [4]. An electroneutral triple associate would form through the association between two substitutional defects and oxygen vacancies, i.e., $(Sr_{La}V_O Sr_{La})^{\times}$. As the substitutional defects increase to a certain high value, these types of defects may order and form microdomains or slabs of new intermediate, perovskite-related phases with lower oxygen content. HRTEM studies have revealed that for $La_{0.3}Sr_{0.7}CoO_{3\text{-}\delta}$ samples equilibrated at 650 °C, 5–50 nm-wide superstructure regions (microdomains) with arrays align along a principal axis. These arrays had a spacing of 0.76–0.77 nm, which corresponds to a doubling of the unit cell in the direction of a principal axis. Figure 2.6 shows the superstructure of oxygen vacancy-ordered $La_{0.3}Sr_{0.7}CoO_{3\text{-}\delta}$ perovskite [4]. In this superstructure, an oxygen anion is missing in every other (001) plane, resulting in a tetragonal $A_2B_2O_5$ superstructure cell with CoO square pyramids of which the nearest neighbors are mirrored with respect to each other. There are two different O-ion positions: one (out of four) in the base of the square pyramid and one in its apex. In Fig. 2.7, a comparison is given of an experimental HRTEM pattern and an HRTEM pattern calculated for a 4-nm-thick slab of the proposed superstructure.

Nano-sized domains were also observed in rare earth-doped ceria. It was found that rare earth cations and oxygen vacancies segregate in the microdomains (n $(Ln_{Ce}V_O Ln_{Ce})^{\times}$ (n is the number of the defect associates, and Ln denotes rare earth elements) by using energy filtering transmission electron microscopy (EFTEM). The nano-sized microdomains and the local ordering develop with increasing doping concentration, and a higher degree of ordering can lead to a more dramatic decrease in ionic conductivity in doped ceria. Figure 2.8a shows the EFTEM elemental maps of the Y-doped ceria samples [5]. The contrast shown in the figure

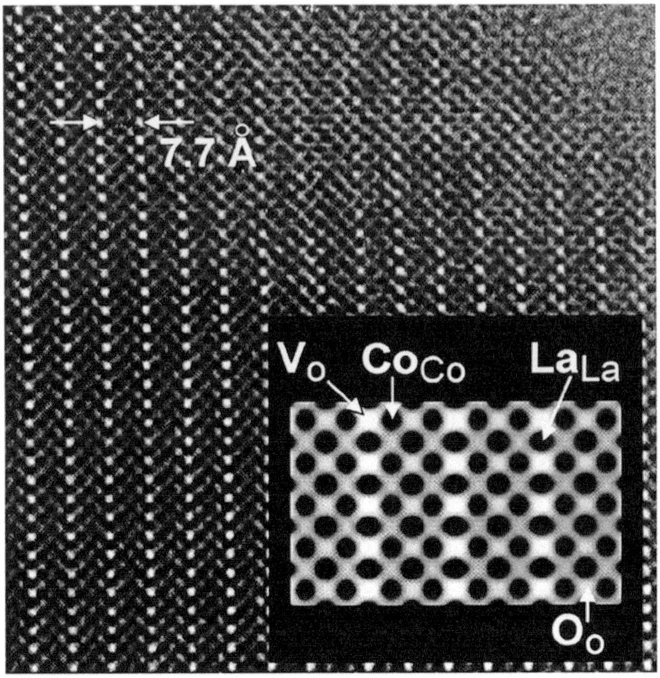

Fig. 2.7 Observed HRTEM pattern, together with a pattern, calculated for a 4-nm slab with the superstructure of Fig. 2.6, both projected along the b-axis (Reproduced from Ref. [4] with permission of Elsevier)

suggests that the compositional distributions of Ce and Y are inhomogeneous in the $Ce_{0.75}Y_{0.25}O_{1.875}$ sample since the used thin slices are uniform. Dark regions in the Ce map with a size about 10–20 nm (as marked by arrows) are brighter in the Y map, indicating that these regions have a higher Y concentration than their surrounding matrix. Figure 2.8b–e show selected area electronic diffraction (SAED) patterns of the Sm, Dy, Y, and Yb samples taken from zone axes [110]. Weak diffuse scattering and extra spots as marked by arrows are related to the microdomain formation, and the microdomains have a locally ordered structure different from the fluorite-structured matrix. Obviously, the microdomains in Sm-doped ceria have a more ordered structure than those in other doped ceria, with the order decreasing in a sequence of Sm > Dy > Y > Yb.

2.1.6 Equilibria of Defect Reactions

The thermodynamic theory of point defects is based on a hypothesis, i.e., a real crystal is regarded as a solution system, in which crystal lattices are the solvent and the point defects are the solute. Since the concentration of point defects is low, this

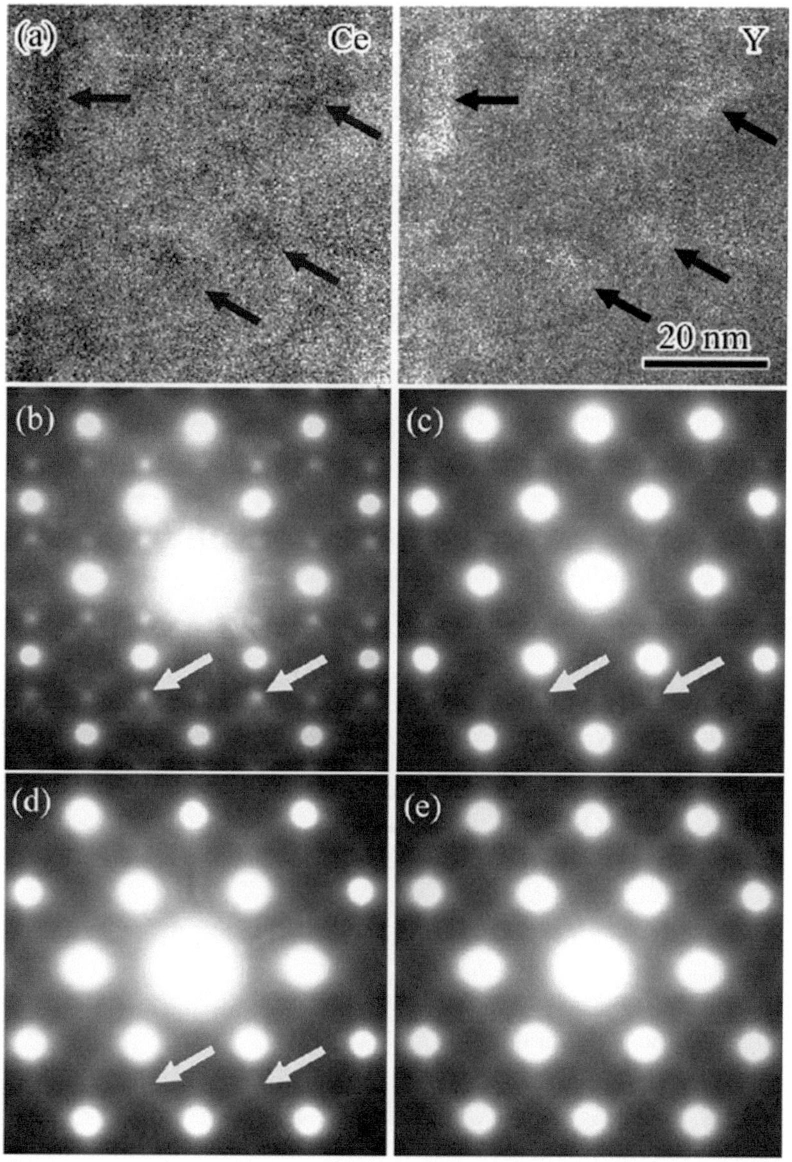

Fig. 2.8 (**a**) Ce and Y maps of Y-doped ceria and SAED patterns of ceria doped with (**b**) Sm, (**c**) Dy, (**d**) Y, and (**e**) Yb (Reproduced from Ref. [5] with permission of AIP Publishing LLC)

kind of "solution" is deemed to be infinitely diluted, and thus, the mass action law can be applied to the defect equilibrium. Electrons, holes, point defects, and defect associates can be treated as atom, ions and molecules, and the reactions that these defects participate in are regarded as quasi-chemical reactions. Therefore, quasi-chemical reaction equations can be written to describe the defect reactions and the principles for chemical reaction equations should be followed. Additionally, the ratios between various crystal lattices are constant before and after the reactions. For example, as oxygen atoms enter the lattices of the MO-type oxides, the reactions can be written as

$$\frac{1}{2}O_2 \rightarrow O_{\mathrm{O}}^{\times} + V_{\mathrm{M}}^{\times} \tag{2.17}$$

Here, the addition of oxygen lattice is accompanied by the addition of M vacancy.

The processes of the ionization of intrinsic and extrinsic semiconductors to release electrons and holes can be regarded as the disassociation of weak electrolytes, such as H_2O. The basic principles in reactions of electrolyte solutions, for instance, mass action law, common ion effect, and equilibrium shifting, are all applicable for the investigation of reactions involved in electrons and holes in solids. In the intrinsic semiconductors, the ionization reaction is

$$0 = e' + h^{\cdot} \tag{2.18}$$

Further, the corresponding equilibrium constant is K_i:

$$K_i = np \tag{2.19}$$

where n and p are the concentrations of electrons and holes in the solid.

In the following text, we use two examples to illustrate the equilibrium of defect reaction. One is a fluorite-type oxide, i.e., ceria, and another is a perovskite-type oxide, i.e., Sr-doped $LaCoO_3$. Pure and doped ceria are MIEC conductors under reducing atmospheres, owing to the reduction of Ce^{4+} to Ce^{3+}, as expressed by Eq. (2.11). The reduction of ceria-based MIEC materials can be described as

$$O_{\mathrm{O}}^{\times} + 2Ce_{\mathrm{Ce}}^{\times} \rightleftharpoons V_{\mathrm{O}}^{\cdot\cdot} + \frac{1}{2}O_2 + 2Ce_{\mathrm{Ce}}' \tag{2.20}$$

The concentrations of $Ce_{\mathrm{Ce}}^{\times}$ and $O_{\mathrm{O}}^{\times}$ are approximately constant, and the equilibrium constant of the above reaction is

$$K_{\mathrm{r}} = [Ce_{\mathrm{Ce}}']^2[V_{\mathrm{O}}^{\cdot\cdot}]P_{\mathrm{O}_2}{}^{\frac{1}{2}} \tag{2.21}$$

In pure ceria, taking the electroneutrality condition into consideration, the following relationship should be approximately fulfilled:

$$2[V_{\mathrm{O}}^{\cdot\cdot}] = [Ce_{\mathrm{Ce}}'] \tag{2.22}$$

The Ce_{Ce}' defects can release electrons to form $Ce_{\mathrm{Ce}}^{\times}$, as indicated by Eq. (2.11). Thus, the concentration of Ce_{Ce}' defects in the solid is

$$[Ce'_{Ce}] = K_e n \tag{2.23}$$

Combining Eqs. (2.21), (2.22), and (2.23),

$$n = (2K_r)^{\frac{1}{3}} K_e^{-1} P_{O_2}^{-\frac{1}{6}} \tag{2.24}$$

The conductivity of a material is proportional to the electron concentration, that is,

$$\sigma \propto P_{O_2}^{-\frac{1}{6}} \tag{2.25}$$

Thus, the electronic conductivity will fall as the oxygen partial pressure increases.

In the doped ceria, the concentration of oxygen vacancies is almost constant because the oxygen vacancies produced according to Eq. (2.20) are too high to be neglected compared to that produced through the following doping reaction:

$$\frac{1}{2} M_2O_3 \overset{CeO_2}{\rightarrow} M'_{Ce} + \frac{1}{2} V_O^{\cdot\cdot} + \frac{3}{2} O_O^{\times} \tag{2.26}$$

Therefore, Eq. (2.21) can be written as

$$K_r = [Ce'_{Ce}]^2 P_{O_2}^{\frac{1}{2}} \tag{2.27}$$

Combining Eqs. (2.23) and (2.27),

$$n = K_r^{\frac{1}{2}} K_e^{-1} P_{O_2}^{-\frac{1}{4}} \tag{2.28}$$

That is,

$$\sigma \propto P_{O_2}^{-\frac{1}{4}} \tag{2.29}$$

The conductivity is now proportional to the $-1/4$ power of the oxygen partial pressure. In the two cases described, electronic conductivity decreases with the increase in the oxygen partial pressure; however, the exact power relationship, $-1/4$ or $-1/6$, depends on the doping level. Figure 2.9 shows the dependence of electronic conductivity on the oxygen partial pressure of Sm-doped ceria in the temperature range of 400–800 °C [6]. The experimental data fit well with the prediction of Eq. (2.29), except at 800 °C. Matsui et al. suggested this deviation is due to the association of oxygen vacancies [6]. Except Sm-doped ceria, other rare earth elements, such as La-, Gd-, and Y-doped ceria, were also investigated to find the dependence of electronic conductivity on the oxygen partial pressure.

At higher oxygen partial pressures, it is assumed that an oxygen atom enters a crystal as oxygen interstitials for pure ceria. The incorporation of neutral oxygen requires the two electrons from the crystal to form oxygen ions, so the corresponding number of holes is generated to maintain the electroneutrality:

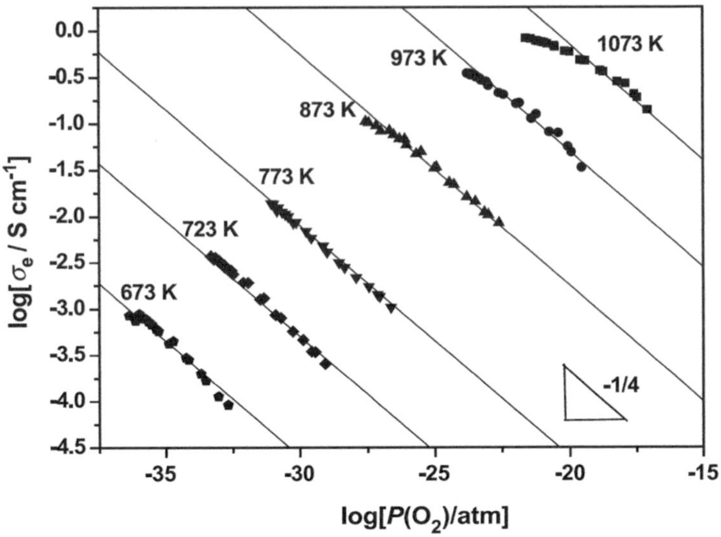

Fig. 2.9 Dependence of electronic conductivity on $P(O_2)$ (Reproduced from Ref. [6] with permission of Elsevier)

$$\frac{1}{2}O_2 \rightleftharpoons O_i'' + 2h^{\cdot} \tag{2.30}$$

The equilibrium constant can be written as:

$$K_r = p^2[O_i'']P_{O_2}{}^{-\frac{1}{2}} \tag{2.31}$$

For every oxygen interstitial introduced into the crystal, two holes form, and thus, Eq. (2.31) is changed to

$$K_r = \frac{1}{2}p^3 P_{O_2}{}^{-\frac{1}{2}} \tag{2.32}$$

Rearrangement gives

$$p = (2K_r)^{\frac{1}{3}} P_{O_2}{}^{\frac{1}{6}} \tag{2.33}$$

Thus, the conductivity is proportional to the $1/6$ power of the oxygen partial pressure for pure ceria. However, for the rare earth-doped ceria, the neutral oxygen atoms incorporate the oxygen vacancies, and this is accompanied by the generation of holes to maintain the electroneutrality:

$$\frac{1}{2}O_2 + V_O^{\cdot\cdot} \rightleftharpoons O_O^{\times} + 2h^{\cdot} \tag{2.34}$$

Considering that the concentrations of oxygen vacancies and lattice oxygen are almost constant, the equilibrium constant is given by

$$K_r = p^2 P_{O_2}{}^{-\frac{1}{2}} \tag{2.35}$$

That is,

$$p = K_r^{\frac{1}{2}} P_{O_2}{}^{\frac{1}{4}} \tag{2.36}$$

Thus, under a higher oxygen partial pressure, the electronic conductivity is proportional to the $1/4$ power of the oxygen partial pressure for rare earth-doped ceria.

Of course, the above relationships are not limited to ceria; the conductivities of other oxides have a similar dependence on oxygen partial pressure. The results obtained can be plotted as log[i] *vs.* $\log P_{O_2}$, i.e., the so-called Brouwer diagram, as illustrated in Fig. 2.10. In the blue area, the concentration of oxygen vacancies is much higher than those of electrons and holes, so the dominant charge carriers are oxygen ions. Good electrolyte materials have a wide blue area. The electronic conductivities of yttria-stabilized zirconia (YSZ) and $La_{0.9}Sr_{0.1}Ga_{0.9}Mg_{0.1}O_{2.9}$ (LSGM), as shown in Fig. 2.11, at 800 °C and oxygen partial pressure of 10^{-15} atm are only 3×10^{-8} S cm^{-1} and 5×10^{-6} S cm^{-1}, respectively [7, 8]. These values are five orders of magnitude lower than their oxygen ionic conductivities, and thus, the two materials are good electrolyte candidates for solid oxide fuel cells. The electronic conductivities are proportional to the $1/4$ power of the oxygen partial pressure at high oxygen pressure region and $-1/4$ at low oxygen pressure region, which fits well with the prediction of Eqs. (2.29) and (2.36). As for rare earth-doped ceria, their electronic conductivities are even higher than their oxygen ionic conductivities at the same temperature and oxygen partial pressure [9–11].

However, as mentioned above, the electronic transfer is inevitable in a solid electrolyte, thus implying that all the oxygen ionic conducting materials in fact are mixed conductors. For materials that can be used as solid electrolyte for SOFCs, the contribution of electronic conductivity to the total conductivity is extremely low, such as YSZ and LSGM. Although the higher electronic conductivity occurs on the rare earth-doped ceria at higher temperatures (>600 °C), they still can be used as electrolytes for SOFCs at low temperatures (<600 °C). The theoretical and experimental open circuit voltages can gradually increase from 1.0 V at 600 °C to 1.15 V at 400 °C for SOFCs based on Sm-doped ceria electrolytes [6, 12]. Therefore, in view of their high ionic conductivity and acceptable open circuit voltage at low temperatures, rare earth-doped ceria oxides are regarded as low-temperature electrolyte materials. Additionally, bilayer electrolytes were suggested for SOFCs to improve the open circuit voltage, in which a thin layer of dense YSZ is inserted between the anode and the ceria layers to block the electronic transport across the electrolyte membrane.

Oxygen ionic conducting materials having high electronic conductivity are good materials as electrodes for fuel cells and batteries, sensors, and dense ceramic membranes for oxygen separation. Perovskite-type (ABO_3) oxides, such as

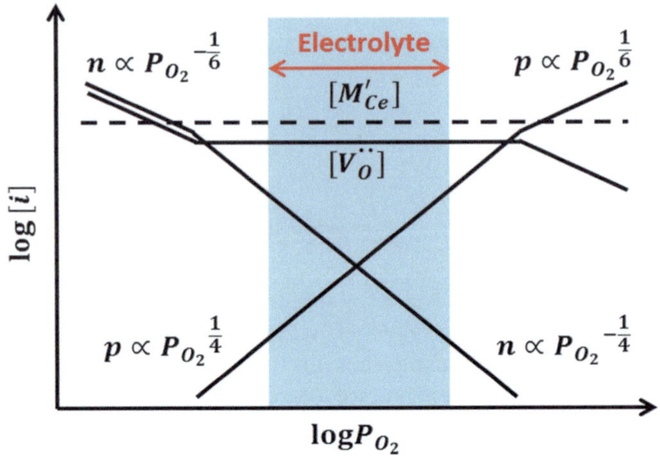

Fig. 2.10 Brouwer diagram of rare earth-doped ceria

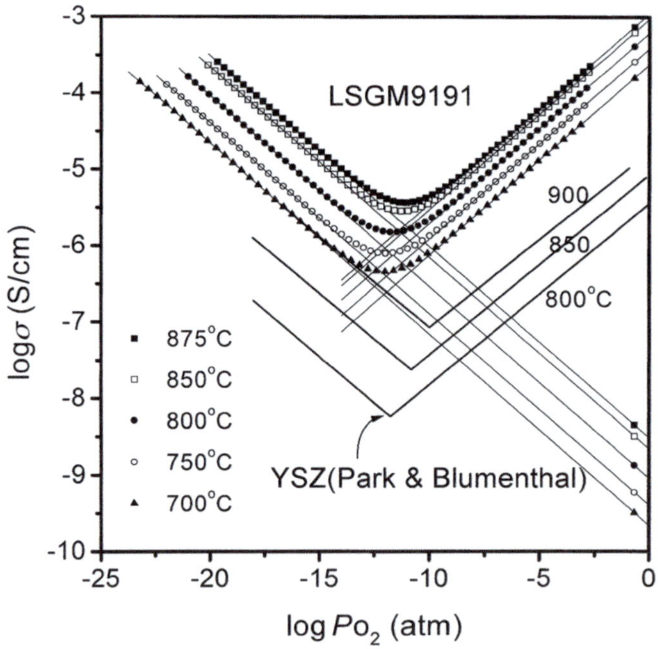

Fig. 2.11 Dependence of the partial electronic conductivity of LSGM9191 on oxygen partial pressure. The partial electronic conductivity of YSZ is plotted for the comparison (Reproduced from Ref. [8] with permission of Elsevier)

$LaCoO_{3-\delta}$, are the most intensively investigated mixed oxygen ionic-electronic conducting oxides, which are widely considered for SOFCs cathodes and membranes for oxygen separation. The introduction into alkali earth cations substitution, such as Sr^{2+}, in the La sites can be compensated by the formation of oxygen vacancies and local holes around cobalt ions, as indicated by Eqs. (2.12), (2.13), and (2.14). A high reducibility of the cobalt ions favors vacancy formation at elevated temperature or lower oxygen partial pressure:

$$O_O^{\times} + 2Co_{Co}^{\cdot} \rightleftharpoons V_O^{\cdot\cdot} + 2Co_{Co}^{\times} + \frac{1}{2}O_2 \tag{2.37}$$

Unlike the case of solid electrolyte materials, the dependences of oxygen vacancy concentration and electronic conductivity on oxygen partial pressure are complicated. Figure 2.12 shows the dependence of oxygen vacancy concentration on oxygen partial pressure at 800 °C for $La_{1-x}Sr_xCoO_{3-\delta}$ with different strontium contents [13]. The exponent of oxygen partial pressure changes from approximately −0.5 to −0.06 as the strontium content increases from 0 to 0.7. This result indicates that in a small oxygen partial pressure range, the oxygen vacancy concentration is slightly influenced by oxygen partial pressure for $LaCoO_{3-\delta}$ perovskite heavily doped with strontium. If the defect associates can be ignored, the oxygen ionic conductivity for perovskite oxides with high oxygen vacancy concentration, such as $La_{0.3}Sr_{0.7}CoO_{3-\delta}$ and $Ba_{0.5}Sr_{0.5}Co_{0.8}Fe_{0.2}O_{3-\delta}$, is almost constant in a small oxygen partial pressure range. This deduction is used as an important hypothesis in the development of the Zhu's oxygen permeation model presented in Chap. 5.

$La_{1-x}Sr_xCoO_{3-\delta}$ shows a typical *p*-type conductivity. Electronic transport in the perovskite oxides (i.e., the perovskite containing reducible B-site cations such as $Fe^{2+,3+,4+}$, $Co^{2+,3+,4+}$, and $Ni^{2+,3+}$) is achieved through a Zener double-exchange mechanism rather than the small polarons hopping mechanism (see Sect. 2.1.3). The Zener double-exchange mechanism is realized according to the following two equations:

$$O_O^{\times} + Co_{Co}^{\cdot} \rightarrow O_O^{\cdot} + Co_{Co}^{\times} \tag{2.38}$$

$$O_O^{\cdot} + Co_{Co}^{\times} \rightarrow O_O^{\times} + Co_{Co}^{\cdot} \tag{2.39}$$

Considering a cubic cobalt-based perovskite oxide, in which one Co^{3+} and one Co^{4+} are bridged by an O^{2-} ion with an intersection angle of 180°, the Co "e_g" orbitals directly interact with the O "2p" orbitals, and one of the Co ions has more electrons than the other. If oxygen ion gives a spin-up electron to the adjacent $Co_{Co}^{\cdot}$, as shown by Eq. (2.38), the resultant vacant orbital can then be filled by an electron from $Co_{Co}^{\times}$, as shown by Eq. (2.39). At the end of the process, an electron has moved between the neighboring metal ions. Figure 2.13 shows the dependence of electrical conductivity on temperature in air for $La_{1-x}Sr_xCoO_{3-\delta}$ with different strontium contents [14]. For the sample with $x = 0.0$, 0.1, and 0.2, the conductivity increases with the temperature, goes through a maximum, and then decreases. For samples with $x \geq 0.3$, the conductivity decreases with the increase in temperature.

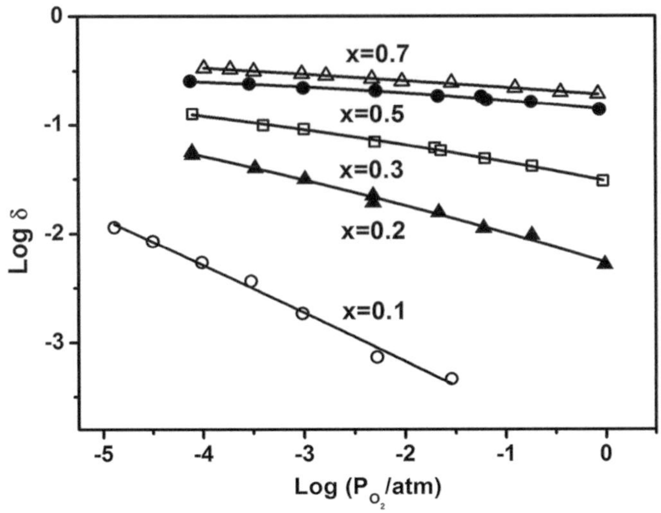

Fig. 2.12 Dependence of oxygen vacancy concentration on oxygen partial pressure at 800 °C for $La_{1-x}Sr_xCoO_{3-\delta}$ with different strontium contents (Reproduced from Ref. [13] with permission of Elsevier)

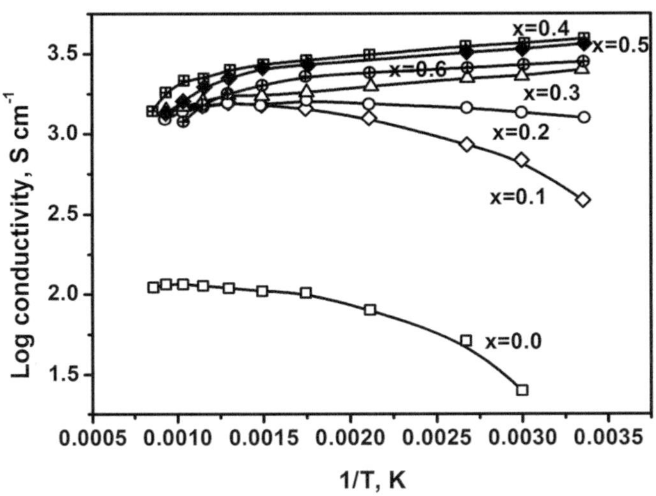

Fig. 2.13 Dependence of electrical conductivity on temperature in air for $La_{1-x}Sr_xCoO_{3-\delta}$ with different strontium contents (Reproduced from Ref. [14] with permission of Elsevier)

Additionally, the conductivity decreases with the decrease in oxygen partial pressure at high temperatures. At high temperature and low oxygen partial pressure, the reduction of $Co_{Co}^{\cdot}$ to $Co_{Co}^{\times}$, as indicated by Eq. (2.37), results in the decrease in electrical conductivity.

2.1.7 Grain Boundaries

Grain boundaries are a type of common 2-D defect in polycrystalline materials, which widely exist in bulk metals, ceramics, and crystalline polymers. Grain boundaries form during the processes of solidification, sintering, etc., and deeply influence the mechanical, electric, magnetic, catalytic, and thermal properties of materials. Grain boundary is a thin 2-D zone corresponding to one crystallographic orientation to another, thus separating one grain from another. The atom in each grain is arranged in an orderly pattern but with irregular junction between two adjacent grains. Grain boundaries are convenient to be separated by the misorientation between the two adjacent grains, i.e., small-angle grain boundaries with a misorientation less than about 15° and large-angle grain boundaries with a misorientation greater than about 11° (the transition angle varies from 10° to 15° depending on the material). The small-angle grain boundaries are composed of an array of dislocations, and there are two limiting cases named small-angle tilt boundaries (which consist of arrays of edge dislocations and can be symmetric or asymmetric) and small-angle twist boundaries (which consist of arrays of screw dislocations). Figure 2.14 shows the HRTEM image of an Fe-doped (0.05 mol%) $SrTiO_3$ bicrystal with 5.4° [001] symmetrical tilt grain boundary and the corresponding structural model [15]. To make the best fit between the adjacent grains, most of the small-angle grain boundaries are of a mixed type, containing dislocations of different types and Burgers vectors. As the misorientation becomes larger, the distance between dislocation cores is gradually shortened to zero; at this point, the small-angle grain boundary is transferred to large-angle grain boundary, and the original grain has been entirely separated into two grains.

For polycrystalline materials, most of the grain boundaries are large-angle grain boundaries, in which atoms are irregularly arranged with complex structures. Figure 2.15 gives a schematic diagram of the large-angle grain boundaries in a polycrystalline material. The boundary between the adjacent two grains is not a smooth curve surface but made of irregular steps. For metal polycrystalline materials, the width of the large-angle grain boundary is usually smaller than the 3-interatomic distance. In recent years, a coincident site lattice (CSL) theory was proposed to describe the large-angle grain boundary, in which the degree of fit (Σ) between the structures of the adjacent two grains is described by the reciprocal of the ratio of coincidence sites to the total number of sites. It means that for a grain boundary, the higher the Σ value, the higher the grain boundary energy. In small-angle boundaries, the distortion is entirely accommodated by dislocations, so Σ is 1.

In the grain boundary area, the irregular arrangement of atoms introduces an excess energy over the matrix energy, and this excess energy is called the grain boundary energy. Therefore, it is thermodynamically favorable for a polycrystalline material to reduce its grain boundary area through grain growth and boundary flatting. During sintering, the grain boundary energy acts as the driving force of ceramic densification and grain growth. At low temperatures ($T < 0.5$ Tm, where

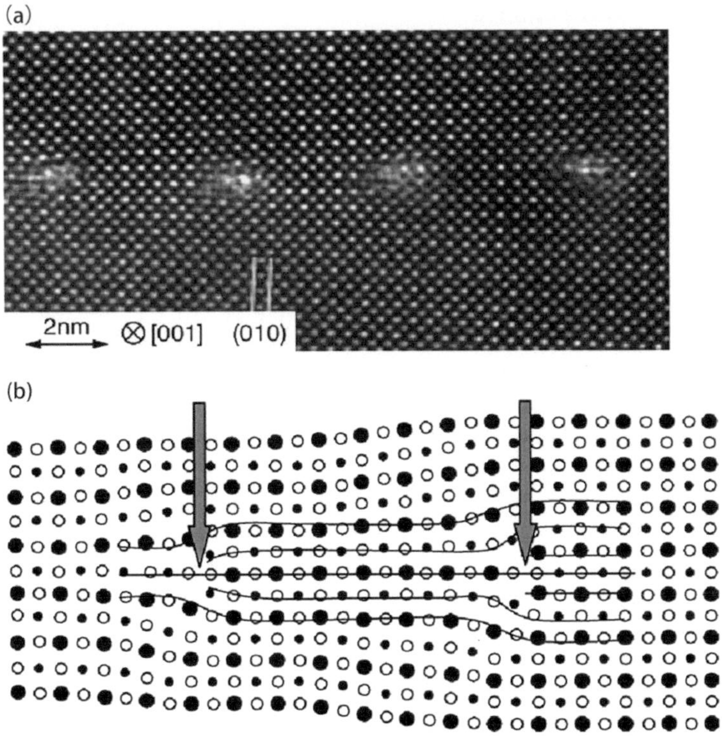

Fig. 2.14 (**a**) HRTEM image of an Fe-doped (0.05 mol%) $SrTiO_3$ bicrystal with 5.48 [001] symmetrical tilt grain boundary. The bright areas are the cores of edge dislocations arranged in an array. (**b**) Corresponding structural model; arrows mark dislocation cores with partially occupied oxygen columns. ●: Sr; •: Ti; ○: O (Reproduced from Ref. [15] with permission of John Wiley and Sons)

Tm is the melting point in K), the grain boundaries act as strong obstacles to dislocation motion, thus inhibiting the plastic deformation of materials. This is reflected by the increase in mechanical strength with a decrease in grain size. Additionally, decreasing grain size, accompanied by an increase in the density of grain boundaries, can improve the fracture toughness and thermal shock tolerance. Atoms in grain boundaries deviating from their equilibrium sites exhibit higher energy than those in grain bulk, and thus, vacancies, impurities, and dislocations are always enriched at grain boundaries; grain boundaries therefore have higher catalytic activity toward chemical reactions, are easier to be corroded, and can transport atoms faster than grain bulk. In the solid-phase transformation, nucleation takes places easier at grain boundaries than in grain bulk owing to the large grain boundary energy and high atomic mobility. Solute segregation and impurity enrichment are frequently observed at grain boundaries for polycrystalline materials, which have a great influence on grain growth and physical properties such as mechanical, electrical, and thermal properties.

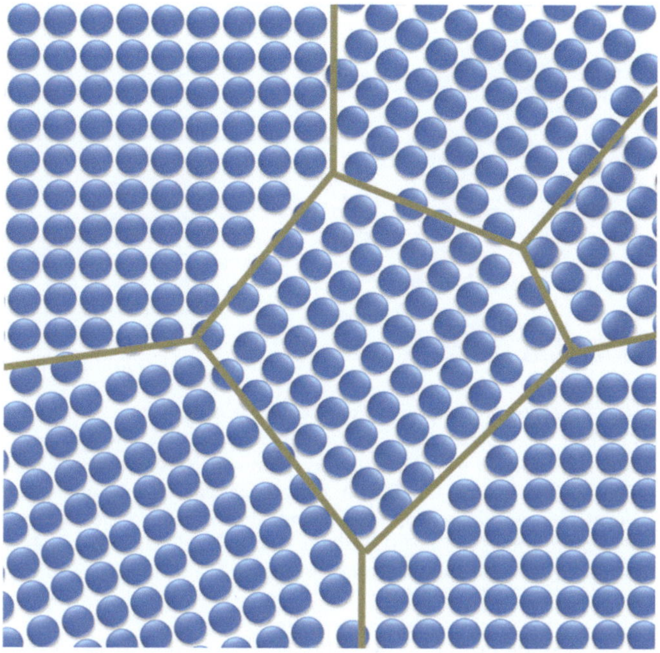

Fig. 2.15 Schematic diagram of the large-angle grain boundaries in a polycrystalline material

For the preparation of dense MIEC membranes, the frequently used method is sintering compacted small particles at elevated temperatures, usually in the range of 1100–1500 °C. Thus, grain boundaries are an inevitable defect feature of MIEC membranes. The most important aspects influenced by grain boundaries are the mechanical and electric properties in the field of MIEC membranes. Although the investigation on grain boundaries is well strengthened for metallic materials, knowledge about the grain boundary effects on ceramic materials is still deficient, especially for perovskite-type MIEC materials. For oxygen ionic conductors and mixed conductors, researchers mainly focus on the oxygen exchange on membrane surfaces and diffusion of oxygen ions in the lattice. The grain boundaries can be fast diffusion paths or diffusion obstacle layers for oxygen ions. Both impurity enrichment and solute segregation take place at grain boundaries during the sintering and subsequent heat treatments. In general, these changes at grain boundaries can abruptly alter the magnitude of oxygen exchange on membrane surfaces and oxygen ionic transport across and along the grain boundaries.

Among oxygen ionic conductors, YSZ is the most intensively investigated material, and there are many studies on the influence of grain boundaries on oxygen activation, ionic transport, and sintering. Shim et al. used high-spatial-resolution (50 nm) secondary ion mass spectrometry (SIMS) to measure the oxygen isotope exchange on a bare YSZ surface [16]. Surface mapping of the ^{18}O distribution by SIMS clearly shows high activity at surface grain boundaries, suggesting that these

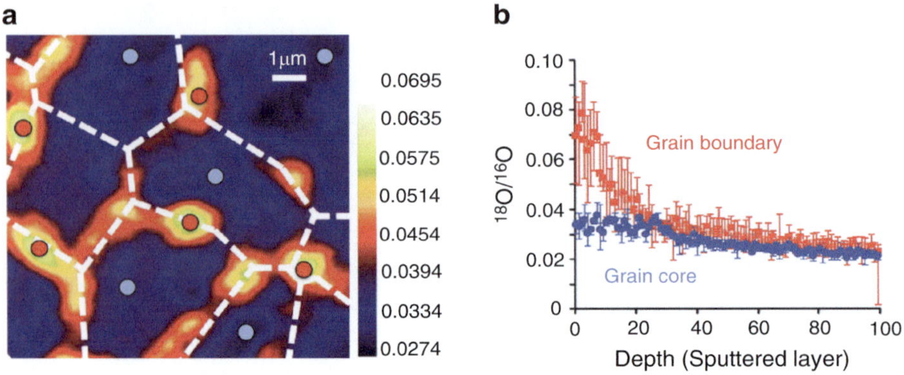

Fig. 2.16 (**a**)$^{18}O/^{16}O$ concentration obtained from nanoSIMS spatial scans. *Yellow* and *red* indicate regions higher$^{18}O/^{16}O$ ratio along the grain boundaries than the *blue regions*. (**b**) Depth profiles of grain boundaries (*red trace*) and grains (*blue trace*) from 100 Cs^+ ion sputter scans, with a sputter rate of approximately 0.2–0.3 nm per sputter scan, of a sample annealed at 600 °C for 3 h in 150 Torr of$^{18}O_2$ gas under an applied cathodic bias of 0.5 V. The *blue spots* (grain core) and the *red spots* (grain boundary) in (a) indicate where the depth profile data were acquired (Reproduced from Ref. [16] with permission of Elsevier)

boundary regions provide preferential pathways for oxygen incorporation into YSZ, as shown in Fig. 2.16 [16]. It was found that the highly active sites for oxygen activation are along the grain boundaries from the nanoscale spatial resolution SIMS image of the YSZ surface. In this study, they found that the$^{18}O/^{16}O$ count ratio along the grain boundaries is about three times higher than that in bulk grain regions. The$^{18}O/^{16}O$ concentration ratio clearly differs for the grain and grain boundary regions near the surface of the sample, but they both decay to a comparable baseline of ~2 %, which is ten times higher than the natural abundance of$^{18}O_2$ deeper in the YSZ substrate. Guo et al. used impedance spectroscopy to investigate the effect of grain size on the electrical properties of 3 mol% Y_2O_3-doped ZrO_2 [17]. The grain boundary conductivity was found to be approximately two orders of magnitude lower than the bulk conductivity, while it increased with decreasing grain size. The low grain boundary conductivity was related to the depletion of oxygen vacancies in the grain boundary space charge layers, in which the space charge potential decreases but the oxygen vacancy concentration increases with the decrease in grain size. Metlenko et al. investigated the oxygen transport along a symmetrical 6° [001] tilt grain boundary (comprised of edge dislocations) of acceptor-doped $SrTiO_3$ perovskite oxide by means of$^{18}O/^{16}O$ exchange experiments and computational techniques [18]. SIMS analysis showed no evidence of fast diffusion along the dislocation array. Two types of dislocation cores in the tilt boundary were predicted, and both cores are the preferred sites for the formation of oxygen vacancies than the bulk, and thus, this was related to the lower migration rate of oxygen ions along both cores than the bulk.

In a study on $LaCoO_3$ perovskite by Royera et al., the authors found that grain boundary diffusion proceeds quickly in comparison with bulk diffusion through isotopic exchange experiments [19]. Diethelm et al. investigated the influence of grain size on the oxygen permeation flux through $La_{0.5}Sr_{0.5}FeO_{3-\delta}$ perovskite membranes [20]. They found the oxygen permeation flux increases from 0.01 to 0.06 μmol cm^{-2} s^{-1} through 1-mm-thick membranes as the average grain sizes decreases from 1.43 to 0.20 μm at 800 °C. The significant improvement in permeability was related to the faster transport of oxygen ions along grain boundaries than through the bulk. However, as shown in other studies, the membrane with smaller grain size has more grain boundaries exposed on the membrane surfaces, which are active sites for oxygen exchange. Thus, there is another possibility that the improvement in permeability is due to the higher catalytic activity of grain boundaries toward oxygen exchange reactions than grain bulk. From the above examples, one can find that the action of grain boundaries on oxygen ionic transport varies from one material to another.

2.2 Ionic Diffusion

Diffusion is a description of the motion of atoms, ions, or molecules in a certain system driven by chemical potential gradient or electrochemical potential gradient. The result of diffusion is to eliminate these gradients. If there is no applied energy field, the final result of diffusion is to realize a uniform distribution of various species in the system. In solids, the motion of hetero atoms or defects is driven by chemical potential or electrochemical potential gradients. The diffusion flux of a species through the unit cross-sectional area and in the unit time is proportional to its concentration gradient and the cross-sectional area, i.e.,

$$j_i = \frac{\partial m_i}{\partial t} = -DA\frac{\partial c_i}{\partial x} \tag{2.40}$$

where $\frac{\partial c_i}{\partial x}$ ((mol cm^{-3}) cm^{-1}) is the concentration gradient of species i along the diffusion direction, $\frac{\partial m_i}{\partial t}$ (mol (cm^2 s)$^{-1}$) is the moles of i passing through the unit cross-sectional area in the unit time, A (cm^2) is the cross-sectional area, and the constant D in Eq. (2.40) is the so-called diffusion coefficient. This equation is the Fick's first law of diffusion. The negative sign in the equation describes a fact that the diffusion is directed to the reduction of the concentration gradient. Ordinarily, the diffusion coefficients of atoms, ions, and vacancies in solids range from 10^{-20} to 10^{-4} cm^2 s^{-1}. Diffusion coefficient is an important physical parameter for solid materials, and it is related to the crystalline structure and microstructures of materials. For MIEC membranes, they usually have a fluorite- or perovskite-type crystalline structure with high oxygen ionic or vacancy diffusion coefficients ranging from 10^{-9} to 10^{-5} cm^2 s^{-1} at elevated temperatures. In a polycrystalline material, besides the diffusion of species in grain bulk (bulk diffusion), they also

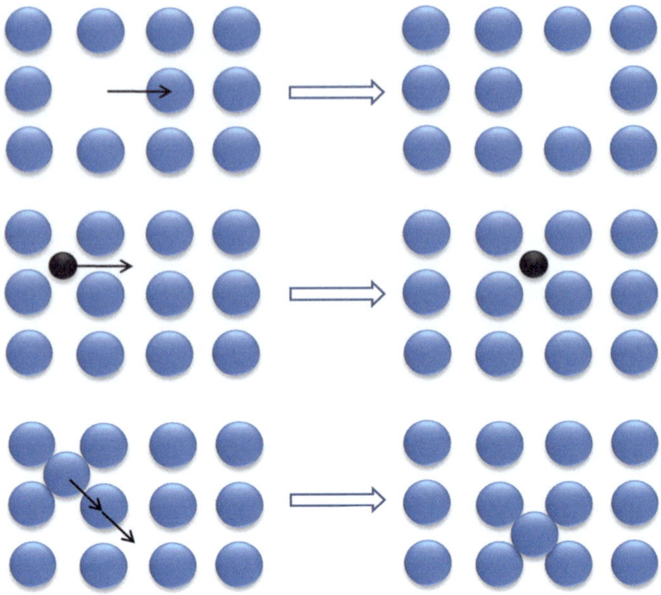

Fig. 2.17 Diffusion in *solid*, (**a**) vacancy diffusion, (**b**) interstitial diffusion (hopping), (**c**) interstitial diffusion (squeezing)

can transport along dislocations, grain boundaries, and inner surfaces. Therefore, the microstructure has significant effect on the diffusion coefficient of a material.

2.2.1 Vacancy Diffusion and Interstitial Diffusion

With vacancy diffusion, atomic motion refers to motion from one lattice site to an adjacent vacancy. Thus, the motion direction of the vacancy is opposite to the diffusion direction of the atom, as shown in Fig. 2.17a. If our eyes are focused on the vacancies, we will find that the diffusion of vacancies is similar to a real particle. The mechanism of vacancy diffusion occurs in most solids. For the MIEC materials, the oxygen ions are the concerned motion species in the material lattice. The diffusion of oxygen ions is dominated by the vacancy mechanism for fluorite- or perovskite-type MIEC materials. However, vacancy diffusion is not the sole mechanism for the MIEC materials. Interstitial diffusion of oxygen ions occurs in some layer-typed MIEC materials. For example, $La_2NiO_{4+\delta}$ has the same crystalline structure as that of K_2NiF_4. In the case of interstitial diffusion, the interstitial atoms can jump to an adjacent interstitial site, as shown in Fig. 2.17b. Another mechanism for interstitial diffusion is that the interstitial atom transports to the normal lattice site by squeezing the occupant into an adjacent interstitial site, as shown in Fig. 2.17c. However, such a diffusion mechanism is infrequently reported for oxygen ionic diffusion in fluorite or perovskite materials. Interstitial diffusion is normally more rapid than vacancy diffusion because bonds between interstitial atoms and the surrounding atoms are normally weaker and the number of empty adjacent interstitial sites is greater than adjacent vacancies for a host atom. Thus, much low diffusion activation energies of oxygen ions and large diffusion coefficients are observed for the $La_2NiO_{4+\delta}$-type materials along the LaO layers.

2.2.2 *Diffusion Path of Oxygen Ions*

Oxide materials with high oxygen diffusion coefficients have attracted considerable attention owing to their diverse applications in SOFCs, MIEC membranes for oxygen separation, sensors, and catalysts as well as batteries. The motion of oxygen ions in crystal lattices is aeolotropic, and thus, there is an optimal path for oxygen ion diffusion for materials, along which the oxygen ion motion has the minimal diffusion activation energy and the maximal diffusion probability. It is important to know the diffusion path of oxygen ions in materials as they are fabricated into devices, such as SOFCs and MIEC membranes. Obviously, materials with different crystalline structures have various diffusion paths for oxygen ions. Fluorite-, perovskite-, and K_2NiF_4-type oxides are the most commonly used materials in the field of MIEC membranes and SOFCs, and thus, their oxygen ionic diffusion paths will be separately introduced in the following sections; additionally, the ionic diffusion paths of other type of materials will also be mentioned.

2.2.2.1 Fluorite-Type Oxides

Cubic fluorite oxides, such as CeO_2, are important ionic conductors with high ionic mobility at elevated temperatures, in which oxygen anions form the eight-coordinated cubic interstices, and half of them are alternately occupied by the cations. Yashima et al. studied the oxygen ionic diffusion path in fluorite oxides by combining the Rietveld refinement, maximum entropy method (MEM), and MEM-based pattern fitting technique together to analyze the data of neutron powder diffraction [21]. They collected the diffraction data of δ-$Bi_{1.4}Yb_{0.6}O_3$ and $Ce_{0.93}Y_{0.07}O_{1.96}$ at elevated temperatures. A fluorite-type structure with space group of $Fm\overline{3}m$ was assigned to the δ-$Bi_{1.4}Yb_{0.6}O_3$ at 384 and 738 °C after the neutron powder diffraction patterns were refined. MEM nuclear density distribution map on the (110) plane is shown in Fig. 2.18a to visualize the oxygen ionic diffusion paths and the structural disorders at 738 °C. As shown in the figure, the oxygen ionic diffusion paths lie along the [001] direction in the (110) plane (white arrows in Fig. 2.18a). According to cubic symmetry of the fluorite oxide, the three directions, i.e., [100], [010], and [001], are identical, and thus, the diffusion paths are a three-dimensional network. Additionally, it was found that the oxygen ions

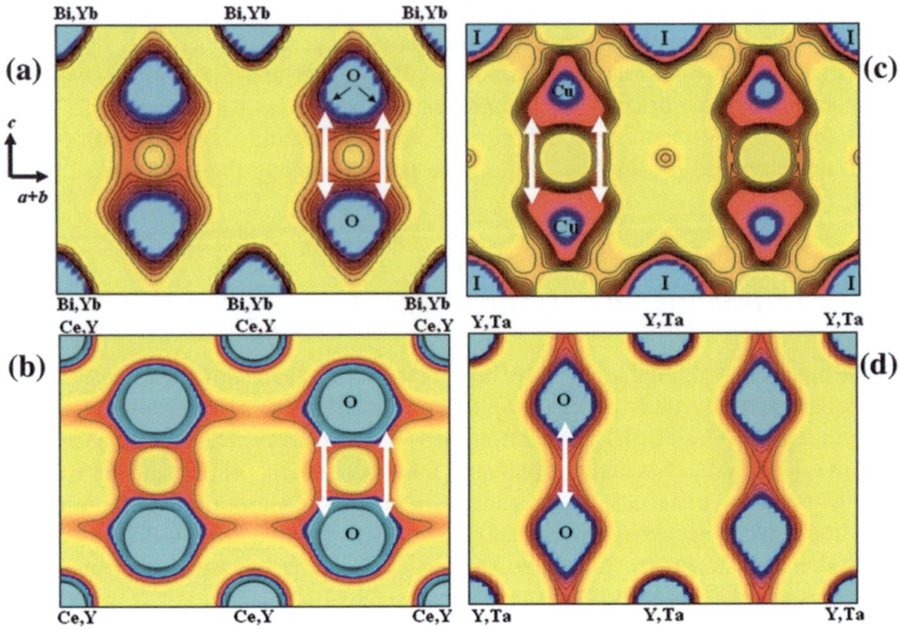

Fig. 2.18 Nuclear density distribution on the (110) plane of the fluorite-type (**a**) δ-$Bi_{1.4}Yb_{0.6}O_3$ at 738 °C, (**b**) $Ce_{0.93}Y_{0.07}O_{1.96}$ at 1434 °C, (**c**) electron density distribution on the (110) plane of the fluorite-type α-CuI at 487 °C, and (**d**) $Y_{0.785}Ta_{0.215}O_{1.715}$ at 535 °C (Reproduced from Ref. [21] with permission of Elsevier)

spread over a wide area showing a complicated disorder compared with the cations. This kind of disorder is strengthened with the increase in temperature, which just reveals that the higher oxygen ionic conductivity, the higher the atomic displacement parameters at higher temperatures. Therefore, the diffusion path along the <100> direction and the disorder of oxygen ions in δ-$Bi_{1.4}Yb_{0.6}O_3$ are the key to its high ionic conductivity.

Yttria-doped ceria was selected as an example to illustrate the oxygen ionic diffusion paths in the ceria-based ionic conductors. Similarly, the Rietveld refinement gives a fluorite-type structure with a space group of $Fm\overline{3}m$. Two types of bulges were observed in the MEM nuclear density distribution map (Fig. 2.18b). One shows the diffusion path along the [100] direction and another along the [110] direction. As illustrated in the figure, the MEM nuclear density is lower along the [110] directions than that along the [100] directions. The most prominent feature in the nuclear density map is the diffusion path along the [100] directions (shown by the white arrows in Fig. 2.18b), while the [100] path is not a straight line but a curved line. Researchers attributed this curve to the repulsion between the cations and anions [22]. Yashima et al. also found the diffusion paths along the [100] directions in other fluorite-type materials, such as α-copper iodide (α-CuI)

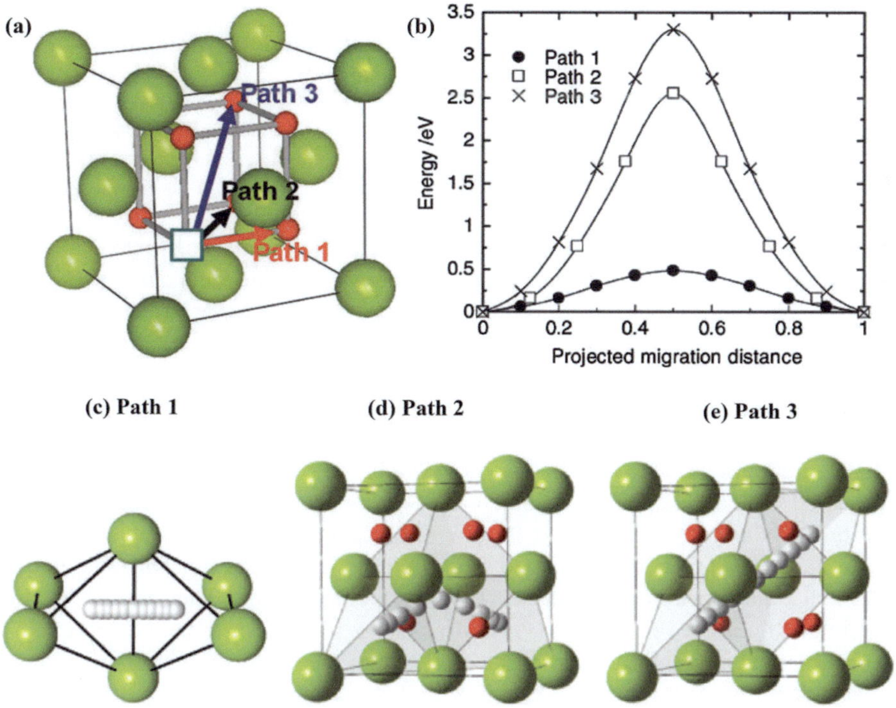

Fig. 2.19 (**a**) Three possible migration pathways of oxygen ions in ceria; (**b**) energy profiles during oxide ion hopping for Paths 1–3, and (**c**)–(**e**) corresponding trajectories of hopping oxygen ions. Large *light-green* spheres indicate cerium, and small *white* and *red* (*brighter* and *darker* small) spheres correspond to hopping and non-hopping oxygen ions, respectively (Reproduced from Ref. [23] with permission of the Royal Society of Chemistry)

(Fig. 2.18c) and $Y_{0.785}Ta_{0.215}O_{1.715}$ (Fig. 2.18d). Thus, the diffusion path along the [100] directions is a common feature of the fluorite-type ionic conductors.

In another research, Nakayama and Martin used a nudged elastic band (NEB) method to study the diffusion paths of oxygen vacancies in the ceria crystal [23]. There are three possible paths for oxygen vacancy migration to the adjacent sites, i.e., along the [100], [110], and [111] directions, as shown in Fig. 2.19a. The computational results of the energy profiles for the above three possible paths are shown in Fig. 2.19b. The horizontal ordinate denotes the migration distance of oxygen vacancy projected onto the straight line connecting the initial site (refers to 0) and the final site (refers to 1). The migration activation energy reaches a maximum at the middle site for all the three paths. Obviously, the activation energies of Paths 2 and 3 are much larger than that of Path 1 (~0.5 eV), so the two paths have negligible contribution to the oxygen vacancy motion in the ceria crystal. Shimojo et al. used a molecular dynamics simulation method to investigate the oxygen ionic migration in yttria-doped ZrO_2 with fluorite structure and found that oxygen ions hop dominantly along the [100] direction between adjacent tetrahedra formed by the cations (i.e., Path 1) [24]. Therefore, one can conclude that for fluorite oxides, the main path for oxygen ionic migration lies along the [100] direction, and it is not a straight line but a curved line.

2.2.2.2 Perovskite-Type and Related Oxides

Perovskite oxides (ABO_3) are another type of oxygen ionic conductors that are widely used in SOFCs, MIEC membranes, and catalysis. The ideal cubic-symmetry structure has the B cation in sixfold coordination, surrounded by an octahedron of oxygen ions, and the A cation in 12-fold cuboctahedral coordination. Oxygen vacancy diffusion is the dominant mechanism for oxygen motion in the crystal. Lanthanum gallate- and $LaGaO_3$-based perovskite oxides have been widely studied because of their possible applications in intermediate-temperature SOFCs and MIEC membranes. Among the series, (Sr,Mg)-doped $LaGaO_3$ has been well addressed because of its higher ionic conductivity than YSZ at intermediate temperatures. Islam et al. computed the migration path of oxygen ions in $LaGaO_3$, $LaCoO_3$, and $CaZrO_3$ crystals based upon energy minimization and interatomic potentials and quantum mechanical (QM) methods [25–27]. They found that the hopping occurs between two oxygen sites along the BO_6 octahedron edge. The migration path is a slightly curved trajectory accompanied by outward relaxation of adjacent cations. Subsequently, Yashima et al. confirmed the above migration path by using MEM based on neutron diffraction data for the composition of $La_{0.8}Sr_{0.2}Ga_{0.8}Mg_{0.15}Co_{0.05}O_{2.8}$, as shown in Fig. 2.20a and b, which is cubic perovskite with space group of $Pm\overline{3}m$ at high temperatures (1198–1392 °C) [21]. In their experiments, they found that the nuclear density in the diffusion path is higher at higher temperatures, which is consistent with the verification that the oxygen ionic motion increases with temperature. In another research, they selected a double perovskite $La_{0.64}(Ti_{0.92}Nb_{0.08})O_{2.99}$, which has lower symmetry P4/mmm space group in the temperature range of 496–1358 °C, as an example to illustrate the oxygen ionic migration path. They disclosed that the nuclear probability density of an O3 atom is connected with that of nearest-neighbor O3 atoms, and thus, this finding indicates that the diffusion of oxygen ions is along a pathway following the [110] and $\left[1\overline{1}0\right]$ directions (path A in Fig. 2.19b). In other words, the diffusion paths of O3 atoms exist along the [110] directions around the center of the path. Similarly, the O3 atom migrates along a curved route as that occurring in cubic perovskite oxides. However, in the present double perovskite, a two-dimensional diffusion pathway is present, i.e., the O3 atoms migrate along the [110] and $\left[1\overline{1}0\right]$ directions (Fig. 2.20c and d). This result indicates that the mobility of the oxygen ions in the ab plane would be much higher than that along the c-axis.

With the powerful MEM, Yashima et al. offered a clear and visual picture on the oxygen ionic migration in the K_2NiF_4-type oxides, as shown in Fig. 2.21 [28]. They selected a composition of $Pr_{1.8}La_{0.2}Ni_{0.74}Cu_{0.21}Ga_{0.05}O_{4+\delta}$ as the example because

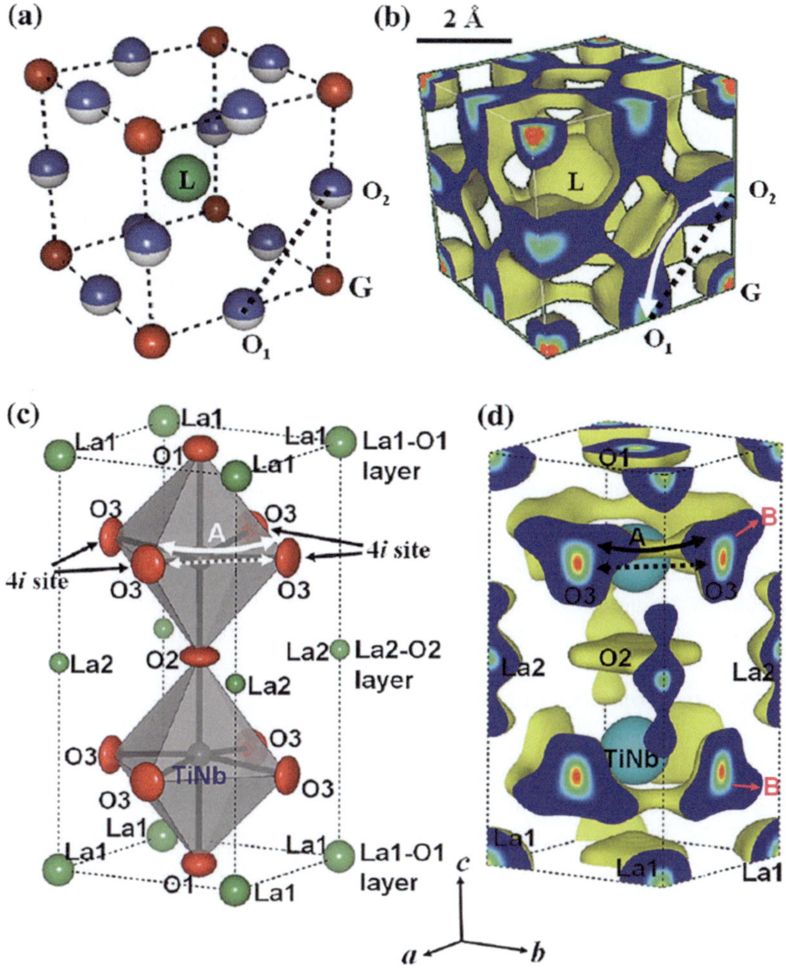

Fig. 2.20 (**a**) Crystal structure and (**b**) isosurface of nuclear density (0.05 fm $Å^{-3}$) in cubic $(La_{0.8}Sr_{0.2})(Ga_{0.8}Mg_{0.15}Co_{0.05})O_{2.8}$ at 1392 °C. (**c**) Crystal structure and (**d**) isosurface of nuclear density (0.08 fm $Å^{-3}$) in the tetragonal $La_{0.64}(Ti_{0.92}Nb_{0.08})O_{2.99}$ with a double perovskite-type structure at 1358 °C (Reproduced from Ref. [21] with permission of Elsevier)

of its high oxygen permeability. This material has a space group of I4/mmm, and its single cell consists of a (Pr,La)(Ni,Cu,Ga)O_3 perovskite layer and a (Pr, La)O rock salt layer. The excess oxygen lies in the rock salt layer as interstitial oxygen ions (O3 in Fig. 2.21). They observed that the oxygen atom at the O2 site has highly anisotropic thermal motion along the [100] directions, and the oxygen ionic migration occurs through the O2 and interstitial O3 sites. However, they did not find the path along the octahedron O–O edges in the perovskite layer, which is the migration paths of oxygen ions in perovskite-type materials mentioned above.

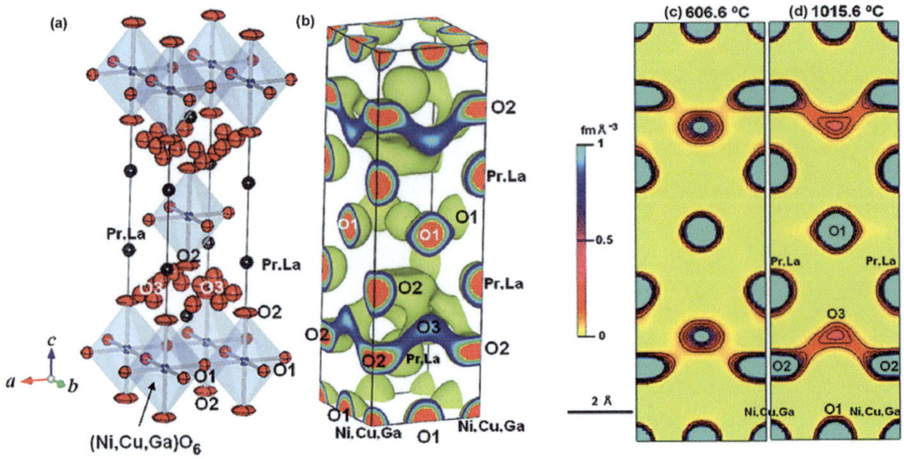

Fig. 2.21 Refined crystal structure (**a**) and isosurface of nuclear density (**b**) at 0.05 fm Å^{-3} of the mixed oxide-ionic and electronic conductor $(Pr_{0.9}\text{-}La_{0.1})_2(Ni_{0.74}Cu_{0.21}Ga_{0.05})O_{4+\delta}$ determined *in situ* at 1015.6 °C. Unit cell: tetragonal *I4/mmm*, $a = 3.875(3)$ Å and $c = 12.738(9)$ Å. Nuclear density distribution on the (100) plane of the mixed conductor $(Pr_{0.9}\text{-}La_{0.1})_2(Ni_{0.74}Cu_{0.21}Ga_{0.05})$ $O_{4+\delta}$ at 606.6 °C (**c**) and 1015.6 °C (**d**). Contour lines from 0.1 to 1.0 by the step of 0.1 fm Å^{-3} (Reproduced from Ref. [28] with permission of the American Chemical Society)

2.2.3 *Diffusion Coefficients*

In a material, under a fixed concentration gradient, if the vacancy diffusion mechanism is dominated, the migration probability of the corresponding atom to vacancies is related to the number of the adjacent vacancies. Additionally, if each migration is independent or not disturbed by other migrations, then for one-dimensional random-walk diffusion, the vacancy diffusion coefficient can be written as

$$D_{\mathrm{V}} = \frac{1}{2} a_0^2 \nu \exp\left(-\frac{\Delta G_{\mathrm{m}}}{RT}\right) \tag{2.41}$$

where a_0, ν, and ΔG_{m} are the distance for each migration, the atomic vibrating frequency, and molar Gibbs energy of migration (i.e., activation energy for the diffusion), respectively. The exponential term denotes the migration probability. In the case of three-dimensional diffusion with identical probability, the corresponding vacancy diffusion coefficient is

$$D_{\mathrm{V}} = \frac{1}{6} a_0^2 \nu \exp\left(-\frac{\Delta G_{\mathrm{m}}}{RT}\right) \tag{2.42}$$

In general, a geometrical factor of γ is included to show the effect of the geometry of the diffusion process, and thus, the above equations become

$$D_{\mathrm{V}} = \gamma a_0^2 v \exp\left(-\frac{\Delta G_{\mathrm{m}}}{RT}\right) \tag{2.43}$$

With regard to the atomic self-diffusion in a material, the migration toward a certain direction is driven by the thermal motion of atoms via a random-walk mechanism but not by the concentration gradients or chemical potential gradients. Accordingly, the self-diffusion coefficient is

$$D_{\mathrm{S}} = \frac{c_{\mathrm{V}}}{c_{\mathrm{L}}} \gamma a_0^2 v \exp\left(-\frac{\Delta G_{\mathrm{m}}}{RT}\right) = \delta D_{\mathrm{V}} \tag{2.44}$$

where c_{V}, c_{L}, and δ are the atomic vacancy concentration in the crystal, atomic lattice site concentration in the crystal, and the atomic vacancy fraction (or nonstoichiometry), respectively. The three parameters are constant and independent of position and time since there is no concentration or chemical potential gradients. For most materials, the δ value is in the range of 0.001–0.1, and thus, the diffusion flux of atoms driven by concentration or chemical potential gradients is much higher than that driven by thermal motion.

To acquire the self-diffusion coefficient, isotope-labeling method is frequently used if the isotopic effect can be ignored. Imagine that for a tracer atom, a new vacancy is produced at the site where tracer atom stays before the transition. As the tracer atom jumps to the adjacent vacancies in the next transition, it has the possibility to jump back to the site where it initially stays. As a result, there is no motion after the two transitions. Furthermore, in another situation, a normal atom jumps to the new vacancy and removes the vacancy around the tracer atom. Therefore, the tracer diffusion coefficient is a fraction of the self-diffusion coefficient:

$$D^* = f D_{\mathrm{S}} \tag{2.45}$$

where f is the correlation coefficient, which is determined by the crystal structure of materials and the diffusion mechanism. Table 2.1 lists the correlation coefficient in different crystal structures with a vacancy diffusion mechanism. As shown in the table, the larger coordinate number, i.e., the number of atoms that the vacancy has the possibility to reach in once transition, results in a correlation coefficient that is

Table 2.1 The correlation coefficient in different crystal structures with a vacancy diffusion mechanism

Structure	Coordinate number	Correlation coefficient (f)
Diamond typed	4	0.5
Simple cubic	6	0.6531
Body-centered cubic	8	0.7272
Face-centered cubic	12	0.7815
Hexagonal close packing	12	$f_x = f_y = 0.7812, f_z = 0.7815$

more close to 1. In the case of interstitial diffusion, in which the interstitial atoms have low concentration in the crystal but many available empty interstitial sites, the diffusion is according to the random-walk mechanism. Thus, the correlation coefficient is expected to be 1.0 regardless of whether the interstitial is a native atom or a tracer atom.

The conductivity relaxation method is convenient to measure the diffusivity (see Chap. 4). The determined diffusion coefficient is the chemical diffusion coefficient, which describes the relaxation of compositional gradients to achieve a homogeneous composition. Under this condition, the diffusion is not driven by a concentration gradient but rather a chemical potential gradient, and thus, relationship between the chemical diffusion coefficient and self-diffusion coefficient for a neutral atom is

$$D_{\text{chem}} = \frac{D_{\text{S}}}{RT}\frac{\partial \mu_i}{\partial \ln c_i} \tag{2.46}$$

where μ_i is the chemical potential of atom i and c_i is the concentration of atom i in the material. The term $\frac{1}{RT}\frac{\partial \mu_i}{\partial c_i}$ is the thermodynamic factor. It is sometimes very large and improves the ionic flux above that would be expected from the concentration gradient alone. If a material is predominantly an electron or hole conductor, the transport of ions is similar to the transport of neutral atom. For MIEC materials, the Eq. (2.46) can be re-written as

$$D_{\text{chem}} = \frac{D_{\text{S}}}{2}\frac{\partial \ln P_{O_2}}{\partial \ln c_{O^{2-}}} \tag{2.47}$$

The thermodynamic factor $\frac{\partial \ln P_{O_2}}{\partial \ln c_{O^{2-}}}$ is measurable by determining the dependence of oxygen vacancy concentration on oxygen partial pressure at a certain temperature. Gravimetric and coulometric methods have been frequently utilized in previous studies. Figure 2.22 shows the dependence of oxygen deficit of $Ba_{0.5}Sr_{0.5}Co_{0.8}Fe_{0.2}O_{3-\delta}$ on oxygen partial pressure and temperature [29]. The thermodynamic factor of oxygen was obtained according to $\gamma_{\text{O}} = \frac{1}{2}\left(\frac{\partial \ln P_{O_2}}{\partial \ln(3-\delta)}\right)_T$ by linear regression.

2.2.4 *Diffusion and Ionic Conductivity*

In ionic crystals, if the vacancies are fully ionized and all contribute to the transport, the diffusion of ions results in the ionic conductivity, which can be described by the Nernst–Einstein equation:

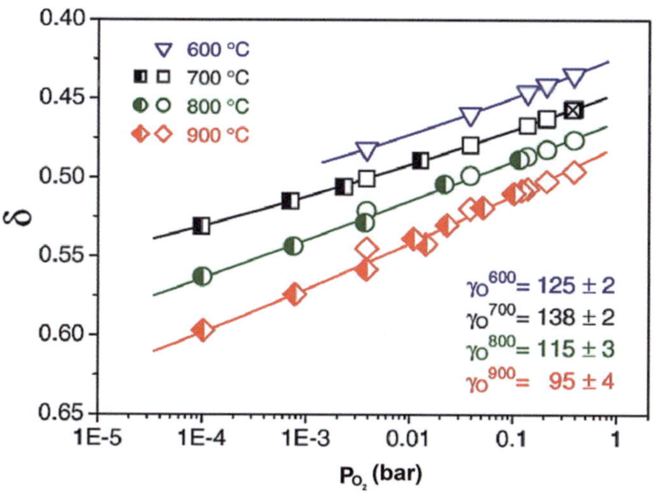

Fig. 2.22 Oxygen nonstoichiometry δ and thermodynamic factor of oxygen γ_O of BSCF5582 at 600–900 °C. Results from isotherms (*half-filled symbols*) are compared with isobars (*open symbols*). The reference point (⊠) was obtained by total reduction in Ar–H_2. The experimental error is estimated to be $\delta \pm 0.007$ (Reproduced from Ref. [29] with permission of Elsevier)

$$\sigma_i = \frac{Z_i^2 F^2 c_V}{RT} D_V = \frac{Z_i^2 F^2 \delta}{RT V_m} D_V = \frac{Z_i^2 F^2 c_L}{RT} D_S = \frac{Z_i^2 F^2 c_L}{RTf} D^* \tag{2.48}$$

where Z_i and V_m are the charge number of the ion and molar volume of the material, respectively. For an ionic conductor like YSZ, its electron or hole conductivity is far lower than the ionic conductivity, and thus, the measured total conductivity is very close to its ionic conductivity. By using Eq. (2.48), the vacancy diffusion coefficient can be calculated if the vacancy concentration is available. However, for MIEC materials, electron or hole conductivity is usually two orders higher than the oxygen ionic conductivity, so the diffusion coefficient of oxygen ions cannot be determined through testing the total conductivity of the material. ^{18}O tracer diffusion combined with SIMS depth profile and conductivity relaxation methods (see Chap. 4) are frequently used to determine the tracer diffusion coefficient and the chemical diffusion coefficient, respectively. By using the Eqs. (2.47 and 2.48), the oxygen ionic conductivity can be calculated. Recently, Zhu et al. developed an oxygen permeation model, and with this model, they could separate the bulk ionic diffusion resistance from surface exchange resistances [30]. Thus, the oxygen ionic conductivity and diffusion coefficient could be obtained through the model regression of the permeation experimental data (see Chap. 5).

2.2.5 *Grain Boundary Diffusion*

The grain boundary is an internal surface between any two grains that have the same crystal structure and composition. The diffusion processes on the internal surfaces of polycrystalline materials are of great importance for many applications, since the grain boundaries and phase interfaces determine many properties of a material. In grain boundaries, the atomic arrangement is not compact as that in the crystal bulk, which results in the reduction of average coordination number and the increase in the bond length of the atoms (or ions) in the grain boundary core. Thus, there is an excess free energy that leads to depletion or accumulation effects for the mobile ionic and/or electronic species. The conventional viewpoint is that elements that strongly segregate in the grain boundaries would retard the grain boundary diffusion, and the greater the magnitude of segregation, the stronger the effect. For example, Park et al. found that grain boundary in YSZ electrolyte showed that the oxide ion diffusion is blocked when it jumps across the grain boundary [31]. However, the diffusion in solids occurring along grain boundaries and over free surfaces is more rapid than through the interiors of the crystals. Ultrafine grained (relative density of 97–99 % and average crystallite sizes of 80 or 300 nm) and undoped monoclinic ZrO_2 materials were selected as examples to explore the effect of grain boundaries on the diffusion of oxygen using ^{18}O as tracer and SIMS profiling. The samples were prepared according to the following steps: Zr sputtering, inert gas condensation, oxidation, in situ consolidation, and pressureless sintering at 950 or 1050 °C in vacuum. Brossmann et al. observed a steep drop of the ^{18}O concentration close to the surface in the ^{18}O diffusion profiles followed by a slower decrease over an extended depth range. Their model investigation attributed this phenomenon to diffusion of ^{18}O directly from the surface and via grain boundaries into the monoclinic ZrO_2 crystallites and disclosed that the diffusion of ^{18}O in grain boundaries in undoped monoclinic ZrO_2 is 10^3–10^4 times faster than in the bulk of the crystallites in the temperature range of 450–950 °C [32].

References

1. Tilley RJD (2008) Defects in solids. Wiley, Hoboken
2. Petrov AN, Cherepanov VA, Kononchuk OF, Gavrilova LY (1990) Oxygen nonstoichiometry of $La_{1-x}Sr_xCoO_{3-\delta}$ ($0 < x \leq 0.6$). J Solid State Chem 87:69–76
3. Cherepanov VA, Gavrilova LY, Aksenova TV, Ananyev MV, Bucher E, Caraman G, Sitte W, Voronin VI (2007) Synthesis, structure and oxygen nonstoichiometry of $La_{0.4}Sr_{0.6}Co_{1-y}Fe_yO_{3-\delta}$. Prog Solid State Chem 35:175–182
4. van Doorn RHE, Burggraaf AJ (2000) Structural aspects of the ionic conductivity of $La_{1-x}Sr_x$ $CoO_{3-\delta}$. Solid State Ionics 128:65–78
5. Ou DR, Mori T, Ye F, Kobayashi T, Zou J, Auchterlonie G, Drennan J (2006) Oxygen vacancy ordering in heavily rare-earth-doped ceria. Appl Phys Lett 89:171911-(1–3)
6. Matsui T, Inaba M, Mineshige A, Ogumi Z (2005) Electrochemical properties of ceria-based oxides for use in intermediate-temperature SOFCs. Solid State Ionics 176:647–654

7. Park JH, Blumenthal RN (1989) Electronic transport in 8 mole percent Y_2O_3-ZrO_2. J Electrochem Soc 136:2867–2876
8. Jang JH, Choi GM (2002) Partial electronic conductivity of Sr and Mg doped $LaGaO_3$. Solid State Ionics 154–155:481–486
9. Shimonosono T, Hirata Y, Ehira Y, Sameshima S, Horita T, Yokokawab H (2004) Electronic conductivity measurement of Sm- and La-doped ceria ceramics by Hebb-Wagner method. Solid State Ionics 174:27–33
10. Yahiro H, Eguchi K, Arai H (1989) Electrical-properties and reducibilities of ceria rare earth oxide systems and their application to solid oxide fuel-cell. Solid State Ionics 36:71–75
11. Steele BCH, Zheng K, Rudkin RA, Kiratzis N, Christitie M (1995) Properties and applications of $Ce(Gd)O_{2-x}$ electrolytes in the temperature range 500–700 °C. In: Dokiya M, Yamonoto O, Tagawa H, Singhal SC (eds) Proceedings of the fourth international symposium on solid oxide fuel cells (SOFC-IV), June 1995. The Electrochemical Society, New Jersey, pp 1028–1038
12. Matsui T, Kosaka T, Inaba M, Mineshige A, Ogumi Z (2005) Effects of mixed conduction on the open-circuit voltage of intermediate-temperature SOFCs based on Sm-doped ceria electrolytes. Solid State Ionics 176:663–668
13. Mizusaki J, Mima Y, Yamauchi S, Fueki K (1989) Nonstoichiometry of the perovskite-type oxides $La_{1-x}Sr_xCoO_{3-\delta}$. J Solid State Chem 80:102–111
14. Petrov AN, Kononchuk OF, Andreev AV, Cherepanov VA, Kofstad P (1995) Crystal structure, electrical and magnetic properties of $La_{1-x}Sr_xCoO_{3-y}$. Solid State Ionics 80:189–199
15. Merkle R, Maier J (2008) How is oxygen incorporated into oxides? A comprehensive kinetic study of a simple solid-state reaction with $SrTiO_3$ as a model material. Angew Chem Int Ed 47:3874–3894
16. Shim JH, Park JS, Holme TP, Crabb K, Lee W, Kim YB, Tian X, Gür TM, Prinz FB (2012) Enhanced oxygen exchange and incorporation at surface grain boundaries on an oxide ion conductor. Acta Mater 60:1–7
17. Guo X, Zhang ZL (2003) Grain size dependent grain boundary defect structure: case of doped zirconia. Acta Mater 51:2539–2547
18. Metlenko V, Ramadan AHH, Gunkel F, Du HC, Schraknepper H, Hoffmann-Eifert S, Dittmann R, Waserb R, De Souza RA (2014) Do dislocations act as atomic autobahns for oxygen in the perovskite oxide $SrTiO_3$? Nanoscale 6:12864–12876
19. Royera S, Duprezb D, Kaliaguinea S (2005) Role of bulk and grain boundary oxygen mobility in the catalytic oxidation activity of $LaCo_{1-x}Fe_xO_3$. J Catal 234:364–375
20. Diethelm S, van Herle J, Sfeir J, Buffat P (2004) Influence of microstructure on oxygen transport in perovskite type membranes. Br Ceram Trans 103:147–152
21. Yashima M (2008) Crystal structures, structural disorders and diffusion paths of ionic conductors from diffraction experiments. Solid State Ionics 179:797–803
22. Yashima M, Kobayashi S, Yasui T (2007) Positional disorder and diffusion path of oxide ions in the yttria-doped ceria $Ce_{0.93}Y_{0.07}O_{1.96}$. Faraday Discuss 134:369–376
23. Nakayama M, Martin M (2009) First-principles study on defect chemistry and migration of oxide ions in ceria doped with rare-earth cations. Phys Chem Chem Phys 11:3241–3249
24. Shimojo F, Okazaki H (1992) Molecular dynamics studies of yttria stabilized zirconia. II Mechanism of oxygen diffusion. J Phys Soc Jpn 61:4106–4118
25. Khan MS, Islam MS, Bates DR (1998) Dopant substitution and ion migration in the $LaGaO_3$-based oxygen ion conductor. J Phys Chem B 102:3099–3104
26. Islam MS, Davies RA (2004) Atomistic study of dopant site-selectivity and defect association in the lanthanum gallate perovskite. J Mater Chem 14:86–93
27. Kuwabara A, Tanaka I (2004) First principles calculation of defect formation energies in Sr- and Mg-doped $LaGaO_3$. J Phys Chem B 108:9168–9172
28. Yashima M, Enoki M, Wakita T, Ali R, Matsushita Y, Izumi F, Ishihara T (2008) Structural disorder and diffusional pathway of oxide ions in a doped Pr_2NiO_4-based mixed conductor. J Am Chem Soc 130:2762–2763

29. Bucher E, Egger A, Ried P, Sitte W, Holtappels P (2008) Oxygen nonstoichiometry and exchange kinetics of $Ba_{0.5}Sr_{0.5}Co_{0.8}Fe_{0.2}O_{3-\delta}$. Solid State Ionics 179:1032–1035
30. Zhu XF, Liu HY, Cong Y, Yang WS (2012) Permeation model and experimental investigation of mixed conducting membranes. AIChE J 58:1744–1754
31. Park JS, Kim YB, An J, Prinz FB (2012) Oxygen diffusion across the grain boundary in bicrystal yttria stabilized zirconia. Solid State Commun 152:2169–2171
32. Brossmann U, Würschum R, Södervall U, Schaefer HE (1999) Oxygen diffusion in ultrafine grained monoclinic ZrO_2. J Appl Phys 85:7646–7654

Chapter 3
Ionic Conductors and Aspects Related to High Temperature

Abstract The ionic conductivity of perovskite-type MIEC membranes is several orders of magnitude lower than their electronic conductivity; however, the ambipolar conductivity determines the diffusion rate of oxygen ions in the membrane bulk. Therefore, the improvement of oxygen ionic conductivity is the key to enhance the oxygen permeability of an MIEC material. In this chapter, the common oxygen ionic conductors with fluorite and perovskite structures are introduced in detail, and other types of oxygen ionic conductors, such as $Bi_4V_2O_{11}$, $La_2Mo_2O_9$, and $La_{10\text{-}x}Si_6O_{26+y}$, are briefly presented. The critical factors influencing the oxygen ionic conductivity are discussed for the perovskite-type oxygen ionic conductors and MIEC conductors. All the oxygen ionic conducting materials can be made into MIEC membranes for oxygen permeation as long as they are properly doped by elements with variable valence states or mixed with a secondary phase with electronic conduction. MIEC membranes are all operated at high temperatures, and thus, the special properties related to high temperature, such as cationic diffusion, kinetic demixing, thermal expansion, chemical expansion, and creep, are briefly introduced to show that several factors should be considered to design or select a practically applicable MIEC membrane material.

Keywords Ionic conductors • Fluorite-type • Zirconia • Ceria • Perovskite • Nonstoichiometric oxygen • Cationic diffusion • Kinetic demixing • Chemical expansion • Creep

3.1 Fluorite-Type Oxygen Ionic Conductors

In terms of oxygen ionic conductors, fluorite-type oxides are first introduced since the best known material, yttria-stabilized zirconia (YSZ), has been widely used as a solid electrolyte for solid oxide fuel cells (SOFCs), sensors, and structural ceramics. This type of oxygen ionic conductor was first investigated in 15 wt% Y_2O_3-doped ZrO_2 by Nernst in the 1890s. This material is a state-of-the-art electrolyte for SOFCs in the temperature range of 600–1000 °C because it exhibits a sufficient oxygen ionic conductivity and an ionic transport number of 1 under the working condition of SOFCs. The cubic fluorite-type structure can be derived from the CsCl

X. Zhu, W. Yang, *Mixed Conducting Ceramic Membranes*, Green Chemistry and Sustainable Technology, DOI 10.1007/978-3-662-53534-9_3

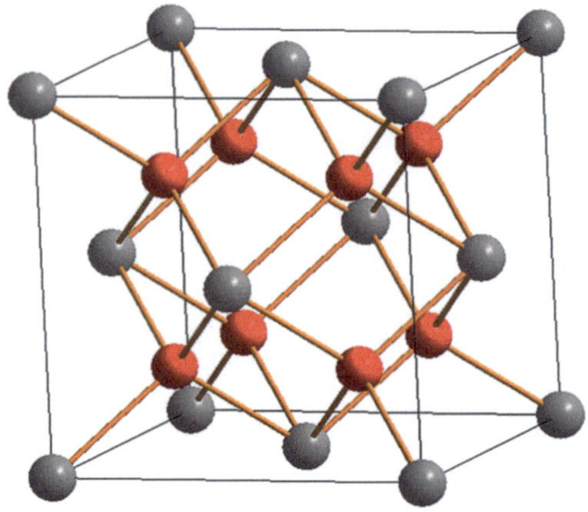

Fig. 3.1 Crystal structure of cubic fluorite-type oxides

structure by taking away the cations alternately; therefore, eight-coordinated cubic interstices are produced. The structure of the cubic fluorite oxides is shown in Fig. 3.1. If the repulsion effect between oxygen ions is weak at a far enough distance between oxygen ions, the cell parameter is determined by the metal–oxygen distance, i.e., the sum of ionic radii for the appropriate coordination numbers. Therefore, to form a stable cubic fluorite structure, the cations should be big enough. Tetravalent cations like Ce^{4+} (0.097 nm), Pr^{4+} (0.096 nm), Th^{4+} (0.105 nm), and U^{4+} (0.100 nm) have large ionic radii, so they can form a stable cubic fluorite structure at ambient conditions, while the ionic radii of Zr^{4+} (0.084 nm) and Hf^{4+} (0.083 nm) are somewhat smaller, and their cubic fluorite structure only appears at high temperatures. For much smaller tetravalent cations like Sn^{4+} and Ti^{4+}, the fluorite structure is replaced by the rutile structure.

3.1.1 Zirconia-Based Ionic Conductors

3.1.1.1 Stabilization of Zirconia by Doping

The phase structure of the undoped zirconia is a function of temperature, i.e., it experiences the phase transition from a monoclinic to tetragonal structure at a temperature of ~1170 °C and the phase transition from tetragonal to cubic at a temperature of ~2370 °C. The theoretic density of the three phases is 5.65, 6.10, and 6.27, respectively. Therefore, as the sintered zirconia is cooled from sintering temperature (1200–1400 °C) to room temperature, a large volume increase accompanies the diffusionless phase transition from a tetragonal to monoclinic structure. The phase transition stems from the fact that the Zr^{4+} ion ($r = 0.084$ nm) is too small for the eight-coordinated cubic interstices formed by oxygen ions. The Ce^{4+} ion ($r = 0.097$ nm) is big enough, so the cubic fluorite structure is stable at room temperature. In contrast, the pure zirconia has extremely low oxygen ionic conductivity at elevated temperatures based on the intrinsic oxygen deficiency. The cations in the fluorite structure can be partially replaced by alkaline earth or rare earth cations in a wide range of compositions. The resultant solid solutions have nonstoichiometric oxygen if the valence state of the doped cations is not +4. The introduction of nonstoichiometric oxygen can be denoted as

$$\frac{1}{2}Y_2O_3 \overset{ZrO_2}{\rightarrow} Y'_{Zr} + \frac{1}{2}V_O^{\cdot\cdot} + \frac{3}{2}O_O^{\times} \tag{3.1}$$

The doping of alkaline earth or rare earth elements can reduce the phase transition temperature and even stabilize the high-temperature cubic phase down to room temperature. Simultaneously, the doping of +2 or +3 cations introduces a large number of oxygen vacancies in the zirconia lattice (see Eq. 3.1) and thus leads to high oxygen ionic conductivity. The conductivity can be written as

$$\sigma = e\mu[V_O^{\cdot\cdot}] \tag{3.2}$$

where μ is the mobility, $[V_O^{\cdot\cdot}]$ is the concentration of oxygen vacancies, and e is the effective charge under the electrical neutrality condition, as illustrated by Eq. (3.1), $2[Y'_{Zr}] = [V_O^{\cdot\cdot}]$. The relationship clearly indicates that the oxygen vacancy concentration is directly dependent on the doping level. Zirconia stabilized by rare earth oxides usually has higher oxygen ionic conductivity than that of alkaline earth oxides. Figure 3.2 shows the influence of doping level on the oxygen ionic conductivity for zirconia doped with different rare earth oxides, including Sc_2O_3, Yb_2O_3, Er_2O_3, Y_2O_3, Dy_2O_3, and Gd_2O_3 [1]. The oxygen ionic conductivity of

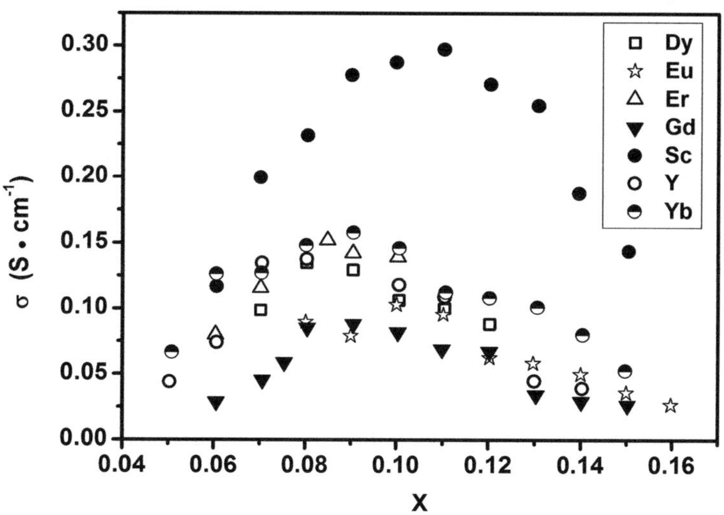

Fig. 3.2 Composition dependence of the electrical conductivity in $(ZrO_2)_{1-x}(Ln_2O_3)_x$ (Ln = lanthanides) (Reproduced from [1] with permission of Springer)

stabilized zirconia is strongly dependent on the dopant element and its doping level. The maximum values appear at the doping amount of ~11 % for Sc_2O_3 and at 8 ~ 9 % for Yb_2O_3, Er_2O_3, Y_2O_3, Dy_2O_3, and Gd_2O_3. At a lower doping level, the conductivity increases monotonically with the increase in the doping level, as illustrated by Eq. (3.2). After reaching the maximum conductivity, further increasing the doping level decreases the oxygen ionic conductivity. Association of point defects (see Chap. 2) occurs between Ln'_{Zr} and $V_O^{\cdot\cdot}$ to form $[Ln'_{Zr} - V_O^{\cdot\cdot}]^{\cdot}$ and/or $[Ln'_{Zr} - V_O^{\cdot\cdot} - Ln'_{Zr}]^{\times}$; as a result, the concentration of free oxygen vacancies decreases with the doping level. Among these oxides, the Sc_2O_3-stabilized zirconia (ScSZ) has the highest conductivity up to 0.3 S cm^{-1} at 1000 °C [2–4]. However, 8 mol% of Zr^{4+} site replaced by Y^{3+}, commonly known as yttria-stabilized zirconia (YSZ) or 8YSZ, is the most widely used electrolyte for SOFC, since it fulfills almost all the criteria for an electrolyte. YSZ has been well studied, and it has exhibited extremely good performance as SOFC electrolyte in the Westinghouse demonstration units over several years [5].

The oxygen ionic conductivity and activation energy for the ZrO_2–Ln_2O_3 system are functions of the ionic radius of Ln^{3+}. Figure 3.3 shows the maximum oxygen ionic conductivity, apparent enthalpy of oxygen ionic diffusion in lattices, and association enthalpy of oxygen defects as a function of the ionic size of dopants [6]. This investigation shows that the ionic conductivity decreases with the increase of rare earth ionic radii. Sc^{3+} cation has the smallest ionic radius, 0.087 nm, which is the closest to that of Zr^{4+} cation ($r = 0.084$ nm); thus the lattice elastic strain is minimal for oxygen ions passing across the potential saddle point. The detailed description of the diffusion path of oxygen ions in fluorite-type oxides is illustrated in Chap. 2. Many published reviews and books have intrduced YSZ electrolyte materials; additionally, great progress has been made on the scandia-stabilized zirconia in recent years, so in the following section, the investigations on the stability of ScSZ are briefly introduced.

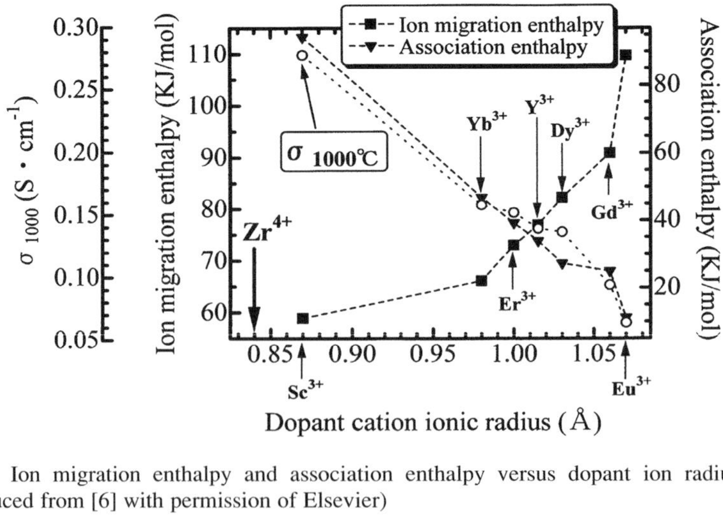

Fig. 3.3 Ion migration enthalpy and association enthalpy versus dopant ion radius curves (Reproduced from [6] with permission of Elsevier)

3.1.1.2 Scandia-Stabilized Zirconia

ScSZ is quite attractive as an electrolyte for SOFCs and MIEC membranes, except for the high cost of scandium. Further, the degradation in oxygen ionic conductivity has been observed in ScSZ as it undergoes annealing treatment at elevated temperatures. For example, Badwal et al. investigated the conductivity stability of ScSZ with different scandia doping levels (7–11 mol%) [7]. They found that all compositions show conductivity degradation with time both at 850 and 1000 °C, but the degradation is slower for the scandia doping amount of 9.0–11.0 mol%, especially at 1000 °C. The fast degradation of conductivity for ScSZ (containing 7.0–9.0 mol % Sc) in the initial stage of annealing was related to the t'-phase, which is a distorted fluorite phase. The t'-phase gradually disappears with time during the annealing and produces coherently grown t-ZrO_2 precipitates in a cubic matrix. The t-ZrO_2 precipitates have a lower Sc doping level than the t'-phase, in addition to a low oxygen ionic conductivity. The slow degradation for samples with scandia doping amount of 9.0–11.0 mol% was related to the formation of microdomains of an ordered phase rich in Sc. This degradation mechanism can be concluded as continual phase transitions to less-conductive phases. Another mechanism is the vacancy trapping or "short-range ordering," whereby attractive coulombic forces between the positive charge field around the cations and the negative field of the vacancies effectively immobilize the charge carriers in the electrolyte. Haering et al. found that the conductivity of the ScSZ samples with higher scandia content (≥11 mol%) was stable after annealing [8]. These specimens were rhombohedral at room temperature but transformed into the cubic phase between 400 and 600 °C. They found that the presence of the rhombohedral phase in the ScSZ system, and not in the YSZ system, is most probably due to the fact that the radius of Sc^{3+} is closest to that of the Zr^{4+}, whereas the radius of the Y^{3+} is more than 10 % larger. Defect association occurs in any stabilized zirconia electrolyte with less than 9 mol % of trivalent dopant. At a low doping level, the main defect association is $[Ln'_{Zr} - V_O^{\cdot\cdot}]^{\cdot}$, which transforms into $[Ln'_{Zr} - V_O^{\cdot\cdot} - Ln'_{Zr}]^{\times}$ during annealing and causes conductivity degradation. However, at a high doping level (>10 %), the main defect association is $[Ln'_{Zr} - V_O^{\cdot\cdot} - Ln'_{Zr}]^{\times}$, which is stable and does not undergo any phase transformation during annealing. Therefore, the conductivity degradation depends mainly on the concentration of dopant but is relatively independent of the nature of the dopant.

To improve the stability of ScSZ and reduce the high cost of scandia, incorporation of other cations in zirconia is regarded as the most effective solution. Many studies have been performed on the ternary system M_xO_y–Sc_2O_3–ZrO_2, where M can be Ce, Y, Yb, Bi, Al, and Ga. Among these dopants, it was reported that a small amount of CeO_2 along with Sc_2O_3 is effective to stabilize ZrO_2 [9, 10]. The resultant material, $Sc_{0.1}Ce_{0.01}Zr_{0.89}O_{1.95}$, no longer exhibits an unfavorable phase transition, making this material a very promising option for intermediate-temperature electrolytes and membranes (see Fig. 3.4).

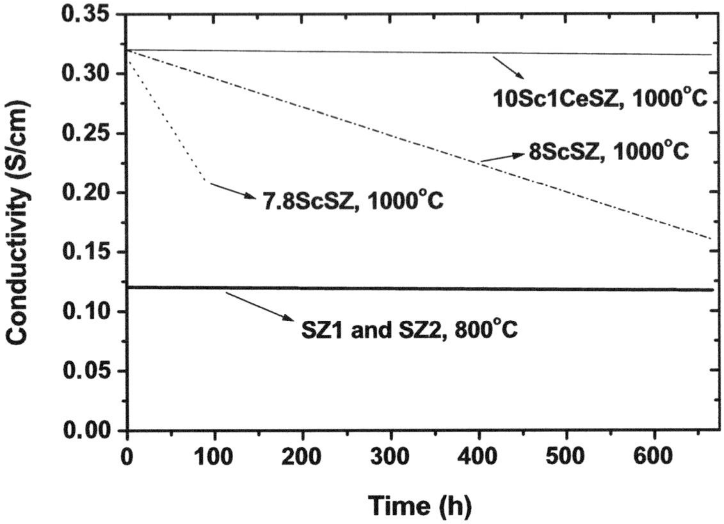

Fig. 3.4 Comparison of long-term stability of SZ samples and other Sc-doped ZrO_2 systems (Reproduced from [9] with permission of Elsevier)

3.1.1.3 Zirconia-Based Membranes for Oxygen Permeation

For oxygen permeation through dense ceramic membranes, except oxygen ionic transport across membranes, electronic transport (including electrons and holes) is also necessary. However, the zirconia-based electrolyte materials have high ionic conductivity but extremely low electronic conductivity because the valence states of Zr^{4+}, Y^{3+}, Sc^{3+}, and other rare earth cations are stable at both oxidizing and reducing atmosphere. However, partial electronic conduction is generated according to the following equations:

$$\text{At low oxygen partial pressure}: O_O^{\times} \rightleftharpoons \frac{1}{2}O_2 + V_O^{\cdot\cdot} + 2e' \tag{3.3}$$

$$\text{At high oxygen partial pressure}: \frac{1}{2}O_2 + V_O^{\cdot\cdot} \rightleftharpoons O_O^{\times} + 2h^{\cdot} \tag{3.4}$$

Assuming that the defect concentrations are low and the interactions between defects can be ignored, according to the mass action law, the concentrations of electrons and holes are

$$[e'] = (K_1[O_O^{\times}]/[V_O^{\cdot\cdot}])^{\frac{1}{2}}P_{O_2}^{-\frac{1}{4}} = K_1'P_{O_2}^{-\frac{1}{4}} \tag{3.5}$$

$$[h^{\cdot}] = (K_2[V_O^{\cdot\cdot}]/[O_O^{\times}])^{\frac{1}{2}}P_{O_2}^{\frac{1}{4}} = K_2'P_{O_2}^{\frac{1}{4}} \tag{3.6}$$

The conductivity of carrier j is proportional to its concentration, as illustrated by Eq. (3.1). Therefore, one can find that the electronic conductivity of YSZ decreases with the decrease in oxygen partial pressure according to $P_{O_2}^{\frac{1}{4}}$ and then increases with the decrease in oxygen partial pressure according to $P_{O_2}^{-\frac{1}{4}}$. The electronic conductivity of YSZ is far lower than its ionic conductivity, so the ionic transport number is very close to 1. For calcium oxide-stabilized zirconia, the ionic transport number can be reduced to 0.5 at 850 °C under oxygen partial pressure of 10^{-30} atm. Obviously, the common chemical method cannot reach such a low value. To introduce electronic conduction into the zirconia, Ce, Pr, Tb, and Ti are doped into the fluorite lattices.

Kawamura et al. investigated the influence of oxygen partial pressure on electrical conductivities of CeO_2-doped CaO–ZrO_2–solid solution system at elevated temperatures [11]. For the $(Zr_{1-x}Ce_xO_2)_{0.9}(CaO)_{0.1}$ system, as the ceria content is low, the total conductivity increases slightly and then decreases with the decrease in oxygen partial pressure (see Fig. 3.5a). Ionic conductivity is primarily proportional to the oxygen vacancy concentration at high oxygen partial pressures. The decrease in ionic conductivity at low oxygen partial pressures is related to the defect association. When the ceria content is in the range of 0.4–0.6, the total conductivity increases remarkably and then decreases with decreasing oxygen partial pressure (see Fig. 3.5b). The remarkable increase in total conductivity is caused by electronic hopping between $Ce_{Zr}^{\times}$ and $Ce_{Zr}^{\prime}$, and electronic conductivity is proportional to the product of $[Ce_{Zr}^{\times}][Ce_{Zr}^{\prime}]$. At high ceria content, the total conductivity increases with the decrease in oxygen partial pressure and is almost independent of Zr content at low oxygen partial pressures (Fig. 3.5c). It was concluded that the ionic conduction is dominated at high oxygen partial pressures, while the electronic conductivity increases with ceria content and exhibits a maximum at $x = 0.6$; however, the

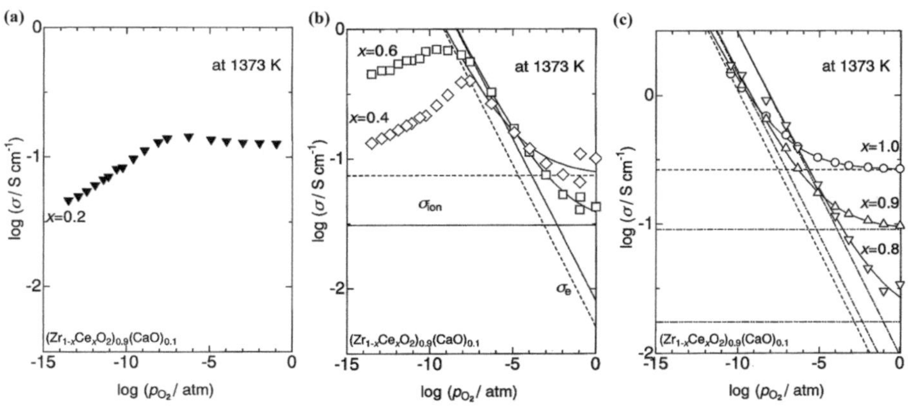

Fig. 3.5 (**a**) The electrical conductivity of $(Zr_{1-x}Ce_xO_2)_{0.9}(CaO)_{0.1}$ at 1373 K. (**a**) $x = 0.1$–0.3, (**b**) $x = 0.4$–0.6, (**c**) $x = 0.8$–1.0 (Reproduced from [11] with permission of Elsevier)

dependence of ionic conductivity on ceria content is opposite to that of electronic conductivity.

Another method to introduce electronic conductivity in membrane materials is to blend a zirconia-based ionic conductor with an electronic conductor to form a dual-phase membrane. Dual-phase membranes will be specially introduced in Chap. 7.

3.1.2 Ceria-Based Ionic Conductors

3.1.2.1 Doped Ceria

Ceria has a cubic fluorite structure with a cell parameter of 0.540 nm and space group of Fm-3 m. Thermal expansion coefficient ($8.6 \times 10^{-6}\ K^{-1}$) of ceria is little smaller than that of stabilized zirconia materials (10–$11 \times 10^{-6}\ K^{-1}$). Pure ceria is an n-type semiconductor, and its oxygen ionic conductivity is extremely low, while the oxygen ionic conductivity can be significantly improved through doping +2 alkaline earth cations or +3 rare earth cations. As in the case of zirconia, the substitution of partial Ce^{4+} by alkaline earth or rare earth cations introduces oxygen vacancies, for instance:

$$\frac{1}{2}Sm_2O_3 \overset{CeO_2}{\rightarrow} Sm'_{Ce} + \frac{1}{2}V_O^{\cdot\cdot} + \frac{3}{2}O_O^{\times} \tag{3.7}$$

Irrespective of whether the ceria is doped or undoped by other oxides, the stable cubic fluorite structure is maintained from room temperature to its melting point. Therefore, unlike MO_x ($x = 1$ or 1.5)-doped zirconia, which is usually shortened as MSZ, the MO_x ($x = 1$ or 1.5)-doped ceria is usually abbreviated as MDC, such as Sm_2O_3-doped ceria (SDC), Gd_2O_3-doped ceria (GDC), and Y_2O_3-doped ceria (YDC). The oxygen ionic conductivity of doped ceria is generally higher than that of stabilized zirconia, especially at low temperatures. The conductivity of doped ceria has a similar dependence on ionic radii and doping levels of dopants as that of stabilized zirconia. Maximum conductivity appears for a certain dopant while changing the doping level. Figure 3.6 shows the dependence of oxygen ionic conductivity of $Ce_{1-x}Sm_xO_{2-\delta}$ on the doping level at various temperatures [12]. The maximum doping level is in the range of 0.1–0.2 at all investigated temperatures. The curves shown in Fig. 3.6 can be interpreted as the interaction between Sm'_{Ce} and $V_O^{\cdot\cdot}$ to form associated defects, i.e., $[Sm'_{Ce} - V_O^{\cdot\cdot}]^{\cdot}$ and $[Sm'_{Ce} - V_O^{\cdot\cdot} - Sm'_{Ce}]^{\times}$, which decrease the free oxygen vacancy concentration at high doping levels. Therefore, additional energy is needed to break the association, which is reflected by a kink as shown by the Arrhenius plot of conductivity. Figure 3.7 shows the dependence of the maximum conductivity on the radii of dopant cations at 800 °C [12]. The ionic radii of Gd^{3+} (0.105 nm) and Sm^{3+} (0.108 nm) are close to that of Ce^{4+} (0.097 nm), and the corresponding doped ceria has the highest ionic conductivity among the rare earth cations owing to the

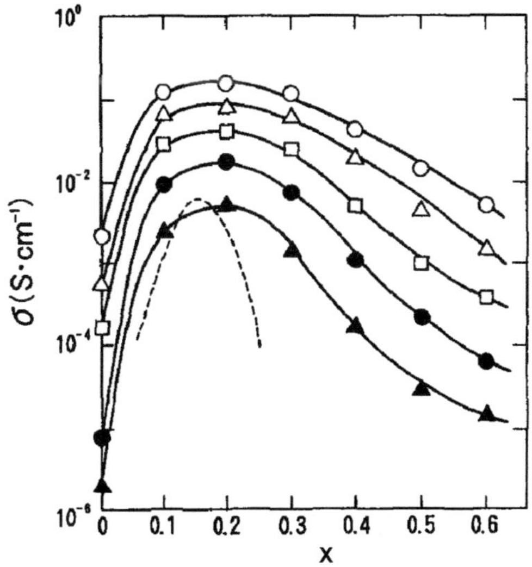

Fig. 3.6 Concentration dependence of conductivities for $(CeO_2)_{1-x}(SmO_{1.5})_x$ at various temperatures: (○) 900 °C, (Δ) 800 °C, (□) 700 °C, (●) 600 °C, (▲) 500 °C, (- - -) $(ZrO_2)_{1-x}(CaO)_x$ at 800 °C (Reproduced from [12] with permission of Elsevier)

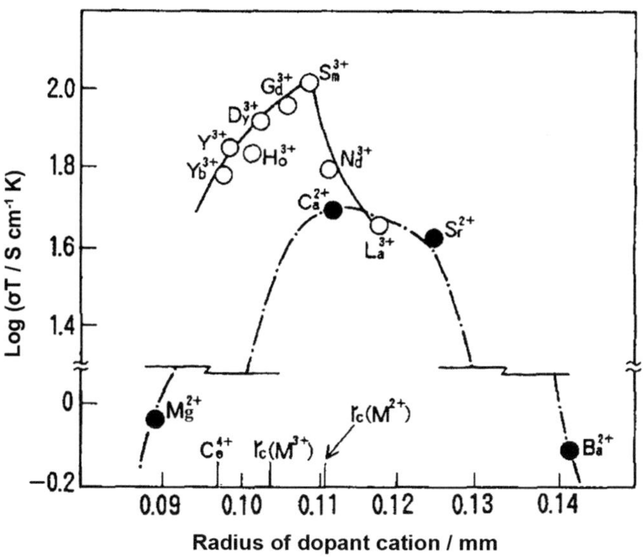

Fig. 3.7 Ionic conductivity of doped ceria at 1073 K against the radius of dopant cation, r_c, shown in the horizontal axis is the radius of the divalent or trivalent cation (Reproduced from [12] with permission of Elsevier)

similar ionic radii as that of the host ion, resulting in the minimum association enthalpy between the dopant ion and the oxygen vacancy. Ceria doped with alkaline earth oxides usually has lower ionic conductivity than ceria doped with rare earth oxides, so researches have mainly focused on the latter in recent years.

Most researchers agree that Gd_2O_3- and Sm_2O_3-doped ceria oxides show higher ionic conductivity than other rare earth oxides but have divergence in the doping level and a better dopant. Yahiro et al. and Eguchi et al. found that SDC shows higher ionic conductivity than that of GDC [13–15]. At 800 °C, the ionic conductivity of $Ce_{0.8}Sm_{0.2}O_{2-\delta}$ is as high as 0.095 S cm^{-1}, which is almost twice that of $Ce_{0.8}Gd_{0.2}O_{2-\delta}$. Zha et al. found that for both Gd- and Sm-doped ceria electrolytes, the maximum electrical conductivity is observed to be at a dopant concentration of 15 mol% in the Ce sublattice, i.e., $Ce_{0.85}Gd_{0.15}O_{1.925}$ and $Ce_{0.85}Sm_{0.15}O_{1.925}$, in the temperature range of 400–850 °C [16]. However, Kudo et al. found the maximum conductivities for GDC were at 20 mol% $GaO_{1.5}$ dopant concentration [17], and Yahiro et al. observed the optimal dopant concentration was 20 mol% $SmO_{1.5}$ for SDC [13]. Steele et al. reported that $Ce_{0.9}Gd_{0.1}O_{2-\delta}$ has the highest lattice conductivity at 500 °C up to about 0.01 S cm^{-1} and $Ce_{0.8}Gd_{0.2}O_{2-\delta}$ has the highest total conductivity because grain boundaries of the latter seem to be more tolerant to impurities [18]. The conductivities reported by different authors include the lattice conductivity and grain boundary conductivity; thus, the conductivities are highly variable from one source to another due to impurities segregation at the boundaries. Steele thought that the divergence is induced by the preparation method of materials and contents of impurities, such as silica, in the chemical regents [18].

3.1.2.2 Co-Doped Ceria

To improve the ionic conductivity of ceria-based oxides, co-doping two or more types of cations in ceria lattices was proposed. Kim et al. studied five trivalent cations (M = Y, Sm, Nd, Pr, and La) to co-dope $Ce_{0.8}(Gd_{0.2-x}M_x)O_{2-\delta}$ with the doping level up to $x = 0.05$ [19]. Only two of the co-dopants, Sm and Y, increase the electrical conductivity of the electrolyte, and the maximum value is observed for Sm at 3 mol%. Wang et al. observed that the co-doping effect was more apparent for the electrolytes with $Ce_{0.85}Gd_{0.15-y}Sm_yO_{1.925}$ than for $Ce_{0.85}Gd_{0.2-y}Sm_yO_{1.9}$, and as the Sm content was in the range of 0.05–0.1, the co-doped ceria of $Ce_{0.85}Gd_{0.15-y}Sm_yO_{1.925}$ exhibited much higher ionic conductivities than that of the singly doped ceria at 500–700 °C [20]. The co-doping strategy can suppress the oxygen vacancies ordering and increase the pre-exponential factor in the system. Other systems like co-doping with Mg, Ca, Y, Nd, etc. with Gd or Sm in ceria sublattices are all reported to be effective for improving the conductivity [21–27]. However, the grain conductivities are improved marginally compared with the grain boundary conductivity. A possible interpretation to the co-doping effect is that the grain boundary conductivity is related to the space charge layer at grain boundaries and the co-doping alters the defect concentrations in the space charge regions.

3.1.2.3 Ceria-Based Membranes for Oxygen Permeation

The doped ceria-based oxides are pure ionic conductors at high oxygen partial pressures, so it needs to introduce an electronic phase to form dual-phase membranes or dope elements with easy changeable valence states in the Ce sublattices. Ceria is chemically compatible with many of perovskite and spinel oxides, so these two types of oxides are frequently employed as the electronic conducting phase for dual-phase membranes. A detailed description of dual-phase membranes, in particular ceria-based dual-phase membranes, is presented in Chap. 7. Therefore, the following section will focus on the incorporation of $Pr^{4+/3+}$ and $Tb^{4+/3+}$ into the Ce sublattices to enhance the electronic conductivity.

The exploration of ceria-based MIEC membranes is motivated by lower thermal expansion coefficients ($10–12 \times 10^{-6}\ K^{-1}$) and higher mechanical strength of the doped ceria-based oxides than that of Co-based perovskite membrane materials (thermal expansion coefficients $>20 \times 10^{-6}\ K^{-1}$), which may be helpful for the practical applications of the MIEC membrane technology. Shuk and Greenblatt investigated the transport properties of $Ce_{1-x}Pr_xO_{2-\delta}$ solid solutions [28]. As shown in Fig. 3.8a, the ionic and electronic conductivities all increase with the doping level of Pr at 700 °C. An electronic transfer number of about 0.5 is reached for a doping amount of 0.25. The electronic transfer number increases with the

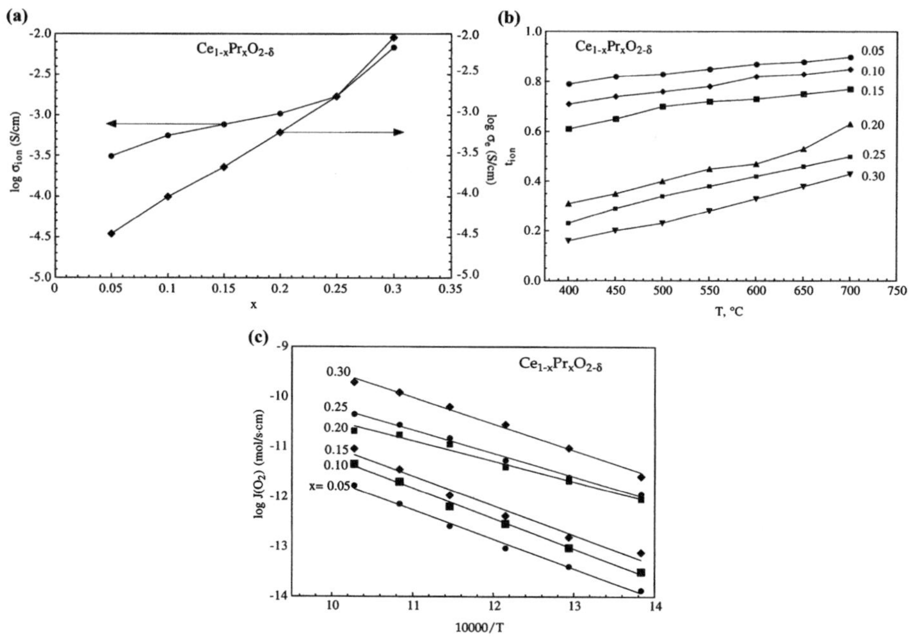

Fig. 3.8 (**a**) Concentration dependence of the ionic and electronic conductivities of $Ce_{1-x}Pr_xO_{2-\delta}$ solid solutions at 700 °C. (**b**) Ionic transfer number of $Ce_{1-x}Pr_xO_{2-\delta}$ solid solutions. (**c**) Temperature dependence of the oxygen permeability of the $Ce_{1-x}Pr_xO_{2-\delta}$ solid solutions (Reproduced from [28] with permission of Elsevier)

decrease in temperature (see Fig. 3.8b), since the electronic conductivity has lower activation energies than the ionic conductivity. The calculated oxygen permeation flux based on the ionic and electronic conductivities increases with the doping level and temperature. The highest oxygen permeation flux that was achieved on $Ce_{0.7}Pr_{0.3}O_{2-\delta}$ is 1.9×10^{-9} mol cm^{-2} s^{-1} through a 1.0-mm-thick membrane at 700 °C (see Fig. 3.8c). Fagg et al. reported an investigation on the oxygen permeability of $Zr_{0.1}Ce_{0.9-x}Pr_xO_{2-\delta}$ and $Ce_{1-x}Pr_xO_{2-\delta}$ ($x = 0.2$–0.3) [29]. They found that Zr substitution leads to a decrease in the total conductivity and an increase in the average grain size. Oxygen permeability in these materials is controlled by both ambipolar bulk diffusion and oxygen surface exchange kinetics, as shown in Fig. 3.9. Typical oxygen permeation flux is approximately 3×10^{-8} mol cm^{-2} s^{-1} through a 1.0-mm-thick membrane at 1000 °C, and this value is comparable to some perovskite membranes, such as $La(Fe,Ni)O_{3-\delta}$. In another study, Fagg et al. improved the electronic conductivity of $Ce_{0.8}Pr_{0.2}O_{2-\delta}$ by using cobalt oxide as the sintering aid (2 mol%) [30]. The addition of cobalt oxide slightly enhances the total conductivity, but the electronic conductivity increases by around two- to threefold, while the ionic conductivity is unaffected. The presence of cobalt oxide can speed up the oxygen surface exchange reactions, so the combination of these effects improves the oxygen permeability of fluorite membranes to compete with some of the best perovskite-type membranes, such as $La_{0.3}Sr_{0.7}FeO_{3-\delta}$, $SrAl_{0.3}Fe_{0.7}O_{3-\delta}$, and $La_2NiO_{4+\delta}$, at temperatures below 850 °C. Similarly, Balaguer et al. investigated the transport properties of $Ce_{1-x}Tb_xO_{2-\delta}$ ($x = 0.1, 0.2$) with 2 mol % cobalt oxide added to improve the sinterability and electronic conductivity

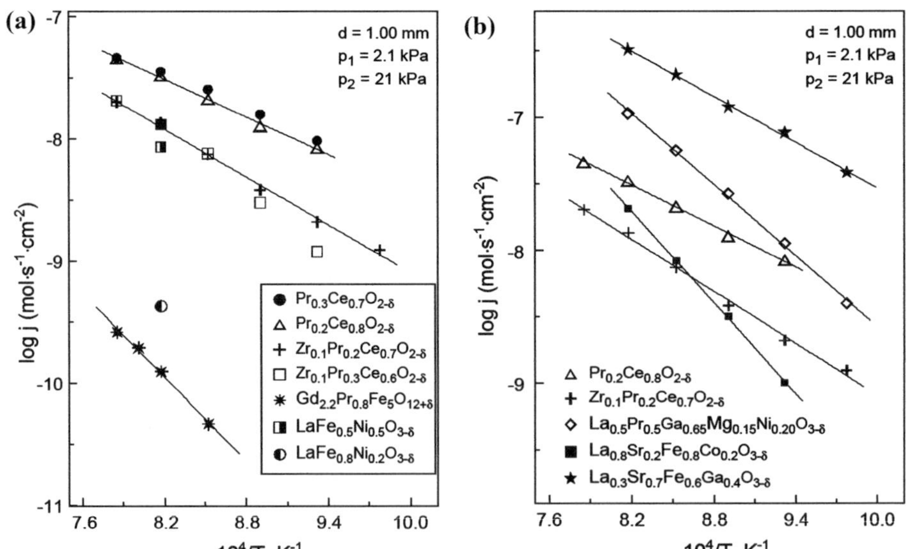

Fig. 3.9 (**a**) Comparison of oxygen permeation flux for $Ce_{1-x}Pr_xO_{2-\delta}$ and $Zr_{0.1}Ce_{0.9-x}Pr_xO_{2-\delta}$ membranes with that of $La(Fe,Ni)O_{3-\delta}$ perovskites and Pr-containing garnets. (**b**) Comparison of oxygen permeation flux for $Ce_{1-x}Pr_xO_{2-\delta}$ and $Zr_{0.1}Ce_{0.9-x}Pr_xO_{2-\delta}$ membranes with data on selected perovskite materials (Reproduced from [29] with permission of Elsevier)

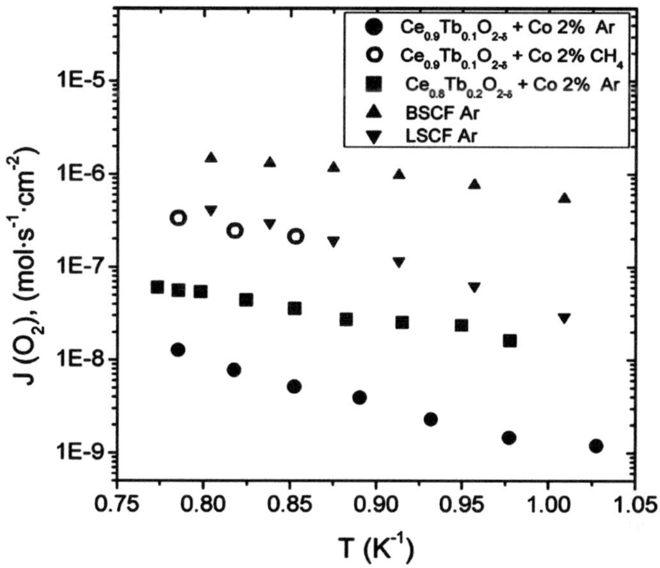

Fig. 3.10 Oxygen permeation flux as a function of inverse temperature for $Ce_{0.9}Tb_{0.1}O_{2-\delta}$ + Co 2 % and $Ce_{0.8}Tb_{0.2}O_{2-\delta}$ + Co 2 % (Reproduced from [31] with permission of American Chemical Society)

[31]. They found that the oxygen permeation is negligible for cobalt-free samples, while that through $Ce_{0.9}Tb_{0.1}O_{2-\delta}$ + Co and $Ce_{0.8}Tb_{0.2}O_{2-\delta}$ + Co 1-mm-thick membranes are about 1.5×10^{-8} mol cm^{-2} s^{-1} and 6.0×10^{-8} mol cm^{-2} s^{-1}, respectively, at 1000 °C, when argon was used as the sweep gas. The permeation flux increased to a high value up to 3.7×10^{-7} mol cm^{-2} s^{-1} at 1000 °C ($Ce_{0.9}Tb_{0.1}O_{2-\delta}$ + Co sample) when methane was used as the sweep gas, as shown in Fig. 3.10.

3.1.3 Bismuth Oxide-Based Ionic Conductors

3.1.3.1 Structure of δ-Bi_2O_3

Four polymorphs of Bi_2O_3 have been reported in the literature, i.e., α, β, γ, and δ phases, but only the δ phase exhibits high oxygen ionic conductivity. The phase transition from the monoclinic phase (α) to the high-temperature cubic phase (δ) is at approximately 730 °C, and the δ phase is stable up to its melting point of about 825 °C. The crystal structure of δ-Bi_2O_3 has a space group of Fm-3 m and is usually described with cations occupying the 4a (0 0 0) position and oxygen anions disordered distribution in 8c (¼ ¼ ¼). Its high ionic conduction is related to the presence of ~25 % oxygen ion vacancies in the structure and disordered oxygen vacancies. However, the locations of oxygen anions in the δ phase were disputed for a long time.

Gattow et al. [32], Willis et al. [33], and Verkerk et al. [34] found that δ phase has an oxygen-defective sublattice and oxygen anions usually occupy the oxygen lattice sites. In other words, the distribution of oxygen vacancies in oxygen lattice sites is disordered. This view of the crystal structure can well explain the extremely high ionic conductivity in δ-Bi_2O_3, since oxygen ions can move from site to site through the bismuth sublattice. However, Sillen [35], Zav'yalova et al. [36], Madernach et al.[37], and Jacobs et al. [38] used different methods, including electronic diffraction, theoretical calculations, and computer simulations, to investigate the oxygen sublattice of δ-Bi_2O_3 and found the vacancies are ordered in the [111] direction. Figure 3.11 shows the structural models for δ-Bi_2O_3. Figure 3.11a shows the ideal cubic fluorite structure, in which Bi^{3+} cations occupy alternate centers of a simple cubic array of oxygen anions on the 8 (c) sites. Harwig et al. suggested that a significant proportion of the structure of oxygen anions are displaced from the 8 (*c*) sites by ~0.1 nm in [111] directions, i.e., entering 32 (*f*) sites at x, x, x, with $x = \frac{1}{4} + \delta = \sim 0.335$, as shown in Fig. 3.11b [39]. Battle et al. proposed a structure model based on their neutron powder diffraction studies of the time-averaged structure of δ-Bi_2O_3 [40], as shown in Fig. 3.11c. Yashima et al. investigated the disorder oxygen anions distribution in δ-Bi_2O_3 by combining Rietveld refinement with a maximum entropy method-based pattern fitting of the neutron powder diffraction data measured at 778 °C [41]. They found the oxygen anions have a complicated disorder spreading over a wide area and shift to the [111] directions from the ideal fluorite site, as shown in Fig. 3.11c. The MEM nuclear density distribution map on the (110) plane is shown in Fig. 3.11d to visualize the structural disorder at 778 °C. Two maxima of the nuclear density are clearly seen at the 32f sites near an ideal 8c position of the oxide ion on the (110) plane. Further, the complicated disorder of oxide ions was regarded to be responsible for the fast oxide ion conduction. Recently, Mohn et al. combined neutron powder diffraction and Born–Oppenheimer molecular dynamics methods to investigate the oxygen anions disorder in δ-Bi_2O_3 [42]. They found that the asymmetric electron density around the Bi^{3+} plays a central role in promoting the anion disorder, i.e., the flexible Bi–O bond and O–Bi–O bond angles under these electronic environments allow the cations to adjust their locations as the oxygen anions migrate, so the oxygen anionic diffusion is promoted by extensive relaxations of both the surrounding anions and a "soft" cation sublattice.

3.1.3.2 Doped Bismuth Oxide

To suppress the phase transition of Bi_2O_3 and keep the high-temperature cubic phase down to room temperature, incorporating rare earth elements into Bi_2O_3 to form a solid solution is the most frequently used method in the literature. Most solid solutions are based on either the face-centered cubic (fcc) or rhombohedral structures, depending on the ionic radius, dopant type, and doping level. In general, large rare earth cations facilitate the formation of rhombohedral phases, while the small rare earth cations facilitate the formation of fcc phases. Iwahara et al. plotted a figure to show the dependence of the phase structures on rare earth cationic radii

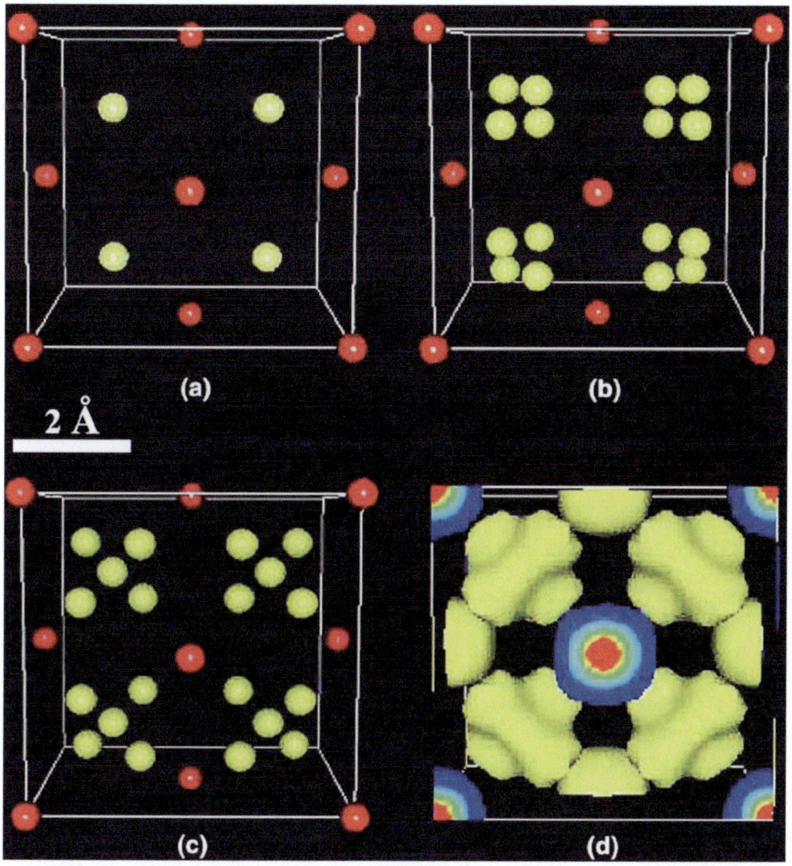

Fig. 3.11 Structural models for δ-Bi_2O_3. (**a**) Gattow model based on an ideal fluorite-type structure: average distribution of six oxygen atoms at the site (8c) of Fm-3 m (xxx, $x = 1/4$); (**b**) Harwig model: average distribution of six oxygen atoms at the site (32f) of Fm-3 m (xxx, $x = 1/4 + \delta$); (**c**) Battle model: six oxygen atoms are distributed at the 8c and 32f sites; and (**d**) scattering amplitude distribution of δ-Bi_2O_3 at 778 °C with equicontour surface at 0.3 fm/Å^3 obtained by the combination technique of Rietveld refinement and the maximum entropy method-based pattern fitting. To simplify only the region of $0 \leq y \leq 1/2$ is drawn. Red and yellow spheres denote the bismuth and oxygen atom sites, respectively (Reproduced from [41] with permission of Elsevier)

[43]. In this series of materials, except for Yb_2O_3, the larger the atomic number of Ln, the lower the content of Ln_2O_3 required to form the fcc solid solution, and the oxygen ionic conductivity of the cubic phase decreases with the increase in the Ln_2O_3 doping level. Similar to ZrO_2 and CeO_2, the dependence of ionic conductivity on cationic radius is also observed in Bi_2O_3-based ionic conductors. Figure 3.12 shows the minimum value of x required to stabilize the fcc structure in $(Bi_2O_3)_{1-x}(Ln_2O_3)_x$ versus the ionic radii of the Ln^{3+} ions. Additionally, the minimum value of x is also the function of cationic radius. Therefore, the best rare earth dopant for Bi_2O_3 is Er_2O_3, and it shows the lowest doping level.

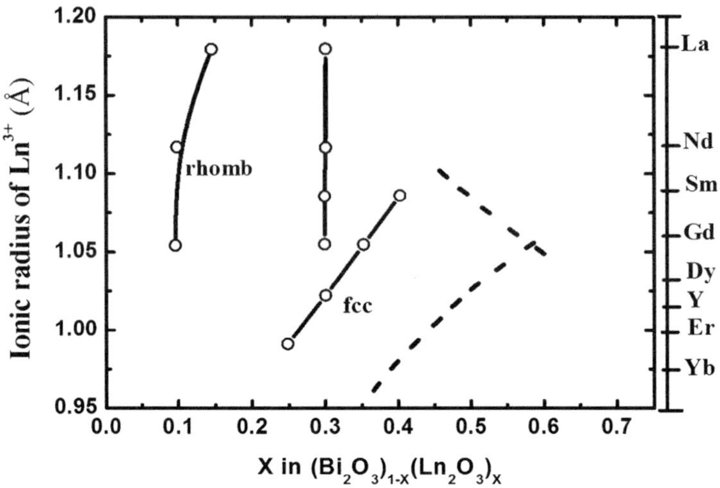

Fig. 3.12 Formation range of the rhombohedral and fcc phases in the ionic radius of Ln^{3+} versus composition diagram; *rhomb* rhombohedral and *fcc* face-centered cubic in relation to δ-Bi_2O_3 (Reproduced from [43] with permission of Elsevier)

However, among the Ln_2O_3-doped Bi_2O_3 materials, Y_2O_3 has been extensively investigated as the dopant since the Bi_2O_3–Y_2O_3 system was reported by Datta and Meehan in 1971 [44]. Takahashi et al. found that a stable fcc phase can be stabilized to lower temperatures as the Y_2O_3 doping level is in the range of 25–43 mol% [45]. Figure 3.13 shows the dependence of oxygen ionic conductivity of the Bi_2O_3–Y_2O_3 system on the temperature and Y_2O_3 doping level. When the doping level is higher than 25 mol%, the phase transition occurring in pure Bi_2O_3 does not appear, which is reflected by a jump in conductivity with the increase in temperature. Hysteresis loops, which indicate phase transitions, were observed as the doping level was lower than 25 mol% during the cooling/heating circles. However, the cubic phase structure of doped Bi_2O_3–Y_2O_3 is not stable during annealing treatment of these materials at 600 °C. Rhombohedral phase formation and a significant decrease in conductivity have been observed frequently in previous studies [46]. It was reported that ZrO_2, ThO_2, and CeO_2 are effective to stabilize the Bi_2O_3–Y_2O_3 system at a low doping amount [47, 48].

In the Bi_2O_3–Er_2O_3 system, the cubic structure can be stable in the doping level range of 17.5–45.5 mol%, and other phases appear for doping levels outside this range. Figure 3.14 shows the dependence of the oxygen ionic conductivity of the Bi_2O_3–Er_2O_3 system on the doping level [49]. However, the cubic phase of Er_2O_3-doped Bi_2O_3 transforms into a hexagonal phase or orthorhombic phase during annealing. In other words, the cubic phase is a metastable quenched phase, not a fully stable phase in thermodynamics. Kruidhof et al. found that the cubic solid solution of the Bi_2O_3–Er_2O_3 system is metastable below 740 °C as the doping level is lower than 27.5 mol% and they gradually transform to the hexagonal phase after long annealing treatment at 650 °C [50]. Watanabe found that at equilibrium state

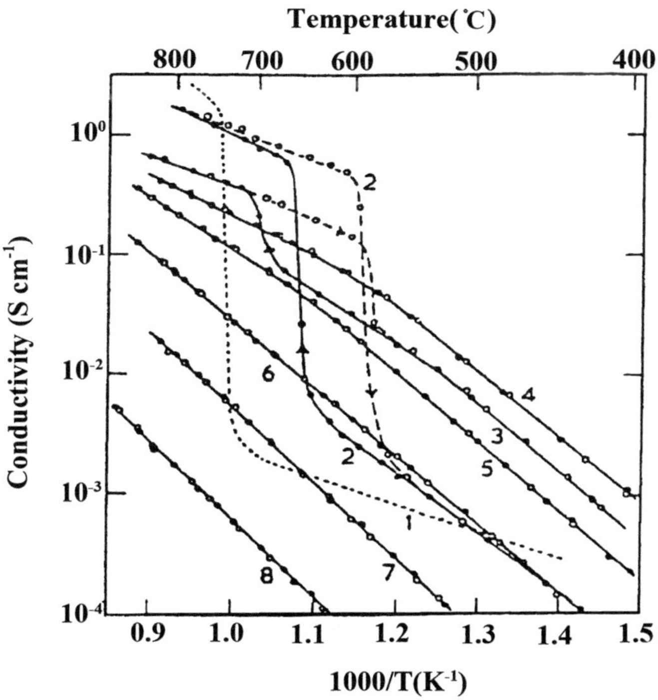

Fig. 3.13 Conductivity of $(Bi_2O_3)_{1-x}(Y_2O_3)_x$ in air; the values for x are 1 = 0, 2 = 0.05, 3 = 0.2, 4 = 0.25, 5 = 0.33, 6 = 0.425, 7 = 0.5, 8 = 0.6 (Reproduced from [45] with permission of Elsevier)

around 650 °C, the metastable hexagonal phase disappears and the low-temperature stable orthorhombic phase appears for Er_2O_3 doping amount of 28 mol% [51]. The orthorhombic phase transforms to the high-temperature stable fcc phase around 825 °C upon heating. This study indicates that it is impossible to stabilize the fcc phase by Er_2O_3 doping or the fcc phase is stable only within a limited high-temperature region. Therefore, the conductivity degradation caused by the sluggish transformation of the fcc phase to orthorhombic phase is inevitable at lower temperatures. Watanabe et al. also found that WO_3 can improve the stability of Bi_2O_3–Er_2O_3 system by replacing 5 mol% Er_2O_3 by WO_3 in the $(Bi_2O_3)_{0.705}(Er_2O_3)_{0.295}$ [52]. The resultant $(Bi_2O_3)_{0.705}(Er_2O_3)_{0.245}(WO_3)_{0.05}$ shows a stable conductivity during 1100 h of annealing at 600 °C, while $(Bi_2O_3)_{0.72}(Er_2O_3)_{0.22}(WO_3)_{0.06}$ exhibits a degradation less than 1.3 % after 830 h of annealing. Therefore, it was thought that the $(Bi_2O_3)_{0.705}(Er_2O_3)_{0.245}(WO_3)_{0.05}$ system is a truly stabilized one.

3.1.3.3 Bi_2O_3-Based Membranes for Oxygen Permeation

Oxides in the Bi_2O_3–Ln_2O_3 system are almost pure ionic conductor in air, so an electronic conducting phase is required. Usually, precious metals, such as Pt, Au,

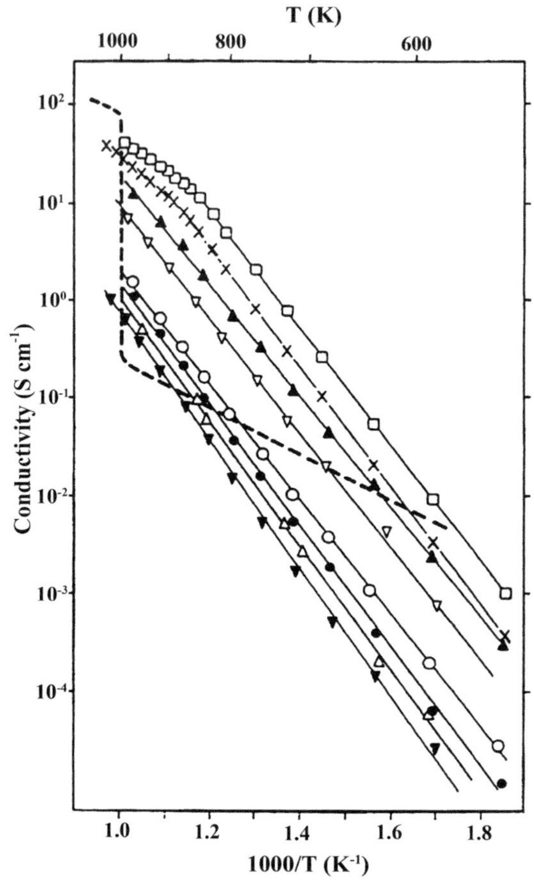

Fig. 3.14 Conductivity of $(Bi_2O_3)_{1-x}(Er_2O_3)_x$ in air; (□) $x=0.2$, (×) $x=0.25$, (▽) $x=0.3$, (○) $x=0.35$, (○) $x=0.4$, (●) $x=0.455$, (Δ) $x=0.5$, (▼) $x=0.6$; the broken line represents the conductivity of pure Bi_2O_3 (Reproduced from [49] with permission of Springer)

Pd, and Ag, are used as the electronic conducting phases to form dual-phase membranes. However, the costs of these dual-phase membranes cannot be accepted in large-scale applications if precious metals used for electronic transport. Unlike ceria-based dual-phase membranes, no chemical compatible electronic conducting oxide has been identified up to now. In the future, if a cheap and chemical compatible electronic conducting phase is developed, it would be promising for the fabrication of Bi_2O_3-dual-phase membranes for oxygen permeation at low temperatures.

3.2 Perovskite-Type Oxygen Ionic Conductors

3.2.1 Structure of Perovskite Oxides

An ideal perovskite-type structure is cubic with a space group of Pm-3m (221). Perovskite-type oxides are frequently denoted as ABO_3, in which A is a larger cation located at a center formed by 12 oxygen anions and B is a smaller cation located at a center formed by 6 oxygen anions. Figure 3.15 shows the structure of an ideal perovskite-type oxide. This structure can be viewed with the A cation located at the apexes of the cube and the B cation at the center of the octahedron (Fig. 3.15a) or the B cation placed at the apexes of the cube and the A cation located at the center of the cube (Fig. 3.15b). In the ideal structure, the relationship between the ionic radii is given by

$$r_{\mathrm{A}} + r_{\mathrm{O}} = \sqrt{2}(r_{\mathrm{B}} + r_{\mathrm{O}}) \tag{3.8}$$

However, it was found that many perovskite-type oxides maintain the perovskite structure even though this equation is not exactly obeyed. Therefore, a tolerance factor (t) was introduced by Goldschmidt to describe the deviation from the ideal situation. The tolerance factor is defined by

$$t = \frac{r_{\mathrm{A}} + r_{\mathrm{O}}}{\sqrt{2}(r_{\mathrm{B}} + r_{\mathrm{O}})} \tag{3.9}$$

For an ideal perovskite oxide, $t = 1$, but as the t value is bigger than 1 and in the range of $0.75 < t < 1.0$, the perovskite-type structure can also be found on many ABO_3 oxides. Usually, with the reduction of t value, the symmetry of perovskite oxides decreases according to an order of cubic ($SrTiO_3$, $t = 0.9992$, Pm-3 m (221)), rhombohedral ($LaCoO_3$, $t = 0.9663$, R-3c (167)), orthorhombic ($LaGaO_3$, $t = 0.9663$, Pnma (62)), and monoclinic ($Pr_{0.6}Sr_{0.4}MnO_3$, $t = 0.931$, I2/c (13)) [53]. A hexagonal structure is formed ($BaMnO_3$, $t = 1.103$, P63/mmc (194)) if the t value is slightly larger than 1. However, the dependence of crystal structure on the tolerance factor is not a strict rule; it also depends on preparation conditions and circumstances. In most cases, various distortions of the BO_6 octahedra appear in the perovskite structure. Deviations from the ideal structure with orthorhombic and rhombohedral symmetry are frequently observed, while perovskites with tetragonal and monoclinic symmetry are scarce. Although the term "perovskite" comes from the naturally occurring mineral $CaTiO_3$, its true symmetry is not cubic but

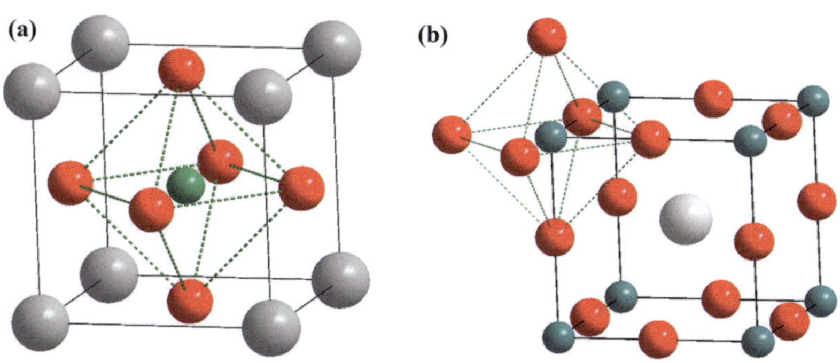

Fig. 3.15 Crystal structure of perovskite oxides

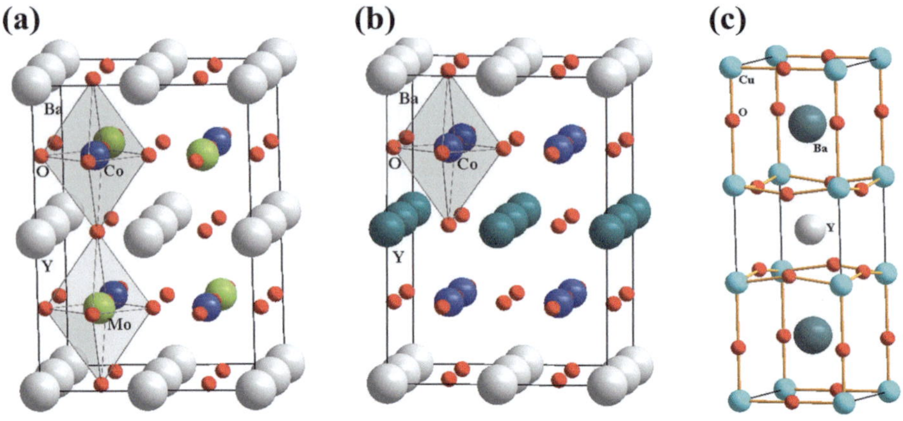

Fig. 3.16 Crystal structures of (**a**)Sr_2CoMoO_6, (**b**)$YBaCo_2O_{5+\delta}$, and (**c**) $YBa_2Cu_3O_{6+\delta}$

orthorhombic. The low symmetric phases can transform to a cubic phase at elevated temperatures, so the cubic phase is frequently called the high-temperature phase.

The perovskite structure allows a strong compositional flexibility, in which the A and B sites can be occupied by many cations with changeable ratios. Almost 90 % of elements in the periodic table can be used to form perovskite-type compounds. Thus, perovskite materials can exhibit many chemical and physical properties or combine many of these properties simultaneously. Partial substitution of A- and/or B-site cations is a very common method for tailoring the chemical and physical properties of perovskite oxides. If the substitution cations occupy sublattices in a disordered manner, the resultant materials are still single-perovskite oxides. However, if the substitution cations occupy sublattices in an ordered manner, the resultant materials could be triple or double perovskite. For example, Sr_2CoMoO_6, $YBaCo_2O_{5+\delta}$, and $YBa_2CuO_{6+\delta}$ are B-site-ordered double perovskites, A-site-ordered double perovskites, and A-site-ordered triple perovskites, respectively. Figure 3.16 shows the crystal structures of these three perovskite-type compounds. These perovskite-type oxides have been investigated extensively in fields of solid oxide fuel cells, dielectrics, ferroelectrics, magnetics, and superconduction, while few works have reported them as MIEC membranes for oxygen permeation. One important reason for this is that their crystal structures do not allow for fast oxygen ionic transport as compared to disordered single perovskite with similar chemical compositions. Oxygen ionic diffusion in the double- or triple-perovskite oxides is along the *ab* plane and is difficult along the *c*-axis, while the oxygen diffusion path in cubic single-perovskite oxides is three dimensional. Obviously, for a polycrystalline MIEC membrane, the three-dimensional diffusion path is much better than the two-dimensional one, only if the membranes were prepared with an *ab* orientation. However, there has been no report yet on the *ab* orientated double- or triple-perovskite membrane. Thus, the following section will focus on single-perovskite oxides if there is no illustration.

3.2.2 Nonstoichiometric Oxygen

Nonstoichiometric oxygen can be oxygen excess or oxygen deficient. The most common apparent oxygen-excess perovskite oxides are $LaMnO_{3+\delta}$-based oxides, which have been well characterized by neutron diffraction. Tofield et al. found oxygen excess in $LaMnO_{3.12}$ accommodated by vacancies at the A and B sites with partial elimination of La (as La_2O_3), and therefore, the composition of the perovskite can be denoted as $(La_{0.94\pm0.02}\square_{0.06\pm0.02})(Mn^{3+}{}_{0.745}Mn^{4+}{}_{0.235}\square_{0.02})O_3$ (open square is a cation vacancy) [54]. Similarly, Roosmalen et al. used neutron diffraction, in combination with chemical analysis and high-resolution transmission electron microscopy, to study a sample with formal composition $LaMnO_{3.158}$ [55]. They found the actual composition is $La_{0.95}Mn_{0.95}O_3$, no defect clustering or crystallographic shear occurs, and the defect chemistry of $LaMnO_{3+\delta}$ must be described with randomly distributed La and Mn vacancies in equal amounts. Therefore, the apparent oxygen excess in $LaMnO_{3+\delta}$-based oxides is a result of cationic vacancies appearing in the perovskite lattice, and it exists as lattice oxygen but not interstitial oxygen. Furthermore, the location of ions and the available space in the perovskite lattice do not allow the diffusion of interstitial oxygen ions since the high-energy migration barrier is > 1.5 eV.

Most often, nonstoichiometric oxygen is oxygen deficient in perovskite oxides. When iron, cobalt, nickel, copper, etc. cations are located in B sites, the oxygen deficiency can reach a high concentration of up to 1/6. Substitution of high valence cations by low valence cations introduces oxygen vacancies. For example, considering the La^{3+} cation in $LaGaO_3$ is partially replaced by Sr^{2+} cation and Ga^{3+} cation is partially replaced by Mg^{2+}:

$$SrO \overset{LaGaO_3}{\rightarrow} Sr'_{La} + \frac{1}{2}V_O^{\cdot\cdot} + O_O^{\times} \tag{3.10}$$

$$MgO \overset{LaGaO_3}{\rightarrow} Mg'_{Ga} + \frac{1}{2}V_O^{\cdot\cdot} + O_O^{\times} \tag{3.11}$$

Therefore, the incorporation of Sr^{2+} and Mg^{2+} cations into $LaGaO_3$ to generate $La_{1-x}Sr_xGa_{1-y}Mg_yO_{3-(x+y)/2}$ introduces oxygen deficiency into the perovskite lattice as oxygen vacancies. This situation is complicated for Sr^{2+} partially replacing La^{3+} cation in $LaFeO_3$. Besides the formation of oxygen vacancies, there is another charge compensation mechanism to neutralize the charge generated by the incorporation of Sr^{2+} in a La^{3+} sublattice:

$$SrO \overset{LaFeO_3}{\rightarrow} Sr'_{La} + \frac{1}{2}V_O^{\cdot\cdot} + O_O^{\times} \tag{3.12}$$

$$SrO \overset{LaFeO_3}{\rightarrow} Sr'_{La} + Fe_{Fe}^{\cdot} + O_O^{\times} \tag{3.13}$$

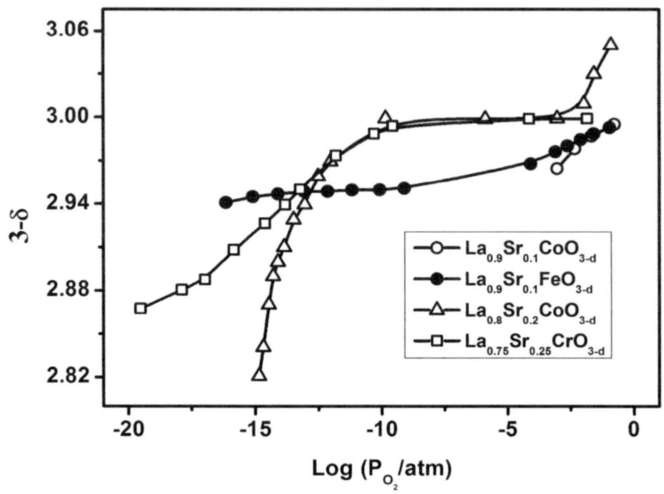

Fig. 3.17 The dependence of oxygen nonstoichiometry on oxygen partial pressure for several SrO-doped $LaMO_3$ (M = Cr, Mn, Fe, and Co) perovskite oxides (Reproduced from [56] with permission of Elsevier)

If the material does not encounter extremely high oxygen pressure or extremely low oxygen pressure, two charge compensation mechanisms are possible for $La_{1-x}Sr_xFeO_{3-\delta}$. Figure 3.17 shows the dependence of oxygen nonstoichiometry on oxygen partial pressure for several SrO-doped $LaMO_3$ (M = Cr, Mn, Fe, and Co) perovskite oxides [56]. Similar to fluorite-type oxygen ionic conductors, the oxygen ionic conduction is accomplished through the diffusion of oxygen ions or oxygen vacancies in perovskite lattices. Of course, the concentration of oxygen vacancies is an important factor in oxygen ionic conductivity. As illustrated in Eqs. 3.10 and 3.11, the doping level has a direct influence on the oxygen vacancy concentration. However, the defect association would take place at a high doping level, and thus, there is an optimal doping level for a certain dopant. This point is similar to that occurring on fluorite-type ionic conductors.

3.2.3 *Critical Radius, Free Volume, and M–O Bonding Energy*

Apart from oxygen vacancy concentration, the critical radius, free volume, and average M–O bonding energy have significant influences on the oxygen ionic diffusion in perovskite oxides. Along the transport path in the <110> direction, oxygen ions need to go through a trigonal window formed by two A-site cations and one B-site cation, as shown in Fig. 3.18. This window is the narrowest position for

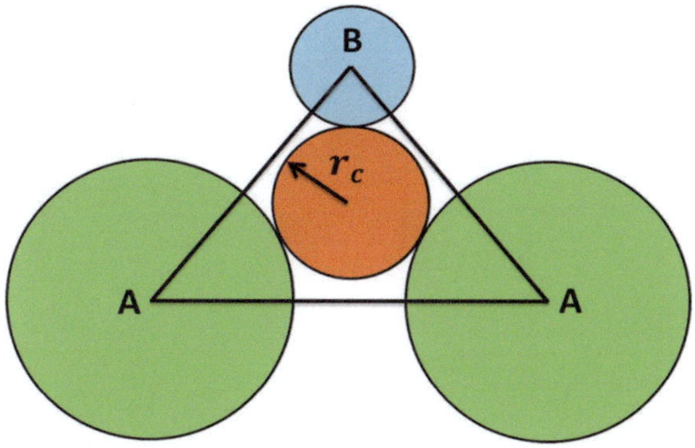

Fig. 3.18 Critical radius (r_c) formed by two A-site cations and one B-site cation in perovskite lattice

oxygen ions along the migration path. The radius of the window is called critical radius (r_c) and is given by Eq. (3.14):

$$r_c = \frac{\frac{3}{4}a^2 - r_A^2 - \sqrt{2}ar_B + r_B^2}{2r_A + \sqrt{2}a - 2r_B} \tag{3.14}$$

where "a" is the cell parameter of a cubic perovskite oxide. For a cubic perovskite oxide, the cell parameter has an approximate relationship with the B-site cationic radius and is given by Eq. (3.15) [57]:

$$a = 2.37r_B + 0.247 - 0.2(t^{-1} - 1) \tag{3.15}$$

where the oxygen ionic radius of 0.14 nm is adopted. Thus, the critical radius can be calculated for given A- and B-site cationic radii. Figure 3.19 shows the dependence of the calculated critical radius on the tolerance factor, assuming that the perovskite maintains the cubic structure. It can be seen that the critical radius decreases with the increase in the tolerance factor and it also decreases with the increase in the B-site cationic radius. Further, the critical radius is far smaller than the oxygen ionic radius, and thus, lattice relaxation is necessary to allow the transport of oxygen ions across the window. The relaxation process is temperature dependent in thermodynamics, and thus, it has a big contribution to the activation energy of oxygen ionic transport in perovskites. Obviously, the larger the critical radius, the easier the transport of oxygen ions. Therefore, in the view of the critical radius, to improve the oxygen ionic conductivity, a smaller B-site cation is preferred on the premise of maintaining the cubic perovskite structure.

Free volume of a cubic perovskite cell is defined by the difference between the crystal volume and the overall ionic volume, as given by Eq. (3.16):

$$F_V = a^3 - \frac{4}{3}\pi\left(r_A^3 + r_B^3 + 3r_O^3\right) \tag{3.16}$$

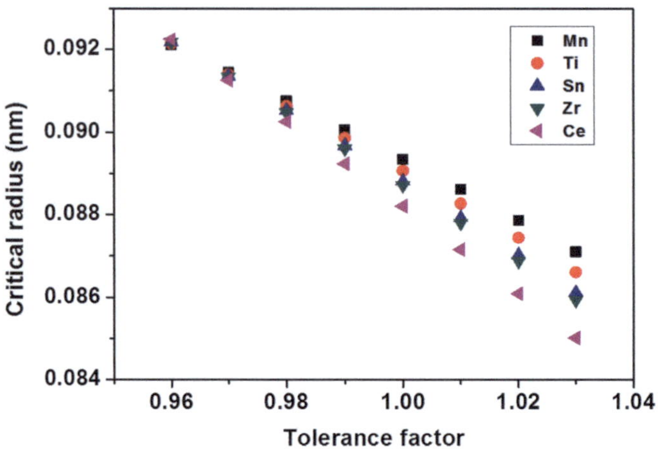

Fig. 3.19 The dependence of the calculated critical radius on the tolerance factor assuming that the perovskite keeps cubic structure

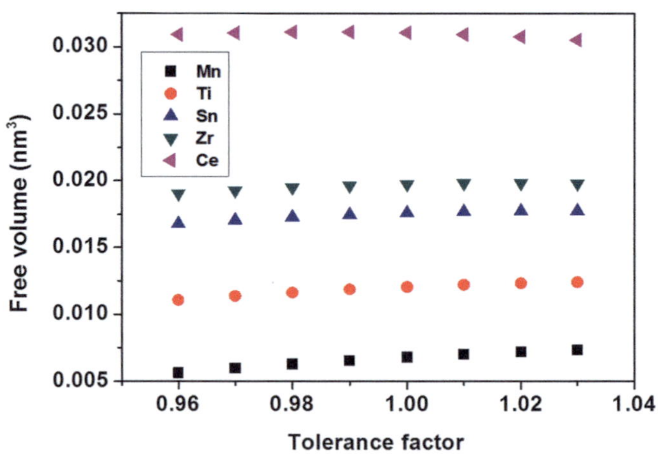

Fig. 3.20 The dependence of the calculated free volume on the tolerance factor assuming that the perovskite keeps cubic structure

Figure 3.20 shows the dependence of the calculated free volume on the tolerance factor assuming that the perovskite maintains the cubic structure. As shown in the figure, free volume is not influenced much by the tolerance factor, but it increases quickly with the increase in the B-site ionic radius. A large free volume means that there is large room allowing the lattice relaxation, thus facilitating the transport of oxygen ions and decreasing the oxygen ionic transport activation energy. In view of the free volume, to improve the oxygen ionic conductivity, a larger B-site cation is preferred on the premise of maintaining the cubic perovskite structure. However, the critical radius has a contrast dependence on the B-site cationic radius, as shown in Fig. 3.19. Therefore, there is an optimal B-site cationic radius for a certain perovskite system.

The average M–O bonding energy in the perovskite system is defined by Eq. (3.17):

$$\overline{E} = \frac{\Delta_f H^0_{\mathrm{A_mO_n}} - m\Delta H^0_{\mathrm{A}} - \frac{1}{2}nD_{\mathrm{O_2}}}{12m} + \frac{\Delta_f H^0_{\mathrm{B_{m'}O_{n'}}} - m'\Delta H^0_{\mathrm{B}} - \frac{1}{2}n'D_{\mathrm{O_2}}}{6m'} \quad (3.17)$$

where $\Delta_f H^0_{\mathrm{A_mO_n}}$ and $\Delta_f H^0_{\mathrm{B_{m'}O_{n'}}}$ are the standard formation enthalpies of A_mO_n and $B_{m'}O_{n'}$ oxides, respectively; ΔH^0_{A} and ΔH^0_{B} are the sublimation heat of A and B elementary substances, respectively; and $D_{\mathrm{O_2}}$ is the dissociation heat of oxygen molecules. A small average M–O bonding energy means the oxygen ions can easily break the bond of the cations. Thus, in view of average M–O bonding energy, to improve the oxygen ionic conductivity, A- and B-site cations with low valence state are preferred on the premise of maintaining the cubic perovskite structure. However, high average M–O bonding energy is good for the chemical stability of a perovskite oxide. The first part in Eq. (3.17) can be regarded as the average A–O bonding energy, and the second part can be regarded as the average B–O bonding energy. Table 3.1 lists the average M–O bonding energies of some frequently used A- and B-site cations. Therefore, to improve the oxygen ionic conductivity, Ba, Sr, Ca, Bi, and Pb are preferred as the A-site cations, and Mg, Fe, Co, Ni, Cu, Zn, and Ga are preferred as the B-site cations. However, in the design of perovskite oxides, the structural stability of the target material under working conditions should be considered, and thus, a balance between stability and performance should be considered. Researchers often face a trade-off between stability and performance when designing a new perovskite material.

3.2.4 $LaGaO_3$-Based Pure Ionic Conductors

In 1994, two groups leaded by Ishihara [58] and Goodenough [59], respectively, reported a perovskite-type oxygen ionic conductor $LaGaO_3$ doped by Sr in the A site and Mg in the B site showing extremely high ionic conductivity at an intermediate to low-temperature range, i.e., 600–800 °C. For example, the ionic conductivity of $La_{0.9}Sr_{0.1}Ga_{0.8}Mg_{0.2}O_{2.85}$ is high up to 0.1 S/cm at 800 °C, which is comparable to ceria-based ionic conductors or one order of magnitude higher than that of YSZ. More importantly, $La_{1-x}Sr_xGa_{1-y}Mg_yO_{3-(x+y)/2}$ (LSGM) shows low hole and electronic conductivity under a wide oxygen partial pressure range (10^{-21}–10^5 Pa). Therefore, it is regarded as a promising electrolyte material for SOFCs. Thus far, LSGM exhibits the highest ionic conductivity among the perovskite-type oxygen ionic electrolytes, which is easily explained from the viewpoint of the critical radius, free volume, and M–O bonding energy. For $LaGaO_3$-based materials, their tolerance factors are in the range of 0.96–0.97, so

Table 3.1 Average M–O bonding energies of some frequently used A- and B-site cations

Oxide	A or B	r_A or r_B (nm)	ΔH_f[a] (kJ/mol)	Q[b] (kJ/mol)	E_{ABE}[c] (kJ/mol)
BaO	Ba^{2+}	0.161	−548.1	185.2	−81.86
SrO	Sr^{2+}	0.144	−591.1	160.6	−83.39
CaO	Ca^{2+}	0.134	−634.9	177.9	−88.48
La_2O_3	La^{3+}	0.136	−1796	430.1	−141.8
Pr_2O_3	Pr^{3+}	0.131	−1810	356.0	−136.2
Nd_2O_3	Nd^{3+}	0.127	−1809	328.4	−133.9
Sm_2O_3	Sm^{3+}	0.124	−1827	205.6	−124.4
Gd_2O_3	Gd^{3+}	–	−1839	398.9	−141.0
Y_2O_3	Y^{3+}	–	−1905	423.8	−145.8
Bi_2O_3	Bi^{3+}	–	−578.5	208.9	−72.64
PbO	Pb^{2+}	0.149	−218.2	195.3	−55.21
MgO	Mg^{2+}	0.072	−601.5	147.2	−166.3
Al_2O_3	Al^{3+}	0.054	−1675	330.1	−256.9
Sc_2O_3	Sc^{3+}	0.075	−1909	377.8	−284.3
TiO_2	Ti^{4+}	0.061	−945.0	473.7	−319.4
V_2O_5	V^{5+}	0.054	−1551	517.3	−319.2
Cr_2O_3	Cr^{3+}	0.062	−1135	397.2	−223.1
CrO_2	Cr^{4+}	0.055	−581.2	397.2	−246.1
Mn_2O_3	Mn^{3+}	0.058	−956.9	283.4	−189.2
MnO_2	Mn^{4+}	0.053	−520.0	283.4	−216.9
FeO	Fe^{2+}	0.061	−267.5	415.3	−155.3
Fe_2O_3	Fe^{3+}	0.055	−823.2	415.3	−200.1
CoO	Co^{2+}	0.065	−238.3	425.6	−152.2
NiO	Ni^{2+}	0.069	−239.7	424.1	−152.1
CuO	Cu^{2+}	0.073	−155.9	338.1	−123.8
ZnO	Zn^{2+}	0.074	−350.5	130.5	−121.7
Ga_2O_3	Ga^{3+}	0.062	−1091	272.0	−198.5
ZrO_2	Zr^{4+}	0.072	−1080	599.0	−362.8
Nb_2O_5	Nb^{5+}	0.064	−1900	723.0	−382.6
MoO_3	Mo^{6+}	0.059	−744.8	657.7	−358.3
In_2O_3	In^{3+}	0.080	−923.0	240.8	−179.3
SnO_2	Sn^{4+}	0.069	−577.5	301.3	−229.5
Sb_2O_5	Sb^{5+}	0.060	−971.7	267.3	−229.3
PbO_2	Pb^{4+}	0.078	−274.6	195.3	−161.3
Y_2O_3	Y^{3+}	0.090	−1905	423.8	−291.6
CeO_2	Ce^{4+}	0.087	−1091	417.2	−334.3
PrO_2	Pr^{4+}	0.085	−949.7	356.0	−300.6

[a]Formation enthalpy of oxides
[b]Sublimation heat of metal elements
[c]Average metal–oxygen bonding energies

large critical radii are achieved; the cationic radius of Ga^{3+} is moderate, which allows a good balance between critical radius and free volume. Hayashi et al. investigated the dependence of ionic conductivity on tolerance factor for single- and double-doped perovskites and found that there is a compromise between the tolerance factor and the lattice free volume. They concluded that the maximum ionic conductivity is always achieved at the tolerance factor of 0.96, though they did not provide any explanation for this [60]. Additionally, the Ga–O bond is weak among M–O bonds formed between trivalent cations and oxygen, as shown in Table 3.1. Other than Sr^{2+}, other alkaline earth cations were tried to partially replace La^{3+} in $LaGaO_3$ perovskite, while only the Sr-doped one shows the highest ionic conductivity and lowest solution energy in the La site [61]. Various crystal structures, such as monoclinic, orthorhombic, and cubic, have been reported in the literature for LSGM, although orthorhombic symmetry has been accepted by most researchers for the parent $LaGaO_3$. The temperature and doping level have a considerable influence on the crystal structure, but a clear correlation has not yet been achieved. To well understand the relationship between crystal structure and the high ionic conductivity of LSGM, Datta et al. investigated the LSGM series in detail [62]. They found that crystal structure depends on the total doping level, i.e., as the sum of Sr and Mg doping level is smaller than 25 at%, the material shows orthorhombic symmetry; it is a mixture of orthorhombic and rhombohedral in the range of 25–30 at%, and cubic when the total doping level is larger than 35 at%, with either the Sr or Mg doping level larger than 20 at%. They also found that the ionic conductivity is structure sensitive, i.e., a LSGM perovskite material with more specific free volume and less lattice distortion has a high oxygen ionic conductivity. When the total doping level is 20 at%, Kajitani et al. found that $La_{0.9}Sr_{0.1}Ga_{0.9}Mg_{0.1}O_{2.9}$ has higher ionic conductivities and lower activation energies than $LaGa_{0.8}Mg_{0.2}O_{2.9}$, although the two materials have the same oxygen vacancy concentrations [63]. They gave three reasons: (1) the former has larger critical radius, (2) double doping was more effective to decrease the GaO_6 octahedral tilt angles, and (3) oxygen displacements in the former was larger than those in the latter. To improve the ionic conductivity, Ishihara et al. doped Co, Fe, Ni, Mn, and Cu into LSGM and found that Co is the most effective and that the resultant material exhibits constant conductivity over a wide oxygen partial pressure range, as shown in Fig. 3.21 [64]. Trofimenko et al. investigated the effects of the incorporation of transition metal cations on conductivity for $La_{0.9}Sr_{0.1}Ga_{0.8}Mg_{0.2}O_{3-\delta}$ to form $La_{0.9}Sr_{0.1}(Ga_{1-y}M_y)_{0.8}Mg_{0.2}O_{3-\delta}$ (M = Cr, Mn, Fe, and Co; y = 0.1–0.3), and the conductivity of $La_{0.9}Sr_{0.1}(Ga_{0.9}M_{0.1})_{0.8}Mg_{0.2}O_{3-\delta}$ is ionic type [65]. There is no change in the activation energies for the Co- and Fe-doped materials, while a change is frequently observed in LSGM at around 600 °C. Additionally, they found that A-site-deficient $La_{0.85}Sr_{0.1}(Ga_{0.9}Co_{0.1})_{0.8}Mg_{0.2}O_{3-\delta}$ shows higher ionic conductivity and it reaches a value twice as high as LSGM at 600 °C. Although Co doping will also slightly increase the electronic conductivity in the material, the ionic transport number is still as high as 0.9 if the doping level is controlled to be below 8.5 at% [64].

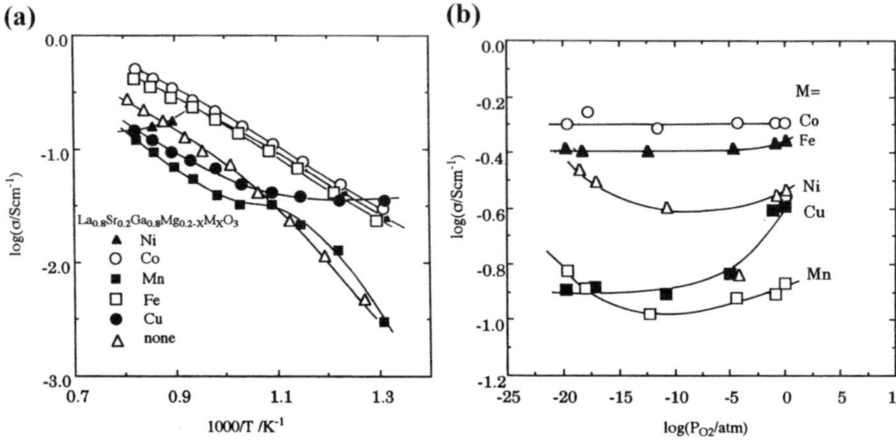

Fig. 3.21 (**a**) Arrhenius plots of electrical conductivity of $La_{0.8}Sr_{0.2}Ga_{0.8}Mg_{0.1}M_{0.1}O_{3-\delta}$ ($M=$ Fe, Co, Ni, Cu, and Mn) and $La_{0.8}Sr_{0.2}Ga_{0.8}Mg_{0.2}O_{3-\delta}$ at $P_{O2}=10^{-5}$ atm. (**b**) Electrical conductivity of $La_{0.8}Sr_{0.2}Ga_{0.8}$ $Mg_{0.1}M_{0.1}O_{3-\delta}$ at 1223 K as a function of oxygen partial pressure (Reproduced from [64] with permission of Elsevier)

Although LSGM has a high ionic conductivity and high ionic transport number over wide oxygen partial pressure range, the drawbacks of this material limit its wide applications as MIEC membranes and electrolyte of SOFCs. In a strong reducing environment, for instance, in anode environments of SOFCs or MIEC membranes for syngas production, the volatilization of gallium in LSGM with the formation of Ga (g) or Ga_2O (g) was observed, and this promoted the formation of a high-resistance phases [66]. Secondary phases, such as $SrLaGa_3O_7$, are frequently detected in the as-prepared LSGM materials, and this stable phase is difficult to be eliminated during post-processing. Further, the high cost of gallium and its compounds lead to a significant increase in material cost, so the economic feasibility of Ga-related technologies is degraded. When LSGM is used as the electrolyte, there is a challenge to ensure the chemical compatibility with other parts of SOFCs under both oxidizing and reducing conditions. There are more elements available to improve the ionic conductivity of perovskite-type MIEC membranes, so the $LaGaO_3$-based perovskite oxides are not well addressed in the field of MIEC membranes.

Apart from $LaGaO_3$-based perovskite oxides, other perovskite oxides, such as $Ln_{1-x}(Sr,Ca)_x(Al,Sc)O_{3-\delta}$, $Ba_2In_2O_5$, and $BaZr_{1-x}M_xO_{3-\delta}$, are oxygen ionic conductors and even have considerable protonic conductivity. However, oxygen ionic conductivities of these perovskite oxides are one order of magnitude lower than that of LSGM. $Ln_{1-x}(Sr,Ca)_x(Al,Sc)O_{3-\delta}$ shows a high level of hole conduction at high oxygen partial pressures. In addition to the low ionic conductivity, the poor sinterability and low grain boundary conductivity of these perovskite oxides also limit their application.

3.2.5 *Perovskite-Type Mixed Ionic and Electronic Conductors*

Perovskite MIECs are the most widely investigated ionic conductors in the field of SOFCs as cathodes and in the field of MIEC membranes for oxygen separation and membrane reactors. Usually, as the B site contains Cr, Mn, Fe, Co, Ni, Cu, etc. cations with variable valences, high electronic or hole conduction is introduced into the materials. In particular, Fe- and Co-doped perovskite oxides with high ionic conductivity show high oxygen permeation flux and good electrocatalytic activity toward oxygen reduction reactions and have been studied intensively over the past 20 years. It is difficult to measure the ionic conductivity of perovskite MIECs since their electronic or hole conductivity is two to five orders of magnitude higher than their ionic conductivity. Therefore, the electromotive force (*emf*) method is invalid to determine the ionic conductivity by multiplying the ionic transport number and total conductivity. Oxygen permeation is based on the motion of oxygen ions or vacancies from one side to another side, and the flux is proportional to the ambipolar conductivity ($\sigma_{\text{amb}} = \sigma_i\sigma_e/(\sigma_i + \sigma_e)$) of the membrane material if oxygen exchange on membrane surfaces can be ignored. For MIEC perovskites, their ambipolar conductivities are very close to their ionic conductivities, so it is possible to calculate the ionic conductivity through oxygen permeation experiments using the Wagner equation or other permeation model (see Chaps. 4 and 5). In Chap. 6, the oxygen permeability of MIEC perovskite and perovskite-related materials, such as Ruddlesden–Popper series will be discussed in detail. When used as a typical MIEC membrane material with the highest oxygen permeability, $Ba_{0.5}Sr_{0.5}Co_{0.8}Fe_{0.2}O_{3-\delta}$ exhibits an ionic conductivity of up to ~1 S cm^{-1} at 800 °C, which is one order of magnitude higher than those of SDC, GDC, and LSGM.

3.3 Other Ionic Conductors

Apart from fluorite-type and perovskite-type oxygen ionic conductors, several types of oxygen ionic conductors, such as $La_2Mo_2O_9$, $Bi_4V_2O_{11}$, and $La_{10-x}Si_6O_{26+y}$, are reported showing considerable ionic conductivity. However, little research work has focused on the application of these conductors in the field of MIEC membranes because of their low stability under reducing atmospheres or lower ionic conductivity compared to perovskite-type oxygen ionic conductors. Further, it is difficult to introduce high electronic conduction in these materials by doping or adding secondary phases. Therefore, in this section, a brief introduction to this type of materials is given below.

3.3.1 $La_2Mo_2O_9$

$La_2Mo_2O_9$ compound, which was first reported by Lacorre et al. in the year 2000, exhibits high oxygen ionic conductivity of up to 0.08 S cm^{-1} at 800 °C [67]. A phase transition occurs in this material from room temperature monoclinic (α) to cubic (β) at around 580 °C accompanied by an increase in conductivity by two orders of magnitude. β-$La_2Mo_2O_9$ has the same crystal structure (space group $P2_13$) as that of β-$SnWO_4$ (Fig. 3.22). In the β-$SnWO_4$ structure, the lone-pair 5 s^2 electrons of Sn^{2+} cation occupy a space similar to the oxygen ion, so its chemical formula can be written as $Sn_2W_2O_8E_2$ (E denotes the lone-pair 5 s^2 electrons). If Sn^{2+} is replaced by La^{3+} and W^{+6} is replaced by Mo^{+6}, two vacancies are produced because La^{3+} has no lone-pair electrons, while to maintain electrical neutrality, one vacancy is occupied by oxygen ions. This process can be described as $Sn_2W_2O_8E_2 \rightarrow La_2Mo_2O_{8+1}$. Therefore, there is one vacancy left in β-$La_2Mo_2O_9$ for oxygen ionic diffusion. This is further favored by the presence of Mo^{6+} cations in the structure that can adopt a distorted environment with various possible coordination of 4–6.

To inhibit the phase transition and improve the stability in a reducing environment, researchers have investigated the effects of cationic substitution on both lanthanum and molybdenum sites. W^{6+} with a similar size to that of Mo^{6+} shows high doping concentration up to 80 % in $La_2Mo_2O_9$. The β-$La_2Mo_2O_9$ phase can be stabilized down to room temperature as the doping level is higher than 25 % [68]. Additionally, the structure stability is improved with the increase in W doping

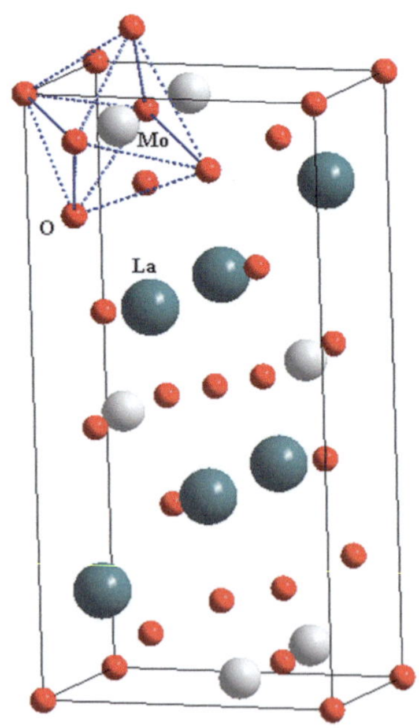

Fig. 3.22 Crystal structure of β-$La_2Mo_2O_9$

level. A maximum conductivity was observed at a doping level of 15 %, and it was close to that of β-$La_2Mo_2O_9$ at 800 °C [69]. Apart from W, the most often used elements to dope Mo sites are V, Nb, and Ta. Doping these V group elements cannot completely inhibit the phase transition, but it decreases the temperature of the phase transition [70]. The 3 %-Nb-doped $La_2Mo_2O_9$ shows the highest ionic conductivity among this series and reaches 0.113 S cm^{-1} at 800 °C, which is comparable to those of LSGM and ceria-based ionic conductors. Rare earth cations with similar size and chemical properties to those of La^{3+} are frequently used as dopants in the La sites. Lacorre et al. reported that incorporation of Nd^{3+}, Gd^{3+}, and Y^{3+} can enhance the conductivity of $La_2Mo_2O_9$ and Gd^{3+} and Y^{3+} can stabilize the cubic phase down to room temperature, while Nd^{3+} cannot. Tsai et al. investigated the 10 % doping of $La_2Mo_2O_9$ by Ce, Nd, Sm, Gd, Dy, Er, and Yb in La sites and found that Dy^{3+} and Er^{3+} cannot only inhibit the phase transition but also significantly improve the ionic conductivity, and the resultant samples show high conductivity values of up to 0.35 and 0.26 S $cm^{-1,}$ respectively, at 700 °C [70]. Further, Er- and Dy-substituted $La_2Mo_2O_9$ exhibit a relatively low thermal expansion coefficient (TEC) 11×10^{-6} K^{-1}, compared with that of the $La_2Mo_2O_9$ parent (TEC of α phase: 14.7×10^{-6} K^{-1}, TEC of β phase: 18.1×10^{-6} K^{-1}). In another paper published by Tsai's group, the conductivity of Dy and W double-substituted $La_2Mo_2O_9$ materials decreases with the increase in the W doping level. At 710 °C, the conductivity of $(La_{1.8}Dy_{0.2})(Mo_{2-x}W_x)O_9$ is 0.20, 0.12, 0.1, and 0.077 S cm^{-1} at $x = 0.2$, 0.4, 0.6, and 1, respectively [71]. However, the positive doping effects of Er and Dy need to be confirmed by other researchers, and the mechanism for the improvement of ionic conductivity by Dy and Er substitution needs to be well disclosed.

3.3.2 $Bi_4V_2O_{11}$

The investigation of the oxygen transport properties of the Bi–V–O system began in the 1980s on $BiVO_4$ compounds [72]. It was found that the $BiVO_4$ is an n-type mixed conductor and its total conductivity was approximately one order of magnitude lower than that of YSZ. Subsequently, Abraham et al. reported $Bi_4V_2O_{11}$ phase showing high ionic conductivity and thoroughly examined the phase structure. They found that the material consists of Bi_2O_2 layers interleaved with V_2O_7 sheets and has three reversible transitions at 447 °C ($\alpha \rightarrow \beta$), 567 °C ($\beta \rightarrow \gamma$), and 877 °C ($\gamma \rightarrow \gamma'$). [73]. However, only the γ phase has high ionic conductivity. γ-$Bi_4V_2O_{11}$ is tetragonal, with cell parameters of $a = 0.3988(2)$ nm and $c = 1.542(1)$ nm at 612 °C and space group of I4/mmm. γ-$Bi_4V_2O_{11}$ belongs to the Aurivillius-layered perovskite family. In its structure, $[Bi_2O_2]^{2+}$ layers are inserted with $(VO_{3.5}(V_O^{\cdot\cdot})_{0.5})^{2-}$ oxygen-deficient perovskite slabs, as shown in Fig. 3.23 [74]. It is believed that the oxygen vacancies are the migration path of oxygen ions and lead to the high ionic conductivity. To suppress the phase transition, Abraham et al. doped Cu and Ni

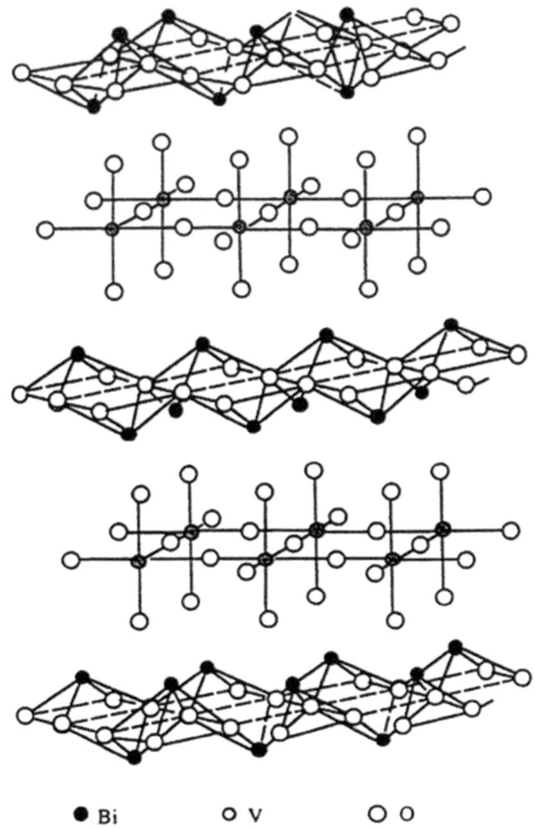

Fig. 3.23 The ideal structure of γ-$Bi_4V_2O_{11}$ phase (Reproduced from [74] with permission of Elsevier)

cations for partial substitution of the vanadium. They found that the resultant $Bi_4V_{2-x}Cu_xO_{11-\delta}$ is orthorhombic and isostructural with α-$Bi_4V_2O_{11}$ as the doping level is in the range of 0–7 % (i.e., $x = 0$–0.14) and is tetragonal and isostructural with γ-$Bi_4V_2O_{11}$ as the doping level in the range of 7–12 % (i.e., $x = 0.14$–0.24) [75]. The conductivity of $Bi_4V_{1.8}Cu_{0.2}O_{11-\delta}$ was found to be as high as 0.003 S cm^{-1} at a low temperature of 237 °C. $Bi_4V_{1.8}Cu_{0.2}O_{11-\delta}$ shows an ionic transport number very close to unity, but the n-type conduction is notable at low oxygen partial pressures. Since then, many cations have been used to partially substitute V^{5+} in the $Bi_4V_2O_{11}$, while only Sb-, Nb-, Zn-, Ni-, and Mg-doped materials show comparable ionic conductivity to that of $Bi_4V_{1.8}Cu_{0.2}O_{11-\delta}$.

3.3.3 $La_{10-x}Si_6O_{26+y}$

The apatite-type phase $Ln_{10-x}M_6O_{26+y}$, where $M = Si$ or Ge, is extensively studied since the 1990s as promising oxygen ionic conductors. Compared with the silicate-based systems, the germanate apatites seem unlikely to find practical applications because of their high volatilization, high cost of GeO_2, and tendency to form glass phase. The oxygen ionic conductivity of the apatite-type phase increases with increasing radius of Ln^{3+} cations; thus, maximum conductivity is achieved for the La-containing phase. The structure of $La_{10-x}Si_6O_{26+y}$ is hexagonal, and its space group is P-3. Figure 3.24 shows the structural sketch of $La_{10-x}Si_6O_{26+y}$ [76]. For the apatite-type phase, its single cell contains six isolated SiO_4 tetrahedra and two extra oxygen anions. The SiO_4 tetrahedra show a laminar distribution along the c-axis without being connected with each other. The extra oxygen anions do not belong to any SiO_4 tetrahedra. La^{3+} cations are located in the two types of cavities, i.e., one sevenfold-coordinate (6h site) and one ninefold-coordinate (4f site). The 4f site La^{3+} cations are located between two SiO_4 tetrahedra and connected with six adjacent SiO_4 tetrahedra. The 6 h site La^{3+} cations are located away from the SiO_4 tetrahedra, located around the sixfold axis (f-axis), and forming channels parallel to the c-axis, while the extra oxygen anions are located in the channel. When $y > 0$ or La^{3+} vacancies ($x > 0$) are present in the lattice, the oxygen anions would be present as interstitial anions in the edge of the channels. There are two paths for the migration of oxygen anions in the apatite-type materials, one is the extra oxygen anion migration in the channels along a straight path in the c-axis direction and the other is the interstitial oxygen anion migration along a curved path in the c-axis direction. The migration of interstitial oxygen anions in the lattice has a very high contribution to the ionic conduction, and thus, apatite-type materials are usually prepared to be La deficient. It is difficult to densify $La_{10-x}Si_6O_{26+y}$, and a much high temperature of up to 1600–1750 °C is required to obtain samples with a density larger than 90 %. To decrease the sintering temperature and improve the ionic conductivity, many cations have been used to dope the Si site. Fe was found to be

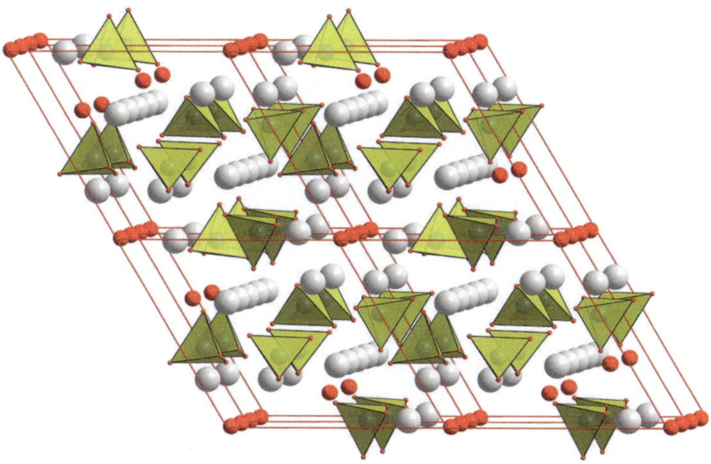

Fig. 3.24 The structure of an apatite-type oxide, $A_{10}(MO_4)_6O_2$ (tetrahedra, MO_4; *white spheres*, A-site cations; *red spheres*, oxygen ions) (Reproduced from [76] with permission of Elsevier)

effective to improve both the ionic conductivity and sinterability. However, one-dimensional transport of oxygen anions in the apatite-type materials limits the fast ionic migration in polycrystalline materials. Its ionic conductivity is comparable to that of YSZ, but if the c-orientated membrane could be prepared in a facile manner, apatite-type materials are promising in the field of SOFCs and MIEC membranes.

3.4 Relevant High-Temperature Ceramic Materials

Similar to SOFCs, MIEC membranes are operated at high temperatures, usually in the range of 500–1000 °C and typically 700–950 °C. At such high temperatures, the problems related to cationic diffusion, thermal expansion, chemical stability, and structural stability must be considered during the design of the membrane materials, preparation of membrane, selection of sealants, design of membrane modules, as well as the optimization of operation conditions. All these factors have a close correlation with the success or failure of the MIEC membrane technology.

3.4.1 Cationic Diffusion

At elevated temperatures, the diffusion of cations in oxygen ionic conducting materials has not received much attention compared with oxygen ions. However, the cationic diffusion does occur in materials and may play a vital role in the stability of materials at high temperatures. Although the cationic diffusion coefficient is several orders of magnitude lower than that of oxygen ions, the long-term degradation effect may be caused by the different cationic diffusivities, which results in demixing of the MIEC membrane materials or segregation of certain elements on membrane surfaces. A common example of the cationic diffusion can be found in the solid-state reaction between two oxides and sintering of oxide ceramics. Cations with a high diffusion rate would lower the reaction temperature of the solid-state reaction and decrease the sintering temperature. Sintering aids, such as Co_3O_4, CuO, and ZnO, are frequently used to decrease the sintering temperature and improve the density of sintering bodies because the liquid phase may form at grain boundaries and thus promote the cationic diffusion.

In the perovskite structure, there is a significant difference in the atomic radius and oxygen coordinate numbers between cations on the A site and B site, and thus, the diffusion properties are different for the two sublattice positions. Figure 3.25 shows some data of the cationic diffusion in air for perovskite oxides, $LaCrO_3$, $LaMnO_3$, $LaCoO_3$, and $LaFeO_3$ [77]. The migration of cations in materials is based on the hopping of cation vacancies. The activation energy of cationic diffusion in perovskite oxides is in the range of 200–500 kJ mol^{-1}, which is much higher than that of oxygen ion diffusion in the same perovskite oxides (usually 50–100 kJ mol^{-1}). The most favorable migration path for A-site cations is along

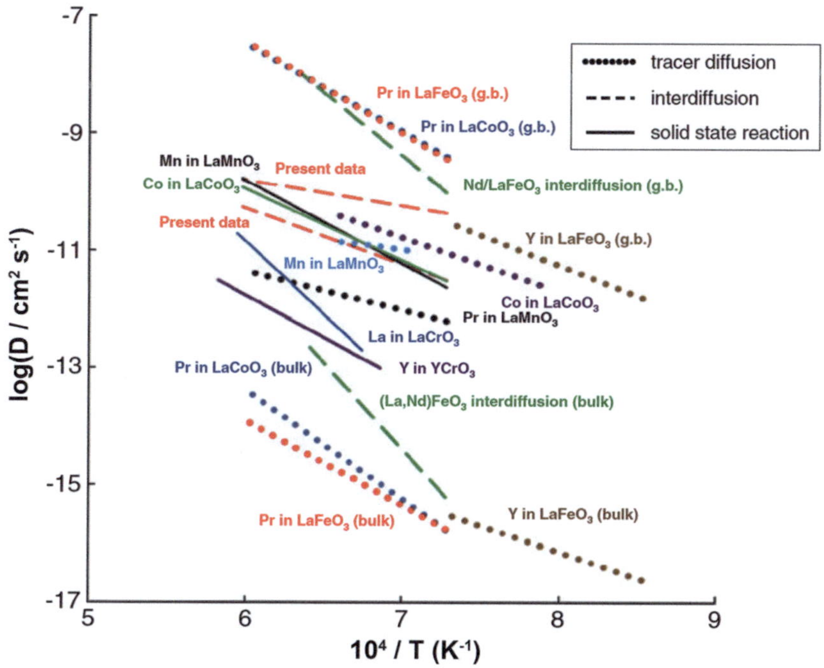

Fig. 3.25 Cations diffusion data in air for perovskite-type oxides $LaCrO_3$, $LaMnO_3$, $LaCoO_3$, and $LaFeO_3$ (Reproduced from [77] with permission of Elsevier)

the [100] direction, while that for B-site cations is more complicated and curved onto the (001) plane.

There are four methods to study the cationic diffusion in solids based on different principles, i.e., tracer or impurity diffusion, interdiffusion, diffusion-controlled solid-state reaction between two binary oxides, and creep measurement, as illustrated in Fig. 3.26. Among the four methods, the tracer or impurity diffusion method is most frequently used to determine the cation diffusion in materials. Kubicek et al. used the tracer and impurity diffusion method to investigate the diffusion of isotopic ^{86}Sr and Fe, Sm impurity elements in $La_{0.6}Sr_{0.4}CoO_{3-\delta}$ thin films below 800 °C [78]. They used time-of-flight secondary ion mass spectrometry (ToF-SIMS) to analyze the profiles and found that the diffusion along grain boundaries is about three orders of magnitude faster than in the grain bulk, as shown in Fig. 3.27. The A- and B-site cations show similar activation energies of ~340 kJ mol^{-1} for bulk and ~400 kJ mol^{-1} for grain boundary diffusion. B-site Fe/Co interdiffusion showed the largest diffusion coefficients, followed by the A-site cations Sr and Sm/La. However, only small differences in diffusion activation energies for all cations were found in their experiments, and thus, a coupled migration mechanism for A- and B-site cations was suggested. Harvey et al. investigated the diffusion of lanthanum and manganese cations in $Ba_{0.5}Sr_{0.5}Co_{0.8}Fe_{0.2}O_{3-\delta}$ polycrystalline ceramics and found that the diffusion

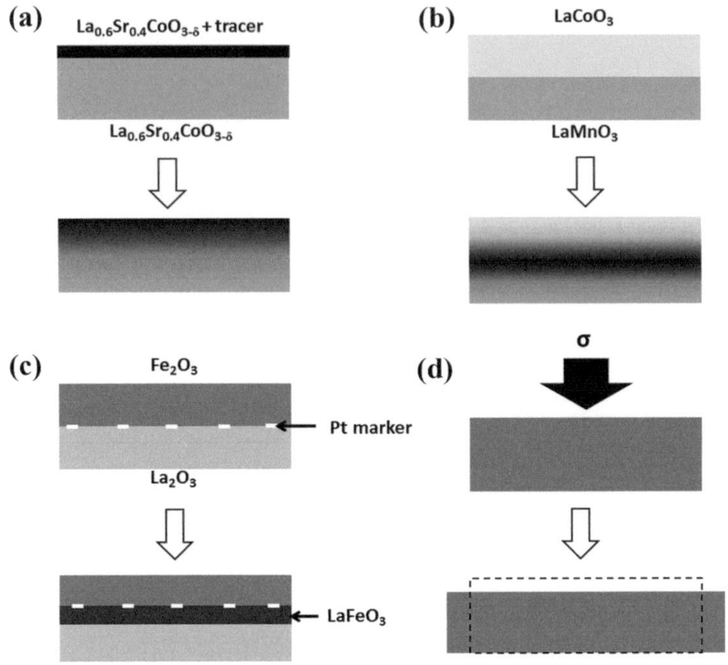

Fig. 3.26 Methods to study the cation diffusion in solids based on different principles, (**a**) tracer or impurity diffusion, (**b**) interdiffusion, (**c**) diffusion-controlled solid-state reaction between two binary oxides, (**d**) creep measurement

activation energies for the two cations are equal, about 150 kJ mol^{-1}, and the diffusion coefficients decrease considerably with decreasing oxygen partial pressure, which indicates a vacancy-mediated migration mechanism [79]. The grain boundary diffusion is 10^3–10^4 times faster than the corresponding diffusion in the grain bulk at each temperature and oxygen pressure investigated. Palcut et al. used the interdiffusion method to investigate the cation diffusion between polycrystalline $LaMnO_3$ and $LaCoO_3$ pellets in air combined with electron microprobe analysis [77]. They found that cation diffusion in $LaMnO_3$ is easier than in $LaCoO_3$ and deduced that larger amount of cation vacancies caused by the high stability of Mn^{4+} occurred in $LaMnO_3$. Čebašek et al. prepared the two interdiffusion couples, i.e., $La_2NiO_{4+\delta}$–$Nd_2NiO_{4+\delta}$ and $La_2NiO_{4+\delta}$–$La_2CuO_{4+\delta}$, to investigate diffusion of A- and B-site cations in the $La_2NiO_{4+\delta}$ [80]. They found the diffusion coefficients of the A- and B-site cations in the grain bulk are comparable but are five orders of magnitude lower than the grain boundary diffusion coefficient of the B-site cations. The activation energies for A-site bulk diffusion are ~275 kJ mol^{-1} and for B-site bulk and grain boundary diffusion are ~450 and 125 kJ mol^{-1}, respectively. Smith et al. utilized the diffusion-controlled solid-state reaction method to investigate the cation diffusion in $LaFeO_3$ with a diffusion couple of Fe_2O_3–La_2O_3 at 950–1350 °C. In their case, Pt was used as the marker to show the motion of the

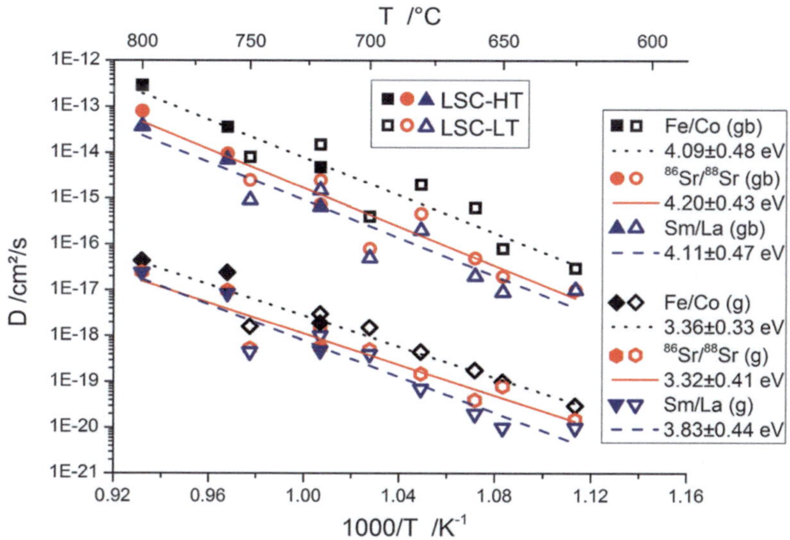

Fig. 3.27 Arrhenius-type plot of tracer diffusion coefficients of grain boundaries and grains. About three orders of magnitude faster grain boundary diffusion and similar activation energies for all cations, ~4.1 eV for grain boundaries (gb) and ~3.5 eV for grains (g), were observed (Reproduced from [78] – Published by the PCCP Owner Societies)

reaction interface. They found the Pt marker was only found at the $Fe_2O_3/LaFeO_3$ interface. This result indicates that the diffusion coefficient of Fe^{3+} is much larger than that of La^{3+} in $LaFeO_3$. Similarly, in the investigation of the Fe_2O_3–Gd_2O_3 diffusion couple, Fe^{3+} diffusion is predominated in the growth of the $GdFeO_3$ perovskite. However, by using the same diffusion-controlled solid-state reaction method to investigate the cation diffusion in $YCrO_3$ [81] and $LaCrO_3$ [82], researchers found the diffusion of A-site cations is the rate limitation step. Suzuki et al. reported an investigation on the Cr and Al diffusion in chromite spinel based on creep measurement in the pressure range of 3–7 GPa and temperature range of 1400–1700 °C [83]. They found the self-diffusion coefficients of Cr is more than one order of magnitude smaller than that of Al in the chromite spinel, which coincides with the patterns of multipolar Cr–Al zoning observed in natural chromite spinel samples deformed by diffusion creep.

3.4.2 *Kinetic Demixing*

When MIEC membranes are used for oxygen permeation, an oxygen chemical potential is applied across the membranes. Cationic diffusion from the low oxygen partial pressure side to the high oxygen partial pressure side cannot be ignored at elevated temperature for those materials with high cationic diffusion coefficients.

For example, perovskite oxide $Ba_{0.5}Sr_{0.5}Co_{0.8}Fe_{0.2}O_{3-\delta}$ contains five elements, and the Gibbs–Duhem relationship can be written as

$$x_{Ba}\mu_{Ba} + x_{Sr}\mu_{Sr} + x_{Co}\mu_{Co} + x_{Fe}\mu_{Fe} + x_{O}\mu_{O} = 0 \tag{3.18}$$

where x_i and μ_i are the molar fraction and chemical potential of species i, respectively. Under a steady-state operation condition, the oxygen chemical potential and molar fraction on the feed-side surface are higher than those on the permeation-side surface, and thus, the cationic chemical potentials on the permeation-side surface are higher than those on the feed-side surface, according to the Gibbs–Duhem relationship. To reach a thermodynamic equilibrium state, the chemical potential difference of cations across the membrane is reduced through the cationic diffusion from the permeation side to the feed side, as illustrated by Fig. 3.28. Since the diffusion rate of the cations is different in the membrane bulk, the fast one will enrich on the feed-side surface, and the slow one will enrich on the permeation-side surface. If cations transport in the membrane bulk through cationic vacancies, a slight movement of the gas–solid interface from the permeation side to the feed side will be observed. For $LaFeO_3$ perovskite, as the cations reach the feed-side surface, the following reaction will take place:

$$La_{La}^{\times} + \frac{3}{2}O_2 \rightarrow \frac{1}{2}La_2O_3 + V_{La}^{'''} + 3h^{\cdot} \tag{3.19}$$

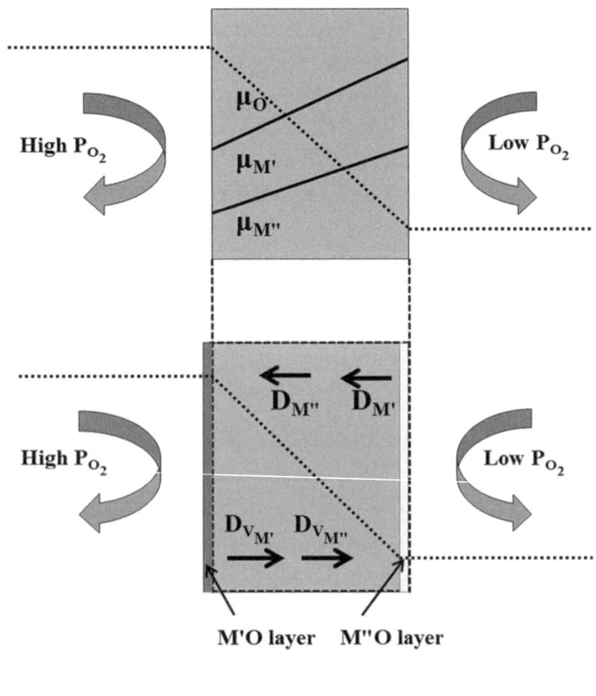

Fig. 3.28 Demixing occurring on the perovskite membrane under a gradient of the chemical potential of oxygen. The kinetic demixing represented by the formation of oxide layer M′ or M″ at the permeation side/feed side (Reproduced from [84] with permission of Wiley)

$$Fe_{\mathrm{Fe}}^{\times} + \frac{3}{2}O_2 \rightarrow \frac{1}{2}Fe_2O_3 + V_{\mathrm{Fe}}^{\prime\prime\prime} + 3h^{\cdot} \tag{3.20}$$

While at the permeation-side surface, the following reaction will take place:

$$LaFeO_3 + V_{\mathrm{La}}^{\prime\prime\prime} + V_{\mathrm{Fe}}^{\prime\prime\prime} + 6h^{\cdot} \rightarrow \frac{3}{2}O_2 + La_{\mathrm{La}}^{\times} + Fe_{\mathrm{Fe}}^{\times} \tag{3.21}$$

Because the diffusion coefficient of Fe^{3+} is much larger than that of La^{3+} in $LaFeO_3$, the expected result is that iron oxide is enriched on the feed-side surface and lanthanum is enriched on the permeation-side surface after long-term running under an oxygen chemical potential gradient at elevated temperatures. However, the phenomenon of kinetic demixing for MIEC materials is easy to be masked by the decomposition of materials and chemical reactions such as the interaction with CO_2, H_2O, etc. Therefore, few reliable experiments have been previously reported on the kinetic demixing in MIEC membranes.

3.4.3 Thermal Expansion and Chemical Expansion

The thermal expansion coefficient of MIEC materials is important when considering the potential problems related to sealing and thermal compatibility with other components in a membrane module. In engineering, the linear thermal expansion coefficient is frequently used, and it is defined by

$$\alpha_{\mathrm{L}} = \frac{1}{L_0}\left(\frac{\partial L}{\partial T}\right) \approx \frac{\Delta L}{L_0 \Delta T} \tag{3.22}$$

where L_0 is the initial length of the sample. For MIEC membranes, the valence states of cations, such as $Co^{3+/4+}$ and Fe^{4+}, decrease with the increase in temperature, while this results in extra expansion due to the increase in the cationic radii. Thus, an increase in the thermal expansion coefficient appears at a certain temperature, above which the reduction of cationic or oxygen loss becomes remarkable. For Co- or Fe-containing MIEC perovskite oxides, the oxygen loss typically occurs as the temperature is higher than 400–600 °C.

Apart from temperature, the oxygen partial pressure and atmosphere can also influence the valence state of cations, and the corresponding expansion is referred to as the chemical expansion. The oxygen chemical potential gradient across an MIEC membrane produces stress because the cationic valence state decreases continuously across the membrane. The maximum stress induced by the chemical expansion in an MIEC membrane can be estimated by

$$\sigma = \frac{E_{\mathrm{YM}}}{1 - \nu_P}\varepsilon_c \tag{3.23}$$

where E_{YM}, ν_P, and $\varepsilon_c = \Delta L/L_0$ are Young's modulus, the Poisson ratio, and the lattice expansion caused by chemical expansion (or linear strain). The linear chemical expansion can be calculated by

$$\alpha_c = \varepsilon_c/\Delta\delta \tag{3.24}$$

where $\Delta\delta$ is the change in the oxygen deficiency due to the change in the oxygen partial pressure. When membranes are used for the conversion of methane to syngas, the side fed with air has an oxygen partial pressure of 0.21 atm, and the side fed with methane has an oxygen partial pressure of 10^{-21} atm. Thus, a large stress is produced across the membrane. Some membranes may break during the operation, such as $SrCo_{0.8}Fe_{0.2}O_{3-\delta}$, and some membranes can protect itself through decomposition of the perovskite structure, which can result in sluggish oxygen evolution kinetics and improve the oxygen partial pressure on the membrane surface, such as $Ba_{0.5}Sr_{0.5}Co_{0.8}Fe_{0.2}O_{3-\delta}$.

3.4.4 Creep

At high temperatures, nonelastic or plastic deformation may take place over time for solid materials if a mechanical stress is applied on the materials. This type deformation is referred to as "creep." Creep describes the plastic deformation at constant stress and is a function of materials properties, temperature, and time. Unlike brittle fracture, creep deformation does not occur suddenly upon the application of stress. As a general guideline, the creep deformation will become noticeable at approximately 30 % of the melting point for metals and at 40–50 % of melting point for ceramics. For MIEC materials, besides the aforementioned factors, oxygen partial pressure has a remarkable influence on the creep. The steady-state creep rate is given by

$$\dot{\varepsilon} = K\left(\frac{1}{d}\right)^p (P_{O_2})^m \sigma^n \exp\left(-\frac{Ea}{RT}\right) \tag{3.25}$$

where K, d, P_{O_2}, σ, and Ea represent a constant, grain size, oxygen partial pressure, applied stress, and activation energy of creep, respectively; p, m, and n are exponents, respectively. The acceptable creep rates in the compressive mode for engineering ceramics are reported to be about 10^{-10} s^{-1}, which permits a strain of about 1 % per year. [84] Fig. 3.29 shows the creep rates of some MIEC perovskite oxides and fluorite oxides [84]. As shown in the figure, no MIEC perovskite oxide can reach a creep rate lower than 10^{-10} s^{-1} in the temperature range of 800–1000 °C at a stress of 10 MPa; in particular, the MIEC perovskite membranes with high oxygen permeability exhibit high creep rate. When MIEC membranes are operated for oxygen permeation, the stress induced by the oxygen chemical potential gradient would lead to creep at high temperatures. The chemically induced

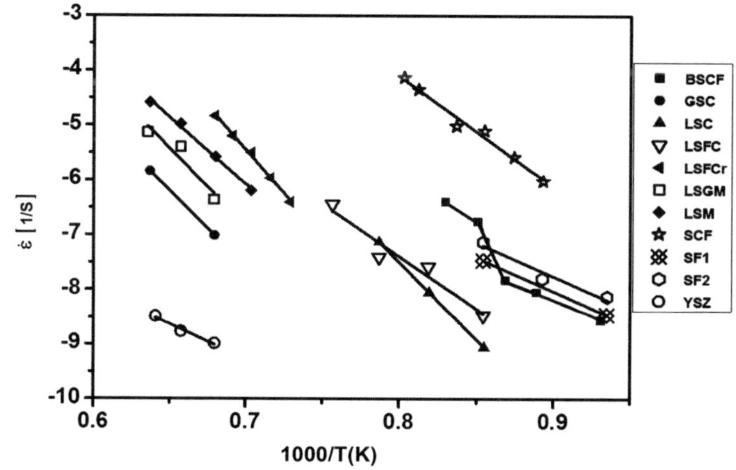

Fig. 3.29 The creep rates of some MIEC perovskite oxides and fluorite oxides (Reproduced from [84] with permission of Wiley)

stress may reach up to 100 MPa as the membrane is operated under a gradient of air or syngas. The MIEC membranes should be operated at a high-pressure gradient across the membranes for oxygen separation, and thus, the applied stress will also lead to the creep deformation. A report from Air Products & Chemicals Inc. showed that creep due to the pressure load on the wafer-type membrane had resulted in a compressive creep strain of the entire stack of wafers of 3.4 % after 15,000-h operation at an air pressure of 1.7 MPa. The creep deformation resulted in the macroscopic wafer displacements [85]. A large fraction of the creep strain was accommodated by decreasing the cross-sectional area of the internal flow channels of the membrane wafers at the central manifold.

References

1. Yamamoto O, Arachi Y, Sakai H, Takeda Y, Imanishi N, Mizutani Y, Kawai M, Nakamura Y (1998) Zirconia based oxide Ion conductors for solid oxide fuel cells. Ionics 4:403–408
2. Dixon M, Lagrange LD, Merten U, Miller CF, Porter JT (1963) Electrical resistivity of stabilized zirconia at elevated temperatures. J Electrochem Soc 110:276–280
3. Strickler DW, Carlson WG (1965) Electrical conductivity in the ZrO_2-rich region of several M_2O_3-ZrO_2 systems. J Am Ceram Soc 48:286–289
4. Hohnke DK (1981) Ionic conduction in doped oxides with the fluorite structure. Solid State Ionics 5:531–534
5. Singhal SC, Kendall K (2002) High temperature solid oxide fuel cells: fundamentals, design and applications. Mater today 5:55
6. Arachi Y, Sakai H, Yamamoto O, Takeda Y, Imanishai N (1999) Electrical conductivity of the ZrO_2-Ln_2O_3 (Ln=lanthanides) system. Solid State Ionics 121:133–139

7. Badwal SPS, Ciacchi FT, Milosevic D (2000) Scandia-zirconia electrolytes for intermediate temperature solid oxide fuel cell operation. Solid State Ionics 136:91–99
8. Haering C, Roosen A, Schichl H, Schnöller M (2005) Degradation of the electrical conductivity in stabilized zirconia system: part II: Scandia-stabilized zirconia. Solid State Ionics 176:261–268
9. Lee DS, Kim WS, Choi SH, Kim J, Lee HW, Lee JH (2005) Characterization of ZrO_2 co-doped with Sc_2O_3 and CeO_2 electrolyte for the application of intermediate temperature SOFCs. Solid State Ionics 176:33–39
10. Liu M, He CG, Wang JX, Wang WG, Wang ZW (2010) Investigation of $(CeO_2)_x(Sc_2O_3)_{0.11-x}(ZrO_2)_{0.89}$ (x = 0.01–0.10) electrolyte materials for intermediate-temperature solid oxide fuel cell. J Alloy Compd 502:319–323
11. Kawamura K, Watanabe K, Hiramatsu T, Kaimai A, Nigara Y, Kawada T, Mizusaki J (2001) Electrical conductivities of CaO doped ZrO_2-CeO_2 solid solution system. Solid State Ionics 144:11–18
12. Inaba H, Tagawa H (1996) Ceria-based solid electrolytes. Solid State Ionics 83:1–16
13. Yahiro H, Eguchi Y, Eguchi K, Arai H (1988) Oxygen ion conductivity of the ceria-samarium oxide system with fluorite structure. J Appl Electrochem 18:527–531
14. Yahiro H, Eguchi K, Arai H (1989) Electrical properties and reducibilities of ceria-rare earth oxide systems and their application to solid oxide fuel cell. Solid State Ionics 36:71–75
15. Eguchi K, Setoguchi T, Inoue T, Arai H (1992) Electrical-properties of ceria-based oxides and their application to solid oxide fuel-cells. Solid State Ionics 52:165–172
16. Zha SW, Xia CR, Gy M (2003) Effect of Gd (Sm) doping on properties of ceria electrolyte for solid oxide fuel cells. J Power Sources 115:44–48
17. Kudo T, Obayashi H (1976) Mixed electrical conduction in the fluorite-type $Ce_{1-x}Gdx\ O_{2-x/2}$. J Electrochem Soc 123:415–419
18. Steele BCH (2000) Appraisal of $Ce_{1-y}Gd_yO_{2-y/2}$ electrolytes for IT-SOFC operation at 500 °C. Solid State Ionics 129:95–110
19. Kim N, Kim BH, Lee D (2000) Effect of co-dopant addition on properties of gadolinia-doped ceria electrolyte. J Power Sources 90:139–143
20. Wang FY, Chen SY, Cheng S (2004) Gd^{3+} and Sm^{3+} co-doped ceria based electrolytes for intermediate temperature solid oxide fuel cells. Electrochem Commun 6:743–746
21. Mori T, Yamamura H, Saito S (1996) Preparation of an alkali-element-doped CeO_2-Sm_2O_3 system and its operation properties as the electrolyte in planar solid oxide fuel cells. J Am Ceram Soc 79:3309–3312
22. Parkash O, Singh N, Singh NK, Kumar D (2012) Preparation and characterization of ceria co-doped with Ca and Mg. Solid State Ionics 212:100–105
23. Kahlaoui M, Chefi S, Inoubli A, Madani A, Chefi C (2013) Synthesis and electrical properties of co-doping with La^{3+}, Nd^{3+}, Y^{3+}, and Eu^{3+} citric acid-nitrate prepared samarium-doped ceria ceramics. Ceram Int 39:3873–3879
24. Guan XF, Zhou HP, Wang YN, Zhang J (2008) Preparation and properties of Gd^{3+} and Y^{3+} co-doped ceria-based electrolytes for intermediate temperature solid oxide fuel cells. J Alloy Compd 464:310–316
25. Sha XQ, Lu Z, Huang XQ, Miao JP, Jia L, Xin XS, Su WH (2006) Preparation and properties of rare earth co-doped $Ce_{0.8}Sm_{0.2-x}Y_xO_{1.9}$ electrolyte materials for SOFC. J Alloy Compd 424:315–321
26. Ma L, Zhao K, Kim BH, Li Q, Huang JL (2015) Electrochemical performance of solid oxide fuel cells with Sm, Nd co-doped $Ce_{0.85}(Sm_xNd_{1-x})_{0.15}O_{2-\delta}$ electrolyte. Ceram Int 41:6391–6397
27. Dikmen S (2010) Effect of co-doping with Sm^{3+}, Bi^{3+}, La^{3+}, and Nd^{3+} on the electrochemical properties of hydrothermally prepared gadolinium-doped ceria ceramics. J Alloy Compd 494:106–112
28. Shuk P, Greenblatt M (1999) Hydrothermal synthesis and properties of mixed conductors based on $Ce_{1-x}Pr_xO_{2-\delta}$ solid solutions. Solid State Ionics 116:217–223

29. Fagga DP, Kharton VV, Shaula A, Marozau IP, Frade JR (2005) Mixed conductivity, thermal expansion, and oxygen permeability of Ce(Pr, Zr)O2−δ. Solid State Ionics 176:1723–1730
30. Fagg DP, Shaula AL, Kharton VV, Frade JR (2007) High oxygen permeability in fluorite-type $Ce_{0.8}Pr_{0.2}O_{2-\delta}$ via the use of sintering aids. J Merbrane Sci 299:1–7
31. Balaguer M, Solís C, Serra JM (2011) Study of the transport properties of the mixed ionic electronic conductor $Ce_{1-x}Tb_xO_{2-\delta}$ + Co ($x = 0.1, 0.2$) and evaluation as oxygen-transport membrane. Chem Mater 23:2333–2343
32. Gattow G, Schroder H (1962) Über Wismutoxide. III. Die kristallsttruker der hochtemperaturemodikation von Wismut (III)-oxid (δ-Bi2O3). Z Anorg Allg Chem 318:176–189
33. Willis BTM (1965) The anomalous behavior of the neutron reflections of fluorite. Acta Crystallogr 18:75–76
34. Verkerk MJ, Burggraaf AJ (1981) High oxygen ion conduction in sintered oxides of Bi_2O_3-Ln_2O_3 system. Solid State Ionics 3(4):463–467
35. Sillén LG (1937) X-ray studies on bismuth trioxide. Ark Kemi Mineral Geol 12A:1–15
36. Zav"yalova AA, Imamov RM (1969) Cubic structure of δ-bismuth sesquioxide. Kristallograya 14:331–333
37. Medernach JW, Snyder RL (1978) Powder diffraction patterns and structure of the bismuth oxides. J Am Ceram Soc 61:494–497
38. Jacobs PWM, Macdonaill DA (1986) Computer simulation of bismuth oxide. Solid State Ionics 18–19:209–213
39. Harwig HA (1978) Structure of bismuthsesquioxide: the α, β, γ, and δ-phase. Z Anorg Allg Chem 444:151–166
40. Battle PD, Catlow RA, Drennan J, Murray AD (1983) The structural properties of the oxygen conducting δ phase of Bi_2O_3. J Phys C Solid State Phys 16:L561–L566
41. Yashima M, Ishimura D (2003) Crystal structure and disorder of the fast oxide-ion conductor cubic Bi_2O_3. Chem Phys Lett 378:395–399
42. Mohn CE, StØlen S, Stefan T, Norberg HS (2009) Oxide-ion disorder within the high temperature δ phase of Bi_2O_3. Phys Rev Lett 102:155502
43. Iwahara H, Esaka T, Sato T, Takahashi T (1981) Formation of high oxide ion conductive phases in the sintered oxides of the system Bi_2O_3-Ln_2O_3 (Ln=La-Yb). J Solid State Chem 39:173–180
44. Datta RK, Meehan JP (1971) The system Bi_2O_3-R_2O_3 (R=Y, Gd). Z Anorg Allg Chem 383:328–337
45. Takahashi T, Iwahara H (1978) Oxide ion conductors based on bismuth sesquioxide. Mater Res Bull 13:1447–1453
46. Fung KZ, Virkar AV (1991) Phase-stability, phase-transformation kinetics, and conductivity of Y_2O_3-Bi_2O_3 solid electrolytes containing aliovalent dopants. J Am Ceram Soc 74:1970–1980
47. Fung KZ, Beak HD, Virkar AV (1992) Thermo-dynamic and kinetic considerations for Bi_2O_3-based electrolytes. Solid State Ionics 52:199–211
48. Huang K, Feng M, Goodenough JB (1996) Bi_2O_3-Y_2O_3-CeO_2 solid solution oxide-ion electrolyte. Solid State Ionics 89:17–24
49. Verkerk MJ, Keizer K, Burggraaf AJ (1980) High oxygen ion conduction in sintered oxides of the Bi_2O_3-Er_2O_3 system. J Appl Electrochem 10:81–90
50. Kruidhof H, Boumeester HJM, de Vries KJ, Gellings PJ, Burggraaf AJ (1992) Thermochemical stability and nonstoichiometry of erbia-stabilized bismuth oxide. Solid State Ionics 50:181–186
51. Watanabe A (2005) Phase relations of Bi_2O_3-rich Bi_2O_3-Er_2O_3 system: the appearance of a new stable orthorhombic phase $(Bi_2O_3)_{0.72}(Er_2O_3)_{0.28}$ against the known oxide-ion conductive hexagonal phase. Solid State Ionics 176:2423–2428
52. Watanabe A, Sekita M (2005) Stabilized δ-Bi_2O_3 phase in the system Bi_2O_3-Er_2O_3-WO_3 and its oxide-ion conduction. Solid State Ionics 176:2429–2433

53. Ritter C, Radaelli PG, Lees MR, Barratt J, Balakrishnan G, Paul DM (1996) A new monoclinic perovskite allotype in $Pr_{0.6}Sr_{0.4}MnO_3$. J Solid State Chem 127:276–282
54. Tofield BC, Scott WR (1974) Oxidative nonstoichiometry in perovskites, an experimental survey: the defect structure of an oxidized lanthanum manganite by powder neutron diffraction. J Solid State Chem 10:183–194
55. Van Roosmalen JAM, Cordfunke EHP, Helmholdt RB, Zandbergen HW (1994) The defect chemistry of $LaMnO_{3\pm\delta}$: 2. Structural aspects of $LaMnO_{3+\delta}$. J Solid State Chem 110:100–105
56. Hassel BAV, Kawada T, Sakai N, Yokokawa H, Doldya M (1993) Oxygen permeation modelling of perovskites. Solid State Ionics 66:295–305
57. Fukunga O, Fujita T (1973) The relation between ionic radii and cell volumes in the perovskite compounds. J Solid State Chem 8:331–338
58. Ishihara T, Matsuda H, Takita Y (1994) Doped $LaGaO_3$ perovskite-type oxide as a new oxide ion conductor. J Am Chem Soc 116:3801–3803
59. Feng M, Goodenough JB (1994) A superior oxide-ion electrolyte. Euro J Solid State Inorg Chem 31:663–672
60. Hayashi H, Inaba H, Matsuyama M, Lan NG, Dokiya M, Tagawa H (1999) Structural consideration on the ionic conductivity of perovskite-type oxides. Solid State Ionics 122:1–15
61. Islam MS, Davies RA (2004) Atomistic study of dopant site-selectivity and defect association in the lanthanum gallate perovskite. J Mater Chem 14:86–93
62. Pradyot D, Peter M, Fritz A (2007) Structural studies of Sr- and Mg-doped $LaGaO_3$. J Alloy Compd 438:232–237
63. Kajitani M, Matsuda M, Hoshikawa A, Harjo S, Kamiyama T, Ishigaki T, Izumi F, Miyake M (2007) Doping effect on crystal structure and conduction property of fast oxide ion conductor $LaGaO_3$-based perovskite. J Phys Chem Solids 68:758–764
64. Ishihara T, Akbay T, Furutani H, Takita Y (1998) Improved oxide ion conductivity of Co doped $La_{0.8}Sr_{0.2}Ga_{0.8}Mg_{0.2}O_3$ perovskite type oxide. Solid State Ionics 113:585–591
65. Trofimenko N, Ullmann H (1999) Transition metal doped lanthanum gallates. Solid State Ionics 118:215–227
66. Yamaji K, Negishi H, Horita T, Sakai N, Yokokawa H (2000) Vaporization process of Ga from doped $LaGaO_3$ electrolytes in reducing atmospheres. Solid State Ionics 135:389–396
67. Lacorre P, Goutenoire F, Bohnke O, Retoux R, Laligant Y (2000) Designing fast oxide-ion conductors based on $La_2Mo_2O_9$. Nature 404:856–858
68. Marrero-Lopez D, Canales-Vazquez J, Zhou WZ, Irvine JTS, Núñez P (2005) Structural studies on W^{6+} and Nd^{3+} substituted $La_2Mo_2O_9$ materials. J Solid State Chem 179:278–288
69. Goutenoire F, Isnard O, Suard E, Bohnke O, Laligant Y, Retoux R, Lacorre PH (2001) Structural and transport characteristics of the LAMOX family of fast oxide-ion conductors, base on lanthanum molybdenumoxide $La_2Mo_2O_9$. J Mater Chem 11:119–124
70. Tsai DS, Hsieh MJ, Tseng JC, Lee HY (2004) Ionic conductivities and phase transitions of lanthanide rare-earth substituted $La_2Mo_2O_9$. J Eur Ceram Soc 25:481–487
71. Jin TY, Rao MMV, Cheng CL, Tsai DS, Hung MH (2007) Structural stability and ion conductivity of the Dy and W substituted $La_2Mo_2O_9$. Solid State Ionics 178:367–374
72. Lu T, Steele BCH (1986) Electrical conductivity of polycrystalline $BiVO_4$ samples having the scheelite structure. Solid State Ionics 21:339–342
73. Abraham F, Debreuille-Gresse MF, Mairesse G, Nowogrocki G (1988) Phase transitions and ionic conductivity in $Bi_4V_2O_{11}$ an oxide with a layered structure. Solid State Ionics 28:529–532
74. Yan J, Greenblatt M (1995) Ionic conductivities of $Bi_4V_{2-x}M_xO_{11-x/2}$ (M=Ti, Zr, Sn, Pb). Solid State Ionics 81:225–233
75. Abraham F, Boivin JC, Mairesse G, Nowogrocki G (1990) The bimevox series: a new family of high performance oxide ion conductors. Solid State Ionics 40–41:934–937
76. Sansom JEH, Richings D, Slater PR (2001) A powder neutron diffraction study of the oxide-ion-conducting apatite-type phases, $La_{9.33}Si_6O_{26}$ and $La_8Sr_2Si_6O_{26}$. Solid State Ionics 139:205–210

77. Palcut M, Knibbe R, Wiik K, Grande T (2011) Cation inter-diffusion between $LaMnO_3$ and $LaCoO_3$ materials. Solid State Ionics 202:6–13
78. Kubicek M, Rupp GM, Huber S, Penn A, Opitz AK, Bernardi J, Stoger-Pollach M, Hutter H, Fleig J (2014) Cation diffusion in $La_{0.6}Sr_{0.4}CoO_{3-\delta}$ below 800 °C and its relevance for Sr segregation. Phys Chem Chem Phys 16:2715–2726
79. Harvey SP, Souza RAD, Martin M (2012) Diffusion of La and Mn in $Ba_{0.5}Sr_{0.5}Co_{0.8}Fe_{0.2}O_3$-$\delta$ polycrystalline ceramics. Energ Environ Sci 5:5803–5813
80. Čebašek N, Haugsrud R, Norby T (2013) Determination of inter-diffusion coefficients for the A- and B-site in the $A_2BO_{4+\delta}$ (A = La, Nd and B = Ni, Cu) system. Solid State Ionics 231:74–80
81. Kawamura K, Saiki A, Maruyama T, Nagato K (1995) Diffusion coefficient of yttrium ion in $YCrO_3$. J Electrochem Soc 142:3073–3077
82. Akashi T, Nanko M, Maruyama T, Shiraishi Y, Tanabe J (1998) Solid-State reaction kinetics of $LaCrO_3$ from the oxides and determination of La^{3+} diffusion coefficient. J Electrochem Soc 145:2090–2094
83. Suzuki AM, Yasuda A, Ozawa K (2008) Cr and Al diffusion in chromite spinel: experimental determination and its implication for diffusion creep. Phys Chem Miner 35:433–445
84. Eriksson A, Einarsrud MA, Grande T (2011) Materials science aspects relevant for high-temperature electrochemistry. In: Kharton VV (ed) Solid state electrochemistry II: electrodes interfaces and ceramic membranes. Wiley, Weinheim, pp 415–466
85. Anderson LL, Armstrong PA, Broekhuis RR, Carolan MF, Chen J, Hutcheon MD, Lewinsohn CA, Miller CF, Repasky JM, Taylor DM, Woods CM (2016) Advances in ion transport membrane technology for oxygen and syngas production. Solid State Ionics 288:331–337

Chapter 4
Fabrication and Characterization of MIEC Membranes

Abstract Preparation method has a significant influence on the performance of MIEC membranes, because the membranes derived from different methods have different microstructures, defects, impurities, etc. In this chapter, the experimental methods related to the MIEC membranes are introduced, including the methods for the synthesis of ceramic powders, for the fabrication of MIEC membranes, and for the performance testing and characterization. The advantages and disadvantages of the methods for preparing ceramic powders and fabricating MIEC membranes are shown in this chapter to facilitate selecting proper methods in experiments.

Keywords Solid-state reaction method • Complexing method • Coprecipitation method • Spray pyrolysis method • Extrusion • Tape casting • Phase inversion • Permeation flux • Electric conductivity • Transport kinetics

4.1 Preparation of Ceramic Powders

The preparation of ceramic membranes usually starts from the synthesis of powders, followed by the shaping process. The synthesis method of the composite powders has a significant influence on the properties of MIEC ceramic membranes, such as permeability, microstructure, sintering, strength, etc. Therefore, many studies have investigated the effects of the preparation method of MIEC powders on the performance of the derived membranes. There are many methods used to prepare MIEC powders: solid-state reactions, the sol–gel method, coprecipitation, spray pyrolysis, hydrothermal, self-propagating combustion, chemical vapor deposition, microemulsion, microwave, etc. The first four methods are frequently used to prepare the composite powders, whereas the other methods have been reported in only a few studies. Here, we introduce the frequently used methods in detail and describe the other methods in brief.

X. Zhu, W. Yang, *Mixed Conducting Ceramic Membranes*, Green Chemistry and Sustainable Technology, DOI 10.1007/978-3-662-53534-9_4

4.1.1 Solid-State Reaction Method

The solid-state reaction method is a standard and traditional method for the preparation of MIEC powders. It is also the most frequently used method for preparing other composite ceramic powders. Cationic transport in solids occurs even at low temperatures such as room temperature, although the diffusion coefficients are too low to be detectable. However, the diffusion coefficients increase exponentially with increasing temperature. The activation energies of cationic transport in oxides, which mainly depend on the metal–oxygen bond energy, are usually in the range of 200–500 kJ mol^{-1}. The reaction between two solids is considered in solid-state synthesis when the temperature increases to the point at which the diffusion of cations becomes remarkable. The solid-state reaction, a heterogeneous reaction process, takes place on the interfaces between two solids, and the reactants need to diffuse across the product layer in order to maintain the reaction. Therefore, the solid-state reaction includes two main processes, i.e., reactions on the interfaces and diffusion across the product layer. Figure 4.1 shows a schematic diagram of a solid-state reaction between two oxides.

Many factors influence the rate of a solid-state reaction, including chemical composition, grain size, structural properties, temperature, pressure, atmosphere, the activities of the reactants, and other factors. These factors promote the inner and outer diffusion of the reactant species. From the viewpoint of thermodynamics, the direction of a chemical reaction is that which decreases the free energy of the reaction system at a certain temperature and pressure. However, a significant change in free energy does not mean the reaction rate is faster than that with a small change. The activation energies of the diffusion of cations play an important role in the solid-state reaction. The activation energy can be decreased by adding a mineralizing agent into the reaction mixture, which acts as a catalyst to speed up the reaction rate by changing the reaction route. The action mechanisms of the mineralizing agent vary for different reaction systems. The commonly accepted mechanisms are (1) the mineralizing agent reacts with one of the reactants and produces a low eutectic mixture at lower temperatures so as to increase the diffusion rate of the reactant species or (2) activated intermediate products are produced. To speed up the reaction rate of a solid-state reaction, the activation energy should be decreased and the pre-exponential factor increased by reducing the particle sizes of the reactants. Reactants with small particle sizes have a high specific surface area, a high surface free energy, and large reaction and diffusion

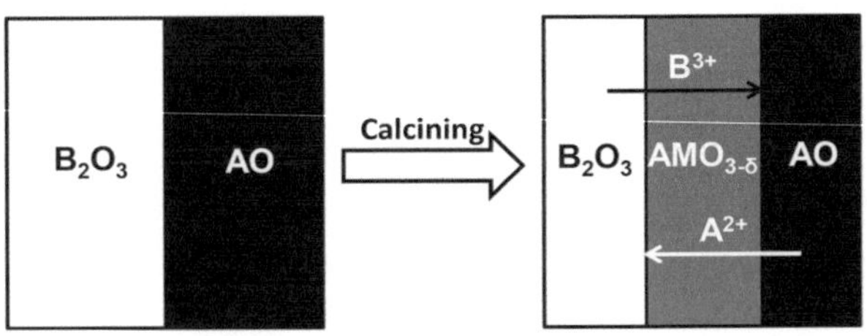

Fig. 4.1 Mechanism of forming perovskite material

interfaces, as well as a thin product layer thickness. All these factors increase the reaction rate. Another choice for accelerating the solid-state reaction is to adopt a reactant with a high activity or precursors that can transform to high activity reactants under the given reaction conditions. For example, for the preparation of a $MgAl_2O_4$ spinel oxide, one can choose MgO and α - Al_2O_3 or $Mg_2(OH)_2CO_3$ and γ - Al_2O_3 as the reactants. $Mg_2(OH)_2CO_3$ decomposes at elevated temperatures and produces MgO nanoparticles with high reaction activities; moreover, γ - Al_2O_3 has an open crystal structure as compared to α - Al_2O_3, resulting in a high reaction activity. As a result, the $MgAl_2O_4$ spinel oxide can be synthesized at lower temperatures when the reactants are $Mg_2(OH)_2CO_3$ and γ - Al_2O_3. In addition to the above factors related to the reactants, external conditions such as temperature, pressure, and atmosphere have a significant influence on the rates of solid-state reactions. Diffusion and reaction rates increase with increasing temperature, and thus the extent of the reaction increases with increasing temperature. However, although the goal is to accelerate the solid-state reaction, the reaction temperature is somewhat restricted because of the volatilization of some reactants and the heavy sintering of the products, as well as the energy consumption. The effects of pressure and atmosphere are also complicated for a solid-state reaction. Speeding up or restricting reactions may happen for different reactant systems. In a word, to accelerate a solid-state reaction, one needs to choose the right reactant precursors with small grain sizes and process them using the proper temperature, pressure, and atmosphere.

For the preparation of MIEC oxide powders, oxides and carbonates are usually used as the precursors. These precursor powders are ball-milled together to reduce their particle sizes and to homogeneously mix them. Ethanol is usually added to the mixed powder to form a slurry, which improves the grinding efficiency. The proper ratio of large balls to small balls and the proper ball diameter and rotation rate are required to obtain fine powders in a short period of time. The selection of mill media needs to be considered because impurities coming from the mill media are introduced during the high-speed and long-term grinding. The use of zirconia as the mill media is recommended for the preparation of MIEC oxide powders owing to its high hardness and density, whereas the use of alumina and hard steels is recommended if the alumina and iron impurities have no significant influence on the properties of the target material. Agate is not recommended for use in ball-milling MIEC oxide materials with high hardness, such as zirconia or ceria-based dual-phase materials, but it can be used to grind Co- or Fe-containing perovskite-type oxides. No detectable silica impurity was introduced by ball-milling $Ba_{0.5}Sr_{0.5}Co_{0.8}Fe_{0.2}O_{3-\delta}$ with agate media in the authors' laboratory. To ensure the complete reaction of the precursors, the mixed powder needs to be ball-milled and calcined repeatedly at elevated temperatures (typically >1000 °C). As mentioned above, the resultant product among the reactant particles limits the diffusion step, and the sizes of the reactant particles increase under high-temperature treatment. All these effects decrease the reaction rate of solid-state reactions. Therefore, multiple calcining and ball-milling steps are required to obtain the target products. Planetary ball-milling machines and rolling-type ball-milling machines,

Fig. 4.2 SEM image of MIEC powder synthesized by solid-state method after being repeatedly ball-milled and calcined at elevated temperatures

which are the most commonly used in laboratories for solid powder preparation, can produce fine powders with diameters down to 1 μm when the proper ratio of large balls to small balls and the proper ball diameter and rotation rate are adopted. Figure 4.2 shows the typical morphology of an MIEC powder that has been repeatedly ball-milled and calcined at elevated temperatures. A wide particle-size distribution is unavoidable if there is no subsequent size-grading process. Fine powders with particle sizes as small as 100 nm can be prepared by the ball-milling method in specially designed machines with balls smaller than 1 mm. MIEC powders derived from the solid-state reaction method may exhibit high sintering activities when the particle size is reduced down to 1 μm because many facets with high surface energies are exposed during the ball-milling process. We found that ceria-based dual-phase powders derived from the solid-state reaction method can be sintered at lower temperatures than those derived from the complexing method via a solution route, although the powders derived from the complexing method have a smaller particle size.

4.1.2 Complexing Method

The solid-state reaction method is the most frequently used method for preparing ceramic powders, and it has many advantages, such as low preparation costs, low raw materials costs, easy scale-up, and environmental friendliness. However, there are some shortcomings that should be considered when selecting this method for the preparation of MIEC powders. These shortcomings include inaccurate stoichiometry owing to volatilization of some elements during repeated calcining at elevated temperatures, low special surface areas, high synthesis temperatures, wide particle-size distributions, introduction of impurities by volatilization of some elements, incomplete solid-state reactions, repeated ball-milling, and reactions with containers at high calcining temperatures. Therefore, there has been a significant effort to improve the preparation method of ceramic powders. Many wet-chemical methods have been developed over the past several decades, such as the complexing method and the coprecipitation method. Composite oxide powders have been successfully prepared with ultrafine particle sizes, high purities, and large special surface areas through wet-chemical methods. The as-prepared powders, which have controllable particle sizes and high sintering activities, are good precursors for the preparation of MIEC membranes.

Among the wet-chemical methods, the complexing method is the most frequently used method for the preparation of MIEC powders. Many investigators call it the sol–gel method. Complexing agents are the core of this method. The most commonly used complexing agents are citric acid, ethylenediamine tetraacetic acid (EDTA), glycine, polyvinyl alcohol (PVA), etc. All these complexing agents have no less than one donor atom. For example, EDTA is a sexadentate ligand that can complex with most of metal ions with high stability constants. An important feature of these complex agents is that the related metal complex solution can transform to a sol during the evaporation of water and form a gel to further remove the water, and so the complexing method is also called the "sol–gel" method. By using the complexing method, one can control the homogeneous mixture of various metal ions on a molecular level, and so the resultant materials show high purity and homogeneous particle sizes, as well as large specific surface areas.

Citric acid, which shows good gelatinization ability under heating, has been used as a complexing agent for the preparation of MIEC powders for several decades. However, the stability constants of citric complexes are not high, especially toward alkali earth elements. As a result, the pH value of a citric acid complex solution needs to be repeatedly adjusted by ammonia water to avoid the precipitation of alkali earth elements during the evaporation of water. For perovskite-type MIEC membranes, alkali earth elements are needed at the A sites to improve oxygen diffusion through the membrane bulk by creating oxygen vacancies. As compared to citric acid complexes, EDTA complexes have high stability constants and are insensitive toward changes in the pH value during the evaporation of water. However, the gelatinization ability of EDTA complexes is not as strong as that of citric acid. Shao et al. combined the two complexing agents, thus combining their advantages [1]. No solid was precipitated during the gelatinization process without adding ammonia water to adjust the pH value of the solution. By using this combined complexing method, they synthesized many perovskite MIEC powders and found that such perovskite structures can be formed at lower calcining temperatures.

Figure 4.3 shows the procedure for preparing MIEC powders using the combined complexing method. Figure 4.4 shows the crystal structure of the dry gel calcined at different temperatures [2]. As shown in the figure, perovskite peaks are

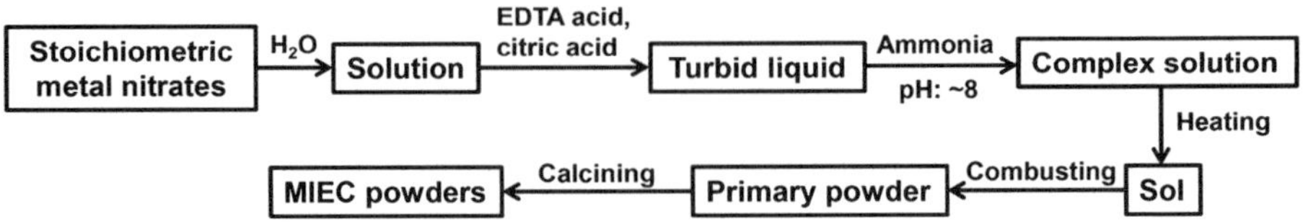

Fig. 4.3 Procedure of preparing MIEC powders through the combined complexing method

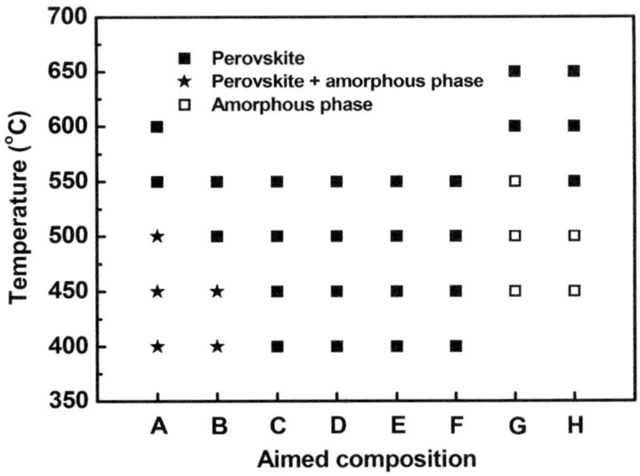

Fig. 4.4 XRD patterns of dry gel calcined at different temperatures

detectable for the powder calcined at 600 °C for 2 h, and the perovskite structure is fully constructed at 800 °C, which is much lower than the calcining temperature used in the solid-state reaction method. The wide XRD diffraction peaks show that the grain size of the as-prepared powder was small, in the range of 20–100 nm. Observing the powder morphology by using SEM and TEM, it was found that the particle size of the as-prepared powder was as large as several microns and small spheroidal grains agglomerated to form large, irregularly shaped particles. The agglomerated grains could only be partly ultrasonically dispersed in a solvent, which shows that hard agglomeration is the major state in powders prepared through the complexing method. Usually, nitrates, citric acids, and ammonia are used in the complexing method, and thus the oxidants and fuels are mixed together in the gel. Violent redox reactions taking place during the calcination of the gel sometimes generate high temperatures up to 1000 °C, especially when the gel calcined in a big batch; thus, the partial sintering of the grains is inevitable, and it results in hard agglomeration. Although the grain size is still on a nanoscale, the sintering activity is lower than that of powders without hard agglomeration. For example, Peng et al. investigated the preparation of $Ce_{0.8}Sm_{0.2}O_{1.9}$ powder by using the complexing method, with glycine as the complex agent [3]. When the glycine-to-metal ratio was approximately the theoretically calculated value (approximately 1.7), highly loose particles with a foam-like morphology and a weak force among them were observed. The resultant powder had a grain size of 10–20 nm, and the

$Ce_{0.8}Sm_{0.2}O_{1.9}$ disks reached a relative density of 95 % after being sintered at 1500 °C for 5 h. However, if the powder is prepared by the coprecipitation method, the sintering temperature could be as low as 1150 °C to reach a relative density of 99 % [4].

4.1.3 Coprecipitation Method

The coprecipitation method is another popular wet-chemical method used for the preparation of composite metal oxides. As compared to the complexing method, it is more feasible for use in preparing composite oxide powders on a large scale because it is easy to control the preparation conditions and it is repeatable, thus, making it possible to obtain homogeneous, ultrafine nano-powders with low production and raw materials costs. Figure 4.5 shows the main steps for preparing composite metal oxide powders using the coprecipitation method. After dissolving all the required metal salts in water or some other solvents to form a homogeneous solution, the mixed metal salt solution is added to a solution containing precipitants. All the metal ions are precipitated in the form of hydroxides, carbonates, oxalates, or compound salts. Composite oxide powders with nano-sized particles are obtained after filtration, washing, drying, and calcining. The key point of the coprecipitation method is the precipitants, which are needed to dissolve out all the metal ions of the target material; otherwise, the chemical composition of the resultant products will deviate from the target. The most frequently used precipitants are ammonium carbonate, ammonium bicarbonate, ammonia water, ammonium oxalate, and organic amines. One important characteristic of these precipitants is that they can be completely removed by subsequent calcination at a proper temperature. Other precipitants, such as NaOH, Na_2CO_3, etc., are also used if the abovementioned chemicals cannot precipitate all the metal ions completely, but it should be noted that thoroughly washing the precipitates is required to eliminate the potential effects of sodium ions on the physicochemical properties of the target material.

Urea is different from the other precipitants referred to above. It cannot precipitate metal ions until its solution is heated higher than 80 °C, at which urea starts to react with water to produce ammonia and carbon dioxide. Because ammonia and

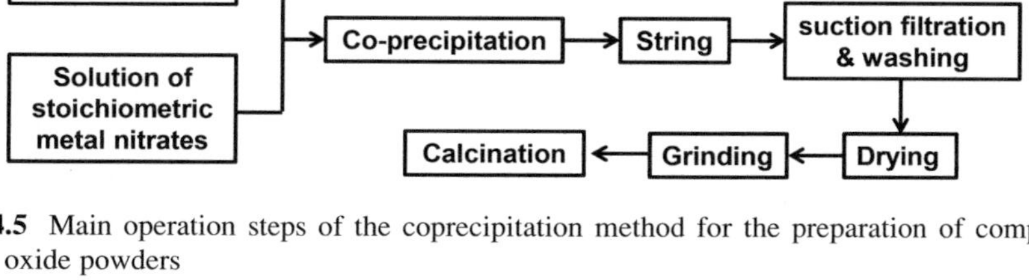

Fig. 4.5 Main operation steps of the coprecipitation method for the preparation of composite metal oxide powders

carbon dioxide are generated homogeneously throughout solution by a hydrolysis reaction, the method utilizing urea as a precipitant is called the homogeneous precipitation method. However, only a few materials systems can make use of the homogeneous precipitation method because the hydrolysis reactions of metal ions usually occur before the release of ammonia and carbon dioxide from urea while heating the solution. Furthermore, some metal ions cannot be completely dissolved out from the solution. These drawbacks limit the widespread use of urea as a precipitant.

Among the abovementioned precipitants, ammonium carbonate is the most frequently used because two kinds of anions, i.e., CO_3^{2-} and OH^-, participate in the precipitation reactions, and they can precipitate out most metal ions. Another important point is that the powders synthesized via the coprecipitation of ammonium carbonate usually have good dispensability, a homogeneous particle size, and a small grain size, as well as good sintering activity. For example, Tok et al. synthesized $Ce_{0.8}Gd_{0.2}O_{1.9}$ powder using ammonium carbonate as the precipitant. After the precipitant was calcined at 700 °C, the resultant $Ce_{0.8}Gd_{0.2}O_{1.9}$ powder had a grain size of 25 nm [4]. The corresponding disk had a high relative density up to 99 % after sintering at a temperature of 1150 °C for 4 h. Liang et al. also used ammonium carbonate as the precipitant for the preparation of $Ce_{0.8}Gd_{0.2}O_{1.9}$ powder and combined it with the supercritical drying technique [5]. The as-prepared powder had a grain size of ~6 nm characterized by TEM and XRD techniques after calcining the precipitate precursor at 400 °C for 2 h. A $Ce_{0.8}Gd_{0.2}O_{1.9}$ ceramic disk with a high relative density up to 97 % was achieved after sintering the compressed nano-powder at 850 °C for 4 h. The grains of the dense disk were small, 100 nm. Figure 4.6 shows the influence of sintering temperature on the grain size and the increase in the relative density of the $Ce_{0.8}Gd_{0.2}O_{1.9}$ ceramic disk.

Few studies have focused on the fabrication of MIEC membranes derived from the coprecipitation method because the cobalt ions frequently used in MIEC materials are difficult to precipitate out. For example, to prepare $Ba_{0.5}Sr_{0.5}Co_{0.8}Fe_{0.2}O_{3-\delta}$ via the coprecipitation method, Na_2CO_3 may be the best precipitant.

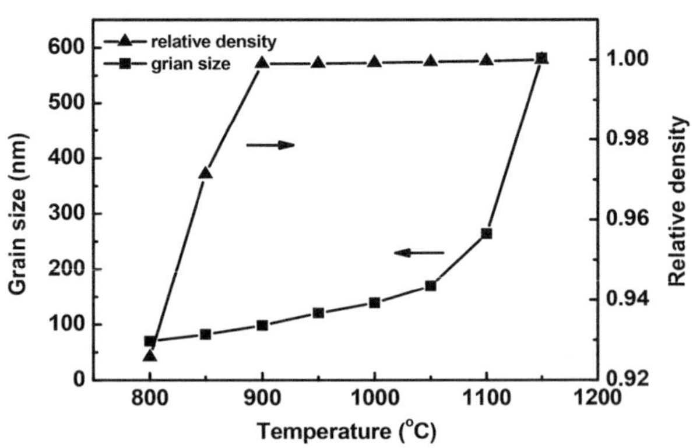

Fig. 4.6 Influences of sintering temperatures on the growth of grain size and the increase of relative density of $Ce_{0.8}Gd_{0.2}O_{1.9}$ ceramic disk (Reproduced from Ref. [5] with permission of Elsevier)

Other precipitants, such as ammonium carbonate and ammonia, as well as ammonium oxalate, cannot completely dissolve the cobalt ions from the solution because of the following reactions:

$$6NH_4^+ + Co^{2+} + 0.25O_2 + 5OH^- = Co(NH_3)_6^{3+} + 5.5H_2O \quad \Delta G(298\ K) = -259\ kJ\ mol^{-1} \tag{4.1}$$

$$6NH_3 + Co^{2+} + 0.25O_2 + 0.5H_2O = Co(NH_3)_6^{3+} + OH^- \quad \Delta G(298\ K) = -104\ kJ\ mol^{-1} \tag{4.2}$$

In the authors' laboratory, the coprecipitation method was used to prepare a dual-phase membrane of $Ce_{0.85}Sm_{0.15}O_{1.925}$–$Sm_{0.6}Sr_{0.4}Al_{0.3}Fe_{0.7}O_3$. In the procedure, metallic nitrates and ammonium carbonate were used as the precursors and the precipitant, respectively. After centrifugation separation and drying at 120 °C, the resultant blocks were crushed into powder and then pressed into disks. A high relative density up to 97 % was achieved after the green disks were sintered at 1000 °C for 3 h. The grain size of the as-sintered membranes was less than 50 nm, as shown in Fig. 4.7.

Fig. 4.7 SEM micrograph of the dual-phase membrane sintered at 1000 °C in air

4.1.4 Spray Pyrolysis Method

The spray pyrolysis method, developed by Aman in 1956 for the preparation of MgO ultrafine powders, now is a popular technique for the preparation of various materials. It developed quickly after its successful use in the industrial production of ceramic powders on a large scale in the 1970s. Now, spray pyrolysis is an important method for the preparation of thin films and fine powders such as metal powders, various ceramic materials, electrode materials, catalysts, etc. The spray pyrolysis method has many merits for the preparation of powders:

1. The pyrolysis reactions finish in seconds at high temperatures, so ultrafine particles are obtained with homogeneous chemical compositions because there is no time for composition segregation.
2. All the precursors are dissolved in solvents, so the chemical composition of the final compounds or materials can be manipulated by adjusting the initial ratios according to the given targets.
3. The morphology, particle size, particle-size distribution, and microstructure of the powders can be easily adjusted by changing the operating conditions, such as the solvent used, pyrolysis temperature, spray speed, concentration, calcining temperature, etc.
4. As compared to the above three methods in terms of batch production, serial production on a large scale is easier to achieve with the spray pyrolysis technique, so powders can be produced at low cost with high repeatability.
5. High-purity products are obtained through this method because filtering, washing, and grinding are unnecessary, because these processes normally result in the introduction of impurities to the target material.

There are many types of setups for the spray pyrolysis method, but usually only five parts are necessary, i.e., an ultrasonic atomizer, carrier gas, reactor, high-temperature furnace, and collector. Figure 4.8 shows a schematic of typical spray pyrolysis equipment used in laboratories. In a typical spray pyrolysis process, a metal salt solution is prepared according to the required stoichiometry [6]. The solution is then ultrasonically atomized and carried by the carrier gas into the high-temperature furnace for pyrolysis. A serial of chemical and physical processes such as solvent evaporation, solute precipitation, particle drying, pyrolysis, solid-state reactions, and sintering are conducted and completed in the reaction furnace before the formation of the final ultrafine powders. Figure 4.9 shows a widely accepted microscopic description of the formation of solid particles from solution droplets in the spray pyrolysis method [7]. It is called the one-droplet-one-particle (ODOP) mechanism. In the first stage, a crusted layer or a solid sphere is formed through solvent evaporation and subsequent dissolving of the solute, depending on the operating conditions and the precursors used. In the second stage, many complex changes take place, which determine the final morphology and microstructure, as demonstrated in Fig. 4.9. More information about the spray pyrolysis method is given in the references for readers who are interested.

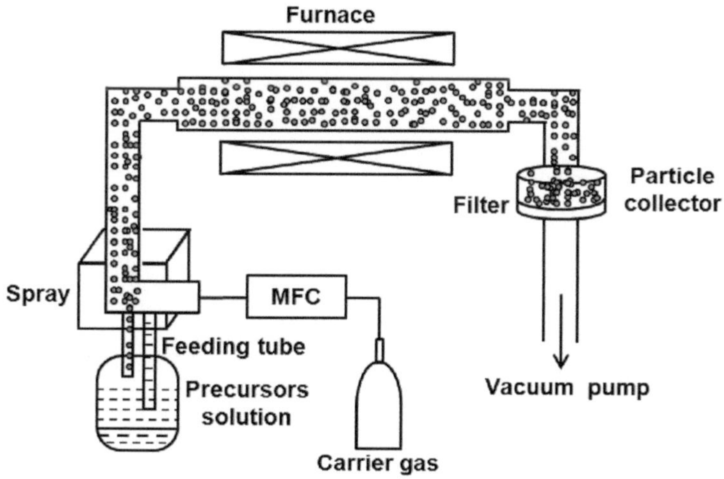

Fig. 4.8 Schematic typical spray pyrolysis equipment used in laboratories

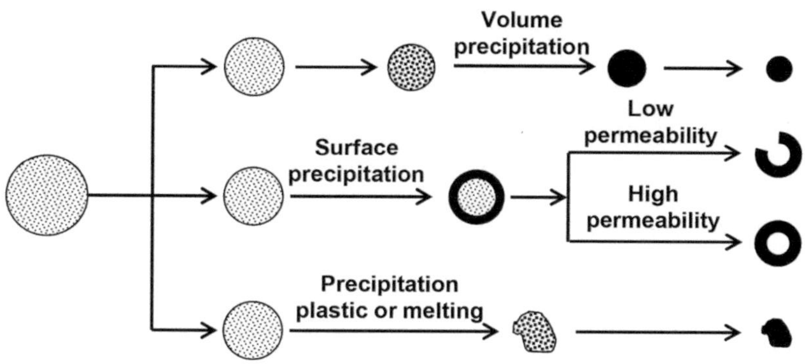

Fig. 4.9 Effect of precipitation conditions on particle morphology in spray pyrolysis (Reproduced from Ref. [7] with permission of Wiley)

Unlike the solid-state reaction and complexing methods, there are few studies on the preparation of MIEC powders through the spray pyrolysis method [8–11]. Meng et al. reported the synthesis of $La_{0.6}Sr_{0.4}Co_{0.2}Fe_{0.8}O_{3-\delta}$ powder via the spray pyrolysis method [8]. A nitrate solution was atomized to 2–5 μm droplets before they were carried into a high-temperature furnace heated to 900 °C. The retention time of the droplets was less than 2 s. Spheroidal particles 0.5–1 μm in size with a perovskite structure formed. The spray pyrolysis technique has been successfully applied to the fabrication of composite powders used in SOFCs, such as cathode powders of $La_{0.8}Sr_{0.2}MnO_3$ [9], $La_{0.6}Sr_{0.4}Co_{0.8}Fe_{0.2}O_{3-\delta}$ [10], $La_{0.6}Sr_{0.4}CoO_3$ [11], and anode powders of $Sm_{0.2}Ce_{0.8}O_{1.9}$–NiO [12] and $La_{0.9}Sr_{0.1}Ga_{0.8}Mg_{0.2}O_3$–NiO [13]. The as-prepared powders with target crystalline structures had particle sizes of 100 nm for $La_{0.6}Sr_{0.4}CoO_3$ and 20 nm for $Sm_{0.2}Ce_{0.8}O_{1.9}$–NiO. However, an amorphous structure was formed for the $La_{0.9}Sr_{0.1}Ga_{0.8}Mg_{0.2}O_3$–NiO composite oxides, and thus an extra calcining step was needed to promote crystallization. A $La_{0.84}Sr_{0.16}MnO_3$ powder was successfully prepared using a polymerized precursor solution containing CA and metal nitrates by an autoignition spray pyrolysis technique [14]. The XRD analysis confirmed the phase formation in the as-sprayed powder itself. $LaFeO_3$, $LaCoO_3$, and $La(M, Pd)O_3$ $(M = Mn, Fe)$ with pure perovskite phases were synthesized in a single step by a modified spray pyrolysis method called ultrasonic spray combustion [15]. The synthesis of perovskite-type oxides was investigated by adjusting the composition of the precursor solution and the synthesis and calcination temperatures. The results showed that the addition of citric acid in sufficient excess to the metal precursors led to micro-sized powders with a hollow spherical morphology and thin porous walls composed of nano-sized perovskite crystallites (30–50 nm). Typical SEM pictures of the resultant powders are shown in Fig. 4.10.

Comparative investigations on solid-state reaction, complexing, coprecipitation, and spray pyrolysis methods for the synthesis of perovskite-type oxides for MIEC membranes or SOFC cathodes have been reported by several groups. Lin et al. used the four methods to prepare $La_{0.8}Sr_{0.2}Co_{0.6}Fe_{0.4}O_{3-\delta}$ perovskite oxides to determine the influence of the preparation method on electron transport and oxygen permeability [16]. In their procedures, oxides and carbonates, nitrates and citric acid, nitrates and KOH, and nitrate solutions were used as the raw materials in the four methods, respectively. All the derived membranes exhibited pure-phase orthorhombic perovskite structures. However, only the one derived from the coprecipitation method had a large strontium deficiency. It showed much lower electronic conductivity and oxygen permeability (ionic conductivity) than the other three membranes with compositions close to the desired stoichiometry. Strontium- or calcium-doped lanthanum manganite ($LaMnO_3$) and lanthanum ferrite ($LaFeO_3$) perovskite cathodes and MIEC membranes were synthesized using different techniques: spray pyrolysis, a modified citrate route, and oxalate and carbonate coprecipitation [17]. Among the three methods, the cyanide and carbonate coprecipitation allowed the lowest temperature for perovskite phase formation, followed by the spray pyrolysis method. The particle-size distributions and surface areas of these materials were significantly affected by the differences in phase formation. The coprecipitation method gave the finest powders on a nanometer scale. From these investigations, it is difficult to tell which method is the best for preparation of MIEC powders because it depends on the chemical compositions of the materials and the procedures used in the experiments. Researchers can choose proper methods in their experiments depending on the material composition and available equipments in their laboratories.

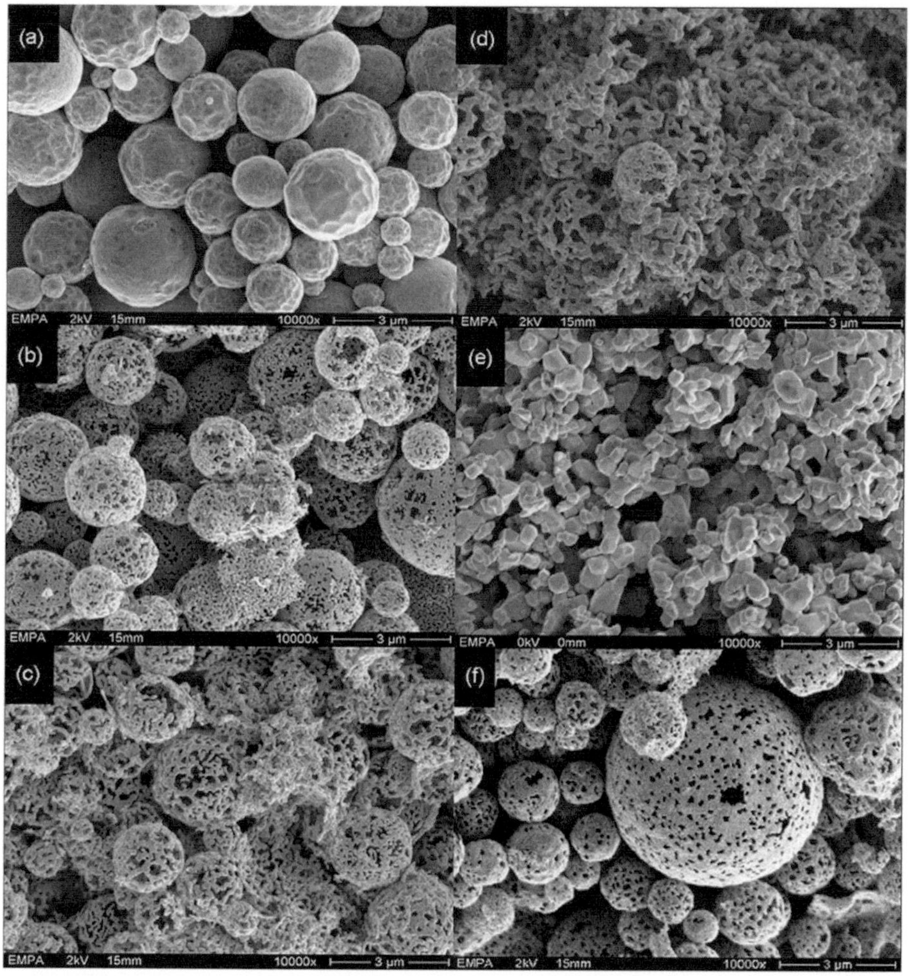

Fig. 4.10 SEM micrographs of $LaMnO_3$ powders generated from precursor solution containing 0.2 mol/L metal nitrates and 0.4 mol/L citric acid. (**a**) $Ts = 800$ °C, as prepared; (**b**) $Ts = Tc = 800$ °C; (**c**) $Ts = Tc = 900$ °C; (**d**) $Ts = Tc = 1000$ °C; (**e**) $Ts = Tc = 1100$ °C; (**f**) $Ts = 1100$ °C, $Tc = 800$ °C. Ts and Tc are the temperatures of the spray pyrolysis process and subsequent calcining step. The calcining time was 2 h (Reproduced from Ref. [15] with permission of Elsevier)

4.2 Preparation of Membranes

The preparation of ceramic membranes requires first forming the ceramic precursor powders into the required shape, which includes three main steps: shaping, sintering, and reprocessing. Among the three steps, shaping is the most important step in the preparation of membranes. Defects such as holes, heterogeneity, and air bubbles formed during the shaping process cannot be removed by the subsequent sintering and reprocessing steps. Thus, shaping is critical for producing high-quality ceramic membranes and other ceramic components. It has a significant influence on the homogeneity of ceramics, production repeatability, rate of finished products, and manufacturing costs. Up to now, MIEC membranes have been fabricated into disks, flat sheets, tubes, multichannel tubes, and hollow fibers, as shown in Fig. 1.4 in Chap. 1. Considering the microstructural symmetry of the membranes, they can be classified as either symmetric (or self-supported) membranes or asymmetric (or supported) membranes, as shown in Fig. 1.5 in Chap. 1. All the shapes mentioned above can be symmetric or asymmetric. To prepare MIEC membranes with the above geometrical structures, different methods are required.

4.2.1 Dry-Pressing

Dry-pressing is one of the most popular shape-forming methods because it involves a relatively simple technique, and thus it is often used in laboratories for preparing ceramic samples and in industrial processing of ceramic workpieces on a large scale. As the term implies, membranes are shaped by compacting dry powders without any visible liquid phase under high applied pressure. High pressures of as much as 1000 MPa can be achieved during the dry-pressing process. The pressure used in the dry-pressing process is the highest among all the methods of preparing ceramic membranes. Ceramic components ranging in size from 1000 mm, such as tiles, down to 1 mm, such as ceramic gears, can be manufactured depending on the design of the ceramic dry-pressing machine and the mold. Before being compacted in a mold, the powder needs to be pretreated by adding a proper amount of organic binders (such as polyvinyl alcohol, polyvinyl butyral, acrylic copolymer) and plasticizers (such as polyethylene glycol, butylbenzyl phthalate, dibutyl phthalate) to modify the flowability and plasticity of the powder. Lubricants (such as stearates) are sometimes necessary to make mold uploading easier. The coarse surfaces of the particles lead to interlocked occlusions in the particles, which results in a bridging effect and significant porosity if the just-received powders are pressed at an elevated pressure. Four main steps are involved in dry-pressing, i.e., powder pretreatment, powder loading into the mold, compaction, and sample unloading. To obtain high-density green samples, the powder pretreatment step is critical because the as-received powders are irregular in shape with poor flowability, which interferes with particle movement during the pressing step. However, for pretreated powders in which the particles are coated by binders and plasticizers, plastic deformation takes place under high pressure, and thus a high initial density is achievable. Detailed discussion about the influence of powder characteristics, binder content, and process parameters on the pressure/density relationship, radial pressure coefficient, wall friction coefficient, etc., are widely available in the literature, and thus they will not be discussed here. One can refer to the reference list for further reading [18].

Two techniques are utilized in dry-pressing. One is isostatic compaction and the other is uniaxial compaction. In isostatic compaction, the pretreated powders are poured into a rubber bag and stress is applied by means of a liquid, which acts as a pressure transmitter, as shown in Fig. 4.11a. Because the rubber bag is immersed in a liquid, this method is also called the wet bag method. After compaction, the rubber bag is taken out of the liquid and opened to remove the green workpieces. Although large and irregularly shaped pieces can be fabricated, the bath mode does not allow for high production rates. To improve the production rate, a dry bag method was developed, in which the rubber bag is part of the equipment. The pressure is transmitted by a liquid on the sides of the sample and a punch on the top and bottom, as shown in Fig. 4.11b. The pouring of powders into the mold and ejection of the as-pressed pieces can be automatically controlled in the dry bag method, which significantly improves production rates when the pieces are small with simple shapes. There are some reports of MIEC membranes prepared by isostatic compaction.

In uniaxial compaction, as the term implies, pressure is applied along the axial direction. This allows for the fabrication of rather complicated shapes, even those with screws or holes perpendicular to the compaction axis. High production rates are offered by industrial pressing machines, such as for the fabrication of tiles. However, the pressure distribution is not homogeneous like that in isostatic compaction because the applied pressure is transmitted from top to bottom through particle contact, where some of the pressure is lost due to the inner friction between particles and the outer friction between the particles and the walls of the mold. Figure 4.12 shows a diagrammatic sketch of the pressure distribution in a disk by

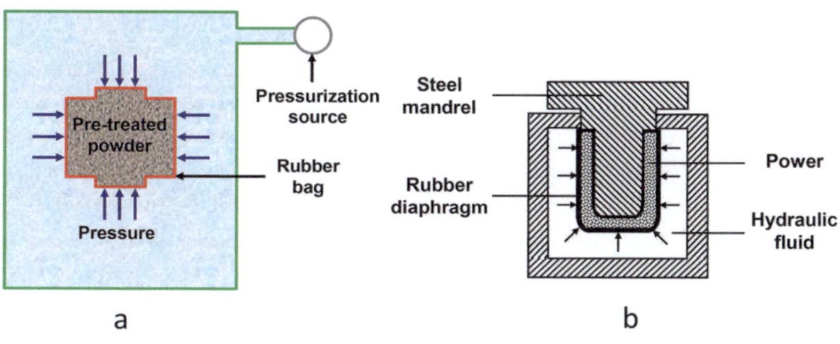

Fig. 4.11 Schematic diagram of the working principle of isostatic compaction, (**a**) wet bag method; (**b**) dry bag method

Fig. 4.12 Inhomogeneous pressure distribution in a green disk prepared by uniaxial compaction

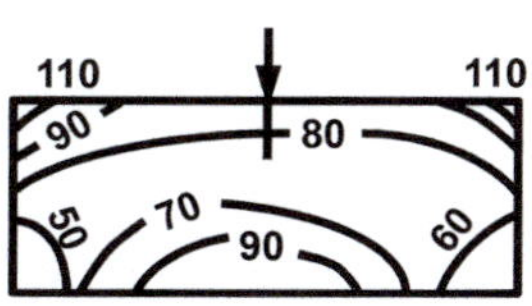

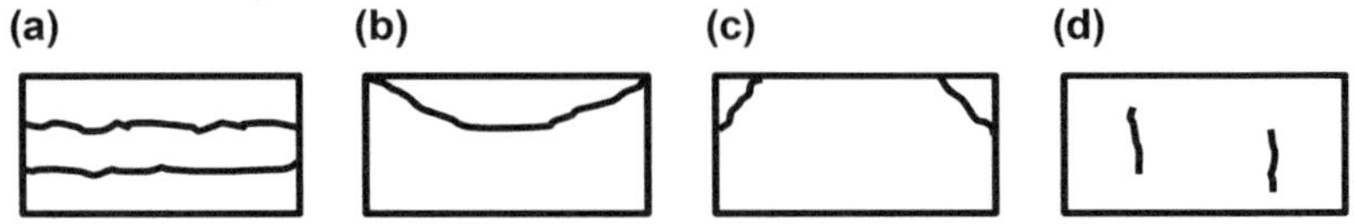

Fig. 4.13 Typical defects of the green disk prepared by uniaxial compaction

uniaxial pressing. Figure 4.13 shows the common defects in compacts formed by uniaxial pressing, which are caused by springback and by friction at the die walls. Therefore, it is important to pretreat the powder by adding binders, plasticizers, and lubricants into the powder to prepare granules with high flowability and plasticity as well as low friction. The treatment can enhance the homogeneity of the pressure distribution, resulting in green disks with high strength, high density, and homogeneous density distribution. As compared to isostatic compaction, uniaxial compaction is more often used for the preparation of small batches of disks in laboratories for the investigation of the permeation performance of MIEC membranes. Uniaxial compaction machines are inexpensive, small, and easily handled. Dense membranes are prepared by pressing powders into disks with diameters in the range of 10–25 mm and thicknesses in the range of 0.5–3 mm. The as-pressed disks are then sintered at elevated temperatures. Most data on oxygen permeation published in journals were measured on disk-type membranes fabricated by the uniaxial compaction method. However, difference in diameters, thicknesses, and oxygen chemical potential gradients make it difficult to compare one report with another. Therefore, a standard testing method should be developed, and this point will be discussed in detail in Sect. 4.3.1.

Uniaxial compaction can also be used for the preparation of asymmetric (or supported) membranes in laboratories. To do that, a powder, a pore-forming agent, and organic additives are first mixed and then pressed in a stainless steel mold under a lower pressure (~50 MPa) to produce the support. A fine and fluffy powder for the dense layer is then homogeneously added onto the as-produced support, followed by pressing under a high compaction pressure up to 300 MPa. Asymmetric membranes are obtained by co-sintering the co-pressed disks at elevated temperatures. The thickness of the dense layer is easily adjusted by changing the loading mass of the fluffy powder, which is usually prepared by the complexing method by adding ammonium nitrate as the combustion improver. This method, also called the co-pressing method, is frequently used in laboratories to prepare anode-supported electrolytes in the field of SOFCs.

4.2.2 Extrusion

The extrusion is an important shaping method for continuously producing ceramic items with a constant cross-sectional area. The lengths of the items can be adjusted according to requirements by cutting the extruded material at right angles to the direction of outflow. Muddy bodies formed from ceramic powders, organic additives, and solvents are used for extrusion shaping because they have high plastic formation ability under squeezing. The body used in this process is completely different from that used in dry-pressing: the extrusion-ready bodies contain many more organic additives and solvents than bodies in dry-pressing. The type and amount of organic additives and solvents depend on the desired properties of the ceramic powders, such as specific surface area, particle size, particle-size distribution, particle shape, materials density, etc. Ceramic membranes in the shape of tubes, capillary tubes, multichannel tubes, flat sheets, etc., are commonly fabricated by the extrusion method. The extrusion process generally consists of passing a column of muddy body through a shaped extrusion die. For a typical extrusion machine, three main parts are needed:

1. A propulsion system, which presses the muddy body at high pressure and forces the muddy body through the extrusion die. There are three types of propellers: piston extruders, expression rolls, and auger extruders. For piston machines, intermittent feeding of the columnar muddy body into barrels leads to the intermittent production of green ceramic items; however, for auger machines, continuous feeding results in highly efficient production.
2. An extrusion die, where the shaped die forms the material when the muddy body is forced through.
3. A cutting device, which cuts the shaped ceramic green body into the required sizes.

Figure 4.14 shows a diagrammatic sketch of an extruder with an auger propulsion system for ceramic tube production [19]. For a continuous extruder, a vacuum deairing pugmill machine is installed to feed the muddy body into the extruder by a rotary propeller under vacuum.

The preparation of muddy bodies from ceramic powders, a solvent, and organic additives is a critical step for manufacturing ceramic bodies with flawless surfaces and a crack-free structure. Two important factors that have a significant influence on the characteristics of the green bodies should be strictly controlled: the rheological characteristics of the muddy body and the extrusion parameters. The rheology of the muddy body is related not only to the contents of the solvent and organic additives but also to the physical and chemical properties of the ceramic powder. A reliable analysis of the extrusion process is necessary for setting optimal extrusion process parameters, including extrusion pressure, rate, die design, etc. Any changes in chemical composition and powder properties would lead to the variations in the rheological characteristics of the muddy body, and so a corresponding adjustment to the extrusion pressure, rate, and die design is needed to produce high-quality ceramic green bodies. The most frequently used solvent for the preparation of muddy bodies is water, and hydrosoluble or water-swelling binders, plasticizers, and lubricants are used to adjust the rheological characteristics. During extrusion shaping, the density of the extruded product increases as the water content in the mass decreases. For the preparation MIEC membranes, the goal

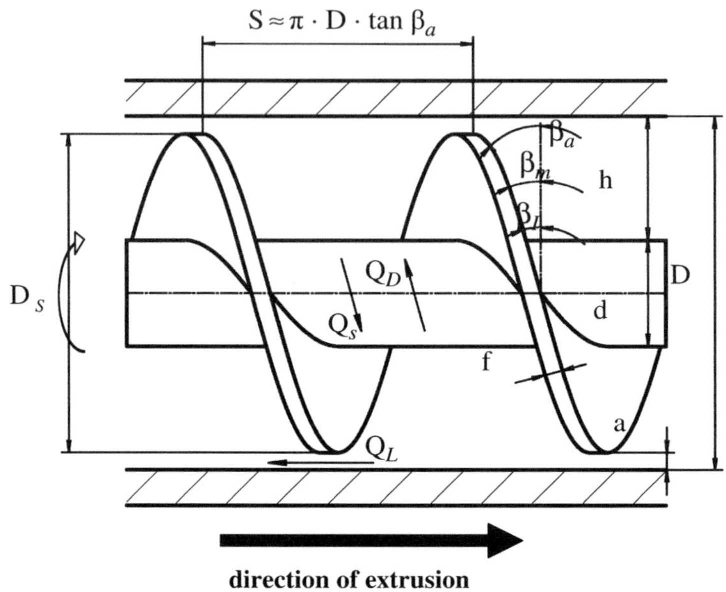

Fig. 4.14 Diagrammatic view of augers with designation of the auger geometry (Reproduced from Ref. [19] with permission of Springer)

is to obtain semi-finished products with an apparent density as high as possible. However, one must bear in mind that a lower body moisture content increases the plastic body attrition forces, resulting in fissures across the direction of extrusion.

In addition to the solvent-based systems, thermoplastic systems are also commonly used for the production of ceramic items through extrusion shaping. Thermoplastic materials are polymers that soften, melt, or become more pliable when heated and harden during cooling in a reversible physical process. Lower abrasivity of the feedstock material is one of the main advantages of using thermoplastic systems for ceramic extrusion as compared to other binder systems, which permits easy fabrication of fine structures such as fibers, hollow fibers, and thin-walled tubes. Because of the very small cross-sectional areas of such fine structures, the extrusion pressures need to reach as high as 80 MPa, resulting in phase separation in a muddy body made of a solvent-based binder, whereas that would not happen in thermoplastic binder systems. For example, as shown in Fig. 4.15, silicon carbide fibers with diameters of approximately 130 μm and alumina hollow fibers with a wall thickness of 30 μm were successfully produced by thermoplastic extrusion [20]. MIEC membrane tubes with a wall thickness of 300 μm were also successfully extruded using a thermoplastic binder system and subsequently sintered and tested. It is thus feasible to produce ceramic sheets and films with thicknesses of less than 200 μm.

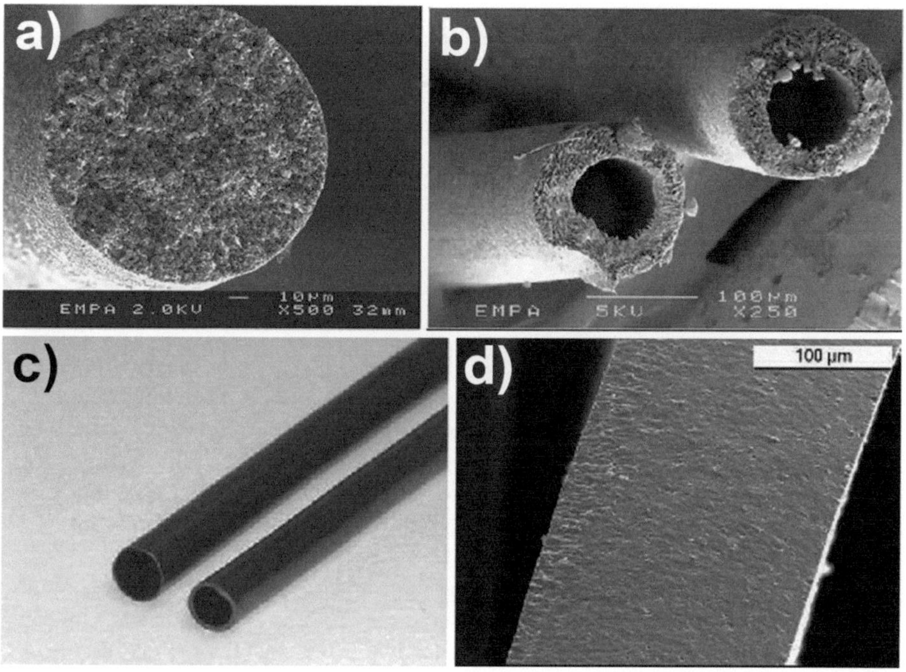

Fig. 4.15 Thermoplastic systems for ceramic extrusion. (**a**) Silicon carbide heat element for micro system devices. (**b**) Alumina hollow fiber for micro system devices. (**c** and **d**): Perovskite tubular membranes for use in the partial oxidation of methane to syngas. (**c**) Green extruded tubes. (**d**) Wall cross section of a sintered membrane (Reproduced from Ref. [20] with permission of Springer)

4.2.3 Slip Casting

Slip casting is a well-known mature ceramic-shaping technique with a long history, and parts of the industrial process can be done on automatic production lines. However, the basic principles and methods of manufacture have not changed over the years. It has been used to produce sanitary ware, artwork, and tableware for a long time, and it is still a vibrant technology for ceramic shaping, although it is occasionally used in the manufacture of advanced (technical) ceramics. The plaster molds used in slip casting are initially made from original models, and they can usually be reused for no more than 60 casting cycles.

Before shaping ceramic items by using the slip-casting method, a slip with a solids content of 30–60 % is made by mixing ceramic powder, water, binders, etc., together to form a slurry with good flowability, proper viscosity, high homogeneity, and stability against deposition. The as-prepared slip is degassed before pouring it into the mold. Some of the water is absorbed due to the capillary force generated by pores in the plaster mold; as a result, a layer of clay with a homogeneous thickness is deposited on the inside of the mold. The thickness of the clay layer increases with

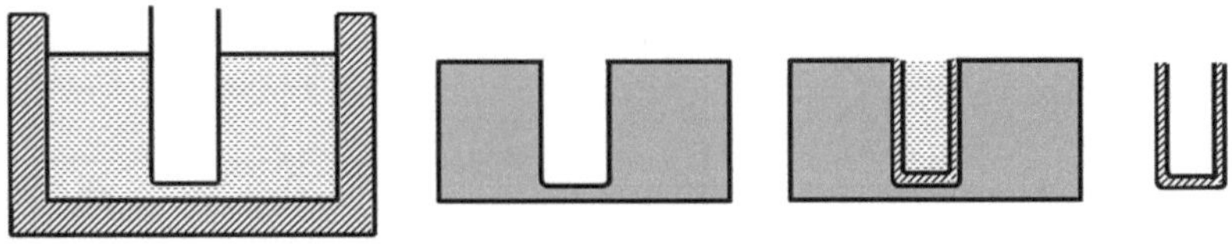

Fig. 4.16 A brief description of operation steps of slip casting. (**a**) A container with a glass test tube in the center; (**b**) a dry tubular plaster mold; (**c**) a plaster mold filled with slurry; (**d**) a tubular substrate

increasing residence time of the slip in the mold. For example, it takes only 3 or 4 min to shape thin translucent China vases, whereas it takes more than 10 min for heavier bodies. After reaching the required thickness, the surplus slip is poured out. The mold continues to absorb the water in the clay layer, drying the layer over several hours or days depending on the size and thickness of the sample. Finally, the green body is obtained by opening the two half-molds and taking it out. Figure 4.16 shows the slip-casting process.

This method is very suitable for producing large ceramic articles with complicated shapes. Although the production process is intermittent, the production rates are still fairly high. However, it also has disadvantages, such as the fact that large-scale production requires many molds and a lot of room and is coupled with the fact that plaster molds have limited durability. These disadvantages are almost equal to the advantages. Modified slip-casting methods such as pressure slip casting and centrifugation slip casting have been developed to improve the production rate. The materials for fabricating molds using polymeric materials are different from those used to fabricate plaster molds. These new molds have large pores and high strength, but they do not generate the same capillary forces as plaster molds. An externally applied pressure to drive the filtration process is required. The pressure can be generated by pressing the slip, vacuuming the mold from the outside, or spinning the mold at high speed. The speed of the casting cycle can be improved by applying a higher pressure ($<$4 MPa) than in traditional slip casting, where the capillary forces generated by the porous plaster mold correspond to pressures less than 0.2 MPa. Furthermore, higher density and strength of the green bodies are obtained by applying higher pressure. Polymeric molds also have a much longer lifetime than plaster molds—40,000 casts per mold are not unusual—and they offer high A-grade recovery over 95 %.

Slip casting is one of the best methods for preparing one-end-closed tubular MIEC membranes. However, the diameter of the tubes cannot be too small or the slip will be difficult to pour into and out of the mold. Ding et al. fabricated $BaCo_{0.7}Fe_{0.2}Nb_{0.1}O_{3-\delta}$ perovskite membranes by slip casting using one-end-closed tubular molds that had lengths of 650 mm, diameters of 20 mm, and thicknesses of 1 mm [21]. Figure 4.17 shows the plaster mold and the as-prepared perovskite membranes. They used the one-end-closed tubular membranes for the partial oxidation of CH_4 in coke oven gas to syngas. The membrane reactor was successfully

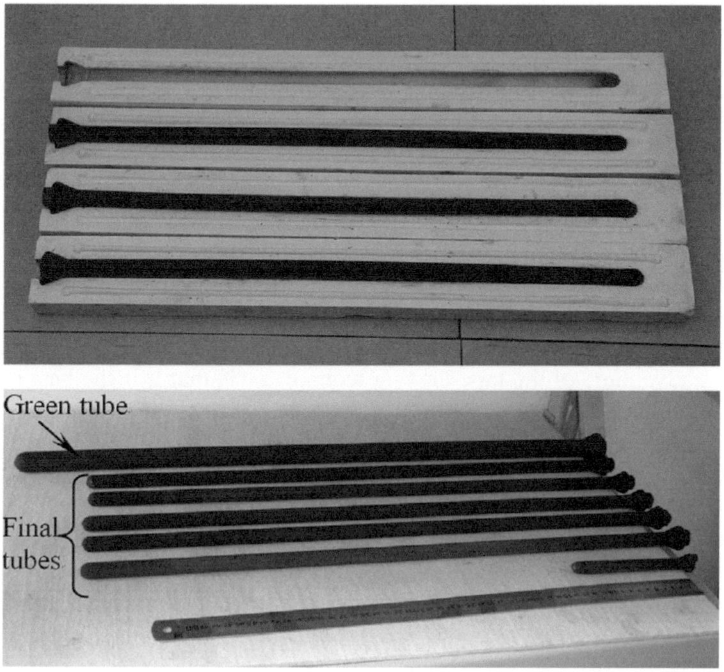

Fig. 4.17 Plaster mold and as-prepared perovskite membranes (Reproduced from Ref. [21] with permission of Elsevier)

operated at 850 °C, and 94 % CH_4 conversion, 93 % H_2, and oxygen permeation flux as high as 11.3 $cm^3\ cm^{-2}\ min^{-1}$ were achieved. Asymmetric membranes with a dense layer and a porous support layer may be prepared via the slip-casting method, in which the slips for the two layers are cast in order. Zyryanov reported a preparation of multilayer mixed conducting oxide membranes on macroporous dead-end tubular composite glass–ceramic substrates through slip casting [22]. However, the dynamics of oxygen loss at high temperatures and the mechanical properties were found to have been affected by the presence of $SrSO_4$ surface inclusions formed as a result of sulfur admixing in the starting reactants. Another possible mechanism to explain the sulfur impurities occurring on membrane surfaces is that sulfate ions coming from the plaster reacted with alkali earth metal ions (Ba^{2+} and Sr^{2+}) to produce sulfates on the surface of the slip-cast membranes. Some papers have reported that alkali earth metal ions can dissolve out from perovskite structures when the powders are dispersed in water. Therefore, polymeric molds are recommended for the preparation of MIEC perovskite membranes to prevent gypsum contamination.

4.2.4 Tape Casting

In 1947, Howatt et al. published a paper introducing a new method for the preparation of ceramic capacitors, which was named "tape casting" [23]. From then on, tape casting developed quickly and gradually matured into an important manufacturing technology for the industrial production of thin sheets. All the powder-type materials can be fabricated into sheets with thicknesses as low as 5 μm. This technology is now widely used in preparing organic and inorganic membranes, electrodes and electrolytes of SOFCs, electrodes of molten carbonate fuel cells, ceramic substrate materials, multilayered ceramic packages and capacitors, separators for batteries, brazing alloys, and intermetallic compounds. The prominent advantage of the tape-casting technology is that it is the best method to produce large-area thin, flat ceramic or metallic items at high production rates. Although it is possible to prepare thin and flat sheets by using other methods, such as dry-pressing or extrusion, tape casting is the most economic method to shape materials with these characteristics. The green tapes are easy to be post-processed into the required shapes and sizes due to their good flexibility, especially in terms of punching holes and slits. They also allow for laminating the sheets into multilayered ceramics, metallic workpieces, or ceramic–metallic composite items. It seems that tape casting is very like the abovementioned slip-casting method for the fabrication of ceramics because high-viscosity slurries are used in both processes. Although the shaping principles and drying mechanisms of these two methods (tape casting and slip casting) are completely different because of the solvent systems used, slurry preparation is similar. In the taping-casting process, the thickness of the tape mainly depends on the gap between the doctor blade and the carrier, the velocity of the carrier, and the solids content in the slurry, whereas the thickness in slip casting is mostly controlled by the casting time. In tape casting, solvents are removed by heating the cast tape, whereas in slip casting, the solvent is removed via absorption by the plaster mold. In terms of the drying method, one can imagine that tape casting offers higher production rates. Figure 4.18 shows the basic principle of the tape-casting process as it exists in most modern factories and laboratories. For a typical tape-casting machine, the four main parts are the blade, the reservoir, and the carrier with a motor and heating system. Other important auxiliary equipment includes a vacuum defoaming machine, automatic delivery pump, filter, wrapping

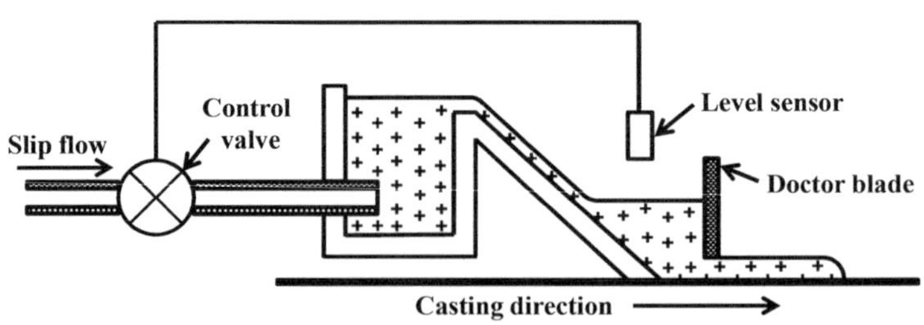

Fig. 4.18 Basic principle of the tape-casting process

setup, etc. These auxiliary setups are integrated with the tape-casting machine for high-speed continuous production.

Figure 4.19 shows the typical procedure for fabrication of ceramic sheets through tape-casting method. In a common operation for tape-casting ceramic sheets, the as-prepared slurry is poured into a reservoir after it is filtrated and degassed under vacuum. The green sheet is then cast onto the carrier, which is driven by a motor. Solvents are evaporated in a drying chamber by heating the baseboard and/or flowing heated air across the sheet. After drying, the as-formed sheet is flexible, as shown in Fig. 4.20. Green tape with a higher density than that derived from the dry-pressing method can be obtained if the starting ceramic powder has the proper particle-size distribution and morphology. For example, the relative density of alumina green tape can be as high as 70 % after tape casting. Powder characteristics, such as morphology, average particle size, particle-size distribution, density, surface area, etc., have a significant influence on the tape-casting technique, so it is important to know the basic information of the powders before preparing the slurry for tape casting. Among the abovementioned properties of powders, the specific surface area is the most important characteristic for tape casting. Fine powders with large specific surface areas are difficult to be shaped into perfect sheets through tape casting, especially those with specific surface areas higher than 20 m^2g^{-1}, because selection of the proper dispersant/solvent combination, as well as the dispersant concentration, are critical factors for very-high-surface-area powders.

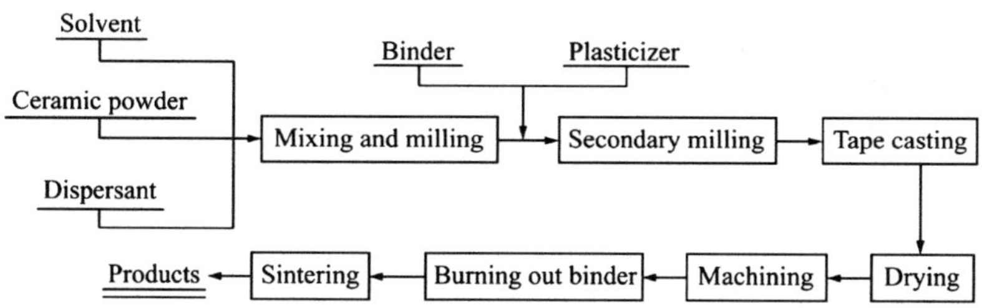

Fig. 4.19 Typical procedure for the fabrication of ceramic sheets through tape-casting method

Fig. 4.20 Sheet prepared by tape casting

Table 4.1 Solvents, dispersants, binders, and plasticizers frequently used in tape casting

Additive type	Additives
Solvents	Water, methyl ethyl ketone (MEK), ethanol, xylenes, toluene, acetone, 1,1,1-trichloroethane, methanol, MIBK, cyclohexanone, butanol, isopropanol, nitropropane, etc.
Dispersants	Polyisobutylene, linoleic acid, oleic acid, citric acid, stearic acid, lanolin fatty acids, salts of polyacrylic acids, salts of methacrylic acids, blown menhaden fish oil, corn oil, safflower oil, linseed oil, glycerol trioleate, synthetic waxy esters, etc.
Binders	Polyvinyl alcohol (PVA), polyvinyl butyral (PVB), polyvinyl chloride (PVC), vinyl chloride-acetate, polyacrylate esters, polymethyl methacrylate, polyethyl methacrylate, petroleum resins, polyethylene, ethylene oxide polymer, polypropylene carbonate, polytetrafluoroethylene (PTFE), poly-alpha-methyl styrene, poly isobutylene, atactic poly(propylene)/poly(butene), polyurethane, cellulose acetate-butyrate, nitrocellulose, methyl cellulose, ethyl cellulose, hydroxyethyl cellulose hydroxypropyl methyl cellulose, etc.
Plasticizers	Dibutyl phthalate, dioctyl phthalate, butyl benzyl phthalate, mixed esters phthalate, dimethyl phthalate, ethyltoluene sulfonamides, glycerine (glycerol), tri-n-butyl phosphate, butyl stearate, methyl abietate, tricresyl phosphate, propylene carbonate, water, polyethylene glycols (PEG), polyalkylene glycols, polypropylene glycols, triethylene glycols, dipropylglycol dibenzoate, etc.

Considering that tape casting is an important method for preparing plate-type MIEC membranes on a large scale, a brief introduction to the whole process is necessary. Table 4.1 lists the solvents, dispersants, binders, and plasticizers frequently used in tape casting. Similar to those used in extrusion processing, solvents, dispersants, binders, plasticizers, and other organic additives are used to preprocess the ceramic powders before shaping [24].

Solvents are used to dissolve dispersants, binders, plasticizers, and other organic additives. They are also used to disperse the ceramic particles in order to adjust the viscosity of slurry. Usually, organic solvents are chosen because of their variety and high volatility, and thus it is easy to find appropriate binders and plasticizers. Although water is an environmentally friendly solvent, its low volatility and the difficulty in finding proper binders and plasticizers lead to its limited application in few ceramic systems. Over the years, some organic solvents have become the standard choices for making a tape-casting slurry. These solvents include ethanol, methanol, toluene, methyl ethyl ketone, xylenes, and 1,1,1 trichloroethylene. To maintain the solubility of solvents to organic solutes, azeotrope mixtures are also used to maintain proper volatility. In some case, cyclohexanone is added into the solvent system to act as a "skin retarder," which slows the formation of a skin on the top layer of the tape owing to the fast evaporation of the solvents [24]. This skin retards the transport of solvent molecules from the bulk to the surface. Because skin retarder works to homogenize the concentration of the solvent throughout the tape by slowing the formation of a solvent-depleted zone (skin) on the drying face, it is also called a "homogenizer." Additionally, the choice of solvent should consider its toxicity, cost, and safety.

Large particles form through agglomeration of small grains in the as-received ceramic powders. These agglomerated particles must be completely dispersed prior to tape casting because these particles make it difficult to obtain a uniform slurry, resulting in green tape with low and inhomogeneous density from the agglomerated powders. Therefore, a dispersion milling step, in which a proper amount of dispersant is added, cannot be omitted. In addition to improving the dispersion of the particles, the function of the dispersant is to enhance the solids content and reduce the solvent amount so as to increase the density of the green tape and decrease the solvent costs. Two different mechanisms are involved in the dispersion effects of dispersants. One is electrostatic repulsion. Particles are changed by the adsorption of ionic dispersant molecules, and a repulsion force forms between particles. It is frequently found that the same mechanisms that serve to disperse the particles also keep the particles in stable suspension in the slurry. Another is steric hindrance. In this case, particles are separated by the adsorption of organic long-chain molecules on their surfaces. Steric hindrance-type dispersants are often utilized when an organic solvent is chosen because of its low dielectric constant. It is worthy of note that most dispersants used in organic solvent systems are blends of various fatty acids and esters. More or less dispersant relative to solid is not in favor of the dispersion of particles. A settlement experiment is an effective method to determine the usage of a dispersant.

After dispersion milling of the powder with the proper amount of dispersant, binders and plasticizers are added for another period of milling. The use of pre-prepared solutions is suggested to avoid incomplete dissolution of the organics. Binders are the most important additives for processing green ceramic tape because they form networks throughout the body, which has a significant influence on the strength and flexibility of the green body. That is to say, the dry green tape comprises a polymer framework filled with solid particles and other organic additives (including a residual solvent, dispersants, and plasticizers). In choosing binders, its solubility in the selected solvent, viscosity, glass transition temperature, cost, strength, burnout atmosphere and temperature, etc., should be considered. Polyvinyl-based organics such as polyvinyl alcohol (PVA) and polyvinyl butyral (PVB) are the most often used binders when water and organic solvents are employed, respectively. Other binders are listed in Table 4.1. The plasticizer allows bending and warping of the tape without breaking. In other words, the addition of a plasticizer improves the flexibility of the dry green tape because most binders result in a tape that is strong, stiff, and brittle after the evaporation of solvents. There are types of plasticizers classed by their plasticization mechanism. Type I is a kind of solvent with very low volatility that can stay in the green tape for a long time so as to dissolve the binder to a certain extent. Type II is a kind of lubricant, which improves the mobility of the long chains of the binders. Sometimes the two types of plasticizers are combined to achieve good plasticization results. The commonly used Type I and Type II plasticizers are listed in Table 4.1.

The organic additives have a complex influence on the properties of slurries. Therefore, optimal compositions and ratios of all the chemicals are obtained based on many experiments, such that the recipes are closely guarded by companies and researchers. No matter how complicated the recipes are, five criteria should be met to obtain a good slurry, i.e., homogeneous dispersion, proper viscosity, high solids content, proper strength, and flexibility after drying. Before being cast, the slurry needs to be filtrated and degassed under vacuum. Then, its viscosity and density should be determined. Usually, slurries with viscosities in the range of 600–6000 mPa s can be easily used for tape casting [24]. To achieve high repeatability from batch to batch, the density of the as-prepared slurry needs to be monitored. An error of less than ± 0.02 g cm^{-3} is required to obtain repeatable physical and chemical characteristics of the resultant green tape. Drying is the last important step in tape casting, and it is influenced by the composition and ratio of the chemicals, so its process conditions are different from one tape to another. There are two steps during the drying of as-cast tapes, i.e., solvent evaporation from the surface of the tape and solvent diffusion through the tape to the drying surface. The second step tends to be the rate-limiting step during drying, in which tape cracking and curling happen frequently. To avoid the formation of these defects, it is important to slow down the drying rate to reduce the heterogeneity of the bulk tape during drying.

Tape casting is the best method for producing plate-type asymmetric membranes on a large scale with high efficiency and low cost. Two possible procedures are involved in the preparation of ceramic plate-type asymmetric membranes through tape casting. In one, a thin layer as the dense layer is cast first, and then a thick layer as the support layer is cast onto the thin layer. In the other procedure, the thick layer is cast first, followed by the deposition of a thin layer through spray coating. This second method has been used to prepare anode-supported electrolyte membranes. Thicknesses of the electrolyte down to 5 μm have been achieved. Air Products & Chemicals Inc. produces plate-type asymmetric MIEC membranes by using tape-casting technology on a large scale, and the thickness of the dense layer is approximately 60 μm, as shown in Fig. 4.21 [25]. Baumann et al. prepared asymmetric BSCF membranes with a dense layer thickness of approximately 70 μm, as shown in Fig. 4.22a, and they reported a high permeation flux of more than 20 mL $cm^{-2}min^{-1}$ [26]. In our group, we prepared a dual-phase asymmetric SDC-SSAF membrane with a dense layer thickness of approximately 40 μm, as shown in Fig. 4.22b, and we achieved a high permeation flux of approximately 4 mL cm^{-2} min^{-1} at 950 °C under an oxygen partial gradient of 21 kPa/0.5 kPa [27].

4.2.5 Phase Inversion

Phase inversion is a mature technique for the preparation of polymeric membranes with asymmetric structure. Most commercially available polymeric membranes are

Fig. 4.21 Cross-sectional SEM of the asymmetric membrane prepared by Air Products & Chemicals Inc. using a tape-casting technique (Reproduced from Ref. [25] with permission of Elsevier)

Fig. 4.22 Fracture cross sections of BSCF membrane (**a**) and dual-phase membrane (**b**) prepared by the tape-casting technique (Reproduced from Refs. [26] and [27] with permission of Elsevier)

prepared by phase inversion, but it is recently being used to fabricate inorganic membranes as compared to the methods mentioned above. To prepare inorganic membranes other than those using ceramic powder, the casting solution comprises a solvent, a nonsolvent, and a polymer. The phase-inversion mechanism can generally be classified into three main categories depending on the parameters that induce demixing, i.e., temperature-, reaction-, and diffusion-induced phase inversion. In diffusion-induced phase inversion, the polymer slurry is first cast to form a thin film or spun into hollow fibers before being immersed in a coagulation bath containing the nonsolvent. The exchange of solvent and nonsolvent between the polymer solution and the coagulation bath induces phase separation in the bulk membrane. Finally, the composition of the polymer-rich phase reaches the glass transition composition and the system solidifies. Figure 4.23 shows the formation

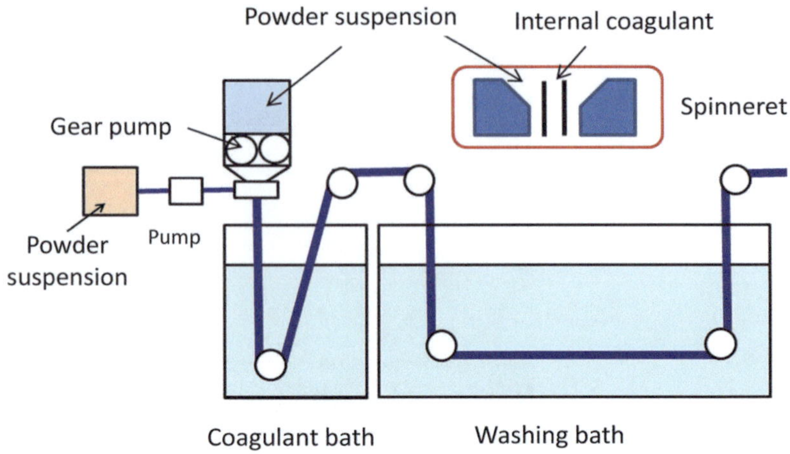

Fig. 4.23 Diagrammatic sketch of the equipment for preparing hollow fibers by the phase-inversion method

process of phase-inversion membranes. Although the process is quite easy to perform, the exact conditions under which a particular membrane will be formed are often derived on empirical grounds. It is clear that membrane morphologies are affected by the original compositions of the polymer solution and coagulation bath, as well as by preparation conditions.

Phase inversion is the best method for preparing ceramic hollow fiber membranes, although there have also been successful attempts to fabricate them using thermoplastic extrusion. Recently, alumina, alumina–titania, perovskite oxides, composite oxides for SOFC anodes, etc., have been fabricated into hollow fibers through the phase-inversion method. Li described the preparation of these ceramic membranes in detail in his book "Ceramic Membranes for Separation and Reaction" [28]. The use of a spinneret for preparing novel dual-layer ceramic hollow fiber membranes, as shown in Fig. 4.24, was reported by Li et al [29]. The developed dual-layer hollow fiber membrane consisted of a thin and dense outer oxygen separation layer (approximately 75 μm in thickness) supported on an inner Ni-based catalytic substrate (approximately 170 μm in thickness), as shown in Fig. 4.25 [30]. Researchers also used this technique to prepare microtubular SOFCs, in which an anode-supported electrolyte dual-layer hollow fiber was formed in one step [31, 32]. To obtain such ceramic hollow fibers, one powder for the dense layer and another powder for the porous support layer are separately mixed with a solvent, a polymer binder, and additives to form the outer and inner layer spinning suspensions, respectively. Then, the two suspensions are simultaneously coextruded through a triple-orifice spinneret, passed through an air gap, and placed into a nonsolvent external coagulation bath. In the meantime, a stream of nonsolvent internal coagulant is supplied through the central bore of the spinneret.

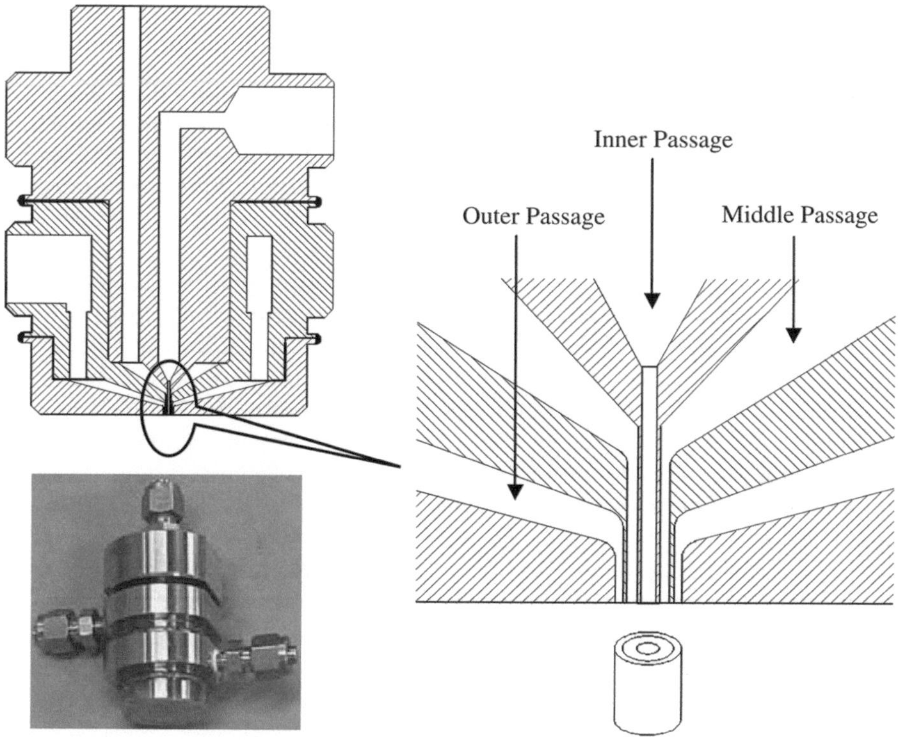

Fig. 4.24 Structure and flow channels of a dual-layer spinneret (Reproduced from Ref. [29] with permission of Elsevier)

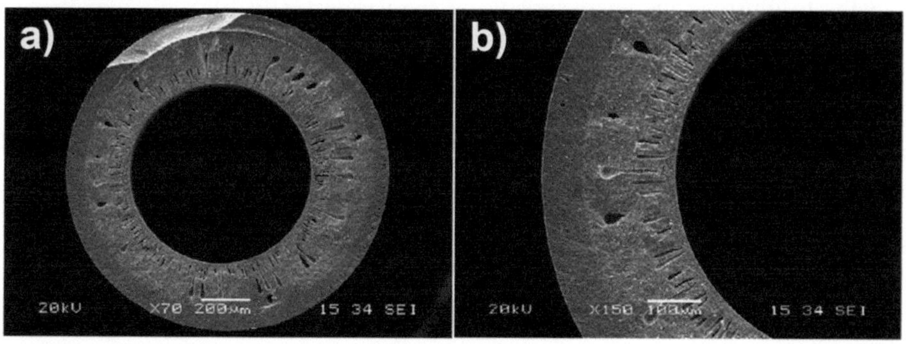

Fig. 4.25 SEM images of dual-layer hollow fibers after co-sintering: (**a**) whole view, (**b**) cross section (Reproduced from Ref. [30] with permission of Elsevier)

The thicknesses of the two layers are easily controlled by the design of the spinneret or by adjusting the extrusion rate, whereas the macrostructure or morphology of the prepared hollow fiber precursor is controlled by adjusting the co-extrusion parameters such as suspension viscosity, air gap, and flow rate of the internal coagulant.

4.2.6 Other Methods

Although several other methods are frequently used to prepare other inorganic membranes, they are occasionally used to prepare MIEC membranes, including dip coating, spin coating, spray coating, chemical vapor deposition (CVD), and pulse laser deposition (PLD). All these methods are used to fabricate asymmetric membranes, i.e., deposition of a dense thin layer on porous supports. Heat treatment to densify the deposited layer is necessary for the first three methods. A much-diluted suspension containing the ceramic powder, dispersant, binder, and plasticizer is prepared as the precursor for dip coating, spin coating, and spray coating. However, in a CVD experiment, volatile metallic salts (organic or inorganic) are carried by a gas into an oven where vapor-phase reactions take place and oxides are produced and deposited onto the pre-placed porous support. The PLD technique is frequently used by physical and materials scientists to prepare epitaxial films with thicknesses down to several nanometers. To obtain high-quality asymmetric membranes through these methods, the quality of the support is the most important factor, and it should be carefully pretreated. For the most part, only repetitious deposition can form the dense thin layer through these methods. In the opinion of the authors, they are somewhat inefficient and uneconomic for the fabrication of MIEC membranes on a large scale if there is no significant improvement in the processing techniques.

4.2.7 Comparison of the Methods

Finally, to illustrate the characteristics of the various MIEC ceramic membrane-shaping methods more clearly, all these methods are included in Table 4.2 to briefly explain and compare their main characteristics and applications.

4.2.8 Sintering

After shaping, the next most important step is sintering the as-prepared green bodies into dense membranes or porous supports. For sintering, a high-temperature furnace is necessary. There are currently many ways to obtain high temperatures, such as traditional combustion, electric, laser, and chemical reactions. For sintering MIEC membranes, furnaces heated by electricity rather than fuel are most often used in laboratories and factories because MIEC membranes consisting of alkali earth and rare earth elements are sensitive to CO_2, SO_2, and steam. Electric furnaces with SiC and $MoSi_2$ as the heating elements, which offer high temperatures up to 1300 and 1700 °C, respectively, are suitable for sintering MIEC membranes in laboratories.

Many physical properties of ceramic membranes such as density, strength, conductivity, microstructure, etc., are significantly influenced by sintering.

Table 4.2 Comparison of the main characteristics and applications of the membrane-shaping methods

	Dry-pressing	Extrusion	Slip casting	Tape casting	Phase inversion
Moisture before shaping	3–8 wt%	15–25 wt %	25–35 wt %	25–40 wt%	25–40 wt%
Moisture after shaping	3–8 wt%	15–25 wt %	15–25 wt %	5–10 wt%	5–10 wt%
Duration of shaping process	Low	Medium	High	Low	Low
Shaping energy consumption	High	Medium	Low	Medium	Low
Green density	High	Medium	Low	High	Low
Green deformability	Low	Medium	Low	High	High
Shrinkage after firing	Low	Medium	High	Low	High
Automation level	Medium	High	Low	High	High
Specific production costs	Medium	Low	High	Medium	High
Productivity	Medium	High	Low	High	High
Mold/die material	Rigid or elastic	Rigid	Rigid	Rigid	Rigid
Shapes	Sheets, tubes	Tubes	Tubes	Sheets	Hollow fibers, sheets
Shaping asymmetric membrane	Easy	Difficult	Difficult	Easiest	Easy
Membrane size	Small/ medium	Small/ medium	Medium/ large	Medium/ large	Small
Typical membrane thickness	0.5–3 mm	0.5–3 mm	1–3 mm	0.02–1 mm	0.1–0.5 mm

Sintering is the result of mass diffusion at high temperatures, which is driven by the surface free energy of the powder. This surface free energy gradually decreases as the gas–solid interface is replaced by a solid–solid interface. Five mass transport mechanisms are involved in the sintering process:

1. Evaporation–condensation
2. Surface diffusion
3. Lattice diffusion from the surface
4. Lattice diffusion from grain boundaries
5. Grain boundary diffusion

Among the five mass transport mechanisms, the first three provide coarsening between grains, but this behavior has little influence on the density of the ceramic body. If the goal of the sintering is to produce porous ceramic membranes or supports, the coarsening effect improves the strength without a significant decrease in porosity. However, the last two provide densification of the ceramic body. The sintering conditions have a significant influence on the predominant mass transport mechanisms, which affect the microstructure of the final ceramic body. These mechanisms can be regulated by changing the temperature, duration, atmosphere, pressure, heating/cooling rate, starting grain size, and additives (pore formers or sintering aids). In addition to the sintering conditions, many factors influence the sintering process and the characteristics of the resulting membranes, such as particle size, particle-size distribution, specific surface area, agglomeration type, shaping method, and the condition of the green bodies. Therefore, the sintering program and method of preparing the powders and membranes need to be carefully selected to obtain a membrane with a thin dense layer and a thick porous support layer through sintering in one step.

There are three stages in sintering, i.e., an initial stage, intermediate stage, and final stage. In the initial stage, evaporation–condensation, surface diffusion, and lattice diffusion from the surface prevail. There is no obvious change in grain size, volume, or density, but a sintering neck forms. To achieve high density, the temperature should be increased quickly in this stage because the formation of necks reduces the driving force for the subsequent densification. In the intermediate stage, with increasing temperature, the sintering necks become big and the distance between grains decreases, accompanied by the formation of a pore network throughout the body; at the same time, the grain size increases and the volume of the body decreases, accompanied by the increase of density. In the final stage, the relative density increases to $>90\,\%$ and more gas pores disappear. Figure 4.26 shows typical SEM images of the three stages.

For all these reasons, it is difficult to acquire ceramic materials with nano-size grains and high density simultaneously. Important progress on ceramic sintering theory was proposed by Chen et al. in 2000 [33]. In their study, they successfully overcame the dilemma by using a two-step sintering method, which greatly promoted the development of nano-size bulk materials. In the first step, the temperature is increased to a high value to complete the initial stage of sintering and reach a relative density of more than 75 %; then, the temperature is decreased to a proper point and the sample is maintained at that temperature to complete densification. As the isolated pores appear in the final period of traditional sintering, boundary diffusion and boundary movement in the bulk occur. Boundary diffusion leads to densification of the ceramic body, whereas boundary movement results in grain growth. The activation energy of boundary movement is higher than that of boundary diffusion; therefore, one can imagine that if a sample is placed in a low-temperature

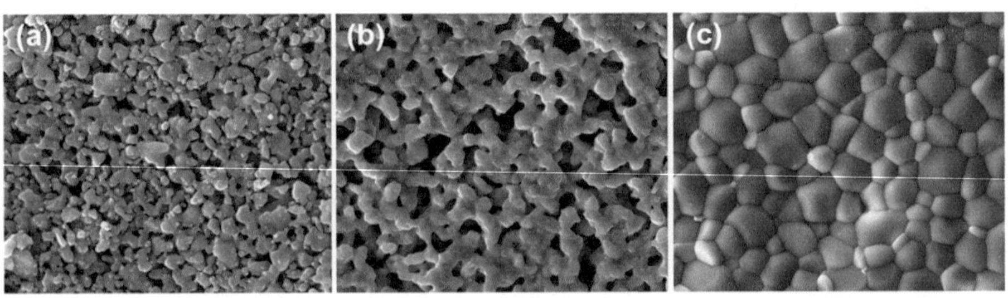

Fig. 4.26 Typical SEM pictures of the three stages in sintering: (**a**) initial stage, (**b**) intermediate stage, (**c**) final stage

environment in its final period of sintering, the movement of the boundary is limited, and thus the increase in grain size is restrained. As a result, the grain size remains almost constant during densification caused by boundary diffusion in the later period of sintering. Many organic additives are utilized while shaping the membrane green body in all the methods mentioned above, and so the temperature should be increased slowly to avoid crack generation during heating. Therefore, combined with sintering mechanisms and the two-step method, sintering curves for dense membranes and porous membranes are plotted in Fig. 4.27.

For MIEC membranes, the preparation of nano-sized membranes is a challenging job. Although the viewpoint is conflicting on the contribution of grain boundaries to oxygen transport, membranes with small grain sizes show high mechanical strength. In the author's group, sintering ceria-based dual-phase membranes via the two-step method was attempted, as shown in Fig. 4.28. Using traditional sintering,

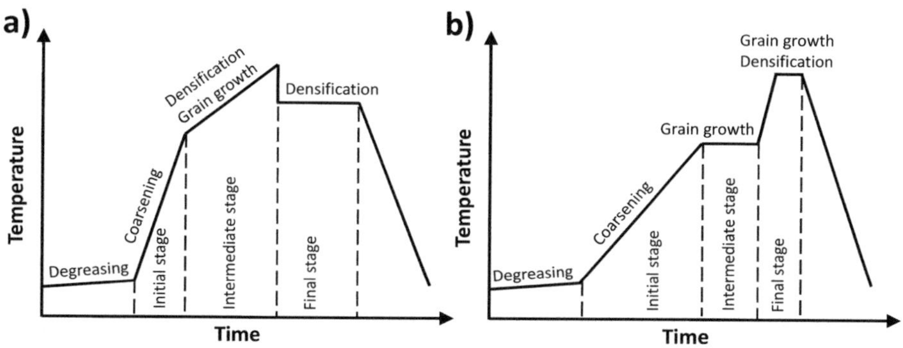

Fig. 4.27 Two different sintering procedures, (**a**) for dense ceramic membranes and (**b**) for porous ceramic membranes

Fig. 4.28 SEM images of dual-phase membranes sintered via the two-step method (**a**) and the traditional method (**b**). For the two-step method, the temperature was first increased to 1450 °C and then decreased to 1250 °C and maintained at 1250 °C for 10 h. For the traditional method, the temperature was increased to 1450 °C and maintained at this temperature for 3 h

the temperature was increased to 1450 °C and maintained at this temperature for 3 h. Using two-step sintering, the temperature was first increased to 1450 °C and then decreased to 1250 °C and maintained at 1250 °C for 10 h. The relative densities of the two disks were all higher than 95 %, but the grain size and microstructure were obviously different. A microstructure of perovskite grains (small grains) and fluorite grains (big grains) was visible on the surface of the two-step sintered membrane, which allowed for facile oxygen exchange and bulk diffusion. However, for the traditional one-step sintered membrane, the surface was covered by fluorite grains because of the lower surface energy of the fluorite phase as compared to the perovskite phase, which does not favor oxygen exchange and electron transport.

4.3 Characterization of MIEC Membranes

4.3.1 Permeation Flux

Permeation flux is one of the most important properties of MIEC membranes. Once a dense MIEC membrane is obtained after sintering, one can test its oxygen permeation flux after sealing it in a permeation setup. There are two operation models used to determine permeation flux, the gas chromatographic (GC) method and the oxygen-pump method. Figure 4.29 shows diagrammatic sketches of the two methods. For the oxygen-pump method, the membrane is sealed on an yttria-stabilized zirconia (YSZ) tube, with both walls coated with porous Pt electrodes. An oxygen electrochemical potential is generated by applying an electric potential between the two electrodes. Driven by this electric potential, oxygen ions are transported out of or in to the YSZ tube so as to adjust the oxygen partial pressure of the permeation side. Under a constant potential difference, the current eventually stabilizes. The stable current value corresponds to the oxygen permeation flux

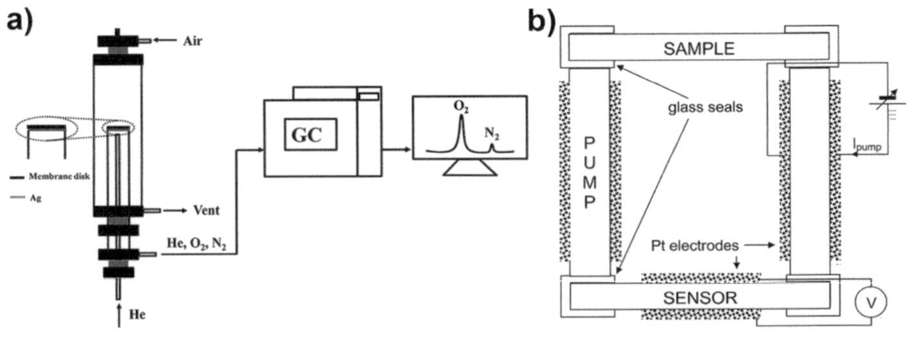

Fig. 4.29 Diagrammatic sketches of the two methods for determining permeation flux. (**a**) Gas chromatographic (GC) method, (**b**) oxygen-pump method (Reproduced from Ref. [34] with permission of Elsevier)

($1\ \mathrm{A\ cm^{-2}} = 3.47\ \mathrm{mL\ cm^{-2} min^{-1}} = 2.59 \times 10^{-6}\ \mathrm{mol\ cm^{-2} s^{-1}}$), whereas the open-circuit voltage (OCV) corresponds to the oxygen chemical potential gradient. Usually, air is fed to the retentate side, so the oxygen partial pressure of the permeation side can be calculated using the Nernst equation. Kharton et al. used this method to test the permeation fluxes of many perovskite membranes [34]. However, the main shortcoming of this method is that the leakage could not be monitored in real time. For the widely utilized GC method, the membrane is sealed on an alumina or quartz tube, and the oxygen chemical potential gradient is created by introducing sweeping gas to the permeation side. The gradient or oxygen partial pressure of the permeation side can be regulated handily by adjusting the flow rate of the sweeping gas. The concentrations of oxygen and nitrogen (due to imperfect sealing) in the effluent are quantitatively analyzed by GC so that the leakage can be monitored in real time during the course of testing. There are four kinds of sealants mentioned in the literature: glass powders or rings, ceramic powders, Ag paste or rings, and Au paste or rings. A detailed discussion of the sealing is given in Chap. 10.

Although there has been continuous research for the past 30 years, no standard testing conditions have been established up to now. Each research group uses different testing conditions to investigate their membranes. Many factors may influence the permeation flux, such as membrane thickness, the oxygen partial pressure gradient across the membrane (Chap. 5), the coarseness of the membrane surface, etc. Most researchers use a sweeping flow rate of $20-50\ \mathrm{mL\ min^{-1}}$, whereas others adopt high flow rates up to $200\ \mathrm{mL\ min^{-1}}$, which may lead to an oxygen partial pressure gradient ranging from 21/5 to 21/0.1 kPa. If bulk diffusion prevails, the difference in driving force would result in a ~3.7-times difference in oxygen permeation flux. This is one of important reasons why the very high permeation flux of a $SrCo_{0.8}Fe_{0.2}O_{3-\delta}$ membrane reported by Toreaka et al. was not achieved by other researchers [35]. Most permeation activation energies (E_a) of membranes reported in the literature are not reliable because the oxygen permeation flux values are usually acquired under a constant sweeping flow rates, in which the oxygen chemical potential difference across membranes varies with temperature. As a result, the calculated E_a is smaller than the actual value. The right way is to fix the oxygen chemical potential difference by changing the sweeping rate with temperature. Of course, different chemical potential differences may give various E_a values because the E_a of surface exchanges is different from that of bulk diffusion. A detailed discussion about how these steps influence the total E_a is given in Chap. 5.

4.3.2 Electric Conductivity

In MIEC ceramic materials, four carriers are considered, i.e., electrons, holes, oxygen ions, and oxygen vacancies. The corresponding electric conductivity for electrons and holes is electronic conductivity and that for oxygen ions and vacancies is oxygen ionic conductivity. The electromotive force (emf) method is used to obtain the ionic transport number when the electronic transport number is small, and the partial conductivity of oxygen ions and electrons can be calculated by multiplying their transport numbers by the total conductivity. However, the electronic transport number of MIEC materials is usually 100–100,000 times higher than the ionic transport number, especially for perovskite-type MIEC materials. Therefore, this method is not suitable for perovskite-type MIEC materials. Another method, called the electronic transport block method, is suggested for the measurement of ionic conductivity. In this method, a thin layer of a pure oxygen ionic conducting oxide such as YSZ is coated on one side of the MIEC sample. The total ohmic resistance across the MIEC sample and the electronic block layer is recorded through alternating current (AC) impedance, and then the ionic conductivity of the MIEC sample is calculated by subtracting the resistance of the electronic block layer from the total ohmic resistance. However, solid-state reactions, which produce new phases with various conductive behaviors, between the MIEC sample and the electronic block layer cannot be prevented at elevated temperatures; as a result, the measured data are inaccurate.

Some researchers suggested that the oxygen ionic conductivity can be calculated by using Wagner's equation through oxygen permeation experiments. To use this method, the following three main assumptions are made:

1. An MIEC membrane has sufficient thickness to make sure that the oxygen transport is controlled by bulk diffusion, and thus oxygen exchange on the membrane surfaces can be totally ignored.
2. The electronic transport number of an MIEC material is much larger (>100 times higher) than its oxygen ionic transport number.
3. Either the ionic conductivity remains almost constant across the MIEC membrane or the ionic conductivity changes little with changing oxygen partial pressure.

Because the electronic conductivity is much closer to the total conductivity, the ambipolar conductivity (Chap. 5) is approximately equal to the ionic conductivity. A porous thin layer (5–20 microns) of a Co-based perovskite catalyst, such as $La_{0.6}Sr_{0.4}CoO_{3-\delta}$, $Sm_{0.6}Sr_{0.4}CoO_{3-\delta}$, etc., with a high oxygen exchange rate was suggested to coat on both sides of the MIEC membrane in order to reduce the oxygen exchange resistances. Because the oxygen exchange resistance cannot be completely eliminated, the obtained ionic conductivity will be lower than the actual value. A new model equation that considers the surface effects in the oxygen permeation process can separate the surface-exchange resistances from the bulk diffusion resistance (see detailed discussion in Chap. 5), and so, accurate ionic conductivity values can be gained through model regression of the permeation data. Another frequently used and convenient way of measuring the ionic conductivity of MIEC materials is to calculate it from the diffusion coefficient obtained from isotropic exchange, conductivity relaxation, coulometric titration techniques, etc. All these methods will be introduced in Sect. 4.3.4.

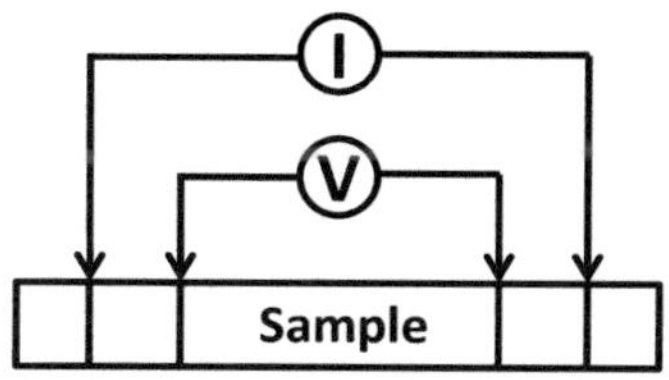

Fig. 4.30 Four-electrode method for conductivity measurement

More often, total conductivities are measured to show the electric transport properties of MIEC materials. There are two common methods employed in the measurement of total conductivity, i.e., the two-electrode method and the four-electrode method, as shown in Fig. 4.30. The resistances of bar-shaped samples are recorded by a multimeter directly for the two-electrode method or indirectly for the four-electrode method, and then total conductivity is calculated through the following equation:

$$\sigma_{\text{total}} = \frac{L_{\text{length}}}{L_{\text{wide}}L_{\text{high}}} \cdot \frac{1}{R_{\text{total}}} \tag{4.3}$$

To avoid introducing a large error, the two-electrode method is suitable for measuring samples with high resistances (for instance, $>1000\ \Omega$). However, the four-electrode method is suitable for measuring samples with low resistances. A constant current is applied by a galvanostat on the outer two electrodes, and the voltage between the inner two electrodes is recorded by a multimeter, as shown in Fig. 4.30.

4.3.3 *Nonstoichiometric Oxygen*

4.3.3.1 Thermogravimetric Analysis

Nonstoichiometric oxygen is an important parameter when studying MIEC oxide materials because it reflects the concentration of oxygen vacancies in the bulk material and because it is closely related to the electronic and ionic transport. Three methods are used to determine the nonstoichiometric oxygen of MIEC materials. One method uses hydrogen to reduce or decompose the MIEC oxide into simple oxides with unambiguous chemical formulas, such as

$$ABO_{3-\delta} + (3 - x - y - \delta)H_2 = AO_x + BO_y + (3 - x - y - \delta)H_2O \tag{4.4}$$

The mass loss ratio (q) due to hydrogen reduction can be determined by thermogravimetric (TG) analysis and the nonstoichiometric oxygen can be calculated by

$$\delta = \frac{M_{\text{O}}(3 - x - y) - qM_{\text{ABO}_3}}{M_{\text{O}}(1 - q)} \tag{4.5}$$

where M_O and M_{ABO_3} are the molar masses of the oxygen atom and the ABO_3 perovskite oxide, respectively, and x and y are the respective stoichiometries of the related oxides. For example, the mass loss of $Ba_{0.5}Sr_{0.5}Co_{0.8}Fe_{0.2}O_{3-\delta}$ was 13.27 % after being reduced by hydrogen.

$$Ba_{0.5}Sr_{0.5}Co_{0.8}Fe_{0.2}O_{3-\delta} + (2-\delta)H_2 \\ = 0.5BaO + 0.5SrO + 0.8Co + 0.2Fe + (2-\delta)H_2O \tag{4.6}$$

Here, $x = 1$, $y = 0$, and the calculated δ is 0.214.

MIEC membranes are operated at elevated temperatures, and so the nonstoichiometric oxygen of the materials at high temperatures and different oxygen partial pressures is more useful than that at room temperature. TG analysis is the most convenient way to determine the change of nonstoichiometric oxygen as a function of temperature and oxygen partial pressure.

4.3.3.2 Iodometry

Not all MIEC oxides can be reduced to products with fixed compositions. For example, the valence states of iron and cerium elements in $BaCe_{0.15}Fe_{0.85}O_{3-\delta}$ are ambiguous after its reduction by hydrogen [36]. Therefore, another method, called iodometry, is used to determine the nonstoichiometric oxygen of these MIEC materials based on the following chemical reactions:

$$ABO_{3-\delta} + (6-2\delta)H^+ + (m-2\delta)I^- \\ = A^{n+} + B^{(6-n-m)+} + (3-\delta)H_2O + (0.5m-\delta)I_2 \tag{4.7}$$

$$2S_2O_3^{2-} + I_2 = S_4O_6^{2-} + 2I^- \tag{4.8}$$

where n is the oxidation state of A-site ions, 6-n-m is the oxidation state of B-site ions after reduction, and δ is the oxygen nonstoichiometry of the perovskite at room temperature. Hence, there is a relationship between the amount of perovskite oxide and the consumption of $S_2O_3^{2-}$, i.e., the molar ratio of $ABO_{3-\delta}$ to $S_2O_3^2$ is 1: (m - 2δ). Assuming the mass used in the reaction is w_0, the molar concentration of $Na_2S_2O_3$ is C, and the consumed volume of the solution is V, one obtains the following equation:

$$\delta = \frac{CVM_{ABO_3} - mw_0}{16CV - 2w_0} \tag{4.9}$$

For $BaCe_{0.15}Fe_{0.85}O_{3-\delta}$, $n = 2$ and $m = 1.85$

However, because the nonstoichiometric oxygen acquired through iodometry is at room temperature, the dependence of the nonstoichiometric oxygen on oxygen partial pressure and temperature needs to be determined by TG or coulometric titration.

4.3.3.3 Coulometric Titration

Another method for determining the amount of nonstoichiometric oxygen at elevated temperatures is coulometric titration. The advantage of this technique is that it allows for direct regulation of the oxygen stoichiometry of a sample, as well as accurate determination of thermodynamic quantities such as the partial energy and entropy associated with the incorporation of oxygen into an oxide [37, 38]. In a typical coulometric titration experiment at an elevated temperature, the oxide sample is sealed inside an electrochemical cell (Fig. 4.31), which comprises an electrochemical oxygen pump equipped with a solid electrolyte (typically YSZ) and three electrodes (typically Pt) [39]. The oxygen stoichiometry of the sample is regulated by electrochemically pumping oxygen into or out of the cell through the solid electrolyte. Two operation models, i.e., galvanostatic and potentiostatic, are used for electrochemically pumping oxygen into or out of the cell. Both methods allow evaluation of the change in nonstoichiometric oxygen of the sample by integration of the current over time. The change in nonstoichiometric oxygen corresponding to a change in the equilibrium value of oxygen partial pressure of the atmosphere surrounding the sample is calculated by numerical integration of the decay current according to the following equation:

$$\Delta\delta = \frac{M}{m}\int_{t=0}^{t=\infty} \frac{I(t) - I_{\mathrm{leak}}}{2F}\,\mathrm{d}t \tag{4.10}$$

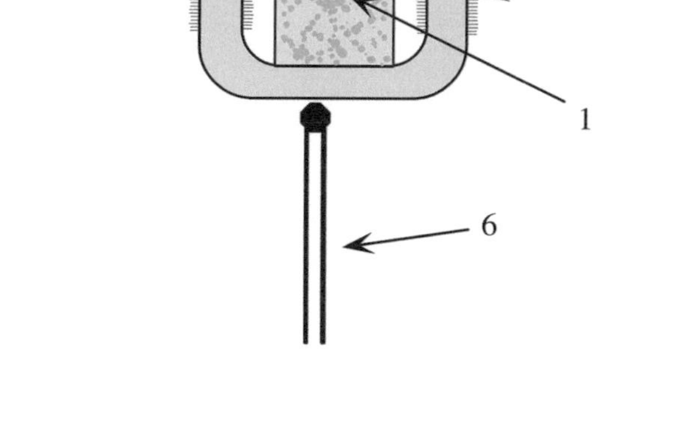

Fig. 4.31 Schematic drawing of the solid-electrolyte cell for coulometric titration: (*1*) porous sample, (*2*) yttria-stabilized zirconia, (*3*) porous Pt electrodes, (*4*) platinum current collectors, (*5*) glass–ceramic sealant, (*6*) thermocouple (Reproduced from Ref. [39] with permission of Elsevier)

where m is the mass of the sample, M is the molar mass of the oxide material, and I_{leak} is the unavoidable leakage current. In the potentiostatic operation model, I_{leak} is determined by measuring the value to which the current decays after a potentiostatic step, whereas in the galvanostatic operation model, I_{leak} can be calculated from the change in emf with time when the sample is in equilibrium with the surrounding gas.

In addition to determining the nonstoichiometric oxygen of oxides, coulometric titration is a powerful method for measuring thermodynamic quantities such as the partial energy and entropy associated with the incorporation of oxygen into the oxides. Therefore, a brief introduction is given here. The oxygen chemical potential of the sealed sample is measured by recording the emf value across an auxiliary solid electrolyte and reference electrode. At an equilibrium state, the oxygen chemical potential of the sealed sample is equal to the oxygen chemical potential of the gas surrounding the sample.

$$\mu_{O_2}^{\text{sample}} = \mu_{O_2}^{\text{gas}} = \mu_{O_2}^{\text{ref.}} - 4F \times emf \tag{4.11}$$

where F is Faraday's constant, $\mu_{O_2}^{\text{ref.}}$ is the oxygen chemical potential of the reference gas (typically air) and *emf* is the electrodynamic force. The value of $\mu_{O_2}^{\text{ref.}}$ is given by the following equations:

$$\mu_{O_2}^{\text{ref.}} = \mu_{O_2}^{0,\text{gas}} + RT\ln(P_{O_2}/P_0) \tag{4.12}$$

$$\mu_{O_2}^{0,\text{gas}} = RT\left[n_1 + n_2/T + n_3\ln T + n_4\ln\left(1 - e^{-n_5/T}\right)\right] \tag{4.13}$$

where $\mu_{O_2}^{0,\text{gas}}$ is the chemical potential of oxygen at 1 atm and $n_1 = -1.225$, $n_2 = -1.045 \times 10^3\text{K}$, $n_3 = -3.500$, $n_4 = -1.013$, and $n_5 = 2.242 \times 10^3\text{K}$. For a sample at a constant nonstoichiometric oxygen, when its oxygen chemical potential, $\mu_{O_2}^{\text{sample}}$, is performed as a function of temperature, the entropy part and energy part are given by the following two equations, respectively:

$$s_{O_2}^{\text{sample}} = -\left(\frac{\partial \mu_{O_2}^{\text{sample}}}{\partial T}\right)_\delta \tag{4.14}$$

$$\mu_{O_2}^{\text{sample}} = \varepsilon_{O_2}^{\text{sample}} - Ts_{O_2}^{\text{sample}} \tag{4.15}$$

4.3.4 Diffusion Coefficients and Exchange Coefficients

The diffusion coefficient and interface exchange coefficient, which can be used to calculate ionic conductivity, membrane characteristic thickness, distribution of permeation resistance, and oxygen permeation flux, are important in investigating mixed conducting membranes. Additionally, proper knowledge of the two coefficients provides a better understanding of oxygen transport across MIEC membranes and the cathode materials of SOFCs. The two parameters can be obtained simultaneously from isotope exchange, relaxation, coulometric titration, impedance spectroscopy, and electrochemical polarization experiments. The most frequently used methods are briefly introduced in this section; detailed discussions of the other methods are found in the references [40–43].

4.3.4.1 Isotope Exchange

One of the direct methods of measuring the oxygen self-diffusion coefficient and interfacial exchange coefficient of oxide materials is through ^{18}O isotope exchange, followed by secondary ion mass spectrometry (SIMS) depth profiling and/or gaseous phase analysis. In the classical implementation, dense samples are evacuated at a given temperature and then equilibrated in an atmosphere with the natural content of oxygen isotopes (0.2 % ^{18}O); once the samples are cooled, the apparatus is evacuated again, and the ^{18}O-labeled gas is introduced into the chamber, followed by rapid heating and maintaining the sample at the target temperature for a certain annealing period. The analysis method depends on the ^{18}O penetration depth. For depths smaller than 10 μm, the profiling method is usually used. For the profilometry mode, the primary ion beam is rastered across a region of the sample surface to produce a square profile while the secondary ion intensities of ^{16}O and ^{18}O are recorded as a function of time. When the depths are relatively large, the annealed samples are cut perpendicular to the exchanged surface and then analyzed by laterally scanning the primary ion beam; the two isotopic secondary ion intensities are recorded. Mass transport during the isotope exchange across the gas–solid interface is given by

$$k^*\left(C_g - C_s\right) = -D^*\left(\frac{\partial C(x)}{\partial x}\right)_{x=0} \tag{4.16}$$

where C_g and C_s are the normalized isotopic fractions of ^{18}O in the gas and solid, respectively, and k^* D^* are the tracer surface-exchange coefficient and tracer oxygen diffusion coefficient, respectively. Neglecting the correlation effects, the latter quantities are often considered essentially equivalent to the equilibrium exchange coefficient, k_{ex}, and the self-diffusion coefficient, respectively. The analytical solution of the one-dimensional diffusion problem is given by Crank [44].

$$\begin{aligned} C'(x) &= \frac{C(x) - C_{bg}}{C_g - C_{bg}} \\ &= \operatorname{erfc}\left(\frac{x}{2\sqrt{D^* t}}\right) - \exp\left(h_i x + h_i^2 D^* t\right) \cdot \operatorname{erfc}\left(\frac{x}{2\sqrt{D^* t}} + h_i\sqrt{D^* t}\right) \end{aligned} \tag{4.17}$$

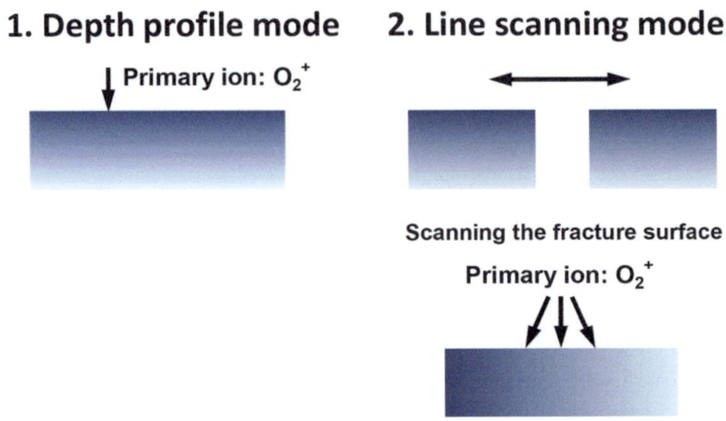

Fig. 4.32 Schematic representation of the SIMS measurement modes to determine the diffusion coefficient ((*1*) depth-profile mode and (*2*) line-scanning mode)

$$\text{with } \mathrm{erfc}(x) = \frac{2}{\sqrt{\pi}} \int_x^{\infty} \exp\left(-x^2\right) dx \tag{4.18}$$

where $C(x)$ is the isotopic fraction obtained from the exchange experiment, C_{bg} is the natural background concentration of ^{18}O in the sample, and t is time. The parameter $h_i = k^*/D^*$ corresponds to the factor describing the oxygen concentration ratio between the sample surface and the gas phase. Figure 4.32 shows the typical SIMS measurement modes to determine the diffusion coefficient. The two transport parameters can be directly obtained by fitting the experimental curves to Eq. (4.17).

Recently, a pulse isotopic exchange technique was developed for rapid evaluation of the oxygen surface-exchange rate under different operation conditions, as shown in Fig. 4.33 [45].

For this technique, the total oxygen surface-exchange rate, j_{ex}^{tot}, is defined by the first-order rate equation under equilibrium conditions:

$$n_g \frac{\partial f_g^{18}}{\partial t} = -j_{ex}^{\mathrm{tot}} S\left(f_g^{18} - f_s^{18}\right) \tag{4.19}$$

where n_g is the total number of oxygen atoms in the gaseous species; f_g^{18} and f_s^{18} are the atomic ^{18}O isotope fractions in the gas and solid phases, respectively; and S is the oxide surface area available for exchange. The exchange reaction involves a sequence of possible elementary steps, and it can be classified into two major steps, namely, dissociative adsorption of oxygen (or oxygen-containing species) and incorporation of oxygen into the oxide lattice. The corresponding rates, designated as j_{ex}^{ad} and j_{ex}^{in}, respectively, have the following relation:

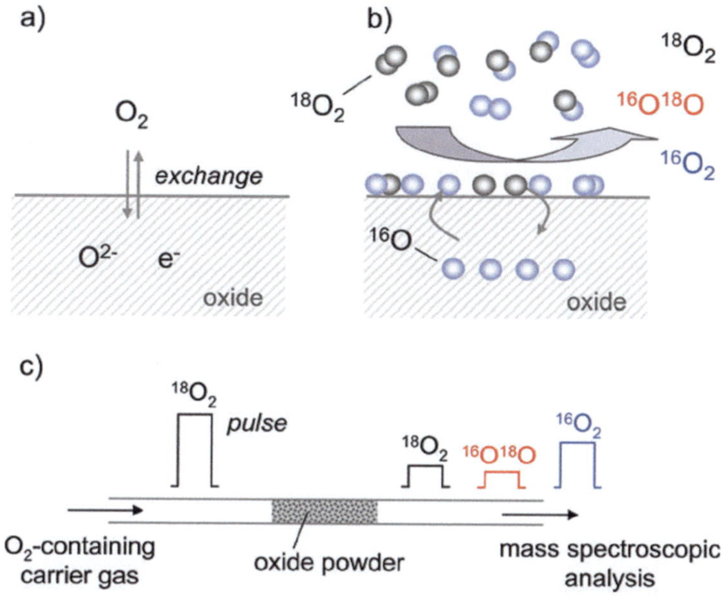

Fig. 4.33 (**a**) Reversible oxygen exchange between gaseous oxygen and oxygen from the oxide, (**b**) two-step oxygen exchange mechanism for the oxygen exchange reaction, with formation of mononuclear (i.e., dissociatively adsorbed) oxygen on the oxide surface as intermediate, and (**c**) principle scheme of the transient isotopic pulse technique (Reproduced from Ref. [45] with permission of RSC)

$$\frac{1}{j_{ex}^{\text{tot}}} = \frac{1}{j_{ex}^{ad}} + \frac{1}{j_{ex}^{in}} \tag{4.20}$$

As for other exchange techniques based on the analysis of $^{18}O/^{16}O$-containing species in the gaseous phase, powder samples are employed in the experiment, and so diffusion coefficients in the solid cannot be acquired. The exchange coefficients collected by the pulse method are similar to those obtained from electrical conductivity relaxation and isotopic exchange depth profiling methods. For mixed conducting perovskites, the total exchange rate is dominated by dissociative adsorption on the surface; the incorporation rate is two to three orders of magnitude higher. For oxygen ionic conductors such as yttria-stabilized zirconia (YSZ), the total exchange rate is jointly controlled by dissociative adsorption and incorporation.

4.3.4.2 Electrical Conductivity Relaxation (ECR)

For mixed conducting oxides, a sudden change of the oxygen partial pressure leads to the change in intrinsic physical parameters until a new equilibrium between the gas and solid phases is achieved. The equilibration kinetics, and thus the oxygen exchange rates, are reflected in the variations of physical parameters, such as electrical conductivity, mass, etc. The time dependence of these parameters is related to the rates of surface oxygen exchange and oxygen chemical diffusion in the bulk. Under relatively small alterations of the oxygen partial pressure, the mobility of change carriers can be considered constant; therefore, one may assume that the electrical conductivity linearly changes with the oxygen vacancy concentration. Again, if the surface reaction kinetics are essentially linear with respect to the oxygen concentration at the surface, the analytical solution for the plane geometry of width $2w_c$, thickness $2h_c$, and length $2l_c$ was given by Crank [44].

$$g(t) = \frac{\sigma(t) - \sigma(0)}{\sigma(\infty) - \sigma(0)} = 1 - \sum_{i=1}^{\infty}\sum_{m=1}^{\infty}\sum_{n=1}^{\infty} \frac{2L_1^2 \exp\left(-\beta_i^2 D_{\mathrm{chem}} t/{h_c}^2\right)}{\beta_i^2\left(\beta_i^2 + L_1^2 + L_1\right)} \times \frac{2L_2^2 \exp\left(-\gamma_m^2 D_{\mathrm{chem}} t/{w_c}^2\right)}{\gamma_m^2\left(\gamma_m^2 + L_2^2 + L_2\right)} \times \frac{2L_3^2 \exp\left(-\delta_n^2 D_{\mathrm{chem}} t/{l_c}^2\right)}{\delta_n^2\left(\delta_n^2 + L_3^2 + L_3\right)} \tag{4.21}$$

where $\sigma(0)$, $\sigma(\infty)$, and $\sigma(t)$ are the conductivity values in the initial state, after achieving a new equilibrium, and during the relaxation process, respectively. Here, D_{chem} is the chemical diffusion coefficient, t is the diffusion time, and L_1, L_2, and L_3 are dimensionless parameters determined by the surface-exchange and chemical diffusion coefficients: $L_1 = h_c k_{\mathrm{chem}}/D_{\mathrm{chem}}$, $L_2 = w_c k_{\mathrm{chem}}/D_{\mathrm{chem}}$, and $L_3 = l_c k_{\mathrm{chem}}/D_{\mathrm{chem}}$, respectively. β_i, γ_m, and δ_n are the nth roots of the equations: $\beta_i \tan \beta_i = L_1, \gamma_m \tan \gamma_m = L_2$, and $\delta_n \tan \delta_n = L_3$, respectively. If the sample is a bar where the ratio of the length to the width is large (for instance, >10), the length can be regarded as infinitely long, and so the last term of Eq. (4.21) can be neglected. Other simplified equations based on Eq. (4.21) are also known in the literature, but it is not possible to give a full account of these models in this brief review. The four-probe method is usually used for the conductivity measurements; the D_{chem} and k_{chem} values are derived by fitting of the relaxation data using Eq. (4.21) or its derivatives as the regression model. The ratio of D_{chem} and k_{chem} is related to the characteristic thickness of the mixed conducting membranes. In simple cases, these can also be converted to the self-diffusion coefficient and equilibrium oxygen exchange coefficient, respectively, via the use of thermodynamic factors. It is necessary to note that the two parameters obtained by the relaxation method are averaged over the given oxygen partial pressure range.

4.3.4.3 In Situ Isothermal Isotope Exchange (IIE)

In a SIMS experiment, dense samples are first annealed in $^{18}O_2$ and then quenched for SIMS depth profiling of the diffused ^{18}O ions. Crank's solution is used to model the depth profile to obtain the tracer diffusion coefficient and surface-exchange coefficient under certain experimental conditions. In an ECR experiment, dense samples are exposed to oxygen partial pressure changes to obtain the time-dependent electrical conductivity changes. Similarly, Crank's solution is used to model the profiles to extract the chemical diffusion coefficient and surface-exchange coefficient. These values are related to the tracer values by a thermodynamic factor. However, the two kinetic parameters are obtained under ex situ analysis (SIMS) or a nonequilibrium state. Furthermore, the surface-exchange coefficient from the two methods is typically determined in the region of diffusion limitation or mixed diffusion and exchange limitation due to the sample thickness on the scale of millimeters. Armstrong et al. proposed a new method, named in situ isothermal isotope exchange (IIE), to overcome these limitations by conducting the isothermal isotope exchange experiments on powder samples [46–48].

For IIE, the same method as that used in SIMS and ECR can be taken to model diffusion in powder particles. If the powder particles are spheres, it is possible to obtain the two kinetic parameters from the accumulation isotope exchange profiles by using Crank's solution to the diffusion in a sphere:

$$\frac{M_t}{M_\infty} = 1 - \sum_{n=1}^{\infty} \frac{6L^2 \exp\left(-\beta_n^2 D^* t/a^2\right)}{\beta_n^2\left(\beta_n^2 + L^2 + L\right)} \tag{4.22}$$

$$\beta_n \cot \beta_n + L - 1 = 0 \tag{4.23}$$

$$L = a \cdot \frac{k^*}{D^*} = \frac{a}{L_c} \tag{4.24}$$

All variables have the same meanings as previously defined. The additional variable a is the radius of the sphere. Because the particle sizes are on the order of or lower than L_c, the oxygen exchange rate between gas phase and solid phase is surface exchange dominated, and meaningful $k*$ values can be extracted. The accumulation of ^{18}O in the powder sample is the desired profile to fit, and it can be described by the following equation:

$$M_t = \int_{t=0}^{t=\infty} r_{\text{net}^{18}\text{O}}\, dt \tag{4.25}$$

where $r_{\text{net}^{18}\text{O}}$ is the rate of incorporation of ^{18}O into the sample, which can be determined by

$$r_{\text{net}^{18}\text{O}} = 2r_{\text{in}^{18}\text{O}} - r_{\text{out}^{18}\text{O}} \tag{4.26}$$

$$r_{\text{out}^{18}\text{O}} = 2r_{\text{out}^{18}\text{O}} + r_{\text{out}}^{^{16}\text{O}^{18}\text{O}} \tag{4.27}$$

where $r_{\text{in}^{18}\text{O}}$ and $r_{\text{out}^{18}\text{O}}$ are the rates of $^{18}O_2$ into and out of the reactor, respectively, and $r_{\text{out}^{18}\text{O}}$ and $r_{\text{out}}^{^{16}\text{O}^{18}\text{O}}$ are the rates of the ^{18}O and $^{16}O^{18}O$ species out of the reactor, respectively. Then the accumulation of ^{18}O in the powder sample can be rewritten as

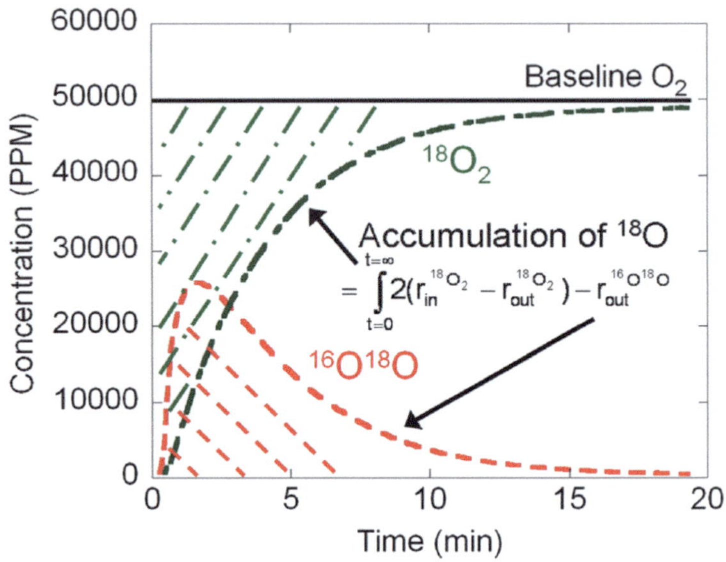

Fig. 4.34 Accumulation profile processing method demonstrated for $(La_{0.6}Sr_{0.4})(Co_{0.2}Fe_{0.8})O_{3-\delta}$ at 800 °C in 0.05 atm O_2 (Reproduced from Ref. [47] with permission of ECS)

$$M_t = \int_{t=0}^{t=\infty} \left(2\left(r_{in^{18}O} - r_{out^{18}O}\right) - r_{out}^{^{16}O^{18}O}\right) dt \tag{4.28}$$

Figure 4.34 gives an example of the calculation of the accumulation of ^{18}O in $La_{0.6}Sr_{0.4}Co_{0.2}Fe_{0.8}O_{3-\delta}$ powder at 800 °C. Once the accumulation profiles were obtained after the processing, the values of the two kinetic parameters, D^* and k^*, can be extracted through regressing Eq. (4.22) and combining Eqs. (4.23) and (4.24).

4.3.4.4 Oxygen Permeation

With the development of the oxygen permeation model, the kinetic parameter, diffusion coefficient and surface-exchange coefficient, can be obtained through some models. In these models, the oxygen permeation through an MIEC membrane is separated into three steps, i.e., oxygen reduction on the retentate side, bulk diffusion, and oxygen evolution on the permeation side. The three steps correspond to oxygen exchange coefficient on the retentate side, the oxygen diffusion coefficient in the bulk, and the oxygen exchange coefficient on the permeation-side surface, respectively. This will be thoroughly introduced in Chap. 5.

4.4 Structure and Morphology Characterizations

Frequently utilized approaches for characterizing the crystal structure, microstructure, valence state, and micro-zone element distribution of MIEC materials or membranes include X-ray techniques [X-ray diffraction (XRD), X-ray photoelectron spectroscopy (XPS), extended X-ray absorption fine structure (EXAFS), etc.], electron microscopic techniques [scanning electron microscopy (SEM), transmission electron microscopy (TEM), high-resolution transmission electron microscopy (HRTEM), atomic force microscopy (AFM), and energy dispersive spectroscopy (EDS)], and spectroscopic techniques [Fourier transform infrared spectroscopy (FTIR), Mössbauer spectroscopy, etc.]. All these techniques are widely discussed in many textbooks and are not particular to the field of solid-state electrochemistry, and so they are not introduced in this book. Several books on these characterization techniques are listed in the reference section.

References

1. Shao ZP, Yang WS, Cong Y, Dong H, Tong JH, Xiong GX (2000) Investigation of the permeation behavior and stability of a $Ba_{0.5}Sr_{0.5}Co_{0.8}Fe_{0.2}O_{3-\delta}$ oxygen membrane. J Membr Sci 172:177–188
2. Shao ZP (2000) Mixed oxygen ionic and electronic conducting membrane and its application in the partial oxidation of methane to synthesis gas process. Dissertation, University of Chinese Academy of Science
3. Peng RR, Xia CR, Fu QX, Meng GY, Peng DK (2002) Sintering and electrical properties of $(CeO_2)_{0.8}(Sm_2O_3)_{0.1}$ powders prepared by glycine-nitrate process. Mater Lett 56:1043–1047
4. Tok AIY, Luo LH, Boey FYC (2006) Consolidation and properties of $Gd_{0.1}Ce_{0.9}O_{1.95}$ nanoparticles for solid-oxide fuel cell electrolytes. J Mater Res 21:119–124
5. Liang JP (2015) Low temperature fabrication of dense gadolinia-doped ceria electrolyte with enhanced electrical conductivity. Electrochim Acta 178:321–328
6. Wei Y (2012) Synthesis and optical properties of self-assembled 2D layered organic-inorganic perovskites for optoelectronics. Dissertation, l'École normale Supérieure de Cachan
7. Messing GL, Zhang S-C, Jayanthi GV (1993) Ceramic powder synthesis by spray pyrolysis. J Am Ceram Soc 76:2707–2726
8. Meng XX, Pang ZB, Tan XY, Meng B, Yang NT (2006) Preparation of $La_{0.6}Sr_{0.4}Co_{0.2}Fe_{0.8}O_{3-\delta}$ ultra fine powder by spray pyrolysis method and its electrical property. Chin J Nonferrous Metals 16:2077–2082
9. Marrero-Lopez D, dos Santos-Gomez L, Canales-Vazquez J, Martin F, Ramos-Barrado JR (2014) Stability and performance of nanostructured $La_{0.8}Sr_{0.2}MnO_3$ cathodes deposited by spray-pyrolysis. Electrochim Acta 134:159–166
10. Marrero-Lopez D, Romero R, Martin F, Ramos-Barrado JR (2014) Effect of the deposition temperature on the electrochemical properties of $La_{0.6}Sr_{0.4}Co_{0.8}Fe_{0.2}O_{3-\delta}$ cathode prepared by conventional spray-pyrolysis. J Power Sources 255:308–317
11. Benel C, Darbandi AJ, Djenadic R, Evans A, Tolke R, Prestat M, Hahn H (2013) Synthesis and characterization of nanoparticulate $La_{0.6}Sr_{0.4}CoO_{3-\delta}$ cathodes for thin-film solid oxide fuel cells. J Power Sources 229:258–264

12. Suda S, Takahashi S, kawana M, Yoshida H, Inagaki T (2006) Effects of atomization conditions on morphology and SOFC anode performance of spray pyrolyzed NiO-$Sm_{0.2}Ce_{0.8}O_{1.9}$-composite particles. Solid State Ionics 177:1219–1225
13. Maric R, Fukui T, Ohara S, Yoshida H, Nishimura M, Inagaki T, Miura K (2000) Powder prepared by spray pyrolysis as an electrode material for solid oxide fuel cells. J Mater Sci 35:1397–1404
14. Kumar A, Devi PS, Das Sharma A, Maiti HS (2005) A novel spray-pyrolysis technique to produce nanocrystalline lanthanum strontium manganite powder. J Am Ceram Soc 88:971–973
15. Wei X, Hug P, Figi R, Trottmann M, Weidenkaff A, Ferri D (2010) Catalytic combustion of methane on nano-structured perovskite-type oxides fabricated by ultrasonic spray combustion. Appl Catal B Environ 94:27–37
16. Qi XW, Lin YS, Swartz SL (2000) Electric transport and oxygen permeation properties of lanthanum cobaltite membranes synthesized by different methods. Ind Eng Chem Res 39:646–653
17. Sfeir J, Vaucher S, Holtappels P, Vogtc U, Schindler H-J, Van Herle J, Suvorova E, Buffat P, Perret D, Xanthopoulos N, Bucheli O (2005) Characterization of perovskite powders for cathode and oxygen membranes made by different synthesis routes. J Eur Ceram Soc 25:1991–1995
18. Rahaman MN (2003) Ceramic processing and sintering. 2nd version, CRC press, Boca Raton
19. Bender W (2007) Types of extrusion units. In: Händle F (ed) Extrusion in ceramics. Springer, Berlin, p p73
20. Clemens F (2007) Thermoplastic extrusion for ceramic bodies. In: Händle F (ed) Extrusion in ceramics. Springer, Berlin, pp p307–p309
21. Zhang YW, Su K, Zeng FL, Ding WZ, Lu XG (2013) A novel tubular oxygen-permeable membrane reactor for partial oxidation of CH_4 in coke oven gas to syngas. Int J Hydrog Energy 38:8783–8789
22. Zyryanov VV (2009) Fabrication of multilayer ceramic membranes. Asia Pac J Chem Eng 4:285–290
23. Howatt GN, Breckenridge RG, Brownlow JM (1947) Fabrication of thin ceramic sheets for capacitors. J Am Ceram Soc 30:237–242
24. Mistler RE, Twiname ER (2000) Tape casting: theory and practice. Wiley, Westerville
25. Miller CF, Chen J, Carolan MF, Foster EP (2014) Advances in ion transport membrane technology for syngas production. Catal Today 228:152–157
26. Baumann S, Serra JM, Lobera MP, Escolástico S, Schulze-Küppers F, Meulenberg WA (2011) Ultrahigh oxygen permeation flux through supported $Ba_{0.5}Sr_{0.5}Co_{0.8}Fe_{0.2}O_{3-\delta}$ membranes. J Membr Sci 377:198–205
27. Cao ZW, Zhu XF, Li WP, Xu B, Yang LN, Yang WS (2015) Asymmetric dual-phase membranes prepared via tape-casting and co-lamination for oxygen permeation. Mater Lett 147:88–91
28. Li K (2007) Ceramic membranes for separation and reaction. Wiley, Chichester
29. Li DF, Chung TS, Wang R, Liu Y (2002) Fabrication of fluoropolyimide/polyethersulfone (PES) dual-layer asymmetric hollow fiber membranes for gas separation. J Membr Sci 198:211–223
30. Wu ZT, Wang B, Li K (2010) A novel dual-layer ceramic hollow fibre membrane reactor for methane conversion. J Membr Sci 352:63–70
31. Li T, Wu ZT, Li K (2014) Single-step fabrication and characterisations of triple-layer ceramic hollow fibres for micro-tubular solid oxide fuel cells (SOFCs). J Membr Sci 449:1–8
32. Othman MHD, Droushiotis N, Wu ZT, Kelsall G, Li K (2011) High-performance, anode-supported, microtubular SOFC prepared from single-step-fabricated, dual-layer hollow fibers. Adv Mater 23:2480–2483
33. Chen IW, Wang XH (2000) Sintering dense nanocrystalline ceramics without final-stage grain growth. Nature 404:168–171
34. Figueiredo FM, Kharton VV, Viskup AP, Frade JR (2004) Surface enhanced oxygen permeation in $CaTi_{1-x}Fe_xO_{3-\delta}$ ceramic membranes. J Membr Sci 236:73–80

35. Teraoka Y, Zhang HM, Furukawa N, Yamazoe N (1985) Oxygen permeation through perovskite-type oxides. Chem Lett 14:1743–1746
36. Zhu XF, Wang HH, Cong Y, Yang WS (2006) Structure and oxygen permeation of cerium light doped $BaFeO_{3-\delta}$ ceramic membranes. Solid State Ionics 117:2917–2921
37. Steele BCH (1968). In: Wachtman JB, Franklin AD (eds) Mass transport in oxides. NBS Special Publication 296. p 165
38. Lankhorst MHR, Bouwmeester HJM, Verweij H (1997) Thermodynamics and transport of ionic and electronic defects in crystalline oxides. J Am Ceram Soc 80:2175–2198
39. Tikhonovich VN, Zharkovskaya OM, Naumovich EN, Bashmakov IA, Kharton VV, Vecher AA (2003) Oxygen nonstoichiometry of Sr(Co, Fe)$O_{3-\delta}$-based perovskites I. Coulometric titration of $SrCo_{0.85}Fe_{0.10}Cr_{0.05}O_{3-\delta}$ by the two-electrode technique. Solid State Ionics 160:259–270
40. Mauvy F, Bassat JM, Boehm E, Manaud JP, Dordor P, Grenier JC (2003) Oxygen electrode reaction on $Nd_2NiO_{4+\delta}$ cathode materials: impedance spectroscopy study. Solid State Ionics 158:17–28
41. Leonide A, Ruger B, Weber A, Meulenberg WA, Ivers-Tiffee E (2010) Impedance study of alternative (La,Sr)$FeO_{3-\delta}$ and (La,Sr)(Co,Fe)$O_{3-\delta}$ MIEC cathode compositions. J Electrochem Soc 157:B234–B239
42. Diethelm S, Van Herle J (2004) Electrochemical characterisation of oxygen nonstoichiometry and transport in mixed conducting oxides-application to $La_{0.4}Ba_{0.6}Fe_{0.8}Co_{0.2}O_{3-\delta}$. Solid State Ionics 174:127–134
43. Adler SB (2004) Factors governing oxygen reduction in solid oxide fuel cell cathodes. Chem Rev 104:4791–4843
44. Crank J (1975) The mathematics of diffusion. Oxford University Press, Oxford
45. Bouwmeester HJM, Song CL, Zhu JJ, Yi JX, van Sint Annaland M, Boukamp BA (2009) A novel pulse isotopic exchange technique for rapid determination of the oxygen surface exchange rate of oxide ion conductors. Phys Chem Chem Phys 11:9640–9643
46. Armstrong EN, Duncan KL, Wachsman ED (2011) Surface exchange coefficients of composite cathode materials using in situ isothermal isotope exchange. J Electrochem Soc 158:B283–B289
47. Armstrong EN, Duncan KL, Oh DJ, Weaver JF, Wachsman ED (2011) Determination of surface exchange coefficients of LSM, LSCF, YSZ, GDC constituent materials in composite SOFC cathodes. J Electrochem Soc 158:B492–B499
48. Armstrong EN, Duncan KL, Wachsman ED (2013) Effect of A and B-site cations on surface exchange coefficient for ABO_3 perovskite materials. Phys Chem Chem Phys 15:2298–2308

Chapter 5
Permeation Principle and Models

Abstract Although the principle of oxygen permeation through an MIEC membrane is simple, it is difficult to develop a proper mathematical equation to describe the processes including oxygen exchange reactions on gas–solid interfaces and ambipolar diffusion in membrane bulk. In this chapter, the classical Wagner equation is introduced firstly. Only the oxygen permeation through an MIEC membrane is dominated by the ambipolar diffusion in membrane bulk; the permeation process can be well described by the Wagner equation. However, most of membranes are limited by the joint control of interfacial exchange and bulk diffusion. Therefore, three permeation models were developed by researchers to include the influence of oxygen exchange reactions on the gas–solid interfaces. The derivation and applications of the three models are separately introduced in detail in this chapter.

Keywords Oxygen permeation model • Wagner equation • Jacobson's model • Xu and Thomson's model • Zhu's model • Permeation resistances • Oxygen interfacial exchange • Transport kinetics

5.1 Introduction

Oxygen permeation through an MIEC membrane involves many chemical and physical processes, which can be classified into three steps: a gas–solid interfacial exchange step on the feed side, a diffusion step in the bulk membrane, and a gas–solid interfacial exchange step on the permeation side. The exchange process between gas-phase oxygen molecules and membrane surfaces includes a series of reaction steps, and every step plays an important role in the oxygen permeation process. Elemental steps, such as adsorption, dissociation, charge transfer, surface diffusion, and oxygen ion incorporation into the lattice, are involved in the interfacial zone of the feed side. The same elemental steps occur in the interfacial zone of the permeation side but in the reverse direction, and the relative roles of the same steps may be different. Interfacial exchange reactions on both sides become the rate-determining steps (r.d.s.) if the bulk transport, including both electronic and ionic transports, is fast enough or if the membranes are thinner than the characteristic thickness (L_c). On the other hand, if bulk diffusion is the r.d.s., oxygen permeation flux through MIEC membranes can be well described by the Wagner equation. If the ambipolar

X. Zhu, W. Yang, *Mixed Conducting Ceramic Membranes*, Green Chemistry and Sustainable Technology, DOI 10.1007/978-3-662-53534-9_5

conductivity is marginally affected by the oxygen partial pressure across membranes, the equation shows a simple relation between the oxygen permeation flux and the operating conditions, i.e., the temperature and oxygen partial pressures on both sides. However, only a few membranes or membranes with sufficient thickness can meet the requirement of bulk diffusion limitation. To achieve a practical and effective description of the oxygen permeation behavior across MIEC membranes, several models and empirical equations have been developed in recent years that consider the effects of oxygen exchange reactions on gas–solid interfaces. For example, considering the influence of the interfacial exchange steps, Boumeester et al. modified the Wagner equation to develop a general permeation model by introducing the important concept of characteristic thickness [1]. Jacobson et al. derived a general equation to describe oxygen permeation through MIEC membranes based on the dependence of a net oxygen interfacial exchange flux under a chemical potential [2–4]. Another general equation was developed by Xu and Thomson, who simplified Lin's model by analyzing the diffusion of oxygen vacancies through membranes and the oxygen exchange reactions at the interfaces [5, 6]. More recently, Zhu et al. developed a permeation model based on the theoretical analysis of the role of interfaces of mixed conducting membranes [7]. In addition to the abovementioned models, several empirical equations were found to fit permeation curves very well. These models were developed from different starting points and based on different assumptions; all have succeeded in modeling some experimental data. Therefore, in this chapter, in addition to the classic Wagner equation, these models will be introduced for a deep understanding of the mechanism of oxygen permeation through MIEC ceramic membranes.

5.2 Wagner Equation and Related Modifications

The basic assumption supporting the Wagner equation is that the bulk diffusion of oxygen or electronic charge carriers through MIEC membranes is the r.d.s. In addition, it is lattice oxygen ions rather than molecules and atoms that are selectively transported through membranes under the driving force of a chemical potential drop of oxygen. External circuitry connecting both surfaces of the membrane such as in SOFCs is unnecessary because continuous oxygen ionic flow is formed by the charge compensation of a simultaneous flow of electronic charge carriers in the bulk. The charge carriers in MIEC membranes are oxygen ions, oxygen vacancies, electrons, and holes. Directional movement of oxygen ions or oxygen vacancies contributes to the oxygen ionic current, whereas that of electrons or holes contributes to the electronic current. There is no applied electric potential across MIEC membranes, and so the sum of the two currents is equal to zero to keep the electric neutrality in the whole membrane. The equilibrium between electrons and holes occurring in MIEC membranes is given as

$$\text{nil} \rightleftharpoons e' + h^{\cdot} \tag{5.1}$$

with an equilibrium constant given as

$$K_e = np \tag{5.2}$$

Here, n and p denote the concentration in molar fraction of electrons and holes, respectively. Oxygen transport in MIEC materials, especially in perovskite oxides, is generally considered to occur through a vacancy transport mechanism. Therefore, in an MIEC conductor, the equilibrium of interest is the following:

$$\frac{1}{2}O_2 + V_O^{\cdot\cdot} \rightleftharpoons O_O^{\times} + 2h^{\cdot} \tag{5.3}$$

At a thermodynamic equilibrium, the oxygen exchange current density from the gas to the solid is equal to that from the solid to the gas. Also the following relationship should be satisfied:

$$\frac{1}{2}\mu_{O_2} + \mu_{V_O^{\cdot\cdot}} = \mu_{O_O^{\times}} + 2\mu_{h^{\cdot}} \tag{5.4}$$

For small deviations from the thermodynamic equilibrium, one can expand Eq. (5.4) to give

$$\frac{1}{2}\delta\mu_{O_2} + \delta\mu_{V_O^{\cdot\cdot}} = 2\delta\mu_{h^{\cdot}} \tag{5.5}$$

where δX denotes a small deviation in the thermodynamic potential X from the equilibrium state. Here, the chemical potential of lattice oxygen is assumed to be constant. The local equilibrium needs to be retained, and this requires that the membrane material can transport at least two kinds of species in and out. For mixed conducting materials, fast transport of both holes and oxide ions makes it easy to maintain local equilibrium. Oxygen ionic and electronic current densities are given by

$$I_{V_O^{\cdot\cdot}} = \frac{\sigma_{V_O^{\cdot\cdot}}}{2F}\nabla\mu_{V_O^{\cdot\cdot}} = \frac{\sigma_{V_O^{\cdot\cdot}}}{2F}\frac{d\mu_{V_O^{\cdot\cdot}}}{dx} \tag{5.6}$$

and

$$I_{h^{\cdot}} = \frac{\sigma_{h^{\cdot}}}{F}\nabla\mu_{h^{\cdot}} = \frac{\sigma_{h^{\cdot}}}{F}\frac{d\mu_{h^{\cdot}}}{dx} \tag{5.7}$$

Substituting Eq. (5.5) into Eq. (5.6), we get the following relationship:

$$I_{V_O^{\cdot\cdot}} = \frac{\sigma_{V_O^{\cdot\cdot}}}{4F}\frac{d\mu_{O_2}}{dx} + \frac{\sigma_{V_O^{\cdot\cdot}}}{F}\frac{d\mu_{h^{\cdot}}}{dx} \tag{5.8}$$

Because no voltage is applied to the MIEC membrane and only an oxygen chemical potential gradient is imposed across the membrane, the net current through the membrane is zero.

$$I_{V_O^{\cdot\cdot}} + I_{h^{\cdot}} = 0 \tag{5.9}$$

and

$$|I_{V_O^{\cdot\cdot}}| = 4Fj_{O_2} \tag{5.10}$$

One can obtain the following equation by combining Eqs. (5.7), (5.8), (5.9), and (5.10):

$$j_{O_2} = -\frac{1}{4^2F^2}\frac{\sigma_{V_O^{\cdot\cdot}}\sigma_{h^{\cdot}}}{\sigma_{V_O^{\cdot\cdot}} + \sigma_{h^{\cdot}}}\frac{d\mu_{O_2}}{dx} \tag{5.11}$$

or, in a more generalized form,

$$j_{O_2} = -\frac{1}{4^2F^2}\frac{\sigma_{\text{ion}}\sigma_{\text{el}}}{\sigma_{\text{ion}} + \sigma_{\text{el}}}\frac{d\mu_{O_2}}{dx} \tag{5.12}$$

where σ_{ion} is the oxygen ionic conductivity and σ_{el} is the electronic conductivity. Integration of Eq. (5.12) across the MIEC membrane thickness, L, yields the Wagner equation in the form of

$$j_{O_2} = -\frac{RT}{4^2F^2L}\int_{\ln P_{O_2}^{I}}^{\ln P_{O_2}^{II}} \frac{\sigma_{\text{ion}}\sigma_{el}}{\sigma_{\text{ion}} + \sigma_{el}}\, d\ln P_{O_2} \tag{5.13}$$

More often, the electronic conductivity of MIEC materials is much higher (>100 times) than their oxygen ionic conductivity. As a result, the ambipolar conductivity is very close to the oxygen ionic conductivity. Furthermore, if the oxygen ionic conductivity remains constant across the membranes, a more simplified Wagner equation is derived in the form of

$$j_{O_2} = -\frac{RT\sigma_{\text{ion}}}{4^2F^2L}\ln\frac{P_{O_2}^{II}}{P_{O_2}^{I}} \tag{5.14}$$

From the above equation, we find that the oxygen permeation flux is directly proportional to $T\sigma_{\text{ion}}$ and $\ln(P_{O_2}^{I}/P_{O_2}^{II})$ and inversely proportional to the membrane thickness. Considering that the activation of conductivity is expressed by the Arrhenius equation, given as

$$\sigma T = Ae^{-\frac{Ea}{RT}} \tag{5.15}$$

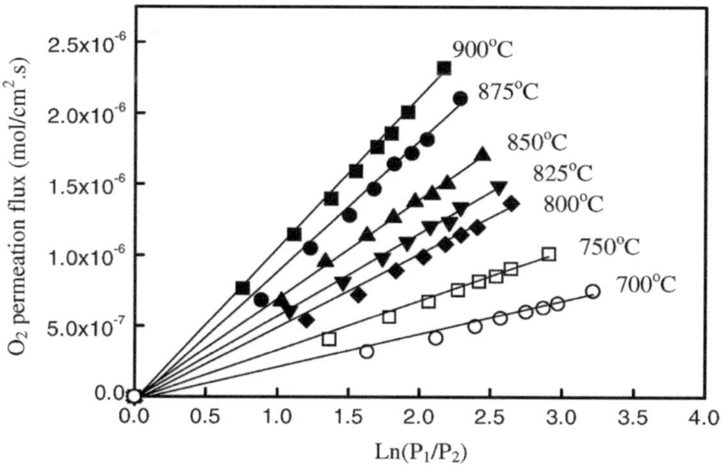

Fig. 5.1 Oxygen permeation flux of the $Ba_{0.5}Sr_{0.5}Co_{0.8}Fe_{0.2}O_{3-\delta}$ membrane tube against $\ln(P_1/P_2)$ at different temperatures. Conditions: P_1 varied from 0.09 to 1 atm; P_2 varied from 0.0093 to 0.1147 atm; $2r_1$=7.96 mm; $2r_2$=4.56 mm; L=17.68 mm; S=2.531 cm^2 (Reproduced from [8] with permission of Elsevier)

the activation energy, E_a, of the oxygen permeation flux is equal to that of the oxygen ionic conductivity. Figure 5.1 shows the dependence of the oxygen permeation flux on $\ln\left(P^I_{O_2}/P^{II}_{O_2}\right)$ performed on a $Ba_{0.5}Sr_{0.5}Co_{0.8}Fe_{0.2}O_{3-\delta}$ (BSCF) tubular membrane with a wall thickness of 1.7 mm [8]. Good straight linear fittings passing the origin of coordinate in the temperature range of 700–900 °C show that the oxygen permeation through the membrane is dominated by either bulk diffusion or oxygen ionic transport across the bulk membrane.

At small oxygen deficiency variations, a limited oxygen pressure difference, or both, acceptor dopants are among the major defects for perovskite-type mixed conducting materials, in which the nonstoichiometric oxygen is a power function of oxygen pressure [9].

$$\delta = \delta_0 P^n_{O_2} \tag{5.16}$$

where δ_0 is the nonstoichiometric oxygen at the reference oxygen pressure (1 atm) and n is an exponent constant. If the oxygen vacancies are fully ionized or not associated with other point defects, that is to say, if all the oxygen vacancies come from nonstoichiometric oxygen and contribute to oxygen transport, the oxygen ionic conductivity can be written as follows:

$$\sigma_{\mathrm{ion}} = \frac{4F^2\delta D_{\mathrm{V}}}{RTV_{\mathrm{m}}} \tag{5.17}$$

where D_V is the vacancy diffusion coefficient and V_m is the molar volume of perovskite oxides. Considering that the electronic conductivity of MIEC materials is several orders of magnitude higher than that of ionic conductivity, after substituting Eqs. (5.16) and (5.17) into Eq. (5.13), the Wagner equation changes to

$$j_{O_2} = -\frac{D_V\delta_0}{4V_mL}\int_{\ln P^I_{O_2}}^{\ln P^{II}_{O_2}} P^n_{O_2} d\ln P_{O_2} = \frac{D_V\delta_0}{4V_mLn}\left[\left(P^I_{O_2}\right)^n - \left(P^{II}_{O_2}\right)^n\right] \quad (5.18)$$

For a membrane material, D_V, V_m, and δ_0 are constants at a certain temperature, and so the above equation can be given by a more simplified form:

$$j_{O_2} = \frac{\alpha}{L}\left[\left(P^I_{O_2}\right)^n - \left(P^{II}_{O_2}\right)^n\right] \quad (5.19)$$

where $\alpha = D_V\delta_0/4V_mn$. Strictly speaking, D_V is not a constant, and its value may decrease slightly with decreasing oxygen partial pressure. According to the classical diffusion theory, D_V is proportional to a factor of $(1-\delta/3)$, which is the probability of vacancy diffusion in perovskite oxides. If this point is considered in deriving Eq. (5.18), the derivation becomes complicated. The value of n can be determined experimentally from the dependence of nonstoichiometric oxygen on the oxygen partial pressure through the thermogravimetric method and the coulometric titration method or from linear fitting of the oxygen flux to $(P^I_{O_2})^n - (P^{II}_{O_2})^n$. Mizusaki et al. investigated the relationship between nonstoichiometric oxygen and the oxygen partial pressure of the perovskite oxides $La_{1-x}Sr_xCoO_{3-\delta}$ ($x=0$, 0.1, 0.2, 0.3, 0.5, 0.7) in the range of $10^{-5} \leq P_{O_2} \leq 1$ atm and $300 \leq T \leq 1000$°C [9]. The n value changes from $-1/2$ to $-1/16$ when the x value increases from 0 to 0.7 at 800 °C. Excellent agreement was achieved on fitting the n values to the oxygen permeation fluxes using Eq. (5.19) when the x value was smaller than 0.6, but it was not good when the x value was larger than 0.6. The success in fitting benefitted from the thick membrane (2 mm) employed at a much high temperature (1000 °C) because bulk diffusion may prevail under those conditions.

Actually, this equation can model the permeation flux dependence on oxygen partial pressure (whenever it is governed by bulk diffusion or surface exchange kinetics) by adopting different values of the exponent n, although the formula is theoretically derived from a bulk diffusion equation. Therefore, the slope coefficient, α/L, is directly related to the vacancy diffusion coefficient only if the permeation process is totally limited by the bulk diffusion; an exchange coefficient can be obtained using α/L as long as the exchange reactions prevail. In the latter case, the physical meaning of n in Eq. (5.19) becomes amphibolous, but some researchers have reported that the permeation is predominantly limited by surface exchange kinetics when n is large and positive and mainly controlled by the bulk processes when n is small. For example, Huang et al. [10] discussed the relationships between n and the oxygen permeation r.d.s. in detail; they argued that the permeation is mainly controlled by bulk diffusion when n is negative and mainly restricted by interfacial exchange when n is close to 0.5. Note that the value of n in Eq. (5.19) is dependent on the thickness and temperature, so n changes from bulk diffusion controlled to interfacial exchange controlled when the thickness is reduced. That is to say, n is not an intrinsic parameter; it varies for different membranes, even if they are prepared via the same route and have the same composition. Thus, it is difficult to use this parameter for extrapolation.

If interfacial exchange is considered, the Wagner equation can be modified by introducing the concept of characteristic thickness, L_c, at which point the transition occurs from predominantly diffusion controlled to control by interfacial exchange. Under a small oxygen partial pressure gradient and the electronic conduction prevailing in MIEC materials, L_c can be expressed as [1]

$$L_c = \frac{D_s}{k_s} = \frac{D^*}{k_s} \tag{5.20}$$

where k_s is the exchange coefficient on a gas–solid interface, D_s is the self-diffusion coefficient of oxygen ions with a valence charge of -2, and D^* is the tracer diffusion coefficient. If the correlation effects are not notable, the two diffusion coefficients are close to each other. The measurement methods for these kinetic parameters are shown in Chap. 4. Under the limitation of a small oxygen partial pressure gradient across an MIEC membrane, the Wagner Eq. (5.14) can be expressed as

$$j_{O_2} = -\frac{1}{1 + (2L_c/L)} \frac{RT\sigma_{ion}}{4^2F^2L} \ln \frac{P_{O_2}^{II}}{P_{O_2}^{I}} \tag{5.21}$$

where the surface exchange rates at the two gas–solid interfaces are assumed to be the same. The factor $1/[1 + (2L_c/L)]$ is the reduction ratio of the oxygen flux as the r.d.s. shifts from the total bulk diffusion limitation to mixed control by both bulk diffusion and interfacial exchange. The diffusion coefficients and exchange coefficient are all dependent on the oxygen partial pressure, and so L_c is a function of process parameters. As a result, the dependence of oxygen permeation flux on the oxygen partial pressure gradient and temperature is not as simple as shown in Eq. (5.21). This equation cannot be directly used for modeling or simulation unless the oxygen partial pressure gradient across the membranes is small or the dependences of L_c on the oxygen partial pressure and temperature are available.

5.3 Jacobson's Model

5.3.1 Model Development

Under the operating conditions of an MIEC membrane, different oxygen partial pressures are applied on the two sides of the membrane. The oxygen partial pressure gradient results in a chemical potential drop across the bulk membrane, as well as a chemical potential drop across the gas–MIEC oxide interface. The chemical potential drop across the bulk gives rise to the diffusion current, whereas the potential drop across the interface gives rise to the interfacial exchange current, as shown in Fig. 5.2. If the interfacial exchange current has a specific dependence on the chemical potential drop at the interface, then the dependence of the permeation behavior on membrane thickness and oxygen partial pressures can be determined as well to obtain the interfacial exchange rate coefficient and bulk diffusion coefficient. In this section, an interfacial exchange ionic flux model is introduced, which was developed by Jacobson and his coworkers [2–4]. Area-specific ionic fluxes (mol cm^{-2} s^{-1}) referred to in this model will be denoted as just ionic fluxes for simplification.

At the gas–solid interface of oxygen molecules and MIEC oxides, the chemical reactions generate a forward exchange ionic flux from the gas phase to the solid phase and a reverse exchange ionic flux from the solid phase to the gas phase. If the overall reaction order of the exchange process is n, then the exchange ionic flux can be given in the form of

$$j_{\text{ion, in}} = k_f c_{\text{O}_2}^0 (P_{\text{O}_2}/P_0)^n \tag{5.22}$$

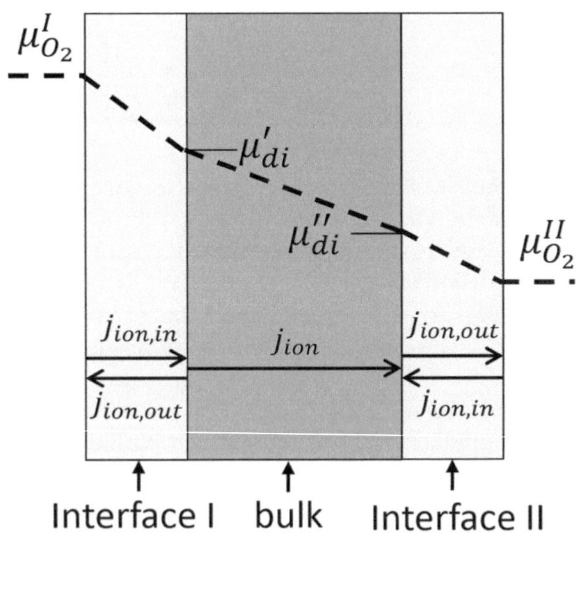

Fig. 5.2 Schematic graph of oxygen chemical potential drop across an MIEC membrane for Jacobson model

where k_f is the interfacial exchange coefficient for the forward reaction, which is assumed to be a constant when the oxygen partial pressure varies, c^0_{O2} is the density of oxygen gas molecules, and the superscript 0 indicates the quantities at 1 atm of oxygen pressure. Considering the oxygen chemical potential relative to that at 1 atm oxygen pressure, we obtain

$$\Delta\mu_{O_2} = RT\ln(P_{O_2}/P_O) \tag{5.23}$$

where T is the absolute temperature, R is the gas constant, and P_O is the oxygen pressure at 1 atm. By substituting Eq. (5.23) into Eq. (5.22), the exchange ionic flux can be written as

$$j_{\text{ion,in}} = k_f c^0_{O_2} e^{n\Delta\mu_{O_2}/RT} \tag{5.24}$$

As the exchange reactions reach thermodynamic equilibrium at the gas–solid interface,

$$\frac{1}{2}\Delta\mu_{O_2} = \Delta\mu_{\text{ambi}} \tag{5.25}$$

and

$$j_{\text{ion},in} = j_{\text{ion,out}} \tag{5.26}$$

where $\Delta\mu_{\text{di}}$ is the chemical potential of the oxygen ion–hole pairs in the MIEC oxide.

To relate the exchange coefficient for the forward reaction to the coefficient for the reverse reaction derived from isotope exchange experiments, the reverse ionic flux is expressed as

$$j_{\text{ion,out}} = k_r c_{O^{2-}} \tag{5.27}$$

where k_r is the interfacial exchange coefficient for the reverse reaction and $C_{O^{2-}}$ is the density of oxygen ions in the MIEC oxide. Substituting Eqs. (5.22) and (5.27) into Eq. (5.26) yields the following equation between the two surface exchange coefficients:

$$k_f c^0_{O_2}(P_{O_2}/P_O)^n = k_r c_{O^{2-}} \tag{5.28}$$

Because the oxygen partial pressure of the gas phase is equal to 1 atm and the density of the oxygen ions is independent on oxygen pressure, then

$$k_f c^0_{O_2} = k^0_r c_{O^{2-}} \tag{5.29}$$

where k_r^0 is the reverse surface exchange coefficient at an oxygen partial pressure of 1 atm. Combining Eqs. (5.24), (5.25), (5.26), and (5.27), the reverse ionic flux can be expressed as

$$j_{\text{ion, out}} = k_r^0 c_{\text{O}^{2-}} e^{2n\Delta\mu_{ambi}/RT} \tag{5.30}$$

All the above discussion is under a thermodynamic equilibrium state, and so the net oxygen ionic flux by summing in and out is zero. However, in the case of oxygen permeation, there is a net exchange ionic flux. If there is a little deviation from the equilibrium state, the chemical potential of oxygen on the feed side is changed to $\Delta\mu_{\text{O}_2}^I$ and that of the oxygen ion–hole pairs in the MIEC oxide is shifted to $\Delta\mu'_{\text{ambi}}$. Then, the chemical potential drop will cause a net exchange ionic flux across the gas–solid interfacial layer I.

$$j_{\text{ion}}^I = j_{\text{ion, in}}^I - j_{\text{ion, out}}^I = k_r^0 c_{\text{O}^{2-}} \left(e^{n\Delta\mu_{\text{O}_2}^I/RT} - e^{2n\Delta\mu'_{ambi}/RT}\right) \tag{5.31}$$

Similarly, a net exchange ionic current across the gas–solid interfacial layer II gives

$$j_{\text{ion}}^{II} = j_{\text{ion, in}}^{II} - j_{\text{ion, out}}^{II} = k_r^0 c_{\text{O}^{2-}} \left(e^{2n\Delta\mu''_{ambi}/RT} - e^{n\Delta\mu_{\text{O}_2}^{II}/RT}\right) \tag{5.32}$$

In the bulk of the MIEC membrane, the ionic flux is given by

$$j_{\text{ion}}^{\text{bulk}} = -\frac{c_{\text{O}^{2-}} D_{ambi}}{RT} \nabla\mu = -\frac{c_{\text{O}^{2-}} D_{ambi}}{RTL} \left(\Delta\mu''_{ambi} - \Delta\mu'_{ambi}\right) \tag{5.33}$$

where L is the thickness of the membrane bulk zone and D_{ambi} is the ambipolar diffusion coefficient of oxygen ion–hole pairs, defined by

$$D_{ambi} = D_{\text{ion}} \frac{\sigma_{\text{el}}}{\sigma_{\text{el}} + \sigma_{\text{ion}}} \tag{5.34}$$

where D_{ion} is the ionic diffusion coefficient. Under a permeation steady state,

$$j_{\text{ion}}^{\text{feed}} = j_{\text{ion}}^{\text{bulk}} = j_{\text{ion}}^{\text{perm.}} = j_{\text{ion}} = 2j_{O_2} \tag{5.35}$$

The overall interfacial exchange reaction on perovskite MIEC oxides can be expressed by Eq. (5.3). If the law of mass action can be used here, the exchange ionic flux is directly proportional to $(P_{\text{O}_2}/P_{\text{O}})^{1/2}$; that is to say, n is equal to 1/2. Furthermore, good fit of the model to the experimental data was found when $n = 1/2$. Therefore, $n = 1/2$ was adopted for the derivation of the following equations.

Combining Eqs. (5.31), (5.32), (5.34), and (5.35) and eliminating $\Delta\mu'_{\mathrm{ambi}}$ and $\Delta\mu''_{\mathrm{ambi}}$, one can obtain the following relationship:

$$\frac{j_{\mathrm{ion}}}{c_{\mathrm{O^{2-}}}k_r^0} = \frac{L_{d0}}{l}\ln\frac{\sqrt{P^I_{\mathrm{O_2}}/P_0} - j_{\mathrm{ion}}/c_{\mathrm{O^{2-}}}k_r^0}{\sqrt{P^{II}_{\mathrm{O_2}}/P_0} + j_{\mathrm{ion}}/c_{\mathrm{O^{2-}}}k_r^0} \tag{5.36}$$

where $L_{\mathrm{d0}} = D_{ambi}/k_{\mathrm{r}}^0$ is the length scale that determines the sample thickness for the r.d.s. of oxygen permeation shifting from diffusion-limited to surface-limited transport. The implication of this parameter is similar to that of L_{c} defined by Bouwmeester et al. [1] However, the self-diffusion coefficient or tracer diffusion coefficient and surface exchange coefficient are used in the latter. For MIEC perovskite membranes, their electronic conductivities are much higher than their ionic conductivities, and so here the ambipolar diffusion coefficient is close to the self-diffusion coefficient. However, here k_r^0 is the surface exchange coefficient for reverse ionic flux at an oxygen partial pressure of 1 atm.

If the oxygen permeation process is limited by interfacial exchange steps, then

$$\Delta\mu'_{ambi} \approx \Delta\mu''_{ambi} \tag{5.37}$$

Therefore, by combining Eqs. (5.31), (5.32), and (5.37), Eq. (5.36) can be simplified as

$$j_{\mathrm{O_2}} = j_{\mathrm{ion}}/2 = \frac{k_r^0 c_{\mathrm{O^{2-}}}}{4}\left(\sqrt{P^I_{\mathrm{O_2}}/P_0} - \sqrt{P^{II}_{\mathrm{O_2}}/P_0}\right) \tag{5.38}$$

If the oxygen permeation process is dominated by bulk diffusion, then

$$\Delta\mu'_{ambi} \approx \Delta\mu^I_{\mathrm{O_2}}/2 \tag{5.39}$$

and

$$\Delta\mu''_{ambi} \approx \Delta\mu^{II}_{\mathrm{O_2}}/2 \tag{5.40}$$

Therefore, by combining Eqs. (5.32), (5.33), (5.39), and (5.40), Eq. (5.36) can be simplified as

$$j_{\mathrm{O_2}} = j_{\mathrm{ion}}/2 = \frac{D_{ambi}c_{\mathrm{O^{2-}}}}{4L}\ln\frac{P^I_{\mathrm{O_2}}}{P^{II}_{\mathrm{O_2}}} \tag{5.41}$$

The dependence of the oxygen permeation flux on the oxygen partial pressure is the same as that shown in the Wagner equation.

For a tubular membrane with an outer diameter of r_1, an inner diameter of r_2, and a length of h, the general Eq. (5.36) changes to

$$\frac{F_{O_2}}{\pi h c_{O^{2-}} D_{ambi}} \ln(r_1/r_2) = \ln \frac{\sqrt{P^{I}_{O_2}/P_0} - F_{O_2}/\pi r_1 h c_{O^{2-}} k_r^0}{\sqrt{P^{II}_{O_2}/P_0} + F_{O_2}/\pi r_2 h c_{O^{2-}} k_r^0} \tag{5.42}$$

where F_{O_2} is the oxygen flow through the membrane. In the interfacial exchange- and bulk diffusion-limited regimes, Eq. (5.42) reduces to

$$F_{O_2} = \frac{\pi r_1 r_2 h c_{O^{2-}} k_r^0}{r_1 + r_2}\left(\sqrt{P^{I}_{O_2}/P_0} - \sqrt{P^{II}_{O_2}/P_0}\right) \tag{5.43}$$

and

$$F_{O_2} = \frac{\pi h c_{O^{2-}} D_{ambi}}{2\ln(r_1/r_2)} \ln(P^{I}_{O_2}/P^{II}_{O_2}) \tag{5.44}$$

5.3.2 Model Application

Jacobson et al. fitted the permeation experimental data of $SrCo_{0.8}Fe_{0.2}O_{3-\delta}$ and $Sm_{0.5}Sr_{0.5}CoO_{3-\delta}$ tubular membranes to this model to deduce the ambipolar diffusion coefficient and the surface exchange rate coefficient from the dependence of oxygen permeation on the oxygen pressure gradient [11]. The results are shown in Fig. 5.3. The kinetic parameters derived from the model fitting agree reasonably well with those obtained from the isotopic tracer experiments. Although there have been some successful applications of this model to simulate experimental data reported by Jacobson and coworkers, few other researchers have utilized this model in their investigations because the complicated expression of Eqs. (5.36) and (5.42) makes it difficult to determine how the process parameters, i.e., oxygen partial pressure and temperature, affect the permeation flux. However, this model shows the clear and simple dependence of the oxygen permeation flux on the oxygen partial pressure when oxygen exchange on the gas–solid interfaces is the limiting step. The plot of $j_{O_2} vs(\sqrt{P^{I}_{O_2}/P_0} - \sqrt{P^{II}_{O_2}/P_0})$ is most frequently used to determine whether the permeation process is dominated by interfacial oxygen exchange or not. Figure 5.4 gives an example of the plots of $j_{O_2} vs(\sqrt{P^{I}_{O_2}/P_0} - \sqrt{P^{II}_{O_2}/P_0})$ and $j_{O_2} vs \ln(P^{I}_{O_2}/P^{II}_{O_2})$ of $SrCo_{0.8}Fe_{0.2}O_{3-\delta}$ and $Sm_{0.5}Sr_{0.5}CoO_{3-\delta}$ tubular membranes [11]. As shown in the figure, the transport in the $Sm_{0.5}Sr_{0.5}CoO_{3-\delta}$ tubular membrane is much closer to being bulk diffusion dominated than that in the $SrCo_{0.8}Fe_{0.2}O_{3-\delta}$ membrane.

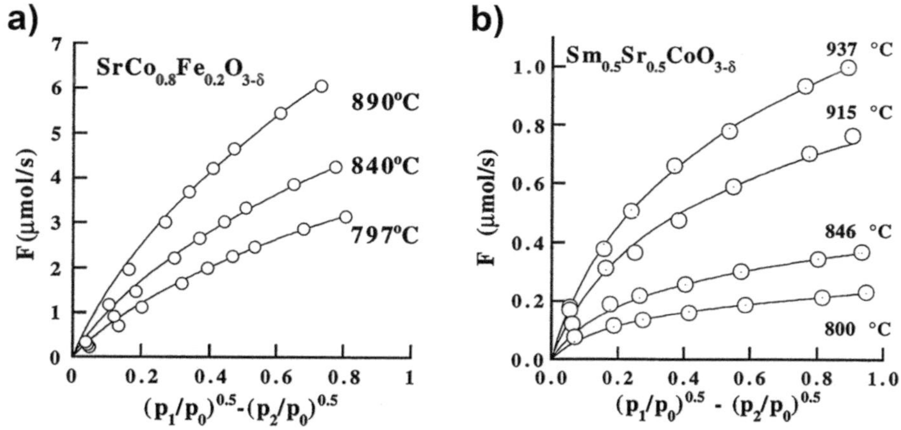

Fig. 5.3 (**a**) Oxygen flow F in $SrCo_{0.8}Fe_{0.2}O_{3-\delta}$ tube measured at the indicated temperatures. The experimental data are given by the open circles. The dimensions of the tube are r_1=0.61 cm, r_2=0.41 cm, and w=1.92 cm. (**b**) Oxygen flow F in $Sm_{0.5}Sr_{0.5}CoO_{3-\delta}$ tube measured at the indicated temperatures. The experimental data are given by the open circles. The dimensions of the tube are r_1=0.64 cm, r_2=0.36 cm, and w=1.48 cm (Reproduced from [11] with permission of Elsevier)

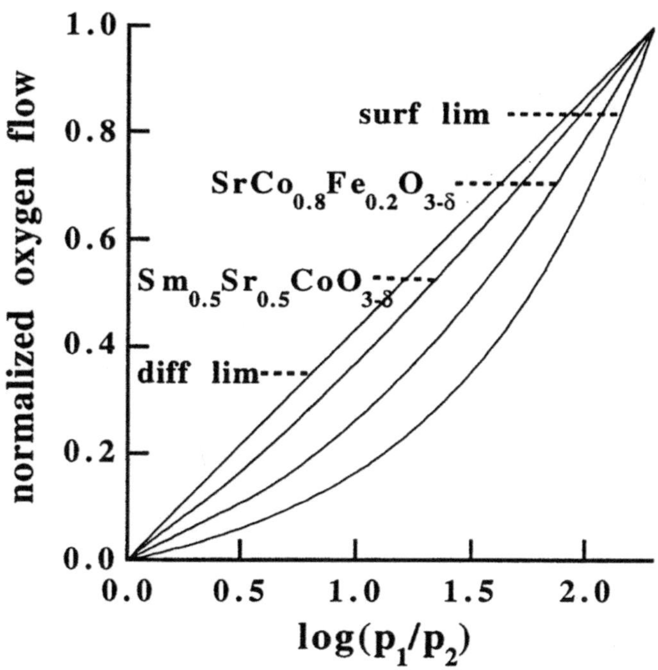

Fig. 5.4 Normalized oxygen flow plotted vs $\log(p_1/p_2)$ in tubular membranes for $SrCo_{0.8}Fe_{0.2}O_{3-\delta}$ at 840 °C and $Sm_{0.5}Sr_{0.5}CoO_{3-\delta}$ at 846 °C. Also shown is the flow in the surface and diffusion-limited regimes (Reproduced from [11] with permission of Elsevier)

5.4 Xu and Thomson's Model

5.4.1 Model Development

As mentioned in the derivation of the Wagner equation, oxygen vacancies and holes are the main carriers in MIEC membranes. Xu and Thomson developed a permeation model based on the fact that oxygen permeation through MIEC membranes under the driving force of the oxygen chemical potential gradient is generated due to the continuous transport of oxygen vacancies from the low oxygen partial pressure side to the high oxygen partial pressure side, as shown in Fig. 5.5 [5]. In brief, it is the transport of oxygen vacancies that gives rise to oxygen flux across the membranes. Therefore, the oxygen exchange reactions on gas–solid interfaces concerned here are at interface I of the feed side:

$$\frac{1}{2}O_2 + V_{\mathrm{O}}^{\cdot\cdot} \underset{k_r}{\overset{k_f}{\rightleftharpoons}} O_{\mathrm{O}}^{\times} + 2h^{\cdot} \tag{5.45}$$

and at interface II of the permeation side:

$$O_{\mathrm{O}}^{\times} + 2h^{\cdot} \underset{k_f}{\overset{k_r}{\rightleftharpoons}} \frac{1}{2}O_2 + V_O^{\cdot\cdot} \tag{5.46}$$

where k_{f} and k_{r} are the rate constants of the forward and reverse reaction Eq. (5.45) or the reverse and forward reaction Eq. (5.46), respectively. To derive the permeation model, the following four main assumptions are proposed:

1. The electronic conductivity is much greater than the ionic conductivity.
2. The diffusion coefficient of the oxygen vacancies does not change with changing oxygen pressure across the membrane.

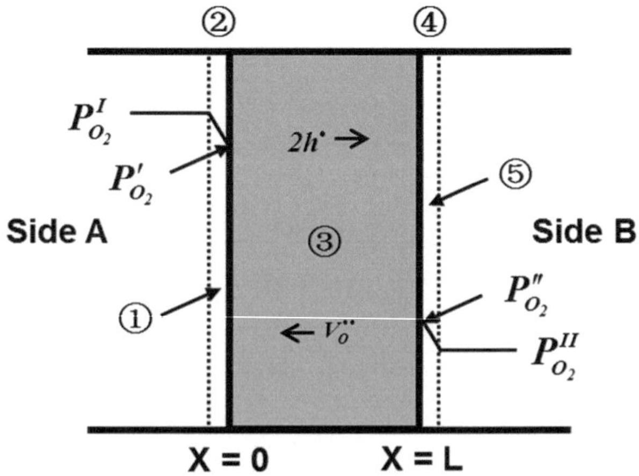

Fig. 5.5 Oxygen permeation through an ion-conducting membrane. Steps 1 and 5: solid mass transfer of gaseous oxygen. Steps 2 and 4: surface exchange between molecular oxygen and oxygen vacancies. Step 3: the oxygen vacancy bulk diffusion

3. The forward and reverse reaction rate constants of reaction Eq. (5.45) are respectively equal to those of reaction Eq. (5.46).
4. The law of mass action can be used for the oxygen exchange reactions on both interfaces.

For MIEC oxides, especially perovskite-type oxides, the electronic conductivity is several orders of magnitude higher than the ionic conductivity; therefore, the transport resistance of holes is much smaller than that of oxygen vacancies. At an ideal status, or when all the oxygen vacancies are ionized and contribute to the mass transport, the oxygen vacancy flux across the MIEC membrane dominates the oxygen permeation flux, which fits with Fick's first law and gives

$$j_{\mathrm{V}} = -D_{\mathrm{V}} \frac{dc_{\mathrm{V}}}{dx} \tag{5.47}$$

where c_{V} is the oxygen vacancy concentration. Integrating the above equation and considering that the value of the vacancy flux is double that of the oxygen permeation flux, then the oxygen permeation flux through membrane bulk is

$$j_{\mathrm{O_2}} = \frac{D_{\mathrm{V}}}{2L}\left(c''_{\mathrm{V}} - c'_{\mathrm{V}}\right) \tag{5.48}$$

where c'_{V} and c''_{V} are the concentrations of oxygen vacancies at the interfaces of the feed and permeation sides, respectively.

It is reasonable to assume that the concentration of holes is constant at both interfaces of the membrane for high electronic conductivities. Therefore, the reverse reaction rate of reaction Eq. (5.45) and the forward reaction rate of reaction Eq. (5.46) are pseudo-zero order at steady state under isothermal conditions if the oxygen ionic concentration remains almost constant. Theoretical and experimental investigations have demonstrated that the two reactions can be considered as elementary reactions and that the law of mass action is therefore applicable, although many elemental steps, such as oxygen adsorption/desorption, dissociation/recombination, incorporation into/evolution out of, and charge transfer, are involved in the two overall reactions. The reaction rates of oxygen permeation fluxes across the two gas–solid interfaces can be written as

across interface I:

$$j_{\mathrm{O_2}} = k_{\mathrm{f}} {P^{I}_{\mathrm{O_2}}}^{0.5} c'_{\mathrm{V}} - k_{\mathrm{r}} \tag{5.49}$$

and

across interface II:

$$j_{\mathrm{O_2}} = k_{\mathrm{r}} - k_{\mathrm{f}} {P^{II}_{\mathrm{O_2}}}^{0.5} c''_{\mathrm{V}} \tag{5.50}$$

At a permeation steady state, oxygen fluxes through interface I, the bulk phase, and interface II are equal to each other. By combining Eqs. (5.48), (5.49) and (5.50) and eliminating c'_V and c''_V, a general oxygen permeation equation that can be readily correlated to oxygen partial pressures is shown in the following:

$$j_{O_2} = \frac{D_V k_r \left({P^I_{O_2}}^{0.5} - {P^{II}_{O_2}}^{0.5} \right)}{2Lk_f \left(P^I_{O_2} P^{II}_{O_2} \right)^{0.5} + D_V \left({P^I_{O_2}}^{0.5} + {P^{II}_{O_2}}^{0.5} \right)} \tag{5.51}$$

where D_V, k_r, and k_f are functions of temperature and the permeation properties of the membrane. All these parameters can be determined by fitting of the experimental oxygen flux data as a function of temperature and oxygen partial pressure gradient.

It is important to know which steps limit the transport process. In this model, for the first time, the concept of permeation resistance is introduced to describe the limitations of gas–solid interfaces and the bulk phase during oxygen transport across MIEC membranes. Interfacial exchange coefficients at both interfaces are correlated to the permeation resistance of exchange kinetics, and they are defined by the following equations:

across interface I:

$$j_{O_2} = k'_{ex} \left(c'_V - c'_{V,eq} \right) \tag{5.52}$$

and

across interface II:

$$j_{O_2} = k''_{ex} \left(c''_{V,eq} - c''_V \right) \tag{5.53}$$

where $c'_{V,eq}$ and $c''_{V,eq}$ are the equilibrium concentrations of oxygen vacancies under the experimental conditions, which can be written as a function of oxygen partial pressure by setting $j_{O_2} = 0$ in Eqs. (5.51) and (5.52).

$$k'_{ex} = k_f P^{I\,0.5}_{O_2} \tag{5.54}$$

$$k''_{ex} = k_f P^{II}_{O_2} 0.5 \tag{5.55}$$

Therefore, the general permeation Eq. (5.51) can be rewritten as

$$j_{O_2} = \frac{\frac{k_r}{k_f} \left({P^{II}_{O_2}}^{-0.5} - {P^{I}_{O_2}}^{-0.5} \right)}{\frac{1}{k'_{ex}} + \frac{2L}{D_V} + \frac{1}{k''_{ex}}} = \frac{\Delta(P_{O_2})}{R' + R_{bulk} + R''} = \frac{\Delta(P_{O_2})}{R^{tot}} \tag{5.56}$$

where $\Delta(P_{O_2}) = (k_r/k_f)\left(P_{O_2}^{II-0.5} - P_{O_2}^{I-0.5}\right)$ is the driving force of the oxygen pressure difference across the membrane and $R^{tot} = R' + R_{bulk} + R''$ is the total permeation resistance. By substituting the relationships shown in Eqs. (5.54) and (5.55), the partial resistances across interface I and the bulk phase, as well as across interface II, can be expressed as

$$R' = \frac{1}{k_f {P_{O_2}^{I}}^{0.5}} \tag{5.57}$$

$$R^{bulk} = \frac{2L}{D_V} \tag{5.58}$$

$$R'' = \frac{1}{k_f {P_{O_2}^{II}}^{0.5}} \tag{5.59}$$

The three parameters, D_V, k_r, and k_f, in Eq. (5.51) can be acquired by fitting the equation to the experimental data. Then, the partial permeation resistances of the three steps can be calculated using the above three equations. As a result, the r.d.s. of oxygen permeation can be valued quantitatively. This is the first model that allows researchers to draw maps of resistance distribution as a function of temperature and oxygen partial pressure.

If the oxygen permeation is dominated by bulk diffusion, i.e., $R_{bulk} \gg R' + R''$, then the general permeation Eq. (5.56) can be simplified as

$$j_{O_2} = \frac{\Delta(P_{O_2})}{R^{bulk}} = \frac{D_V k_r}{2Lk_f}\left({P_{O_2}^{II}}^{-0.5} - {P_{O_2}^{I}}^{-0.5}\right) \tag{5.60}$$

If the oxygen permeation is dominated by interfacial exchange, i.e., $R' + R'' \gg R^{bulk}$, then the general permeation Eq. (5.56) can be simplified as

$$j_{O_2} = \frac{\Delta(P_{O_2})}{R' + R''} = k_r \frac{{P_{O_2}^{II}}^{-0.5} - {P_{O_2}^{I}}^{-0.5}}{{P_{O_2}^{II}}^{-0.5} + {P_{O_2}^{I}}^{-0.5}} \tag{5.61}$$

The oxygen permeation flux linearly increases with decreasing thickness in the bulk diffusion limitation regime until the interfacial exchange resistance cannot be neglected as compared to the reduced bulk diffusion resistance. L_c is defined as the point at which the transition occurs from predominantly diffusion-controlled to control by surface exchange. In other words, as the point at which the exchange resistance is equal to the bulk diffusion resistance, thus, L_c can be written as

$$L_c = \frac{D_V}{2k_f}\left({P_{O_2}^{I}}^{-0.5} + {P_{O_2}^{II}}^{-0.5}\right) \tag{5.62}$$

If the vacancy diffusion coefficient and reaction rate constant remain constant as the oxygen partial pressure changes, the above equation shows the dependence of the characteristic thickness on the oxygen partial pressure of both sides of the membrane. Clearly, this expression of L_c is different from that shown by Eq. (5.20).

5.4.2 Model Application

Xu and Thomson fitted the permeation experimental data of $La_{0.6}Sr_{0.4}Co_{0.2}Fe_{0.8}O_{3-\delta}$ disk-type membranes to their model to acquire D_V, k_r, and k_f in Eq. (5.51) from the dependence of the oxygen permeation on the oxygen pressure gradient and temperature [5]. The experimental measurements of oxygen flux well matched the model results, with a correlation coefficient of 0.9966. Under a small oxygen partial pressure difference, the oxygen exchange coefficients at both interfaces are functions of oxygen partial pressure, and they have the relation $k'_{ex}/k''_{ex} = \left(P^I_{O_2}/P^{II}_{O_2}\right)^{0.5}$. Because $P^I_{O_2}$ is usually much higher than $P^{II}_{O_2}$, the contribution of interfacial exchange to total permeation resistance at the feed side is often negligible as compared to that at the permeation side. Figure 5.6 shows permeation resistance distributions under different conditions: (a) membrane thickness, (b) oxygen partial pressure of the feed side, and (c) oxygen partial pressure of the permeate side combined with temperature. The oxygen exchange activation energy of the permeation side is higher than that of oxygen vacancy diffusion, and so R'' increases quickly with decreasing temperature, as shown in Fig. 5.7a. The activation energy values are lower at higher temperatures and higher at lower temperatures, even if neither the change of crystal phase nor the ordering of oxygen vacancies happens. Xu and Thomson thought that the two-stage characteristic of the permeation activation energy of the $La_{0.6}Sr_{0.4}Co_{0.2}Fe_{0.8}O_{3-\delta}$ membrane was induced by the change of the r.d.s. from bulk diffusion to interfacial oxygen exchange kinetics. For the $La_{0.6}Sr_{0.4}Co_{0.2}Fe_{0.8}O_{3-\delta}$ membrane, the interfacial exchange at the feed side was not the r.d.s., and so the distribution of these resistances holds the line as the oxygen pressure at the feed side increases, as shown in Fig. 5.7b. However, R'' increases quickly with decreasing oxygen partial pressure on the permeation side, as shown in Fig. 5.7c. This can be used to explain why in some cases the oxygen flux linearly increases with $\ln P^I_{O_2}/P^{II}_{O_2}$ at a smaller oxygen pressure gradient across the membrane but gradually deviates from the straight line as the oxygen pressure gradient increases.

Among these three newly introduced models, the vacancy diffusion model developed by Xu and Thomson is especially useful in identifying the relative roles of bulk diffusion and interfacial exchange kinetics in a permeation process, which is why this model is widely used [12–16]. Tan and coworkers simulated their $La_{0.6}Sr_{0.4}Co_{0.2}Fe_{0.8}O_{3-\delta}$ hollow fiber membranes using this model and found that the kinetic parameters of the $La_{0.6}Sr_{0.4}Co_{0.2}Fe_{0.8}O_{3-\delta}$ membrane reported by Xu

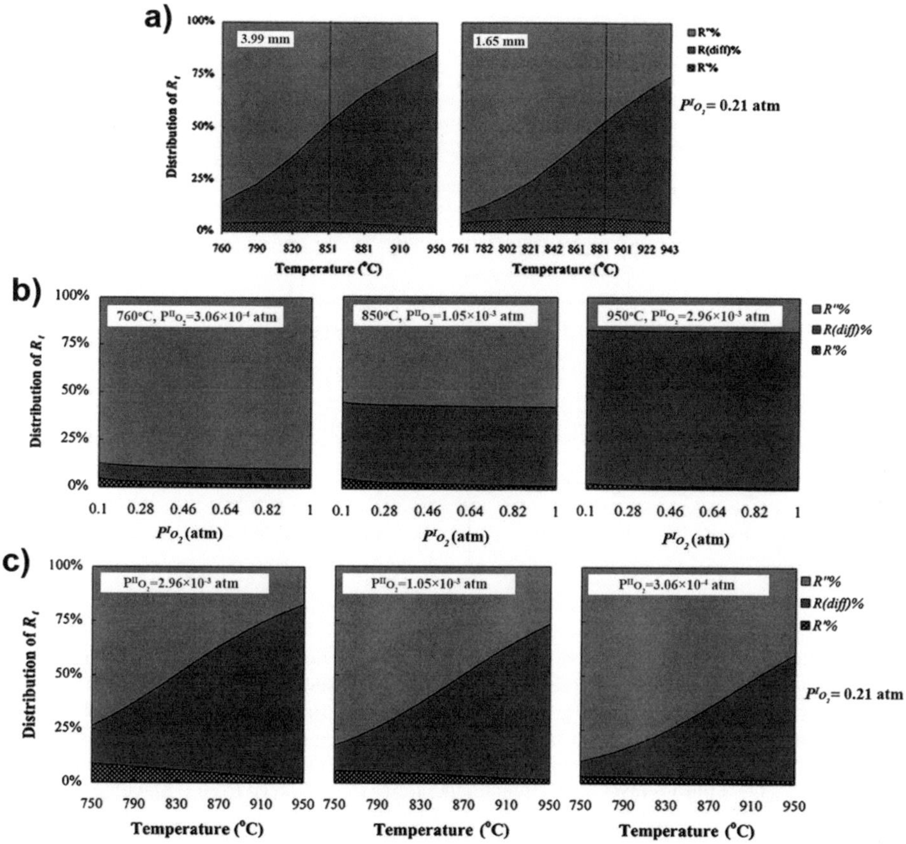

Fig. 5.6 (**a**) Effect of temperature on distribution of total permeation resistance. (**b**) Effect of $P^{I}_{O_2}$ on distribution of total permeation resistance under constant $P^{II}_{O_2}$ and T over the disk with thickness of 3.99 mm. (**c**) Effect of $P^{II}_{O_2}$ on distribution of total permeation resistance (Reproduced from [5] with permission of Elsevier)

and Thomson agreed well with their experimental results on the hollow fiber $La_{0.6}Sr_{0.4}Co_{0.2}Fe_{0.8}O_{3-\delta}$ membranes for oxygen separation from air [12]. The performance of the hollow fiber module at various operating conditions was studied theoretically, and the simulation results revealed that the cocurrent was a better operating flow pattern than the countercurrent flow pattern. Similarly, they found that the interfacial exchange reaction of the permeation side was more important than that of the feed side, especially at lower operating temperatures. In a comparison of the two interfacial exchange resistances, that on the permeation side was much larger than that on the feed side. Furthermore, to derive a much more precise model, Asadi et al. modified the Xu and Thomson model by considering the effect of oxygen partial pressure on each side of the membrane on the gas flow rate and assuming a nonelementary rate of reaction with respect to the oxygen partial

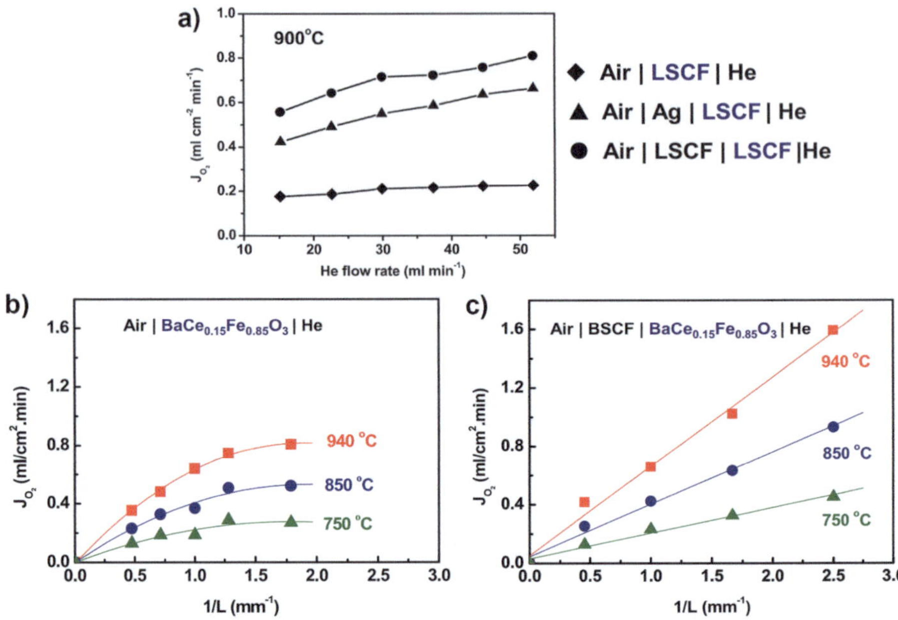

Fig. 5.7 (**a**) Comparison of the modified and the unmodified hollow fiber membranes in oxygen permeation flux. Reproduced from [17] with permission of Elsevier. (**b**) The relationship between oxygen permeation fluxes and reciprocals of thicknesses of $BaCe_{0.15}Ce_{0.85}O_{3-\delta}$ membranes at different temperatures. Reproduced from [18] with permission of Elsevier. (**c**) The relationship between oxygen permeation fluxes and reciprocals of thicknesses of membranes with $BaCe_{0.15}Ce_{0.85}O_{3-\delta}$ porous layer at different temperatures (Reproduced from [18] with permission of Elsevier)

pressure. The predictions of the improved mathematical model showed excellent agreement with the experimental data [14].

Although there have been many successful cases applying the model to other membranes or LSCF membranes with different configurations, one assumption proposed during the derivation of the model equation should be reevaluated, i.e., whether the forward and reverse reaction rate constants of one gas–solid interfacial reaction are identical to those of other gas–solid interfacial reactions. The ratio between partial permeation resistances on gas–solid interfaces I and II is obtained by comparing Eqs. (5.57) and (5.59), which gives

$$\frac{R'}{R''} = \sqrt{\frac{P_{O_2}^{II}}{P_{O_2}^{I}}} \tag{5.63}$$

In typical oxygen permeation operations, the oxygen partial pressure of the feed side is 10–10^3 times higher than that of the permeation side. As a result, the partial permeation resistance on gas–solid interface II is always 3–30 times larger than that on interface I. In other words, the interfacial exchange reaction on interface I cannot be the r.d.s. Coating an MIEC membrane surface with a porous catalyst layer is an effective method for speeding up the exchange kinetics on both interfaces. From the above analysis, one can infer that the permeation flux of a membrane can only be marginally improved by coating porous catalyst layer on the surface of the feed side for the improvement of oxygen exchange on the feed side. However, experimental investigations have shown that the permeation flux of MIEC membranes can be significantly improved by coating catalysts on the feed side. For example, the oxygen permeation fluxes of LSCF hollow fiber membranes were improved by ~200 % by coating a porous LSCF layer on the feed side surface [17]. In our group, we found that the oxygen permeation flux of a $BaCe_{0.15}Fe_{0.85}O_{3-\delta}$ perovskite membrane was improved by 40–100 % by coating a 10-μm-thick $Ba_{0.5}Sr_{0.5}Co_{0.8}Fe_{0.2}O_{3-\delta}$ porous layer on the feed side of the membrane [18]. As a result, the permeation process was shifted from being jointly limited by both interfacial exchange and bulk diffusion to being limited predominantly by bulk diffusion, as shown in Fig. 5.7. The above results show that the relationship between the two partial permeation resistances is not as expressed by Eq. (5.63). Furthermore, if the feed gas or the sweep gas contains impurities, such as H_2O and CO_2, the phase structure on the membrane surface and the elemental composition will change significantly during oxygen permeation, and so the forward and reverse reaction rate constants will vary accordingly. Therefore, strictly speaking, the model given in Eq. (5.56) is unsuitable for modeling because the reaction rate constants are identical on both interfaces.

5.5 Zhu's Model

5.5.1 Model Development

In the models described above, the focus is only on the role of defect chemistry (for instance, the Wagner equation) or on taking into account either ion or vacancy transport across interfaces (for instance, the surface exchange ionic flux and vacancy diffusion models) but not both. Here, we developed a permeation model that takes into account both ionic and electronic transports across gas–solid interfaces and the bulk phase of MIEC membranes [7]. Appropriate assumptions are necessary for the theoretical analysis, the development of simple equations, and an understanding of the nature of interface effects. The following three assumptions were therefore used in this model:

1. The transport properties of electrons/holes and oxygen ions in a given zone are constant, regardless of position and oxygen chemical potentials.
2. The diffusion resistance or concentration polarization resistance from oxygen molecules from the environmental atmosphere to the membrane interfaces on both sides can be totally ignored. It should be noted that this hypothesis may be invalid for thin dense membranes on thick supports.

3. The law of mass action can be used for the oxygen exchange reactions on both interfaces.

The best-known one-dimensional diffusion equation describing a continuous flux of species through a unit surface is Fick's first law.

$$j_i = -D_i \frac{dc_i}{dx} \tag{5.64}$$

where j_i is the flux of the species being transporting through the unit area (mol cm^{-2} s^{-1}), D_i is the diffusion coefficient of the species (cm^2 s^{-1}), and c_i is the concentration of the species i (cm^{-3}). The equation can be rewritten by combining classical thermodynamics with the chemical potential of species i, which gives

$$j_i = -\frac{D_i c_i}{k_B T} \frac{d\mu_i}{dx} \tag{5.65}$$

where μ_i is the electrochemical potential of component i, k_B is the Boltzmann constant, and T is the temperature. For a mixed conducting membrane, it is unnecessary to apply an electric potential due to the inner short circuit, and so here, the electrochemical potential is equal to the chemical potential aroused by the oxygen partial pressure gradient across the MIEC membrane.

The current density I_i (A cm^{-2}) due to the directional transport of species i for a one-dimensional case is given by

$$I_i = z_i e j_i = -\frac{z_i e D_i c_i}{k_B T} \frac{d\mu_i}{dx} = -\frac{\sigma_i}{z_i e} \frac{d\mu_i}{dx} \tag{5.66}$$

where $\sigma_i = z_i^2 e^2 c_i D_i / k_B T$ is the conductivity of species i (S cm^{-1}) transported in the mixed conducting material, z_i is the electrical charge of species i, and e is the electronic charge (C). It should be kept in mind that no net current flows through the membranes once the stationary state has been reached. Accordingly, equivalent amounts of electrons and oxygen ions are oppositely transported across the membrane. For a mixed ionic–electronic conductor, the equilibrium of interest is the following:

$$\frac{1}{2} O_2 + 2e' \rightleftharpoons O^{2-} \tag{5.67}$$

At thermodynamic equilibrium, the oxygen exchange current density from the gas to the solid is equal to that from the solid to the gas. Also, the following relationship should be satisfied:

$$\frac{1}{2} \mu_{O_2} + 2\mu_e = \mu_{O^{2-}} \tag{5.68}$$

For small deviations from the thermodynamic equilibrium, one can expand Eq. (5.68) to give

$$\frac{1}{2}\delta\mu_{O_2} + 2\delta\mu_e = \delta\mu_{O^{2-}} \tag{5.69}$$

where δX denotes a small deviation in the thermodynamic potential X from the equilibrium state. Local equilibrium needs to be retained, and this requires that the membrane material transport at least two kinds of species in and out. For mixed conducting materials, fast transport of both electrons and oxide ions makes it easy to maintain local equilibrium. Oxygen ionic and electronic current densities are given by

$$I_{O^{2-}} = \frac{\sigma_{O^{2-}}}{2F}\nabla\mu_{O^{2-}} = \frac{\sigma_{O^{2-}}}{2F}\frac{d\mu_{O^{2-}}}{dx} \tag{5.70}$$

and

$$I_e = \frac{\sigma_e}{F}\nabla\mu_e = \frac{\sigma_e}{F}\frac{d\mu_e}{dx} \tag{5.71}$$

Substituting Eq. (5.69) into Eq. (5.70), we obtain the following relationship:

$$I_{O^{2-}} = \frac{\sigma_{O^{2-}}}{4F}\frac{d\mu_{O_2}}{dx} + \frac{\sigma_{O^{2-}}}{F}\frac{d\mu_e}{dx} \tag{5.72}$$

Figure 5.8 shows a schematic illustration of a mixed conducting membrane across which there is an oxygen chemical potential difference ($\mu^I_{O_2} - \mu^{II}_{O_2}$) and an electronic

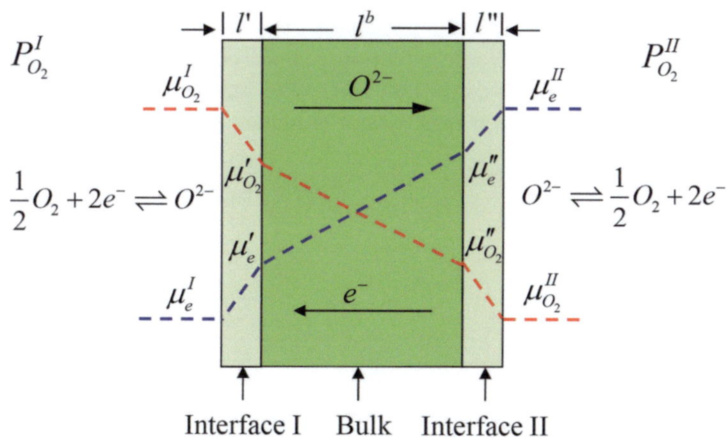

Fig. 5.8 Schematic graph of oxygen chemical potential drop across an MIEC membrane for Zhu's model

chemical potential difference ($\mu_e^I - \mu_e^{II}$). The transport of oxygen ions and electrons occurs across the interfacial zones, which can be greatly affected by external conditions. Therefore, the thickness of the interfacial zones is difficult to determine and not very well defined. It is assumed that the "interfacial zones" are of a certain thickness, which is used to describe transport properties through both interfacial zones in terms of conductivity and thickness. $\sigma'_{\mathrm{O}^{2-}}$, σ'_e, $\sigma^b_{\mathrm{O}^{2-}}$, σ^b_e, $\sigma''_{\mathrm{O}^{2-}}$, and σ''_e are the oxygen ionic and electronic conductivities at interfacial zone I, bulk zone, and interfacial II, respectively. l', l^b, and l'' are the thicknesses of interfacial zone I, the bulk zone, and interfacial zone II, respectively. The area-specific conductance (S cm^{-2}) and area-specific resistance (Ω cm^2) are introduced because it is impossible to separately determine the conductivities and thicknesses of the interfacial zones. However, the area-specific conductance and resistance can be measured experimentally. The following relationships describe the area-specific oxygen ionic conductance and resistance of interfacial zone I, the bulk zone, and interfacial zone II, respectively:

$$\kappa'_{\mathrm{O}^{2-}} = \frac{\sigma'_{\mathrm{O}^{2-}}}{l'} = \frac{1}{r'_{\mathrm{O}^{2-}}} \tag{5.73}$$

$$\kappa^b_{\mathrm{O}^{2-}} = \frac{\sigma^b_{\mathrm{O}^{2-}}}{l^b} = \frac{1}{r^b_{\mathrm{O}^{2-}}} \tag{5.74}$$

$$\kappa''_{\mathrm{O}^{2-}} = \frac{\sigma''_{\mathrm{O}^{2-}}}{l''} = \frac{1}{r''_{\mathrm{O}^{2-}}} \tag{5.75}$$

There are relationships comparable to Eqs. (5.73), (5.74), and (5.75) that express the area-specific electronic conductance and resistance of interfacial zone I, the bulk zone, and interfacial zone II, respectively. The interfacial oxygen ionic or electronic conductivity is not equal to the conductivity in the bulk membrane because the membrane is greatly influenced by its microstructure, interfacial elemental composition, etc. These equations are based on the simple assumption that the transport properties are constant across the interfacial and bulk zones, and thus the oxygen ionic and electronic current densities are given by

$$I'_{\mathrm{O}^{2-}} = \frac{\kappa'_{\mathrm{O}^{2-}}}{4F}\left(\mu'_{\mathrm{O}_2} - \mu^I_{\mathrm{O}_2}\right) + \frac{\kappa'_{\mathrm{O}^{2-}}}{F}\left(\mu'_e - \mu^I_e\right) = \frac{\mu'_{\mathrm{O}_2} - \mu^I_{\mathrm{O}_2}}{4Fr'_{\mathrm{O}^{2-}}} + \frac{\mu'_e - \mu^I_e}{Fr'_{\mathrm{O}^{2-}}} \tag{5.76}$$

$$I^b_{\mathrm{O}^{2-}} = \frac{\kappa^b_{\mathrm{O}^{2-}}}{4F}\left(\mu''_{\mathrm{O}_2} - \mu'_{\mathrm{O}_2}\right) + \frac{\kappa^b_{\mathrm{O}^{2-}}}{F}\left(\mu''_e - \mu'_e\right) = \frac{\mu''_{\mathrm{O}_2} - \mu'_{\mathrm{O}_2}}{4Fr^b_{\mathrm{O}^{2-}}} + \frac{\mu''_e - \mu'_e}{Fr^b_{\mathrm{O}^{2-}}} \tag{5.77}$$

$$I''_{\mathrm{O}^{2-}} = \frac{\kappa''_{\mathrm{O}^{2-}}}{4F}\left(\mu^{II}_{\mathrm{O}_2} - \mu''_{\mathrm{O}_2}\right) + \frac{\kappa''_{\mathrm{O}^{2-}}}{F}\left(\mu^{II}_e - \mu''_e\right) = \frac{\mu^{II}_{\mathrm{O}_2} - \mu''_{\mathrm{O}_2}}{4Fr''_{\mathrm{O}^{2-}}} + \frac{\mu^{II}_e - \mu''_e}{Fr''_{\mathrm{O}^{2-}}} \tag{5.78}$$

By analogy with the oxygen ionic current densities through the two interfacial zones and the bulk zone, the electronic current densities can be written as

$$I'_e = \frac{\kappa'_e}{F}\left(\mu'_e - \mu^I_e\right) = \frac{\mu'_e - \mu^I_e}{Fr'_e} \tag{5.79}$$

$$I^b_e = \frac{\kappa^b_e}{F}\left(\mu''_e - \mu'_e\right) = \frac{\mu''_e - \mu'_e}{Fr^b_e} \tag{5.80}$$

$$I''_e = \frac{\kappa''_e}{F}\left(\mu^{II}_e - \mu''_e\right) = \frac{\mu^{II}_e - \mu''_e}{Fr''_e} \tag{5.81}$$

In the above equations, μ'_{O_2}, μ'_e, μ''_{O_2}, and μ''_e are the oxygen chemical potential and electronic chemical potential at interfaces I and II, respectively. For mixed conducting membranes, when the oxygen permeation reaches a steady state, the oxygen ionic current densities in all zones are equal to each other, so do the electronic current densities. There is no longer any differentiation; therefore, $I_{O^{2-}}$ and I_e will be used to denote the oxygen ionic and electronic current densities, respectively. Because no voltage is applied across the MIEC membrane except an oxygen chemical potential gradient, the net current through the membrane is zero, that is, $I_{O^{2-}} + I_e = 0$. Furthermore, $I_{O^{2-}} < 0$, and $I_e > 0$. By combining Eqs. (5.76) and (5.79) and Eqs. (5.77) and (5.80), as well as Eqs. (5.78) and (5.81), and rewriting the equations, we obtain expressions for the oxygen chemical potentials at the interfaces and the bulk.

$$\Delta\mu'_{O_2} = \mu'_{O_2} - \mu^I_{O_2} = -4F(r'_{O^{2-}} + r'_e)|I_{O^{2-}}| \tag{5.82}$$

$$\Delta\mu^b_{O_2} = \mu''_{O_2} - \mu'_{O_2} = -4F(r^b_{O^{2-}} + r^b_e)|I_{O^{2-}}| \tag{5.83}$$

$$\Delta\mu''_{O_2} = \mu^{II}_{O_2} - \mu''_{O_2} = -4F(r''_{O^{2-}} + r''_e)|I_{O^{2-}}| \tag{5.84}$$

Combining the above three equations gives

$$\begin{aligned}\left(\mu^{II}_{O_2} - \mu^I_{O_2}\right) &= -4F|I_{O^{2-}}|(r'_{O^{2-}} + r'_e) - 4F|I_{O^{2-}}|(r^b_{O^{2-}} + r^b_e) \\ &\quad - 4F|I_{O^{2-}}|(r''_{O^{2-}} + r''_e)\end{aligned} \tag{5.85}$$

where the term on the left is the total oxygen chemical potential difference across the membrane. The terms on the right are the reduced oxygen chemical potentials at interface I, the bulk diffusion zone, and interface II. Therefore, the available total driving force is distributed in three parts across the membrane.

$$\Delta\mu^{tot}_{O_2} = \mu^{II}_{O_2} - \mu^I_{O_2} = \Delta\mu'_{O_2} + \Delta\mu^b_{O_2} + \Delta\mu''_{O_2} \tag{5.86}$$

Considering $|I_{O^{2-}}| = 4Fj_{O_2}$ and substituting this equation into Eq. (5.85), a general transport equation can be expressed as

$$j_{O_2} = -\frac{1}{4^2F^2}\frac{1}{r' + r^b + r''}\nabla\mu_{O_2}^{tot} = -\frac{1}{4^2F^2 r^{tot}}\nabla\mu_{O_2}^{tot} \tag{5.87}$$

where $r' = r'_{O^{2-}} + r'_e$, $r^b = r^b_{O^{2-}} + r^b_e$, and $r'' = r''_{O^{2-}} + r''_e$. $\Delta\mu_{O_2}^{tot}$ is the total driving force, and $r^{tot} = r' + r^b + r''$ is the total permeation resistance across the membrane. According to the relationship between the oxygen chemical potential and the oxygen partial pressure shown in Eq. (5.23), (5.87) can be rewritten as

$$j_{O_2} = -\frac{RT}{4^2F^2}\frac{1}{r' + r^b + r''}\ln\frac{P_{O_2}^{II}}{P_{O_2}^{I}} = -\frac{RT}{4^2F^2 r^{tot}}\ln\frac{P_{O_2}^{II}}{P_{O_2}^{I}} \tag{5.88}$$

and the total resistance can be calculated from the following equation:

$$r^{tot} = -\frac{RT}{4^2F^2 j_{O_2}}\ln\frac{P_{O_2}^{II}}{P_{O_2}^{I}} \tag{5.89}$$

In general, the area-specific resistances of the interfacial regions are functions of the oxygen partial pressure. Usually, a simple power function can be used to describe the relationship between the interface specific resistance and the oxygen partial pressure [19]; therefore, we obtain the following relationships:

$$r' = r'_0\left(P_{O_2}^{I}/P_0\right)^n \tag{5.90}$$

$$r'' = r''_0\left(P_{O_2}^{II}/P_0\right)^n \tag{5.91}$$

where P_0, r'_0, and r''_0 are the oxygen partial pressure of 1 atm and the area-specific resistances of the feed side and the permeation side at an oxygen partial pressure of 1 atm, respectively. n is the exponent. Under conditions close to thermodynamic equilibrium, the permeation flux through an interface can be described as [1]

$$j_{O_2} = -j_{ex}\frac{\Delta\mu_{O_2}}{RT} \tag{5.92}$$

where j_{ex} is the balance exchange rate in the absence of oxygen potential gradients at the interface. In the simplest cases, the latter quantity is directly related to the equilibrium interfacial exchange coefficient k_{ex}, which can be obtained from experimental data of ^{18}O–^{16}O isotopic exchange as follows [1]:

$$j_{ex} = \frac{1}{4}k_{ex}c_{O^{2-}} \tag{5.93}$$

where $c_{O^{2-}}$ (mol m^{-3}) is the volume concentration of oxygen anions at equilibrium. Therefore, by combining Eqs. (5.82), (5.92), and (5.93) and the relationship $|I_{O^{2-}}| = 4Fj_{O_2}$, the real exchange coefficients at interfaces I and II give

$$k'_{\mathrm{ex}} = \frac{RT}{4F^2 c'_{O^{2-}}} \frac{1}{r'} \tag{5.94}$$

$$k''_{\mathrm{ex}} = \frac{RT}{4F^2 c''_{O^{2-}}} \frac{1}{r''} \tag{5.95}$$

Experimental and theoretical investigations have revealed that the equilibrium interfacial exchange coefficient is a function of oxygen partial pressure:

$$k_{\mathrm{ex}} = k_0 (P_{O_2}/P_0)^{0.5} \tag{5.96}$$

where k_0 is the equilibrium interfacial exchange coefficient at an oxygen partial pressure of 1 atm. Therefore, combining Eqs. (5.94), (5.95), and (5.96) and considering that the concentration of oxygen ions at the interfacial zone is almost constant over a small oxygen pressure range, we can determine that the value of n in Eqs. (5.90) and (5.91) is $-1/2$. Therefore, we obtain the following two relationships:

$$r^{\mathrm{tot}} = r'_0 \left(P^I_{O_2}/P_0\right)^{-\frac{1}{2}} + \left(r^b + r''\right) \tag{5.97}$$

$$r^{\mathrm{tot}} = r''_0 \left(P^{II}_{O_2}/P_0\right)^{-\frac{1}{2}} + \left(r^b + r'\right) \tag{5.98}$$

The total specific resistance can be calculated from Eq. (5.89), and if the oxygen pressure on the permeation side is fixed and that on the feed side is varied, we obtain r'_0 and $r^b + r''$ through linear regression of Eq. (5.97). Similarly, r''_0 and $r^b + r'$ can be obtained through linear regression of Eq. (5.98). Another more professional method to fit the permeation data is to use a combined algorithm, i.e., the Levenberg–Marquardt method and the universal global optimization method.

Considering that both interfacial exchange processes contribute to the characteristic thickness, a combined oxygen exchange coefficient is defined as

$$k^{\mathrm{tot}}_{\mathrm{ex}} = k'_{\mathrm{ex}} k''_{\mathrm{ex}} / \left(k'_{\mathrm{ex}} + k''_{\mathrm{ex}}\right) \tag{5.99}$$

Under a small oxygen partial pressure gradient, the concentration of oxygen ions in every zone of an MIEC membrane can be regarded as a constant. Therefore, by substituting Eqs. (5.94) and (5.95) into Eq. (5.99), we obtain the following equation:

$$k^{\mathrm{tot}}_{\mathrm{ex}} = \frac{RT}{4F^2 c_{O^{2-}}} \frac{1}{r' + r''} \tag{5.100}$$

Bouwmeester et al. introduced L_c to define the membrane thickness corresponding to the transition from predominantly bulk diffusion controlled to the state where permeation is mainly governed by the interfacial exchange, as expressed by Eq. (5.20) [1]. Here, the characteristic thickness can be expressed as

$$L_c = \frac{D_{\mathrm{ambi}}}{k_{\mathrm{ex}}^{\mathrm{tot}}} \tag{5.101}$$

where D_{ambi} is the ambipolar diffusion coefficient of oxygen ionic–electronic pairs derived from the classical Nernst–Einstein relationship.

$$D_{\mathrm{ambi}} = \frac{\sigma_{\mathrm{ambi}} RT}{4F^2 c_{O^{2-}}} = \frac{RT}{4F^2 c_{O^{2-}}} \frac{l^b}{r^b} \tag{5.102}$$

The above equations are simplified and only valid for small oxygen partial pressure gradients applied across membranes and under the condition that the concentration of oxygen ions in oxides is treated as a constant. Because the electronic conductivity of an MIEC material is much larger than its ionic conductivity, the ambipolar diffusion coefficient is close to the self-diffusion coefficient. Combining Eqs. (5.100–5.102) and assuming $l', l'' \ll l^b \approx L$, one can obtain the following relationship:

$$L_c = \frac{r' + r''}{r^b} L = \frac{r'_0 + r''_0}{r^b} L \left[\left(P^I_{O_2} / P_0 \right)^{-\frac{1}{2}} + \left(P^{II}_{O_2} / P_0 \right)^{-\frac{1}{2}} \right] \tag{5.103}$$

Therefore, it is found that L_c is a process parameter that changes with oxygen partial pressure on both sides of a membrane. Also, the relationship reveals that if the membrane thickness is equal to the characteristic thickness, the bulk diffusion resistance is equal to the total interfacial exchange resistance.

5.5.2 *Model Applications*

5.5.2.1 Experimental Verification

BSCF, developed by Shao and Yang et al. in our group, has attracted significant attention in recent years, and it has been investigated thoroughly as an oxygen permeable membrane [20] and as a cathode of solid oxide fuel cells [21]. In order to determine r'_0, r^b, and r''_0 in Eqs. (5.97) and (5.98), the oxygen partial pressure of one side was kept constant, whereas that of the other side was changed. Then, the two interfacial exchange constants were obtained through linear regression of the two equations. Zhu and Yang et al. employed this permeation model to fit the permeation experimental data of BSCF membranes [7]. The relative errors of the interfacial exchange resistance constants were found to be less than 2 %, and

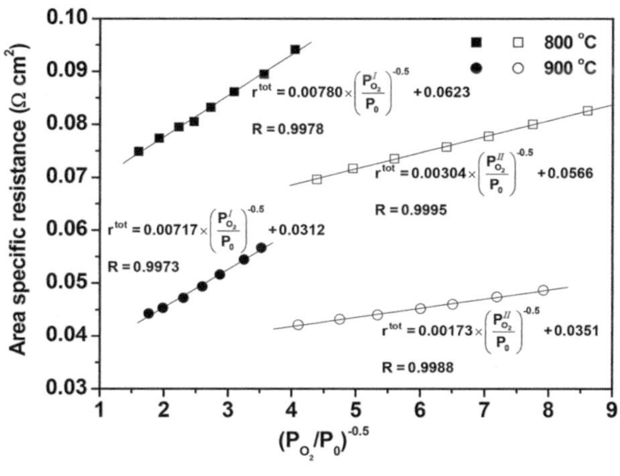

Fig. 5.9 Dependence of total permeation resistance on oxygen partial pressure of each side for a 0.5-mm BSCF membrane. Solid symbols: varying the oxygen partial pressure of feed side; open symbols: varying the oxygen partial pressure of permeation side (Reproduced from [7] with permission of Wiley)

the bulk permeation resistance constant on a 0.5-m-thick membrane calculated from the two equations showed small relative average errors in the range of 0.1–2.7 %. These average values were then used for the modeling investigation. Figure 5.9 shows the total permeation resistance, r^{tot}, of a 0.5-mm-thick BSCF membrane plotted as a linear function of $(P_{O_2}/P_0)^{-0.5}$ on each side at different temperatures. The figure reveals that model Eqs. (5.97) and (5.98) closely match the experimental data, with a correlation coefficient greater than 0.997. To examine the applicability of these parameters in modeling other membranes, the dependences of the oxygen permeation flux of a 1.0-mm-thick BSCF membrane on the oxygen partial pressure were tested. The results showed that the model data were somewhat higher than the test values by a factor of 1.05. This difference was much smaller than that obtained in Xu's model, in which a factor of 1.67 was found when the parameters obtained for a 3.99-mm-thick membrane were applied to a 1.68-mm-thick membrane [5]. Surface and bulk microstructural variations from one sample to another were considered to have caused the difference. Although the same procedures were used for the preparation of the membranes, it is impossible to produce two membranes with the exact same surfaces and bulk microstructures.

In addition to BSCF, the model was employed to fit the permeation experimental results of a $BaCe_{0.05}Fe_{0.95}O_{3-\delta}$ (BCF) perovskite membrane and SDC–SSAF dual-phase membranes. Linear dependences of the total permeation resistances on $(P_{O_2}/P_0)^{-0.5}$ were found for the two MIEC membranes, with a correlation coefficient greater than 0.995. The relative errors of all the interfacial exchange resistance constants were found to be less than 2 %, and the relative average errors of the bulk permeation resistance constants on a 0.5-mm-thick membrane were less than 8 %. These experimental results revealed that the model fit not only both cobalt-containing and cobalt-free perovskite membranes but also dual-phase membranes with a fluorite-type ionic conductor as the major phase and a perovskite mixed conductor as the minor phase.

5.5.2.2 Kinetic Parameters

The electronic conductivities of BSCF and BCF perovskite oxides are far higher than their ionic conductivities. Therefore, the bulk-transport resistance, r^b, is dominated by the oxygen ionic transport resistance, $r^b_{O^{2-}}$. The oxygen nonstoichiometry of BSCF is often found in the range of 0.35–0.65, typically around 0.5. The oxygen nonstoichiometry of BCF at room temperature and high temperature was determined by iodometry and thermogravimetry according to the literature [22]. The oxygen nonstoichiometry of BCF is 0.30 at room temperature and 0.50–0.57 in the temperature range of 750–950 °C. The kinetic parameters k'_0, k''_0, and D_s of the BCF membrane can be calculated according to Eqs. (5.94), (5.95), and (5.102). Both the permeation resistance constants and the kinetic parameters at different temperatures are listed in Tables 5.1 and 5.2. It can be seen that $r'_0 > r''_0$, corresponding to $k'_0 > k''_0$, indicating that the oxygen exchange rate on interface II is faster than that on interface I under the same oxygen partial pressure. A comparison of the kinetic parameters of BSCF and BCF obtained from the permeation model to those of other common SOFC cathode materials measured by tracer diffusion is presented in Fig. 5.10. D^* and k^* are usually obtained through tracer diffusion experiments under equilibrium states (Chap. 4). However, in this case, k'_0, k''_0, and

Table 5.1 Permeation resistance constants and kinetic parameters of BSCF membrane at various temperatures

Temperature (°C)	r'_0 (Ω cm^2) $\times 10^{-3}$	r''_0 (Ω cm^2) $\times 10^{-3}$	r^b (Ω cm^2) $\times 10^{-3}$	k'_0 (cm s^{-1}) $\times 10^{-4}$	k''_0 (cm s^{-1}) $\times 10^{-4}$	D_s (cm^2 s^{-1}) $\times 10^{-6}$
940	6.86	1.69	15.7	6.08	24.7	13.3
900	7.17	1.73	19.3	5.63	23.3	10.5
850	7.56	2.05	27.9	5.11	18.9	6.92
800	7.80	3.04	40.2	4.73	12.1	4.59
750	10.2	4.01	62.3	3.45	8.78	2.83
715	12.3	6.41	81.7	2.76	5.30	2.08
680	16.5	8.08	114	1.99	4.06	1.44

Reproduced from [7] with permission of Wiley

Table 5.2 Permeation resistance constants and kinetic parameters of BCF membrane at different temperatures

Temperature (°C)	r'_0 (Ω cm^2) $\times 10^{-3}$	r''_0 (Ω cm^2) $\times 10^{-3}$	r^b (Ω cm^2) $\times 10^{-3}$	k'_0 (cm s^{-1}) $\times 10^{-4}$	k''_0 (cm s^{-1}) $\times 10^{-4}$	D_s (cm^2 s^{-1}) $\times 10^{-6}$
950	18.7	3.22	32.9	2.34	13.7	6.71
900	19.7	3.39	37.8	2.12	12.3	5.59
850	19.3	4.48	47.2	2.06	8.89	4.26
800	20.7	5.19	58.3	1.82	7.29	3.26
750	23.0	6.43	69.6	1.55	5.59	2.57

Reproduced from [23] with permission of Wiley

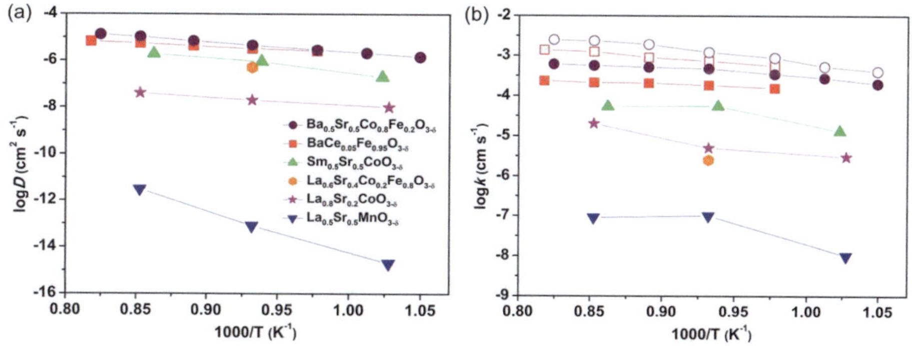

Fig. 5.10 (**a**) Comparison of oxygen diffusion coefficient and (**b**) oxygen surface exchange coefficient of various SOFCs cathode materials with BCF. Kinetics parameters of $Ln_{1-x}Sr_x$(Mn, Fe,Co)$O_{3-\delta}$ (Ln = lanthanide elements) MIEC materials are tracer diffusion and surface exchange coefficients (D^* and k^*), while the kinetics parameters of BCF and BSCF are obtained through the permeation model. Solid and open symbols in (**b**) for BSCF and BCF denote the interfacial exchange coefficients of feed and permeation side, respectively. Oxygen partial pressure for $La_{0.6}Sr_{0.4}Co_{0.8}O_{3-\delta}$ is 10.1 kPa, and those for other materials are 101 kPa in (**b**). All the data are adopted from [23–27]. This figure is reproduced from [23] with permission of Wiley

D_s were obtained by the oxygen permeation model under permeation steady state. Chemical diffusion and exchange coefficients are more often reported in the literature through the conductivity relaxation method (Chap. 4); however, the data measured under nonsteady and non-equilibrium states may be significantly different from those under actual steady states. Therefore, among the three method methods, the observed oxygen diffusion and exchange coefficients of MIEC materials through the permeation model under a permeation steady state are closest to the actual values in their working states.

5.5.2.3 Permeation Resistance Distributions

As mentioned above, one of the strong points of Xu's model is that the permeation resistance distribution across an MIEC membrane can be drawn. Obviously, it is easy to draw graphs of the permeation resistance distribution based on the basic constants listed in Tables 5.1 and 5.2. The influences of temperature, oxygen partial pressure of each side, and membrane thickness on the permeation resistance distribution across the BSCF membrane are shown in Fig. 5.11. The r.d.s. is easily found on the graphs. For example, as the oxygen partial pressure of the permeation side decreases from 0.1 bar to 0.07 bar, the contribution of the bulk resistance to total resistance decreases from ~70 % to ~47 % at a feed air pressure of 5 bar, which corresponds to a change in the permeation rate from being limited by bulk diffusion to being jointly controlled by bulk diffusion and interfacial exchange.

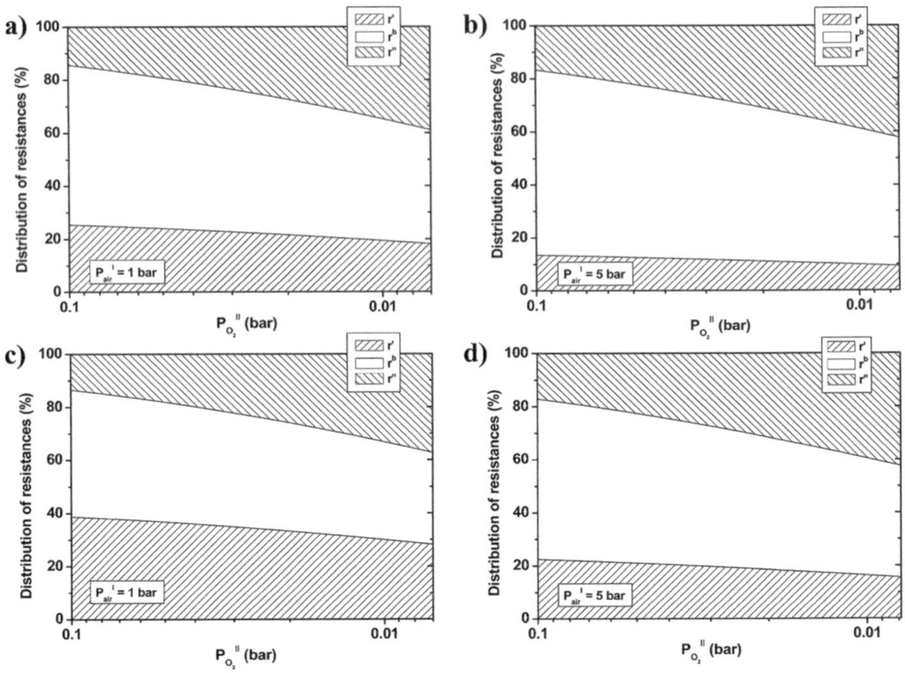

Fig. 5.11 Distribution of permeation resistances for the 0.5-mm-thick BSCF membrane under various oxygen partial pressures of permeation side. For (**a**) and (**b**): 800 °C; for (**c**) and (**d**): 900 °C (Reproduced from [7] with permission of Wiley)

5.5.2.4 Degradation Mechanism Analysis

The degradation of the permeation flux of BSCF membranes has typically been attributed to the phase transformation of the material at intermediate-low temperatures (see Chap. 6). The changes in the interfacial exchange resistances and bulk resistance were investigated by the oxygen permeation model, as shown in Fig. 5.12. We found that the interfacial oxygen exchange resistances increased more quickly with time than bulk resistance [28]. The degradation of oxygen permeation flux of BSCF is predominated by the increase of interfacial oxygen exchange resistances induced by the enrichment of $BaSO_4$ particles on membrane surfaces and is slightly affected by the bulk phase transformation. After the membrane surfaces were coated by porous $Sm_{0.5}Sr_{0.5}CoO_{3-\delta}$ (SSC) catalyst layers, the interfacial oxygen exchange resistances kept constant while the bulk diffusion resistance increased gradually. The increase of the bulk diffusion resistance for the SSC-coated BSCF membranes with time confirms the occurrence of a bulk phase transformation at intermediate-low temperatures. The resistance distribution reveals that the interfacial exchange kinetics is the rate-limiting step for the BSCF membranes at intermediate-low temperatures; therefore, changes in the oxygen exchange kinetics have a significant influence on the oxygen permeation

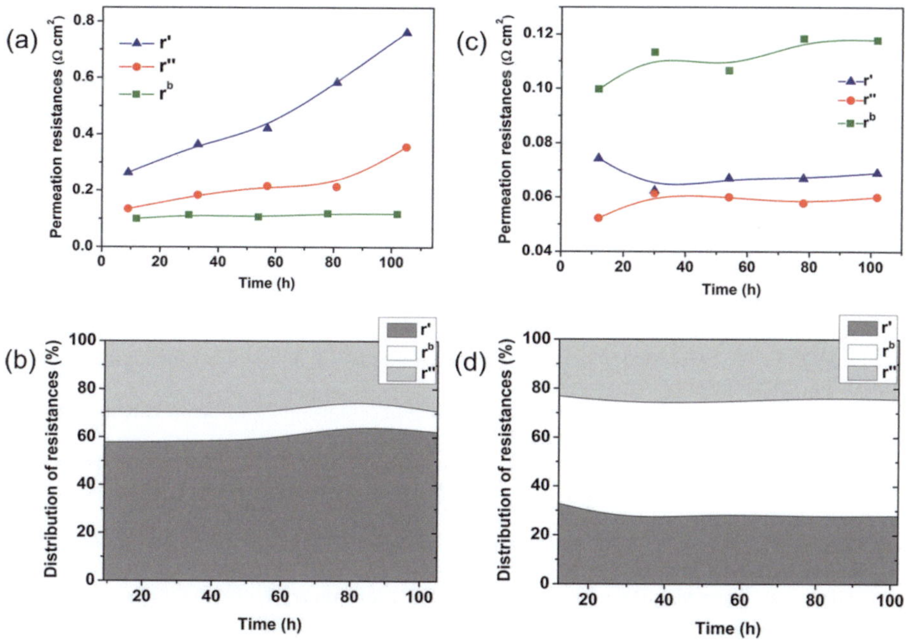

Fig. 5.12 Time-dependent changes of the permeation resistances (**a** and **c**) and resistance distribution (**b** and **d**) for the 0.5-mm-thick BSCF membranes at 600 °C. (**a** and **b**) without SSC porous layers, and (**c** and **d**) with SSC porous layers on both sides (Reproduced from [28] with permission of Wiley)

flux. The bulk phase transformation, which increases the bulk diffusion resistance, is a minor contributor to the degradation of the BSCF membrane at intermediate-low temperatures.

References

1. Bouwmeester HJM, Burggraaf AJ (1997) Dense ceramic membranes for oxygen separation. In: Gellings PJ, Bouwmeester HJM (eds) The CRC handbook of solid state electrochemistry. CRC Press, Boca Raton, pp 481–553
2. Kim S, Yang YL, Jacobson AJ, Abeles B (1999) Oxygen surface exchange in mixed ionic electronic conductor membranes. Solid State Ionics 121:31–36
3. Lee TH, Yang YL, Jacobson AJ, Abeles B, Zhou M (1997) Oxygen permeation in dense $SrCo_{0.8}Fe_{0.2}O_{3-\delta}$ membranes: surface exchange kinetics versus bulk diffusion. Solid State Ionics 100:77–85
4. Kim S, Yang YL, Jacobson AJ, Abeles B (1999) Diffusion and surface exchange coefficients in mixed ionic electronic conducting oxides from the pressure dependence of oxygen permeation. Solid State Ionics 106:189–195
5. Xu SJ, Thomson WJ (1999) Oxygen permeation rates through ion-conducting perovskite membranes. Chem Eng Sci 54:3839–3850

6. Lin YS, Wang WJ, Han J (1994) Oxygen permeation through thin mixed-conducting solid oxide membranes. AIChE J 40:786–798
7. Zhu XF, Liu HY, Cong Y, Yang WS (2012) Permeation model and experimental investigation of mixed conducting membranes. AIChE J 58:1744–1754
8. Wang HH, Cong Y, Yang WS (2002) Oxygen permeation study in a tubular $Ba_{0.5}Sr_{0.5}Co_{0.8}Fe_{0.2}O_{3-\delta}$ oxygen-permeable membrane. J Membr Sci 210:259–271
9. Mizusaki J, Mima Y, Yamauchi S, Fueki K, Tagawa H (1989) Nonstoichiometry of the perovskite–type oxides $La_{1-x}Sr_xCoO_{3-\delta}$. J Solid State Chem 80:102–111
10. Huang K, Goodenough JB (2001) Oxygen permeation through cobalt–containing perovskites surface oxygen exchange vs. lattice oxygen diffusion. J Electrochem Soc 148:E203–E214
11. Kim S, Yang YL, Jacobson AJ, Abeles B (1998) Diffusion and surface exchange coefficients in mixed ionic electronic conducting oxides from the pressure dependence of oxygen permeation. Solid State Ionics 106:189–195
12. Tan XY, Li K (2002) Modeling of air separation in a LSCF hollow-fiber membrane module. AIChE J 48:1469–1477
13. Tan XY, Li K (2007) Oxygen production using dense ceramic hollow fiber membrane modules with different operating modes. AIChE J 53:838–845
14. Asadi AA, Behrouzifar A, Iravaninia M, Mohammadi T, Pak A (2012) Preparation and oxygen permeation of $La_{0.6}Sr_{0.4}Co_{0.2}Fe_{0.8}O_{3-\delta}$ (LSCF) perovskite-type membranes: experimental study and mathematical modeling. Ind Eng Chem Res 51:3069–3080
15. Mancini ND, Mitsos A (2011) Ion transport membrane reactors for oxy-combustion – Part I: intermediate-fidelity modeling. Energy 36:4701–4720
16. Ghadimi A, Alaee MA, Behrouzifar A, Asadi AA, Mohammadi T (2011) Oxygen permeation of $Ba_xSr_{1-x}Co_{0.8}Fe_{0.2}O_{3-\delta}$ perovskite-type membrane: experimental and modeling. Desalination 270:64–75
17. Tan XY, Wang ZG, Liu H, Liu SM (2008) Enhancement of oxygen permeation through $La_{0.6}Sr_{0.4}Co_{0.2}Fe_{0.8}O_{3-\delta}$ hollow fibre membranes by surface modifications. J Membr Sci 324:128–135
18. Zhu XF, Cong Y, Yang WS (2006) Oxygen permeability and structural stability of $BaCe_{0.15}Fe_{0.85}O_{3-\delta}$ membranes. J Membr Sci 283:38–44
19. Virkar AV (2005) Theoretical analysis of the role of interfaces in transport through oxygen ion and electron conducting membranes. J Power Sources 147:8–31
20. Shao ZP, Yang WS, Cong Y, Dong H, Tong JH, Xiong GX (2000) Investigation of the permeation behavior and stability of a $Ba_{0.5}Sr_{0.5}Co_{0.8}Fe_{0.2}O_{3-\delta}$ oxygen membrane. J Membr Sci 172:177–188
21. Shao ZP, Haile SM (2004) A high-performance cathode for the next generation of solid-oxide fuel cells. Nature 431:170–173
22. Bucher E, Egger A, Ried P, Sitte W, Holtappels P (2008) Oxygen nonstoichiometry and exchange kinetics of $Ba_{0.5}Sr_{0.5}Co_{0.8}Fe_{0.2}O_{3-\delta}$. Solid State Ionics 179:1032–1035
23. Liu Y, Zhu Y, Li MR, Zhu XF, Yang WS (2016) Oxygen transport kinetics of MIEC membrane with different catalysts. AIChE J. doi:10.1002/aic.15239
24. Han N, Zhang SG, Meng XX, Yang NT, Meng B, Tan XY, Liu SM (2016) Effect of enhanced oxygen reduction activity on oxygen permeation of $La_{0.6}Sr_{0.4}Co_{0.2}Fe_{0.8}O_{3-\delta}$ membrane decorated by K_2NiF_4-type oxide. J Alloys Compd 654:280–289
25. Fuilarton IC, Jacobs JP, van Benthem HE, Kilner JA, Brongersma HH, Scanlon P, Steele BCH (1995) Study of oxygen ion transport in acceptor doped samarium cobalt oxide. Ionics 1:51–58
26. Li Y, Gerdes K, Horita T, Liu X (2013) Surface exchange and bulk diffusivity of LSCF as SOFC cathode: electrical conductivity relaxation and isotope exchange characterizations. J Electrochem Soc 160:343–350
27. Carter S, Selcuk A, Chater RJ, Kajda J, Kilner JA, Steele BCH (1992) Oxygen transport in selected nonstoichiometric perovskite-structure oxides. Solid State Ionics 53:597–605
28. Liu Y, Zhu XF, Yang WS (2015) Degradation mechanism analysis of $Ba_{0.5}Sr_{0.5}Co_{0.8}Fe_{0.2}O_{3-\delta}$ membranes at intermediate-low temperatures. AIChE J 61:3879–3888

Chapter 6
Perovskite-Type MIEC Membranes

Abstract Perovskite-type MIEC oxides are the most widely investigated membrane materials since the pioneering works of Teraoka and coworkers in 1985. In the early stage, cobalt-based perovskite oxides attracted much attention of researchers due to their high oxygen permeability. $Ba_{0.5}Sr_{0.5}Co_{0.8}Fe_{0.2}O_{3-\delta}$ and $La_{0.6}Sr_{0.4}Co_{0.2}Fe_{0.8}O_{3-\delta}$ are the two materials which have been thoroughly investigated in the past decades. Subsequently, the stability of the materials was regarded more important than the permeability for a practical application, and then cobalt-free perovskite membranes were extensively investigated in the past several years. In this chapter, the development of the cobalt-based and cobalt-free perovskite membranes is introduced in detail, and the influence of doping on oxygen permeability and stability is concerned.

Keywords Perovskite • Co-containing membranes • $LnCoO_{3-\delta}$ • $Ba_{0.5}Sr_{0.5}Co_{0.8}Fe_{0.2}O_{3-\delta}$ • Co-free membranes • $LaGaO_3$ • $BaFeO_{3-\delta}$ • Oxygen permeation • Phase transformation • Stability

6.1 Perovskite Structure

Perovskites are multifunctional materials that successfully act as dielectric materials, piezoelectric materials, photocatalysts, electrocatalysts, ionic conductors, and membrane materials. All materials with a crystal structure similar to that of calcium titanium oxide, e.g., $CaTiO_3$, are called perovskite materials. The term "perovskite" is not limited to oxides; it applies to any material with a similar atomic arrangement in the crystal structure. For perovskite oxides, the formula ABO_3 is usually used to denote the perovskite structure, where A is the larger cation and B is the smaller cation. Perovskite is a framework structure constructed from BO_6 octahedra, with A ions located in 12-coordinated interstices, in which the A ion is in the cell center. In another layout, the B ion is in the cell center, and A ions are located at the apexes of the hexahedron, as shown in Fig. 3.15.

From a geometric perspective, the ionic radius of B ions (sixfold-coordination) should be bigger than 0.051 nm to achieve stable octahedra, and that of A ions should be bigger than 0.09 nm (twelvefold-coordination) to minimize the distortion

X. Zhu, W. Yang, *Mixed Conducting Ceramic Membranes*, Green Chemistry and Sustainable Technology, DOI 10.1007/978-3-662-53534-9_6

of the BO_6 octahedra. The relationship between the tolerance limits and ionic radii is defined as follows:

$$t = (R_A + R_O)/\sqrt{2}(R_B + R_O) \tag{6.1}$$

where $R_A, R_B,$ and R_O are the ionic radii of A cations (twelvefold-coordination), B cations (sixfold-coordination), and oxygen ions (0.14 nm), respectively. A stable perovskite structure is formed when the tolerance factor t is in the range of 0.75–1. If it is smaller than 0.75, the crystal structure transitions to that of ilmenite ($FeTiO_3$); if it is much bigger than 1, a crystal structure such as calcite or aragonite ($CaCO_3$) is formed. Ideally, t should be equal to 1 for cubic perovskite oxides. If the tolerance factor is in the range of $0.75 < t < 1$, the perovskite oxide can transition to an anorthic, monoclinic, orthorhombic, trigonal (or rhombohedral), or cubic structure, in that order, with increasing t value. If the tolerance factor is slightly greater than 1, the oxide usually has a hexagonal structure. However, in practice, the cubic structure can exist across a wide range of t values. For example, $BaFeO_3$ perovskite has a hexagonal structure because its t value is larger than 1.09, whereas after doping 5 % Ce^{4+} ions in the B sites, the structure changes to cubic, even though the t value is still larger than 1.08. In addition to the tolerance factor, the metal–oxygen bonding energy, ionic doping at A sites, B sites, or both, as well as the oxygen vacancy concentration in the lattice, influence the crystal structure.

The stability of BO_6 octahedra largely depends on the B–O bonding energy, while it has a significant influence on the structural stability of the perovskite oxides. A-site cations with proper ionic radii can retain the cubic structure of the perovskite crystal; meanwhile, the proper A-site cation will improve the structural stability of the perovskite oxide. A typical example is the $SrCo_{0.8}Fe_{0.2}O_{3-\delta}$ perovskite. This oxide is unstable when the oxygen partial pressure decreases to 0.1 atm at ~790 °C because oxygen vacancy ordering takes place, causing a phase transformation from cubic to orthorhombic [1]. At low oxygen partial pressures, the tolerance factor gradually decreases with the decreasing oxidation state of cobalt and iron ions. However, because some Sr^{2+} is replaced by Ba^{2+}, the resultant $Ba_{0.5}Sr_{0.5}Co_{0.8}Fe_{0.2}O_{3-\delta}$ can sustain the cubic structure, even with an oxygen partial pressure of 10^{-5} atm at ~800 °C [2].

For stable perovskite oxides, the BO_6 octahedra are connected by corner-sharing. According to the Pauling connect rule of coordination polyhedron, the corner-sharing connection is more stable than edge- or facet-sharing connections. The corner-sharing connection of the BO_6 octahedra allows for large cations, large defect concentrations, and fast oxygen ionic diffusion in the lattice. The changeable crystal structure, element composition, defects, and valence state facilitate the adjustment of the physical and chemical properties of perovskite oxides according to the requirements of various applications. Therefore, perovskite oxides are often called omnipotent materials.

6.2 Defect Chemistry in Perovskite Oxides

To maintain electrical neutrality, the total charge of A and B cations should be equal to the total charge of the oxygen ions. $A^{2+/3+}B^{3+/4+}O_{3-\delta}$ is the most common formula for perovskite oxides, where δ is usually in the range of 0–0.5. It is possible to retain the perovskite structure when the A and B cations are partially or completely substituted by other cations with different ionic radii and oxidation states. Approximately 90 % of the elements in the periodic table can form a perovskite structure. Doping elements in A or B sites with different oxidation states can be the same as or different from the mother materials. For example, a $LaCoO_3$ perovskite oxide is doped by $SrFeO_3$ both in the A sites with Sr^{2+} and in the B sites with Fe^{4+}. This kind of replacement can be denoted as

$$SrFeO_3 \overset{\mathrm{LaCoO_3}}{\longrightarrow} Sr'_{\mathrm{La}} + Fe^{\cdot}_{\mathrm{Co}} + 3O^{\times}_{\mathrm{O}} \tag{6.2}$$

At high temperatures, low oxygen partial pressures, or both, oxygen vacancies and electronic holes are formed according to the following equations:

$$2Fe^{\cdot}_{\mathrm{Co}} + O^{\times}_{\mathrm{O}} \rightleftharpoons 2Fe^{\times}_{\mathrm{Co}} + V^{\cdot\cdot}_{\mathrm{O}} + \frac{1}{2}O_2 \tag{6.3}$$

$$Co^{\times}_{\mathrm{Co}} \rightleftharpoons Co'_{\mathrm{Co}} + h^{\cdot} \tag{6.4}$$

The doping of $SrFeO_3$ in $LaCoO_3$ perovskite oxide can significantly change both the ionic and electronic conductivities. The electronic conductivity is maximized at a Sr^{2+} doping amount of approximately 50 %; however, the ionic conductivity increases monotonically with increasing Sr^{2+} doping amount. Therefore, perovskite oxides demonstrate both high oxygen ionic conductivity and high electronic conductivity because the structure allows both high oxygen vacancy concentrations and high electronic conductivities. The structure also allows the same elements with different valence states to occupy the B sites.

If both the A and B cations have fixed valence states, the introduction of an aliovalent cation in the A or B site produces oxygen vacancies or interstitial oxygen ions in the lattice, such as with a $BaZrO_3$ perovskite doped by Y^{3+} and Nb^{5+} in the Zr site [Eqs. (6.5) and (6.6)], respectively.

$$Y_2O_3 \overset{\mathrm{BaZrO_3}}{\longrightarrow} 2Y'_{\mathrm{Zr}} + 3O^{\times}_{\mathrm{O}} + V^{\cdot\cdot}_{\mathrm{O}} \tag{6.5}$$

$$Nb_2O_5 \overset{\mathrm{BaZrO_3}}{\longrightarrow} 2Nb^{\cdot}_{\mathrm{Zr}} + 4O^{\times}_{\mathrm{O}} + \mathrm{O}''_{i} \tag{6.6}$$

As a result, the two resultant materials are predominantly oxygen ionic conductors, but the charge carriers are different. Partial electronic conductivity will be produced in the yttrium- and niobium-doped $BaZrO_3$ perovskite oxides. At high oxygen partial pressures, electronic holes appear, shown as

$$\frac{1}{2}O_2 + V_O^{\cdot\cdot} \rightleftharpoons O_O^{\times} + 2h^{\cdot} \tag{6.7}$$

and at low oxygen partial pressures, electrons appear, shown as

$$O_i'' \rightleftharpoons \frac{1}{2}O_2 + 2e' \tag{6.8}$$

Therefore, the materials may change from pure oxygen ionic conductors to mixed conductors under certain conditions.

A cationic deficiency may be encountered on perovskite oxides owing to stoichiometric deviation during the preparation of the materials. In the perovskite structure, B-site cationic vacancies are not energetically favorable owing to the large positive charge and the small size of B cations in the perovskite lattice. A-site cations may be partially empty without breaking down the perovskite structure. An A-site cationic deficiency influences the oxygen defects and the valence state of B-site cations, as well as the oxygen ionic and electronic conductivities. Here, $LaFeO_3$ perovskite oxide is used as an example to show the influence of A-site cationic deficiency. A La^{3+} deficiency is introduced into the lattice if a small excess of Fe_2O_3 is added during the preparation of the material, so that A-site cationic vacancies are formed according to the following equations:

$$3Fe_2O_3 \xrightarrow{LaFeO_3} 2V_{La}''' + 6Fe_{Fe}^{\cdot} + 9O_O^{\times} \tag{6.9}$$

$$2Fe_{Fe}^{\cdot} + O_O^{\times} \rightleftharpoons 2Fe_{Fe}^{\times} + V_O^{\cdot\cdot} + \frac{1}{2}O_2 \tag{6.10a}$$

$$Fe_{Fe}^{\cdot} \rightleftharpoons Fe_{Fe}^{\times} + h^{\cdot} \tag{6.10b}$$

Obviously, the A-site cationic deficiency would raise the oxidation state of B-site cations according to Eq. (6.9) and increase the oxygen vacancies according to Eq. (6.10a) and electronic holes according to Eq. (6.10b).

6.3 An Introduction to the Pioneering Works of Teraoka and Coworkers

Perovskite membranes for oxygen permeation were first reported by Teraoka et al. in 1985 [3]. They found that $La_{1-x}Sr_xCo_{1-y}Fe_yO_{3-\delta}$ perovskite membranes showed oxygen permeability at elevated temperatures. The permeation flux increased with increasing operation temperature and increasing doping amounts of strontium and cobalt. A membrane with a composition of $SrCo_{0.8}Fe_{0.2}O_{3-\delta}$ exhibited the highest

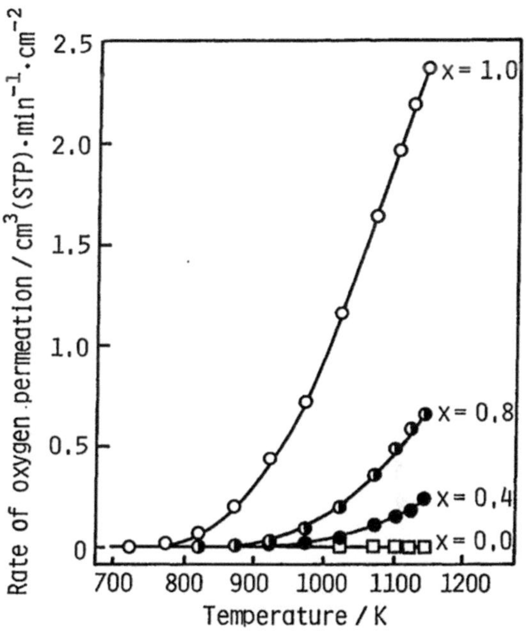

Fig. 6.1 Influences of temperature and composition on oxygen permeation through $La_{1-x}Sr_xCo_{0.4}Fe_{0.6}O_{3-\delta}$ ($x = 0.0$, 0.4, 0.8, and 1.0) (Reproduced from [3] with permission of CSJ Journals)

oxygen permeation flux in this series. Figure 6.1 shows the influence of chemical composition on the oxygen permeation flux of the $La_{1-x}Sr_xCo_{1-y}Fe_yO_{3-\delta}$ perovskite membranes. In their subsequent investigation on the substitution effects, they found that the oxygen permeation flux decreased in the following order [4]:

B-site substitution, Cu > Ni > Co > Fe > Cr > Mn
A-site substitution, Ba > Ca > Sr > Na
Rare earth element substitution, Gd > Sm > Nd > Pr > La

According to the substitution effect on the oxygen permeation flux, they designed a perovskite membrane material with a chemical composition of $Gd_{0.2}Ba_{0.8}Co_{0.7}Cu_{0.2}Fe_{0.1}O_{3-\delta}$, in which the iron element was thought to be important to keep the perovskite-type structure at a high alkaline earth metal content. An abrupt increase in oxygen permeation flux was found on the membrane when the temperature was higher than 700 °C. Finally, an oxygen permeation flux of 5.5 mL cm^{-2} min^{-1} was achieved at 870 °C on a 1.5-mm-thick disk-type membrane under an oxygen chemical potential gradient created by ambient air and He flow (30 mL min^{-1}). At that time, Teraoka et al. believed that the perovskite membranes were superior to polymer membranes for oxygen separation in terms of selectivity and thermal stability, even though they operated at relatively high temperatures. It should be pointed out that the experimental methods developed by Teraoka et al. in 1985, including the preparation and testing of the permeation performance of the membranes, are still used today.

Thereafter, perovskite membranes for oxygen separation at elevated temperatures started to attract wide attention. Perovskite membranes have been explored for use as catalytic membrane reactors for partial oxidation, selective oxidation, and water-splitting reactions, and they are promising materials in combination with oxyfuel and integrated gasification combined cycle (IGCC) processes for CO_2 capture and energy saving. Although researchers have not been able to achieve the high permeation fluxes reported by Teraoka and coworkers, they have verified that Co-containing perovskite materials show reliably high permeation fluxes. Although 30 years have passed since the first report on perovskite membranes by Teraoka and coworkers, and since then hundreds of membrane materials have been developed by different groups, the substitution effects found by Teraoka and coworkers are still correct. With broadening application of oxygen permeable membranes, many new membrane materials are being developed to meet the requirements of various applications in terms of permeability and stability. Therefore, except for the $LnCoO_{3-\delta}$ system first studied by Teraoka and coworkers, other materials systems such as $(Ba,Sr)CoO_{3-\delta}$, $LaGaO_{3-\delta}$, $LaFeO_{3-\delta}$, and $BaFeO_{3-\delta}$ have been widely investigated in recent years. Therefore, all these material systems will be discussed in detail in the following sections.

6.4 Co-containing Perovskite Membranes

6.4.1 $LnCoO_{3-\delta}$ System

6.4.1.1 Substitution in B Sites

Lanthanum cobaltite, $LaCoO_{3-\delta}$, a perovskite-type oxide exhibiting oxygen ionic conductivity with high electronic conductivity, is a promising basis for high-performance membrane materials after proper doping in A and B sites. Oxygen permeation through the $LaCoO_{3-\delta}$ membrane is low and limited by its low oxygen ionic conductivity because the low nonstoichiometric oxygen (or oxygen vacancy, $\delta = 0.001–0.03$) under oxygen permeation conditions cannot supply a sufficient path for the transport of oxygen ions [5]. Therefore, to enhance the oxygen ionic conductivity and permeability of the membrane materials, it is effective to increase the concentration of oxygen vacancies in the perovskite lattice. Divalent metal ions such as Ni^{2+} and Mg^{2+} were used to partially replace cobalt ions in the B sites [6, 7]. However, the oxygen ionic conductivity was only slightly improved and the permeability remained low. Other trivalent and tetravalent metal ions such as Ga^{3+}, $Fe^{3+/4+}$, and $Cr^{3+/4+}$ were also incorporated into the $LaCoO_{3-\delta}$ lattice [6, 7]. Introducing divalent metal ions into the perovskite lattice led to increases in the oxygen nonstoichiometry and electrical conductivity and a decrease in the thermal expansion coefficients; however, doping with trivalent and tetravalent metal ions has been demonstrated to have the opposite effect. Although the incorporation of divalent metal ions in the cobalt sites can improve the ionic conductivity, they cannot produce enough oxygen vacancies because of their limited solubility, and thus they cannot significantly enhance the ionic conductivity. Up to now, there have been no reports showing that the permeability of $LaCoO_{3-\delta}$ membranes can be significantly improved just by doping with cations in the B sites.

6.4.1.2 Substitution in A Sites

An alternative method for increasing the number of oxygen vacancies is to dope divalent metal ions in the A sites. The most frequently used divalent metal ions are alkaline metal ions, i.e., Ba^{2+}, Sr^{2+}, and Ca^{2+}. Almost all the experimental data reveal that the permeation flux of $La_{1-x}A'_xCoO_{3-\delta}$ ($A' = Ba^{2+}$, Sr^{2+}, and Ca^{2+}) membranes increases with increasing doping amount of alkaline metal ions as long as the membrane materials do not change to a hexagonal phase. Islam et al. calculated the solution energy of the three alkaline ions in the La^{3+} sites in perovskite and found that the solution energy increased in the order of $Sr^{2+} < Ca^{2+} < Ba^{2+}$ [8–10]. The favorable incorporation of dopant ions (such as Sr^{2+}) enhanced the oxygen diffusivity owing to the increase in the number of oxygen vacancies. However, this finding is opposite to the oxygen permeation data from the A-site cationic substitution experiments reported by Teraoka and coworkers [3, 4].

Tsai et al. found that the steady-state oxygen permeation fluxes of $La_{1-x}A'_xCo_{0.2}Fe_{0.8}O_{3-\delta}$ perovskite membranes with air on one side of the membrane and helium on the other decreased in the order of $A(x)' =$ Ba-0.8 > Ba-0.6 > Ca-0.6 > Sr-0.6 [11]. Perovskite crystallographic-related parameters were used to explain the order, including (1) the average metal–oxygen bond energy (ABE) of the perovskite lattice, (2) the free volume of the lattice, and (3) the critical radius (r_c). It was found that except for the Ca-substitution membranes, a higher ABE leads to a higher permeation activation energy and a lower permeation flux. The Ca-substitution material had the highest ABE and free volume, and thus the comprehensive results were that the membrane exhibited moderate permeation activation energy and permeation flux. However, Stevenson et al. showed that the oxygen permeation fluxes of $La_{1-x}A'_xCo_{0.2}Fe_{0.8}O_{3-\delta}$ perovskite membranes were in the order of $A(x)' =$ Sr-0.6 > Ba-0.6 > Ca-0.6 [12]. They attributed the superior ionic conduction in the Sr-doped perovskite to the higher mobility of oxygen vacancies rather than to a higher concentration of vacancies. The tolerance factors ranged from 0.97 to 0.98, from 0.99 to 1.02, and from 0.95 to 0.96 for the $La_{0.4}Sr_{0.6}Co_{0.2}Fe_{0.8}O_{3-\delta}$, $La_{0.4}Ba_{0.6}Co_{0.2}Fe_{0.8}O_{3-\delta}$, and $La_{0.4}Ca_{0.6}Co_{0.2}Fe_{0.8}O_{3-\delta}$ materials, respectively, as calculated using the Shannon and Prewitt ionic radii. The authors suggested that a slight negative deviation from ideality in the tolerance factor may have enhanced the oxygen ion vacancy mobility. This supposition was also presented by Hayashi et al. after they surveyed many perovskite-type oxygen ionic conducting materials [13]. They found that the optimum tolerance factor for $A^{3+}B^{3+}O_{3-\delta}$-type perovskites to achieve the maximum oxygen ionic conductivity was ~0.96 owing to the balance between the specific free volume and the tolerance factor. Soon after, Li et al. [14] investigated the substitution effects of alkaline metal ions on the oxygen permeation fluxes of $La_{0.2}A'_{0.8}Co_{0.2}Fe_{0.8}O_{3-\delta}$ perovskite membranes and found the same order as that reported by Stevenson et al. Among the three materials, it was found that the Sr-doped material had the maximum free volume and critical radius.

In the above four experimental investigations on the substitution effects of alkaline ions in $LaCoO_{3-\delta}$ perovskite, different conclusions were drawn by the different authors, and different explanations were adopted to describe the order they observed. These conflicting conclusions well exhibit the complexity of substitution influences in perovskite A sites. Except for the doping effects, there are many factors that could potentially affect the oxygen permeation flux. First of all, the influence of oxygen exchange on the membrane surfaces was not discussed in these investigations. It was found that the three alkaline ion-doped $LaCoO_{3-\delta}$ perovskite oxides showed different activities toward the oxygen reduction/evolution reactions at elevated temperatures. Other factors such as membrane surface microstructure, grain boundaries, impurities, grain size, synthesis method, etc., all have a significant influence on the permeation flux of membranes. Therefore, it is difficult to tell which alkaline ion is the best choice for $LaCoO_{3-\delta}$ perovskite. However, Sr^{2+} is the most frequently used cation to partially substitute for A-site cations in $LnCoO_{3-\delta}$ perovskite.

With the exception of lanthanum, other rare earth elements have been used as A-site cations in $Ln_{1-x}A'_xCoO_{3-\delta}$ ($Ln = Pr^{3+}$, Nd^{3+}, Sm^{3+}, Gd^{3+}, Y^{3+}, etc.; $A' = Ba^{2+}$, Sr^{2+}, and Ca^{2+}) perovskite oxides. Kharton et al. found that the oxygen permeation flux through $Ln_{0.3}Sr_{0.7}CoO_{3-\delta}$ perovskite membranes decreased in the order of La > Nd > Sm > Gd [15]. This order is in accord with the order of shrinkage of the radii of the rare earth cations. The authors thought this showed dependence of the oxygen ion mobility in oxides having a perovskite ABO_3 structure on the mean radius of cations in the A-site sublattice. Most $LnCoO_{3-\delta}$-based perovskite membranes are double doped in both the A and B sites. Main-group metallic elements such as Mg, Al, and Ga and subgroup metallic elements in the fourth period have been doped in the B sites of $La_{1-x}Sr_xCoO_{3-\delta}$ to alter the permeability and stability of the perovskite membranes. Usually, the incorporation of Mg^{2+}, Ni^{2+}, Cu^{2+}, Zn^{2+}, etc., divalent cations can improve the oxygen permeability, but it also decreases the mechanical strength and chemical stability. One important reason for this is that the doping reduces the ABE of the perovskite lattice. However, doping Al^{3+}, Sc^{3+}, Ti^{4+}, $V^{3+/4+}$, $Cr^{3+/4+}$, $Mn^{3+/4+}$, Ga^{3+}, Zr^{4+}, Nb^{5+}, etc., cations in the B sites of $La_{1-x}Sr_xCoO_{3-\delta}$ decreases the oxygen permeability but improves the mechanical strength and chemical stability of the materials.

6.4.1.3 Typical $LnCoO_{3-\delta}$-Based Perovskite Membrane Materials

The most popular $LnCoO_{3-\delta}$-based perovskite material is $La_{0.6}Sr_{0.4}Co_{0.2}Fe_{0.8}O_{3-\delta}$ (LSCF), which has been well investigated in the field of solid electrochemistry, including solid oxide fuel cells and MIEC membranes, owing to its high electronic conductivity, moderate oxygen ionic conductivity, moderate thermal expansion coefficient, and high mechanical strength as compared to other perovskite oxides of this series. Almost all available methods have been used to investigate the oxygen surface exchange coefficient and bulk diffusion coefficient of LSCF, i.e., isotropic tracer, conductivity relaxation, and oxygen permeation model methods. The kinetic parameters obtained by the above methods have different physical meanings (see Chap. 5), but they can be interconverted by multiple special factors i.e. thermodynamic factor, γ_O, and the Haven ratio, H_R. The Haven ratio equals the well-known tracer correlation factor (f) for a vacancy diffusion mechanism. Lane et al. investigated the two kinetic parameters of LSCF perovskite and compared the data with those acquired from ^{18}O isotope exchange/secondary ion mass spectrometry (SIMS) [16]. They found that the bulk diffusion coefficients obtained by the conductivity relaxation method fit very well with those determined by the isotropic tracer method after they were multiplied by H_R/γ_O. However, the oxygen surface exchange coefficients obtained by the isotropic tracer method were one order of magnitude higher than those obtained from the conductivity relaxation method, although the thermodynamic factor was considered. Further investigation is required to understand this difference.

Bouwmeester et al. used the conductivity relaxation technique to investigate the surface exchange coefficient and chemical diffusion coefficient of oxygen for the perovskite oxides of $La_{0.6}Sr_{0.4}Co_{1-y}Fe_yO_{3-\delta}$ ($y = 0.2$, 0.5, and 0.8) between 600 and 800 °C in an oxygen partial pressure range between 10^{-4} and 1 bar [17]. They found that the ordering of oxygen vacancies led to a remarkable decrease of the two kinetic coefficients when the oxygen partial pressure was lower than 10^{-2} bar, as shown in Fig. 6.2, but that the chemical diffusion coefficient was almost constant at oxygen partial pressures larger than 10^{-2} bar. In other words, as the oxygen partial pressure was reduced to a certain value, the increased oxygen vacancy concentration enhanced the interactions between oxygen vacancies and other point defects, and thus not all oxygen vacancies were involved in the ionic transport.

Xu and Thomson developed an oxygen permeation model based on oxygen vacancy diffusion and exchange on membrane surfaces (see Chap. 5) [18]. They used the model equation to obtain the vacancy diffusion coefficient and surface exchange coefficient through nonlinear regression of the experimental permeation fluxes and oxygen partial pressures. The results of their model showed that the bulk diffusion of oxygen vacancies for LSCF had an activation energy of 74 kJ mol^{-1}, which is much lower than those for surface exchange kinetics. It was found that the oxygen permeation was limited by the surface exchange of the permeation side at lower temperatures but controlled by bulk diffusion at higher temperatures. This was the first attempt to acquire the kinetic parameters of perovskite materials through oxygen permeation experiments.

For utilization in catalytic membrane reactors, perovskite membranes must exhibit long-term stability under reducing environments. In general, the decomposition of ABO_3 perovskites is the result of the heavy loss of lattice oxygen under reducing environments:

$$ABO_3 \rightarrow \frac{1}{2}A_2O_3 + BO + \frac{1}{4}O_2 \tag{6.11a}$$

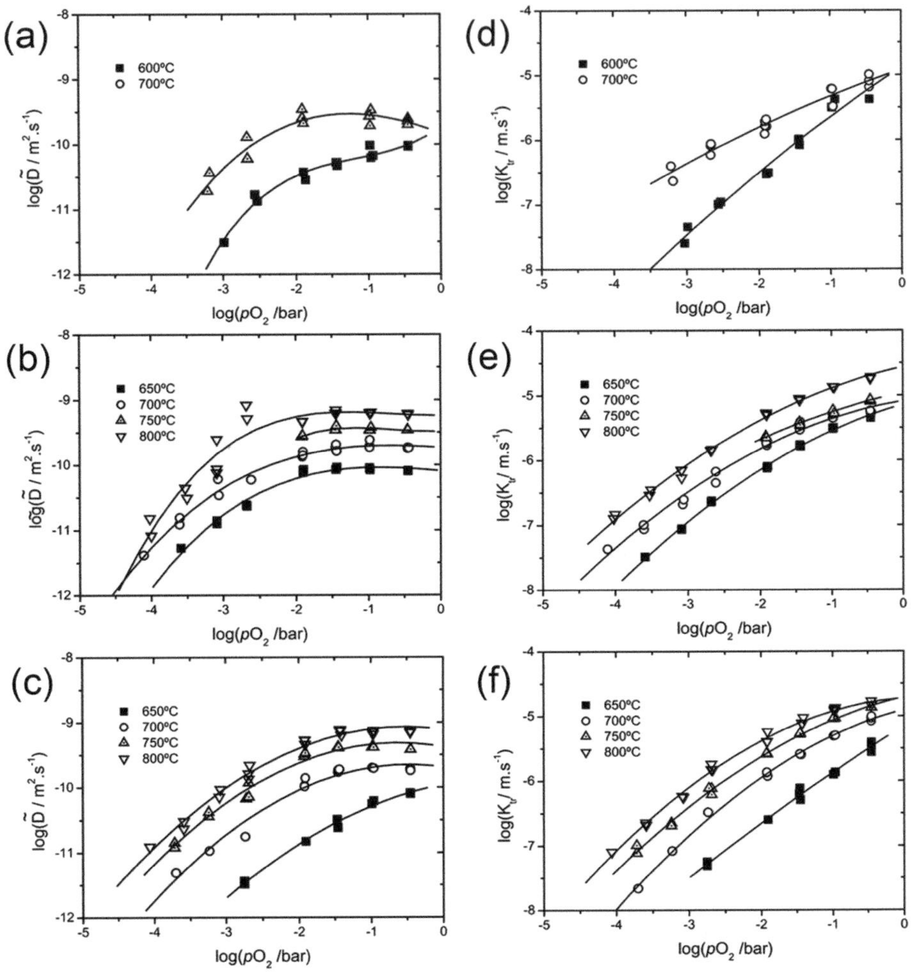

Fig. 6.2 Chemical diffusion coefficient (**a**, **b**, and **c**) and surface exchange coefficient (**d**, **e**, and **f**) of $La_{0.6}Sr_{0.4}Co_{1-y}Fe_yO_{3-\delta}$ plotted as a function of oxygen partial pressure at different temperatures: (**a** and **d**) $y = 0.2$, (**b** and **e**) $y = 0.5$, and (**c** and **f**) $y = 0.8$ (Reproduced from [17] with permission of Springer)

and

$$ABO_3 \rightarrow \frac{1}{2}A_2O_3 + B + \frac{3}{4}O_2 \quad (6.11b)$$

Factors that influence these reactions include the bulk structure and composition of ABO_3, temperature and the reducing capability of the environment. The high-temperature phase stability of LSCF as a function of oxygen activity was reported by Tai et al. [19]. They found that partial replacement of La by Sr had little

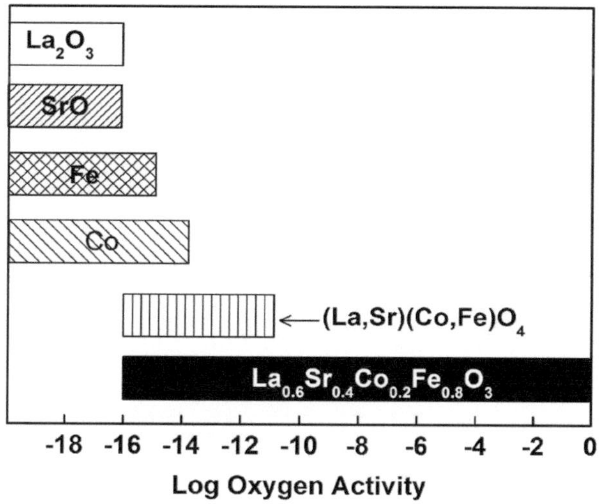

Fig. 6.3 The equilibrium phases of $La_{0.6}Sr_{0.4}Co_{0.2}Fe_{0.8}O_{3-\delta}$ at 1000 °C as a function of oxygen activity (from XRD analyses) (Reproduced from [19] with permission of Elsevier)

effect on the high-temperature phase stability of LSCF. Under a strong reducing environment, dissociation of LSCF occurred gradually by the formation of transient oxides. (La,Sr)(Fe,Co)O_4 is the major transient compound formed during decomposition under a reducing atmosphere, which would be reduced to La_2O_3, SrO, Co, and Fe under further reduction. Figure 6.3 shows the equilibrium phases of LSCF at 1000 °C as a function of oxygen activity from the analysis of XRD data. However, for catalytic membrane reactors, the situation regarding the structural stability of perovskite membranes becomes much complicated. Usually, one side of the membrane is exposed to air, and the other side is exposed to a reducing atmosphere. The membrane surface reduced by the reducing gases can be reoxidized by the permeated oxygen from the air side. As a result, the oxygen partial pressure around the membrane surface is higher than that of the gas bulk area. Surface exchange kinetics of the permeation side determine the oxygen partial pressure around the membrane surface area, and slow surface exchange kinetics will protect the membrane. That is why many perovskite membranes with poor stability under reducing environments can be successfully used to construct membrane reactors for the partial oxidation of methane to syngas over hundreds to thousands of hours. If other gases, such as steam, CO_2, etc., are involved in the catalytic reactions, the influence of oxygen partial pressure on the structural stability of perovskite becomes complicated.

6.4.2 (Ba,Sr)CoO_3 System

For (Ba,Sr)CoO_3-type perovskite membranes, the A sites are occupied by the divalent alkaline earth metallic ions Sr^{2+}, Ba^{2+}, or a mixture of the two cations; Ca^{2+} ions are also sometimes included. (Ba,Sr)$CoO_{3-\delta}$ oxides are hexagonal

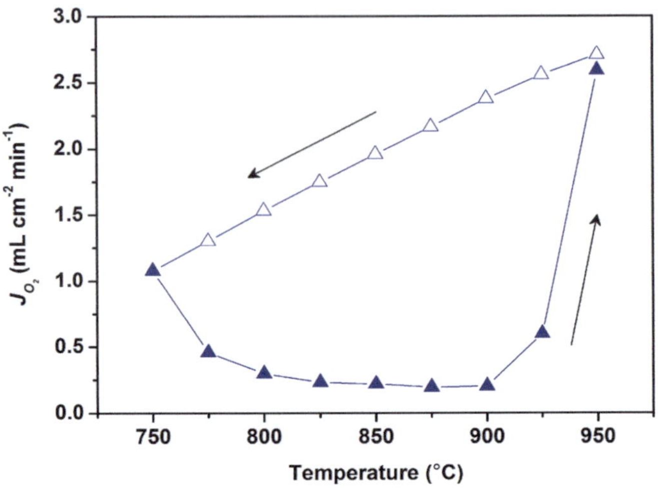

Fig. 6.4 The dependence of the oxygen permeation flux of $SrCoO_{3-\delta}$ on the temperature (Reproduced from [20] with permission of Elsevier)

whenever the ratio of the two alkaline earth metallic ions is changed in the range of $0 - \infty$ under normal air pressure and at a temperature lower than 800 °C. The ionic conductivity of the hexagonal phase is extremely low, but it increases sharply as the temperature increases higher than 850–900 °C owing to the phase transformation from hexagonal to cubic. Figure 6.4 shows the dependence of the oxygen permeation flux of $SrCoO_{3-\delta}$ on the temperature [20]. A hysteresis loop was observed during the temperature cycle experiments owing to the slow phase-change kinetics. Usually, significant volume and strength changes are induced by the phase transformation; therefore, this type of material is not suitable for any applications.

6.4.2.1 $SrCo_{1-x}M_xO_{3-\delta}$

Partial substitution of cobalt by other elements has been identified as an efficient method to stabilize the high-temperature cubic phase to room temperature. $SrCo_{0.8}Fe_{0.2}O_{3-\delta}$ (SCF) was the first membrane material among the $SrCoO_{3-\delta}$ series reported by Teraoka and coworkers in 1985 [3]. It showed the highest oxygen permeation flux in the series of $La_{1-x}Sr_xCo_{1-y}Fe_yO_{3-\delta}$ perovskite membranes. It was found that Fe in the B sites was necessary to keep the cubic structure to room temperature. However, the cubic structure of SCF perovskite was not stable; a phase transformation from the cubic perovskite structure to an orthorhombic brownmillerite structure ($A_2B_2O_5$) with ordered oxygen vacancies was observed at oxygen partial pressures lower than ~0.1 bar and temperatures lower than ~790 °C [1]. The crystalline structure of the orthorhombic brownmillerite is shown in Fig. 6.5. The structure of brownmillerite consists of alternating perovskite layers of corner-sharing BO_6 octahedra and layers of BO_4 tetrahedra, so it is usually

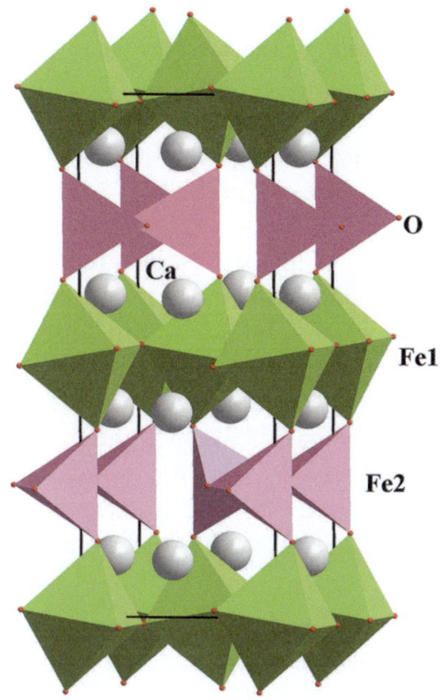

Fig. 6.5 Crystal structure of the brownmillerite phase $Ca_2Fe_2O_5$

considered as an oxygen-deficient perovskite, where the oxygen vacancies are ordered along the (010) planes. These vacancies can contribute to the transport of oxygen ions, but they only offer a one-dimensional diffusion pathway for oxygen ion migration in perovskite layers along the *c* axis. Although the diffusion in tetrahedral layers of the brownmillerite structure is also possible, its contribution to the entire transport is negligible. As a result, the phase transformation leads to the decline of oxygen permeation flux over time.

To inhibit the phase transformation, many researchers have investigated the phase structure stability and oxygen permeability of the SC-based materials with Co-site doped by various cations [21–26]. For example, Nagai et al. investigated the substitution effects of metallic elements, such as Al, Sc, Ti, V, Cr, Mn, Fe, Ni, Cu, Zn, Ga, Zr, Nb, In, Sn, and Ce, in the B sites of $SrCo_{0.9}M_{0.1}O_{3-\delta}$ oxides [21]. It was found that Ni, Cu, Zn, In, and Ce could not stabilize the $SrCoO_{3-\delta}$ oxide to a cubic perovskite structure. The samples doped by Al, Ti, V, Cr, Fe, Ga, Zr, Sn, and Nb did not contain the hexagonal phase but rather the perovskite or perovskite-related phase. Among the second group of samples, the XRD patterns of Fe-, Al-, and Ga-doped samples showed that the samples mainly contained brownmillerite-type oxides as the major phase. Detailed analysis of the samples doped by Ti, V, Cr, Zr, Nb, and Sn showed that the resultant compounds were tetragonal, with oxygen vacancy ordering rather than cubic perovskite with disordered vacancies. For example, the lattice parameters of $SrCo_{0.9}Nb_{0.1}O_{3-\delta}$ were $a = 0.3878$ nm and $c = 0.7799$ nm. However, hexagonal phases were detected after the powders were treated by pure oxygen at 850 °C for 24 h. Especially for the samples doped by

Al, V, Cr, Ga, Zr, and Sn, their strong XRD peaks were assigned to the $SrCoO_{2.52}$ hexagonal phase, and no peaks belonged to the perovskite phase. In contrast, only weak peaks assigned to hexagonal phases could be detected for Fe-, Ti-, and Nb-doped samples. Therefore, the stability order of the $SrCo_{0.9}M_{0.1}O_{3-\delta}$ oxides according to the substitution cations is given as Ni, Cu, Zn, In, Ce < Al, V, Cr, Ga, Zr, Sn < Fe, Ti, and Nb.

It is well known that the tolerance factor [Eq. (6.1)] is an important index of the stability of the perovskite structure. A cubic perovskite structure is easily formed at tolerance factors close to 1. Table 6.1 shows the calculated tolerance factors and cation radii (sixfold-coordination) of the $SrCo_{0.9}M_{0.1}O_{3-\delta}$ oxides. All the tolerance factors are in the range of 0.993–1.003, values that are all very close to 1. However, not all of the samples retained a stable cubic perovskite structure at room temperature. Nagai et al. thought that the stability of the perovskite structure in the $SrCo_{0.9}M_{0.1}O_{3-\delta}$ mixed conductors was closely related to the electrostatic repulsion between B-site cations and the larger impact of the repulsion on the stability of the hexagonal structure as compared to perovskite [21].

Table 6.1 Cations radii (r) (6-coordination) and tolerance factors (t) of the $SrCo_{0.9}M_{0.1}O_{3-\delta}$ (SCM) oxides

M^{n+}	r (Å)	t
Al^{3+}	0.54	1.003
Sc^{3+}	0.75	0.992
Ti^{4+}	0.605	0.999
V^{5+}	0.54	1.003
Cr^{4+}	0.55	1.002
Mn^{4+}	0.53	1.003
Fe^{3+}	0.645 (HS)	0.997
Co^{3+}	0.61 (HS)	0.999
Ni^{2+}	0.69	0.995
Cu^{2+}	0.73	0.993
Zn^{2+}	0.74	0.993
Ga^{3+}	0.62	0.999
Zr^{4+}	0.72	0.994
Nb^{5+}	0.64	0.998
Mo^{6+}	0.59	1.000
In^{3+}	0.8	0.990
Sn^{4+}	0.69	0.995
Sb^{5+}	0.6	1.000
Ta^{5+}	0.64	0.998
W^{6+}	0.60	1.000
Ce^{4+}	0.87	0.986

Tolerance factor was calculated using ionic radii of ions. The effect of the oxygen vacancy was ignored in the calculation
HS high spin

6.4.2.2 $SrCo_{1-x}Fe_xO_{3-\delta}$

McIntosh and coworkers investigated the phase stability and oxygen nonstoichiometry of $SrCo_{0.8}Fe_{0.2}O_{3-\delta}$ by in situ neutron diffraction, and they found that $SrCo_{0.8}Fe_{0.2}O_{3-\delta}$ adopted a cubic perovskite structure in their studied temperature range of 600–900 °C under a higher oxygen partial pressure (i.e., 1 atm) [27]. However, it became a mixture comprising a cubic perovskite phase and a vacancy-ordered brownmillerite phase as the temperature decreased below 649 °C under a medium oxygen partial pressure (i.e., 0.1 atm) and presented as a pure brownmillerite phase in the oxygen partial pressure range of 10^{-2}–5×10^{-4} atm at temperatures lower than 747 °C. The influence of the doping amount of Fe on the stability of $SrCo_{1-x}Fe_xO_{3-\delta}$ ($x = 0.2, 0.4, 0.6$) oxides was investigated by Wang and coworkers [28]. All the samples were cubic perovskite oxides with a small amount of impurities assigned to the $Sr_2Co_2O_5$ brownmillerite phase for the $x = 0.2$ sample. When the samples were treated in a pure Ar atmosphere at 900 °C, their structures changed from cubic to orthorhombic brownmillerite. In other words, the incorporation of Fe only marginally increased the structural stability of $SrCo_{1-x}Fe_xO_{3-\delta}$ in an atmosphere with a higher oxygen partial pressure (air), but it had no positive effect on the stability in an atmosphere with a lower oxygen partial pressure (Ar).

Shin et al. found that the nonstoichiometry of A-site cations had an important influence on the structural stability of $SrCo_{1-x}Fe_xO_{3-\delta}$ [29]. For $Sr_{1-y}Co_{0.5}Fe_{0.5}O_{3-\delta}$, a slight Sr deficiency ($y = 0.05$–0.1) enhanced the stability of the perovskite structure at high temperatures in an inert atmosphere. However, a heavy Sr deficiency ($y = 0.15$–0.3) degraded the stability according to the in situ XRD analysis when the sample was heated from room temperature to 1000 °C in He, as shown in Fig. 6.6. TG/DSC analysis by He and coworkers showed that in addition to a Sr deficiency, a slight Sr excess also inhibited the phase transformation in an inert atmosphere at elevated temperatures, as revealed by Fig. 6.7 [30]. No endothermic peak was detected for the $Sr_{1.1}Co_{0.8}Fe_{0.2}O_{3-\delta}$ sample, but two endothermic peaks at ~720 °C and 800 °C were recorded for the $Sr_{0.8}Co_{0.8}Fe_{0.2}O_{3-\delta}$ sample. As compared to the endothermic peak of the $SrCo_{0.8}Fe_{0.2}O_{3-\delta}$ sample, the peak at ~720 °C was assigned to the phase transformation from cubic perovskite to the vacancy-ordered orthorhombic brownmillerite structure. The much stronger endothermic peak indicated that a heavy Sr deficiency promoted the phase transformation. The second endothermic peak of the $Sr_{0.8}Co_{0.8}Fe_{0.2}O_{3-\delta}$ sample was assigned to the decomposition of $(Co,Fe)_3O_4$ spinel oxides to (Co,Fe)O. The stabilization mechanism of the slight Sr excess for the $SrCo_{0.8}Fe_{0.2}O_{3-\delta}$ sample is unclear. Because dense membranes are prepared by pressing and sintering the powder, the membranes were pulverized quickly when they were exposed to ambient air. XRD analysis showed that the impurity $Sr_3(Co,Fe)_2O_{7-\delta}$ adsorbed water quickly to form $Sr_3(Co,Fe)_2O_{7-\delta}\cdot 2H_2O$, accompanied by a great volume expansion [31].

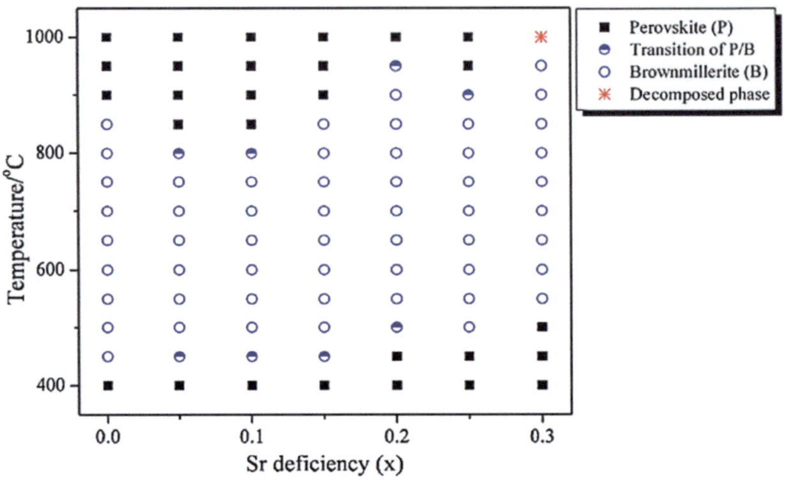

Fig. 6.6 Phase diagram of $Sr_{1-x}Fe_{0.5}Co_{0.5}O_{3-\delta}$ (x = 0–0.3) as determined by in situ XRD analysis when heated from room temperature to 1000 °C in He (Reproduced from [29] with permission of Elsevier)

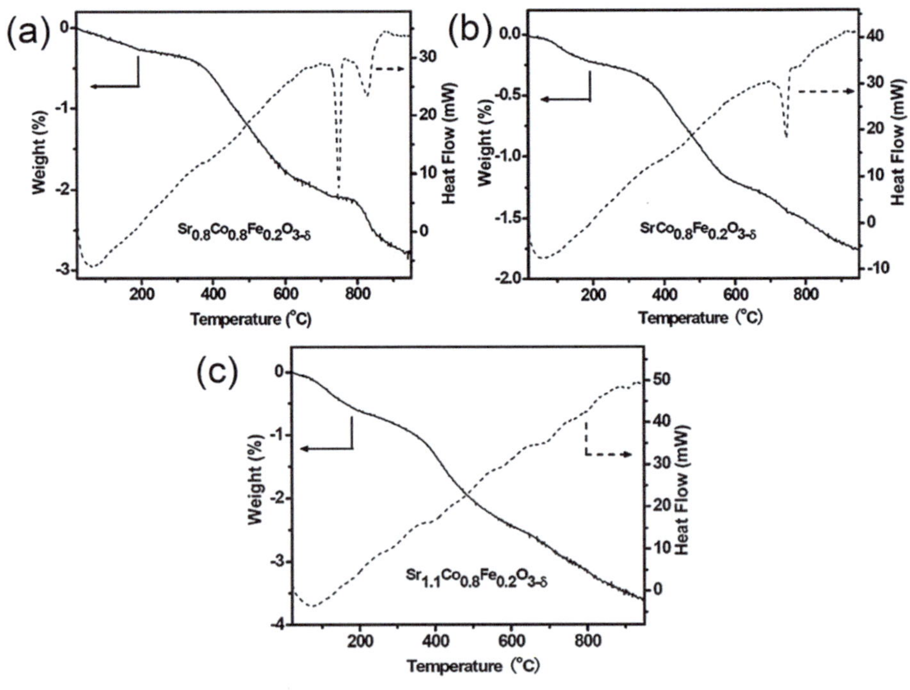

Fig. 6.7 TGA/DSC profiles of $Sr_{1+x}Co_{0.8}Fe_{0.2}O_{3-\delta}$ oxides (**a**) $x = -0.2$, (**b**) $x = 0$, (**c**) $x = 0.1$ (Reproduced from [30] with permission of Elsevier)

As $SrCo_{1-x}Fe_xO_{3-\delta}$ membranes are employed for oxygen permeation experiments, usually one side is exposed to air, and the other side is swept by inert gases. The oxygen partial pressure of the sweeping side is usually lower than 0.05 atm, which is below the stabilization limitation of the cubic phase. Therefore, the phase structure of the sweeping side surface may change to an orthorhombic brownmillerite structure with ordered oxygen vacancies. Additionally, the A-site stoichiometry has a remarkable influence on the stability of the materials. As a result, different groups have reported that oxygen permeation fluxes for $SrCo_{1-x}Fe_xO_{3-\delta}$ membranes change in the range of one order of magnitude, even under similar testing conditions, as listed in Table 6.2.

The high oxygen permeability of $SrCo_{1-x}Fe_xO_{3-\delta}$ membranes encourages researchers to improve membrane stability through substitution of A-site cations, B-site cations, or both. Ti, Zr, and Nb elements have been verified to be effective for stabilizing the cubic perovskite phase. Li et al. found that 5 mol% Zr in the B sites of $SrCo_{0.95-x}Fe_xZr_{0.05}O_{3-\delta}$ ($x = 0.1$–0.8) can inhibit the phase transformation when these powders are treated with helium at 900 °C [35]. The DSC profile of $SrCo_{0.8}Fe_{0.15}Zr_{0.05}O_{3-\delta}$ showed no endothermic or exothermic peak throughout the heating and cooling processes, which indicates that no structural transition occurs for the oxides. Similarly, Lu et al. identified a 10 mol% Nb-substituted

Table 6.2 Oxygen permeation fluxes for the $SrCo_{1-x}Fe_xO_{3-\delta}$ membranes reported by different groups

Materials	T (°C)	Thickness (mm)	Conditions	J_{O_2} (mL cm^{-2} min^{-1})	Ref.
$SrCo_{0.4}Fe_{0.6}O_{3-\delta}$	900	1	Air (0.21 atm)/He (0.001 atm)	2.4	[3]
$SrCo_{0.8}Fe_{0.2}O_{3-\delta}$	850	1	Air (0.21 atm)/He (0.001 atm)	3.1	[3]
$SrCo_{0.8}Fe_{0.2}O_{3-\delta}$	870	1	Air (0.2 atm)/He (20 mL min^{-1})	0.64	[32]
$SrCo_{0.8}Fe_{0.2}O_{3-\delta}$	870	2	Air (0.2 atm)/He (40 mL min^{-1})	0.93	[32]
$SrCo_{0.8}Fe_{0.2}O_{3-\delta}$	870	1	Air (0.2 atm)/He (0.0149 atm)	1.09	[33]
$Sr_{1.1}Co_{0.8}Fe_{0.2}O_{3-\delta}$	900	1.2	Air (100 mL min^{-1})/He (0.02 atm)	0.98	[30]
$SrCo_{0.8}Fe_{0.2}O_{3-\delta}$	900	1.2	Air (100 mL min^{-1})/He (0.02 atm)	1.54	[30]
$Sr_{0.8}Co_{0.8}Fe_{0.2}O_{3-\delta}$	900	1.2	Air (100 mL min^{-1})/He (0.02 atm)	1.47	[30]
$SrCo_{0.5}Fe_{0.5}O_{3-\delta}$	900	1	Air/He	0.78	[29]
$Sr_{0.9}Co_{0.5}Fe_{0.5}O_{3-\delta}$	900	1	Air/He	0.81	[29]
$Sr_{0.7}Co_{0.5}Fe_{0.5}O_{3-\delta}$	900	1	Air/He	0.52	[29]
$SrCo_{0.95}Fe_{0.05}O_{3-\delta}$	910	1.2–1.3	Air/He	2.18	[34]
$SrCo_{0.9}Fe_{0.1}O_{3-\delta}$	910	1.2–1.3	Air/He	1.88	[34]
$SrCo_{0.8}Fe_{0.2}O_{3-\delta}$	910	1.2–1.3	Air/He	1.48	[34]

$SrCo_{0.8}Fe_{0.1}Nb_{0.1}O_{3-\delta}$ membrane with high structural stability as a result of in situ XRD analysis (room temperature to 1100 °C) [36]. A stable oxygen permeation flux at 800 °C indicated that there was no phase transformation under the conditions of oxygen permeation testing.

6.4.2.3 $BaCo_{1-x-y}Fe_xM_yO_{3-\delta}$

Unlike $SrCo_{1-x}Fe_xO_{3-\delta}$, the $BaCo_{1-x}Fe_xO_{3-\delta}$ ($x = 0$–1) family comprises hexagonal phases under ambient oxygen partial pressure, but they transform to a cubic phase as the temperature increases beyond a certain temperature, typically 850 °C, or with decreasing oxygen partial pressure. Therefore, to avoid the phase transformation across the whole range of operating temperatures, another metal ion that can form strong metal–oxygen bonds or that has a larger ionic radius or both needs to be incorporated into the B sites of the perovskite. Up to now, Ti^{4+}, Zr^{4+}, Nb^{5+}, Ta^{5+}, Ce^{4+}, Bi^{5+}, etc., metal ions have been successfully used to stabilize $BaCo_{1-x}Fe_xO_{3-\delta}$ perovskite to a cubic structure across the whole temperature range [37–44]. Usually, a doping amount of $y = 0.1$–0.2 for these metallic ions in the B sites of perovskite is enough to stabilize the perovskite structure. Much higher doping would decrease the permeability or cause formation of a second phase. Additionally, the cobalt content has a significant influence on the solubility of the doping metallic ions, but for the same cobalt content, different metallic ions have various solid-solution limitations in $BaCo_{1-x-y}Fe_xM_yO_{3-\delta}$ perovskite. For the $BaCo_{0.8-x}Fe_xTi_{0.2}O_{3-\delta}$ system [40], single-phase cubic perovskite oxides can be synthesized in an x range of 0.3–0.5; however, if the materials are treated with high-purity nitrogen under elevated temperatures, only $BaCo_{0.5}Fe_{0.3}Ti_{0.2}O_{3-\delta}$ can retain its cubic structure, whereas its Zr- and Nb-doped counterparts have higher stability than the Ti-doped ones [37, 40]. The two ions have high solid-solution limitations in $BaCo_{1-x-y}Fe_xM_yO_{3-\delta}$ perovskite. Much more doping of the two ions is possible, but the oxygen permeation fluxes of the resultant perovskite membranes are reduced significantly. Ce^{4+} has a much bigger ionic radius than cobalt and iron ions, and so limited solubility in the perovskite lattice was reported for $y = 0.1$ [43]. However, $y = 0.1$ was found to be enough to stabilize the cubic structure in both air and inert atmospheres. The doping effects of Ta^{5+} in perovskite are similar to those of Nb^{5+} because they are in the same group and have the same valence state and similar ionic radii [38, 44]. The stabilization of the $BaCo_{1-x}Fe_xO_{3-\delta}$ perovskite to a cubic structure by Zr^{4+}, Nb^{5+}, Ta^{5+}, and Ce^{4+} is realized based on two kinds of effects: one increases the average metal–oxygen bond energy, and the other increases the average B-site ionic radii to obtain a tolerance factor close to 1. However, for the $BaCo_{0.8-x}Fe_xTi_{0.2}O_{3-\delta}$ system, Ti^{4+} has an ionic radius similar to those of cobalt and iron ions, but it can form stronger metal–oxygen bonds. As a result, its stability is lower than the $BaCo_{1-x-y}Fe_xM_yO_{3-\delta}$ (M = Zr, Nb, Ta, or Ce) counterparts. For the Bi^{5+}-doped $BaCo_{1-x-y}Fe_xBi_yO_{3-\delta}$ materials, when the Bi-doping amount is $y = 0.2$, the maximum cobalt content must be 0.4 to retain a pure cubic phase, whereas that of iron can be as high as 0.8 [41]. As the cobalt

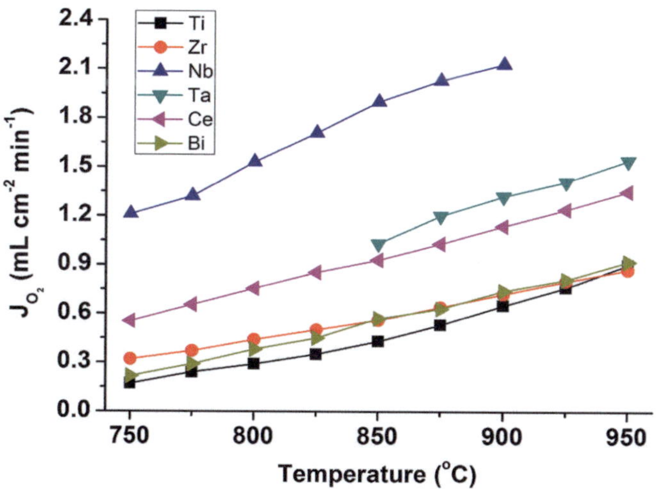

Fig. 6.8 The dependence of the oxygen permeation flux on the temperature for the optimized compositions for Ti^{4+}-, Zr^{4+}-, Nb^{5+}-, Ta^{5+}-, Ce^{4+}-, and Bi^{5+}-doped $BaCo_{1-x-y}Fe_xM_yO_{3-\delta}$ perovskite membranes. $L = 1.0$ mm

content is fixed at 0.2, in the investigated Bi-doping range of $y = 0.1$–0.5, the perovskite can maintain the cubic structure. It is interesting to find that Bi-doped perovskite oxides have high structural stability in both air and inert atmospheres. Unlike the abovementioned metallic ion-doped perovskite membranes, the oxygen permeability is only slight influenced by the doping amount for $BaCo_{0.2}Fe_{0.8-y}Bi_yO_{3-\delta}$ ($y = 0.1$–0.5). One possible reason is that the Bi–O bonds are weak as compared to the Fe–O bonds, and so the increase of the Bi-doping amount would decrease the oxygen vacancies in the materials; however, it decreases the average metal–oxygen bond energy facilitating the diffusion of oxygen ions.

Figure 6.8 compares the oxygen permeation flux of the optimized compositions for Ti^{4+}-, Zr^{4+}-, Nb^{5+}-, Ta^{5+}-, Ce^{4+}-, and Bi^{5+}-doped $BaCo_{1-x-y}Fe_xM_yO_{3-\delta}$ perovskite membranes. For membranes with a thickness of 1.0 mm operated at 900 °C under an air/inert gradient, $BaCo_{0.7}Fe_{0.2}Nb_{0.1}O_{3-\delta}$, which had the highest cobalt content, also showed the highest oxygen permeation flux, followed by $BaCo_{0.7}Fe_{0.2}Ta_{0.1}O_{3-\delta} > BaCo_{0.4}Fe_{0.5}Ce_{0.1}O_{3-\delta} > BaCo_{0.4}Fe_{0.4}Zr_{0.2}O_{3-\delta} \approx BaCo_{0.2}Fe_{0.5}Bi_{0.3}O_{3-\delta} > BaCo_{0.5}Fe_{0.3}Ti_{0.2}O_{3-\delta}$. The high oxygen permeation flux of the $BaCo_{0.7}Fe_{0.2}Nb_{0.1}O_{3-\delta}$ membrane makes it useful for the partial oxidation of methane to syngas [45]. The $BaCo_{0.4}Fe_{0.4}Zr_{0.2}O_{3-\delta}$ membrane has shown higher stability than other materials for methane conversion to syngas [46]. The operation lasts up to 2200 h, which is the longest experiment for the conversion of methane to syngas in a cobalt-based perovskite membrane reactor. All in all, $BaCo_{1-x-y}Fe_xM_yO_{3-\delta}$ perovskite membranes have good oxygen permeability and considerable stability for syngas production.

6.4.3 $Ba_{0.5}Sr_{0.5}Co_{0.8}Fe_{0.2}O_{3-\delta}$

Among all the perovskite oxides, $Ba_{0.5}Sr_{0.5}Co_{0.8}Fe_{0.2}O_{3-\delta}$ (BSCF) is the most famous, not only in terms of oxygen permeation but also for oxygen activation as a cathode of solid oxide fuel cells. BSCF was first reported by Shao and his coworkers in 2000 as an oxygen permeable membrane material [2]. The introduction of barium into $SrCo_{0.8}Fe_{0.2}O_{3-\delta}$ stabilizes the cubic perovskite structure under lower oxygen partial pressures, and it improves the oxygen permeation flux as compared to $SrCo_{0.8}Fe_{0.2}O_{3-\delta}$ under an air/He gradient. After a series of experiments, Shao et al. found that $Ba_{0.5}Sr_{0.5}Co_{0.8}Fe_{0.2}O_{3-\delta}$ was the optimum composition in the $Ba_{1-x}Sr_xCo_{1-y}Fe_yO_{3-\delta}$ system in terms of permeability and stability [41]. Figure 6.9 shows the long-term operation of BSCF membranes at 850, 825, and 750 °C [2]. The membrane shows stable oxygen permeation during the 1045 h on-stream as long as the operation temperature is higher than 850 °C. However, degradation of the permeation flux cannot be avoided if the membrane is operated below 850 °C. In the pioneer research of Shao et al., they found that the degradation had no relationship with CO_2 poisoning and the enrichment of Ba and Sr on the membrane surfaces; rather, it was a direct result of the phase transformation of the material at intermediate temperatures. They found that BSCF transformed to a hexagonal phase and a new perovskite phase as the powder was treated in air at 750 °C [41]. The reduced oxygen permeation flux at intermediate temperatures can return quickly to its initial value as long as the operating temperature is higher than 850 °C. The following sections will show the details on the oxygen exchange kinetics, bulk diffusion kinetics, oxygen permeation, and phase stability of BSCF.

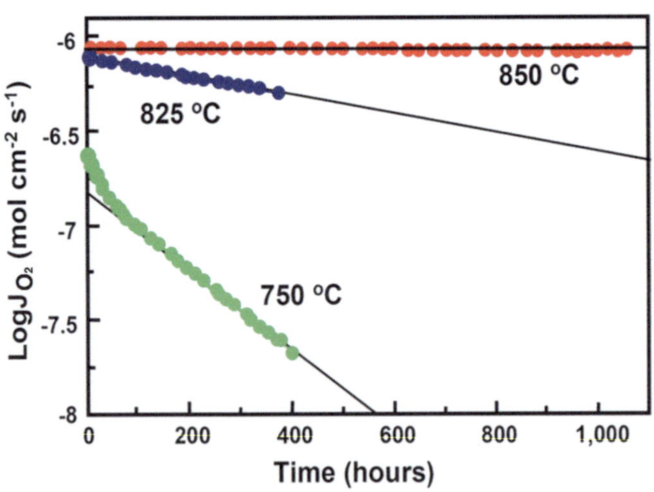

Fig. 6.9 The long-term operation of the BSCF membranes at 850, 825, and 750 °C with the thicknesses of 1.46, 1.40, and 1.76 mm, respectively (Reproduced from [2] with permission of Elsevier)

6.4.3.1 Oxygen Exchange and Diffusion Kinetics

The high oxygen permeability of BSCF membranes indicates that the material has fast oxygen exchange kinetics at the gas–solid interfaces and large oxygen ionic diffusion coefficients. The possible oxygen permeation-limiting steps are oxygen reduction reactions on the gas–solid interface of the feed side (usually air is used as the feed gas), bulk phase oxygen ionic diffusion, and oxygen evolution reactions on the gas–solid interface of the permeation side (usually an inert gas is used to sweep the permeated oxygen in laboratories) for MIEC membranes. Therefore, only materials with both high oxygen interfacial exchange and high bulk diffusion kinetics can show high oxygen permeability. Additionally, the BSCF material is suitable for use as cathodes of solid oxide fuel cells (SOFCs) to catalyze the oxygen reduction reactions and the subsequent transfer of oxygen ions from the cathode to the electrolyte. Shao et al. applied the BSCF material to ceria-based SOFCs and achieved a much higher peak power density of 1010 mW cm^{-2} and an extremely low polarization resistance at 600 °C [47]. Since then, BSCF has been widely and deeply investigated by researchers in the field of fuel cells and MIEC membranes all over the world. To understand the high performance of the BSCF material, its oxygen interfacial exchange and bulk diffusion kinetics are two important keys.

The oxygen exchange reactions that happen on the gas–solid interface are complex and comprise a series of elemental steps, such as diffusion, adsorption, dissociation, ionization, and, finally, incorporation of oxygen into the crystal lattice of the materials. Many techniques can be used to either directly or indirectly determine the interfacial exchange and bulk diffusion kinetics of MIEC materials, such as oxygen permeation experiments combined with modeling, oxygen isotope exchange depth profiling (IEDP) combined with secondary ion mass spectrometry (SIMS), pulse isotope exchange combined with mass spectrometry analysis, coulometric titration, electrical conductivity or mass relaxation techniques, and electrochemical impedances. These techniques were introduced in detail in Chap. 4.

Bucher et al. investigated the oxygen nonstoichiometry and the interfacial exchange and bulk diffusion kinetic parameters of BSCF by thermogravimetric and electrical conductivity relaxation techniques [48]. The nonstoichiometric oxygen of BSCF is 0.43–0.6 in the temperature range of 600–900 °C and an oxygen partial pressure range of 1×10^{-4}–0.4 bar. The thermodynamic factors of oxygen were obtained at different temperatures by thermogravimetric techniques. The chemical diffusion coefficients (D_{chem}) and interfacial exchange coefficients (k_{chem}) were in the range of 1×10^{-6}–3×10^{-5} $cm^2\ s^{-1}$ and 2×10^{-4}–3×10^{-3} cm s^{-1} in the temperature range of 550–725 °C. Their activation energies were 86 ± 8 and 64 ± 12 kJ mol^{-1}, respectively. The self-diffusion and interfacial exchange coefficients were calculated by dividing the chemical diffusion and interfacial exchange coefficients, respectively, by the thermodynamic factors. They derived the ionic conductivity of BSCF at 600 and 700 °C; however, the values were obviously lower than those obtained through the oxygen permeation method [49, 50]. Similarly, Chen et al. investigated the oxygen exchange and transport properties of BSCF by using the electrical conductivity relaxation technique [51]. They reported much larger activation energies for D_{chem} (111 ± 5 kJ mol^{-1}) and k_{chem} (110 ± 6 kJ mol^{-1}) in the temperature range of 600–800 °C but ionic conductivities comparable to those derived from the oxygen permeation testing.

Girdauskaite et al. reported an investigation of the oxygen exchange and transport properties of BSCF by the oxygen permeation and electrical conductivity relaxation techniques [52]. Much higher activation energies were found for the kinetic parameters obtained through the electrical conductivity relaxation technique than for those acquired from oxygen permeation. They reported that a comparison of kinetic parameters and activation energies obtained from the two methods was difficult owing to highly different gradients of the oxygen chemical potential in the two methods. D_{chem} and k_{chem} are determined from conductivity relaxation experiments under nonsteady-state oxygen chemical potential in MIEC materials; in contrast, the kinetic parameters acquired under steady-state oxygen permeation are more useful for modeling or simulating the permeation behaviors of membranes and membrane modules.

An oxygen permeation model that included three permeation resistances (r', r^b, r'') was developed by Zhu et al. [50]. The three permeation resistances corresponded to the interfacial oxygen exchange coefficient (k') on the feed side, the bulk diffusion coefficient (D_{ambi}, ambipolar diffusion coefficient) in the bulk membrane, and the interfacial oxygen exchange coefficient (k'') on the permeation side, respectively. Their corrections can be described by the following equations:

$$k' = \frac{RT}{4F^2 C'_{O^{2-}}} \frac{1}{r'} \tag{6.12}$$

$$D_{ambi} = \frac{RT}{4F^2 C^b_{O^{2-}}} \frac{L}{r^b} \tag{6.13}$$

$$k'' = \frac{RT}{4F^2 C''_{O^{2-}}} \frac{1}{r''} \tag{6.14}$$

where $C'_{O^{2-}}$, $C^b_{O^{2-}}$, and $C''_{O^{2-}}$ are the oxygen ion concentrations at the gas–solid interface of the feed side, the bulk zone, and the gas–solid interface of the permeation side, respectively. R, T, and F have their usual meanings. The self-diffusion coefficient (D_s) of the oxygen ions is approximate to the ambipolar diffusion coefficient if the electronic conductivity is much larger than the oxygen ionic conductivity. For BSCF, its electronic conductivity was higher than 30 S cm^{-1} in its working temperature range, so $r^b \approx r^b_{O^{2-}}$. By using Zhu's permeation model (see Chap. 5), the three kinetic parameters were obtained through regressing the permeation experimental data. Figure 6.10 shows the activation energies of the interfacial exchange coefficients (k' and k'') and the bulk diffusion coefficient ($D_{ambi} \approx D_s$) in the temperature range of 680–940 °C. As shown in the figure, the two interfacial

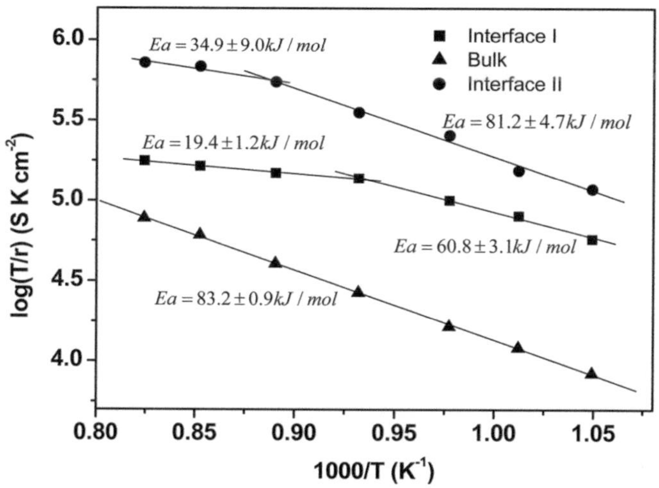

Fig. 6.10 Activation energies of interfacial resistance constant r'_0, r''_0, and bulk-resistance constant r^b under oxygen partial pressure gradient of 0.21 bar/0.02 bar (Reproduced from [50] with permission of Wiley)

exchange coefficients had lower activation energies at high temperatures and higher values at low temperatures. The change of the activation energies indicates that the r.d.s. varied as the temperature varied in the investigated temperature range. However, the activation energy of the bulk diffusion coefficient (83.2 ± 0.9 kJ mol^{-1}) did not change over the wide temperature range. The change of apparent activation energy for oxygen permeation has been found for BSCF membranes by many researchers. Typically, the transition temperature is in the range of 750–800 °C. For example, Shao et al. found that the permeation activation energy was 40.9 kJ mol^{-1} in the temperature range of 775–950 °C and 72.6 kJ mol^{-1} in the temperature range of 600–775 °C [2]. The transition temperature of 750–800 °C coincided with the bulk phase transformation temperature of the materials (a detailed discussion is given in Sect. 6.4.3.3); therefore, some researchers thought that the change of apparent activation energy at approximately 750–800 °C was a reflection of the phase transformation. However, the results shown in Fig. 6.10 reveal that the change of apparent activation energy is due to the r.d.s shifting from predominantly interfacial exchange to predominantly bulk diffusion. Investigation of the phase transformation of BSCF showed that the transformation kinetics were slow and took hundreds of hours. However, it took just tens of minutes to accomplish the permeation testing. Through a series of experiments, Hong et al. concluded that the activation energy for bulk diffusion was greater than that for interfacial exchange at high temperatures [53]. Figure 6.11 compares the interfacial exchange and bulk diffusion kinetic parameters of some MIEC perovskite materials [51, 54–58] with those of BSCF. Clearly, BSCF shows both higher interfacial exchange and bulk diffusion coefficients across a wide temperature range. It has thus gradually

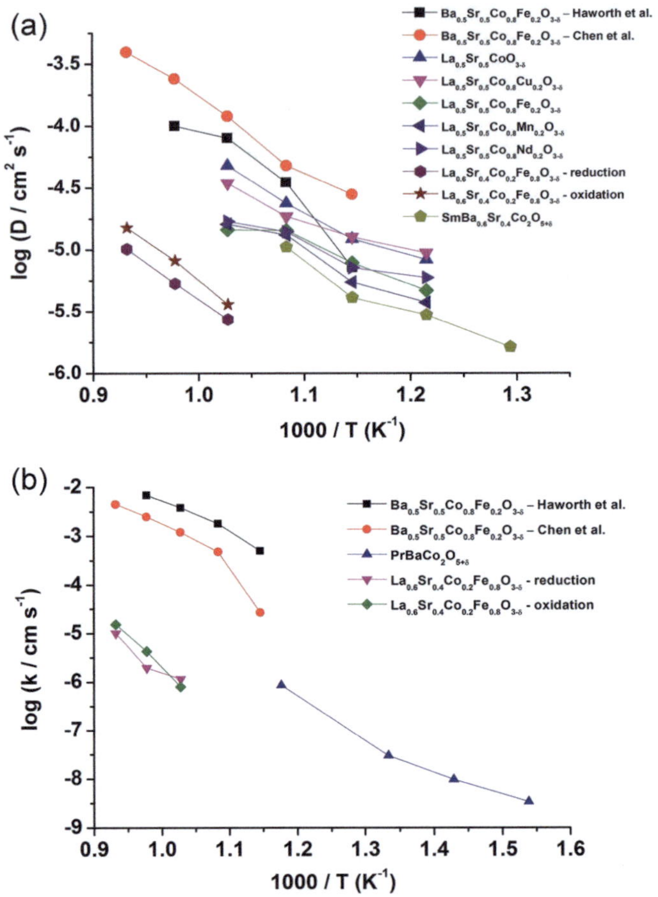

Fig. 6.11 The (**a**) bulk diffusion and (**b**) interfacial exchange kinetic parameters of some MIEC perovskite materials. Oxygen partial pressure change: $Ba_{0.5}Sr_{0.5}Co_{0.8}Fe_{0.2}O_{3-\delta}$, Haworth et al. [54], 0.21–1 atm; $Ba_{0.5}Sr_{0.5}Co_{0.8}Fe_{0.2}O_{3-\delta}$, Chen et al. [51], 0.21–0.1 atm; $La_{0.5}Sr_{0.5}CoO_{3-\delta}$ and $La_{0.5}Sr_{0.5}Co_{0.8}M_{0.2}O_{3-\delta}$ (M = Mn, Fe, Ni, Cu) [55], 0.21–0.05 atm; $La_{0.5}Sr_{0.5}Co_{0.2}Fe_{0.8}O_{3-\delta}$, reduction and oxidation [56], 0.50–0.35 atm and 0.35–0.50 atm, respectively; $SmBa_{0.6}Sr_{0.4}Co_2O_{5+\delta}$ [57], 0.21–0.05 atm; $PrBaCo_2O_{5+\delta}$ [58], 0.02–0.1 atm

become the benchmark in the field of the oxygen permeation membranes and SOFC cathodes.

6.4.3.2 Oxygen Permeation

BSCF has been fabricated into disks, tubes, hollow fibers, and asymmetric plates. Although BSCF has the fastest interfacial exchange kinetics, researchers have found that its permeation flux is still limited by interfacial exchange steps as the membrane is thinner than 1 mm. Permeation model investigations have shown that the characteristic thickness of a BSCF membrane is approximately 0.7 mm at 900 °C and under an oxygen partial pressure gradient of 21/2 kPa [50]. It is impossible to exceed a permeation flux of 10 mL cm^{-2} min^{-1} under the above operating conditions by decreasing the membrane thickness. Zhu et al. prepared a capillary BSCF membrane with OD = 2.38 mm and ID = 1.40 mm and operated it under high-pressure air and vacuum conditions. The relationship between the oxygen recovery and oxygen permeation flux was especially addressed, as shown in Fig. 6.12 [59]. The oxygen permeation flux and oxygen recovery both increased with increasing feed pressure; however, it gradually leveled off. The increase of oxygen permeation flux is attributed to the improved driving force across the membrane. The oxygen permeation flux reaches 6.5 and 9.5 mL cm^{-2} min^{-1} under an air pressure of 7 atm and oxygen recoveries of 70 % and 48 %, respectively.

Leo et al. reported the fabrication of a BSCF hollow fiber membrane with a thickness of ~200 μm, which provided an oxygen permeation flux of 7 mL cm^{-2} min^{-1} at 900 °C and an oxygen partial pressure gradient of 21/2.2 kPa [60]. A flux higher than 10 mL cm^{-2} min^{-1} can only be achieved by depositing catalysts such as Pd particles on both surfaces of the hollow fiber membranes to catalyze the oxygen exchange reactions at the gas–solid interfaces. Baumann et al. reported an asymmetric BSCF membrane with a 70-μm-thick dense layer and an 830-μm-thick porous support layer prepared by using a tape-casting technique, as shown in Fig. 6.13 [61]. An oxygen permeation flux of approximately 5 mL cm^{-2} min^{-1} was achieved at 900 °C and under an oxygen partial pressure gradient of air/Ar. The oxygen permeation flux was improved to 10 mL cm^{-2} min^{-1} by coating a 17-μm-thick porous BSCF layer on the dense layer. When pure oxygen was used as the

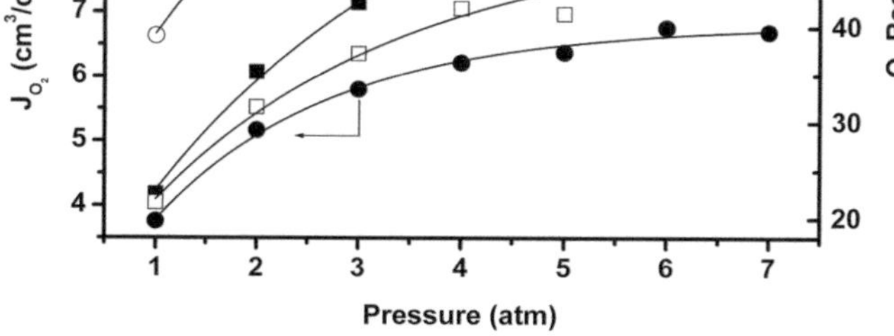

Fig. 6.12 Dependence of oxygen permeation flux and oxygen recovery on temperatures. *Solid symbols*: 3 atm; *open symbols*: 1 atm. Airflow rate: 200 mL/min for *square* and 100 mL/min for *circle*. Length: 4.0 cm, $S = 2.3$ cm^2 (Reproduced from [59] with permission of Elsevier)

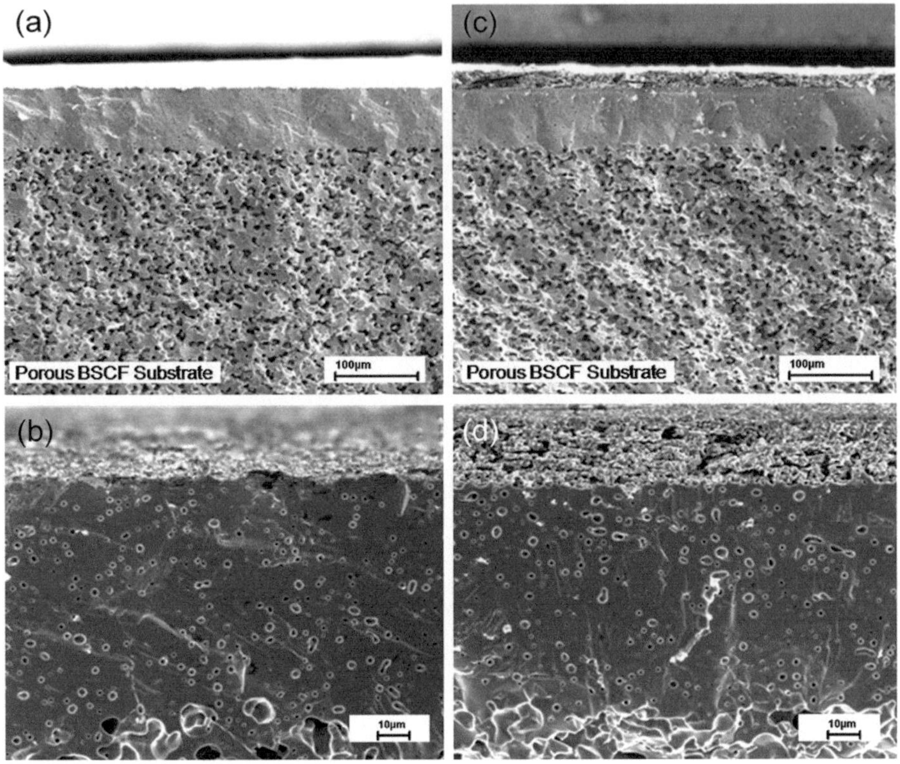

Fig. 6.13 Fracture cross sections (SEM pictures) of two BSCF membranes after oxygen permeation measurements: (**a** and **b**) bare and (**c** and **d**) with an oxygen activation layer (Reproduced from [61] with permission of Elsevier)

feed gas, an oxygen permeation flux higher than 50 mL cm^{-2} min^{-1} was achieved at 900 °C and under an oxygen partial pressure gradient of 21/2.7 kPa, as shown in Fig. 6.14. This amazing flux value was higher than that of many of zeolite membranes used in gas separation.

For asymmetric membranes, concentration polarization cannot be ignored, especially for membranes exhibiting high permeation fluxes. In Baumann's experiments, air or oxygen was fed to the porous support layer. The effect of concentration polarization induced by the porous support can be excluded by using pure oxygen as the feed gas. A five-time increase in oxygen permeation flux was observed after the feed gas was switched from air to pure oxygen, as shown in Fig. 6.14. Correspondingly, the oxygen partial pressure gradient across the BSCF asymmetric membrane changed from 21/2.7 to 101/13.8 kPa. However, the driving force term ($\ln \frac{P^{I}_{O_2}}{P^{II}_{O_2}}$) in the Wagner equation slightly decreased from 2.05 to 1.99. Based on this experiment, Baumann et al. thought that the concentration polarization of the support was very severe. Strictly speaking, the interfacial exchange

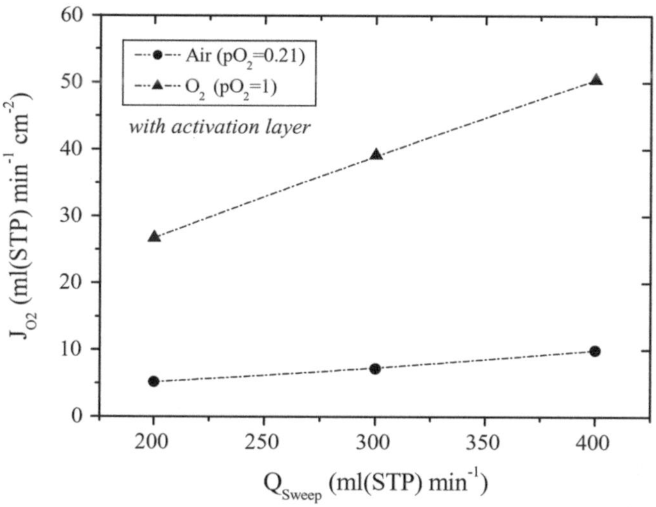

Fig. 6.14 Effect of the porous support. BSCF supported thin-film membrane with an O_2 activation layer. $T = 900$ °C; $Q_{\mathrm{Feed}} = 300$ mL (STP) min^{-1} (Reproduced from [61] with permission of Elsevier)

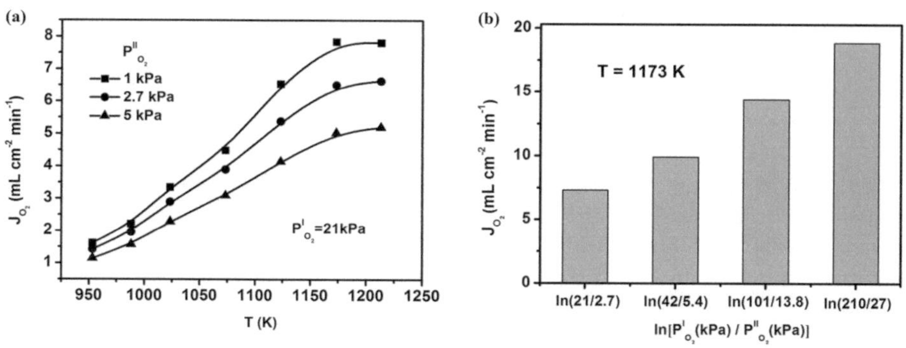

Fig. 6.15 Predicted oxygen permeation fluxes based on the kinetic parameters acquired by Zhu's model for a 70-μm-thick BSCF membrane without porous support and catalyst layer for oxygen activation

resistances decreased when the feed gas was switched from air to pure oxygen because they are functions of oxygen partial pressure. The decrease of interfacial exchange resistances leads to a reduction of the total permeation resistance across the membrane. The predicted oxygen permeation fluxes based on the kinetic parameters acquired by Zhu's model are plotted in Fig. 6.15 for a 70-μm-thick BSCF membrane without a porous support for oxygen activation. Under an oxygen partial pressure gradient of 21/2.7 kPa, its oxygen permeation flux is ~6 mL cm^{-2} min^{-1}, and it increases to ~12 mL cm^{-2} min^{-1} when the oxygen partial pressure gradient is changed from 21/2.7 to 101/13.8 kPa. Therefore, for Baumann's membrane, the significant increase in oxygen permeation flux after the feed gas was shifted from air to pure oxygen is ascribed to the concentration polarization effect and the decrease of interfacial exchange resistances. If the interfacial exchange kinetics are fast enough and can be ignored as compared to the bulk diffusion kinetics, an oxygen permeation flux of ~65 mL cm^{-2} min^{-1} could be achieved on the 70-μm-thick BSCF membrane at 900 °C and under an oxygen partial pressure gradient of 101/13.8 kPa. The BSCF asymmetric membrane with BSCF catalyst layer gives a flux of ~50 mL cm^{-2} min^{-1} under the same conditions [61]. These results indicate that the interfacial exchange resistance of the membrane is extremely small as compared to its bulk resistance, and thus the BSCF material is a good catalyst for oxygen activation.

6.4.3.3 Phase Transformation

Recently, many researchers have focused on the degradation mechanism of BSCF membranes at intermediate temperatures. Here, some important findings are introduced to show the complexity of the phase transformation of BSCF. Švarcová et al. found that extended annealing of the sintered powders for 10 days at 750 °C resulted in the partial conversion of the cubic parent phase to a 2H-type hexagonal perovskite structure [62]. They suggested that the formation of the hexagonal phase occurs via a nucleation and growth mechanism and that it is inhibited by slow diffusion of the Ba or Sr ions. They also reported that the transformation of BSCF from a cubic to a hexagonal polymorph can be rationalized by the Goldschmidt tolerance factor. Additionally, they suggested that the transformation can be suppressed by appropriate substitutions in the A sites, the B sites, or both. However, Arnold et al. suggested that the real driving force to form the hexagonal perovskite structure is the oxidation of Co^{2+} to Co^{3+}, which is accompanied by a change in the preferred spin state ($r_{Co^{2+}} = 0.0745$ nm (HS), $r_{Co^{3+}} = 0.0545$ nm (*LS*) as compared to $r_{Fe^{2+}} = 0.0780$ nm (*HS*), $r_{Fe^{3+}} = 0.0645$ nm (*HS*)) [63]. Under the driving force, the formation of BSCF, as well as its decomposition, is a process of reversible reordering of the hexagonal AO_3-layer stacking sequence in the cubic perovskite, which happens if the cubic BSCF is kept at temperatures below 900 °C for a long period, thereby leading to the decomposition of BSCF into a three-phase mixture, i.e., a mixture containing cubic, hexagonal, and monoclinic phases. The monoclinic phase was regarded as an intermediate step toward the formation of the hexagonal phase.

By combining XRD and TEM, Mueller et al. investigated the phase transformation of the BSCF powder and sintered ceramic at 800 °C [64]. The dense BSCF ceramic annealed for 8 days at 800 °C showed a grain boundary phase and a lamellar phase. The grain boundary phase was described as 2H-$Ba_{0.5-x}Sr_{0.5-x}CoO_{3-\delta}$ based on energy dispersive X-ray spectroscopy (EDXS) and selected area electron diffraction (SAED) along the *c* axis; there was no further investigation of the lamellar phase. The phase transformation kinetics of the BSCF cubic phase to the hexagonal phase were investigated by a modified Johnson–Mehl–Avrami (JMA) model.

$$\alpha_{\text{hex}}(t) = \alpha_{\text{hex}}(t = \infty)\left(1 - e^{-kt^n}\right) \tag{6.15}$$

where $\alpha_{\text{hex}}(t)$ and $\alpha_{\text{hex}}(t = \infty)$ are the volume fractions of the hexagonal phase at an annealing time t and an equilibrium state, respectively, and k and n are the rate constants of formation and the Avrami exponent, respectively. Table 6.3 lists the fitting parameters obtained from the curves shown in Fig. 6.16. It was found that the Ba/Sr ratio and the forms (powder vs. sintered ceramic) had no influence on the volume phase fraction of the hexagonal phase in an equilibrium state. However, the Avrami exponents for powder samples were close to 0.5; for the sintered ceramic, the Avrami exponent was close to 1. A value of $n = 0.5$ is usually regarded as being related to nucleation and one-dimensional growth. For this mechanism, the nucleation is very fast, but the growth of the phase is limited by the slow diffusion of the involved ions. The results suggest that all nucleation sites are occupied with nuclei before the growth of the phase begins. However, because the Avrami exponent is

Table 6.3 Fitting parameters of the Johnson–Mehl–Avrami (JMA) analysis of the kinetic data on the formation of the hexagonal phase in air at $T = 1073$ K

Composition	Form	$\alpha_{\text{hex}}(t=\infty)$	k/d^{-n}	n
BSCF5582	Powder	0.53 ± 0.03	0.248 ± 0.0018	0.58 ± 0.05
BSCF5582	Ceramic	0.53 ± 0.05	0.078 ± 0.019	0.86 ± 0.11
BSCF1982	Powder	0.49 ± 0.13	0.229 ± 0.0058	0.43 ± 0.07

Reproduced from [59] with permission of Elsevier

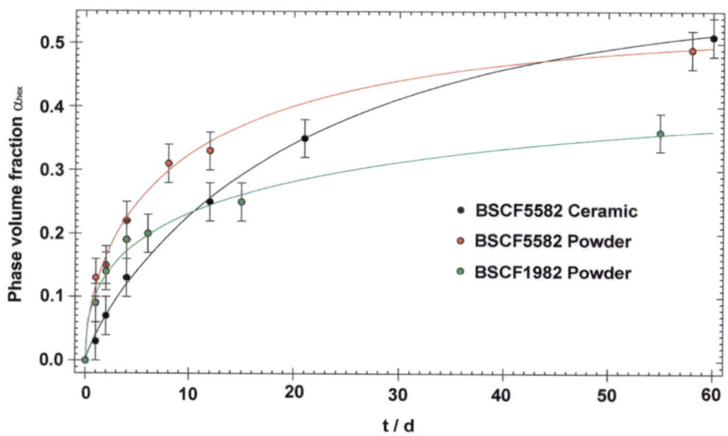

Fig. 6.16 Volume fraction of the hexagonal phase as a function of time for the three different samples, BSCF5582 powder (*red*), BSCF5582 ceramic (*green*), and BSCF5582 dense ceramics (*black*). *Lines* are fits to the JMA model (Reproduced from [64] with permission of the Royal Society of Chemistry)

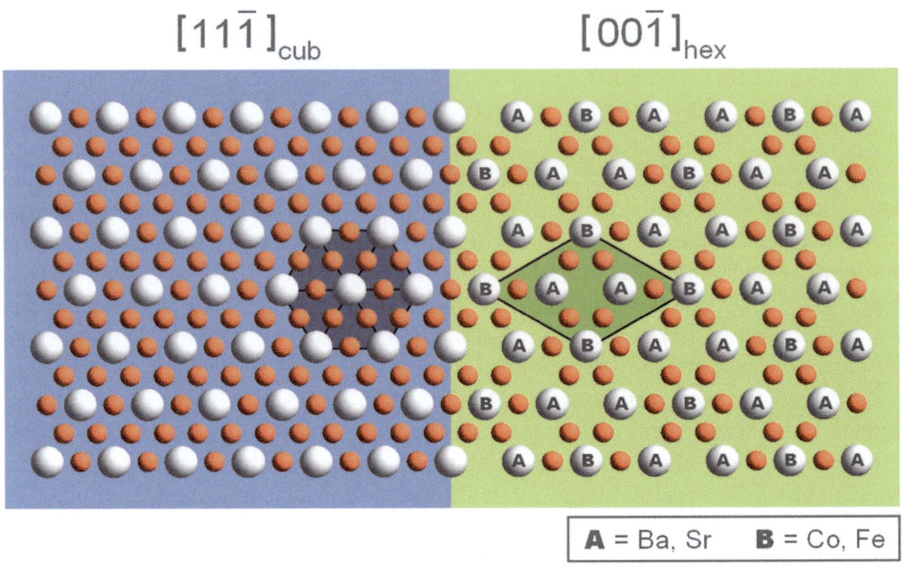

Fig. 6.17 Representation of the orientation of the hexagonal (*right*) and cubic lattice (*left*) at the contact plane determined from the SAED patterns. The unit cells are *highlighted* (Reproduced from [64] with permission of the Royal Society of Chemistry)

equal to 1, nucleation occurs at the grain boundaries, with fast saturation of the nucleation sites. The difference in the nucleation behavior of the powders and ceramics is because dense ceramics do not offer as many nucleation sites as powders. Additionally, Mueller et al. found a certain orientation relationship between the cubic phase and the hexagonal phase at the interface, i.e., $\left[11\bar{1}\right]_{\mathrm{cub}}$ $\left[00\bar{1}\right]_{\mathrm{hex}}$ and $\left\{\bar{1}21\right\}\left\{\bar{1}10\right\}_{\mathrm{hex}}$, which indicates that the lattice planes also represent the contact planes of the interface, as shown in Fig. 6.17. Therefore, it can be expected that the contact planes between the cubic and the hexagonal grains are the {111} planes of the cubic polymorph and the {001} planes of the hexagonal polymorph.

Aside from hexagonal 2H-$Ba_{0.5-x}Sr_{0.5-x}CoO_{3-\delta}$ perovskite, the formation of lamellar noncubic phases was observed in both the powder and the sintered ceramic samples by Efimov et al. [65]. The lamellar structure with a composition close to $Ba_{1-x}Sr_xCo_{2-y}Fe_yO_{5-\delta}$ was found to be related to the 15R hexagonal perovskite polytype. They found that the growth of lamellae often occurred in various directions in the cubic perovskite grain, as shown in Fig. 6.18a. It is interesting to note that the lamella marked by a dashed line in Fig. 6.18a is located at an angle of approximately 25°. The angle between the (100) and $\left(3\bar{1}1\right)$ cubic planes coincidentally equals 25.24°, and so the growth direction of the lamella is parallel to the $\left(3\bar{1}1\right)$ cubic plane. Similarly, the growth direction of the lamella marked by a dashed–dotted line is parallel to the $\left(2\bar{1}1\right)$ cubic direction, and the angle between the lamellae is approximately 35°, whereas the angle between the (100) and $\left(2\bar{1}1\right)$

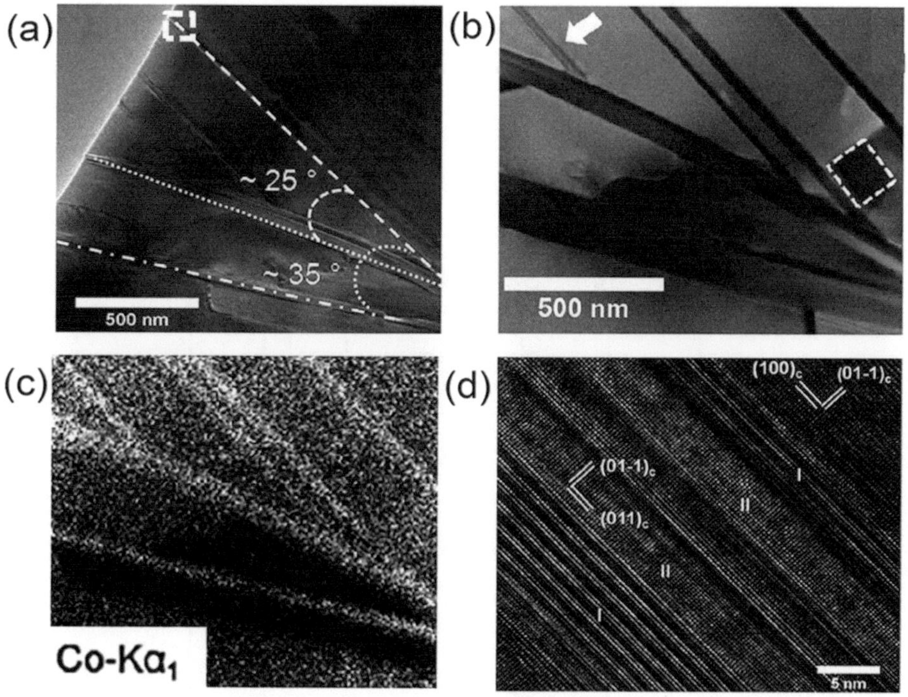

Fig. 6.18 (**a**) TEM bright-field micrograph of the annealed BSCF ceramic sample showing lamellae grown along different directions. (**b**) STEM annular dark-field micrograph showing platelike phases adjacent to lamellae. *Arrow* marks the lamella shown in Fig. 6.18(**b**). *Rectangle* marks the area of the $Ba_{0.6}Sr_{0.4}CoO_{3-\delta}$ composition. (**c**) Cobalt distribution by EDX spectroscopy using the Co $K\alpha_1$ line. (**d**) HRTEM micrograph showing the presence of cubic phase inside the lamella (Reproduced from [65] with permission of the American Chemical Society)

cubic directions is 35.26°. Around the cobalt-enriched lamellae, platelike phases adjacent to the lamellae, shown in Fig. 6.18b, were found. The cobalt elemental distribution and the EDXS results in Fig. 6.18c show that the platelike phases were a mixed barium/strontium oxide, with a stoichiometry of $Ba_{0.4}Sr_{0.6}O$. Based on the above findings, a summary of the phase transformation of BSCF is illustrated in Fig. 6.19. The products include a cobalt-enriched hexagonal perovskite, $Ba_{0.6}Sr_{0.4}CoO_{3-\delta}$, a 15R-related $Ba_{1-x}Sr_xCo_{2-y}Fe_yO_{5-\delta}$ complex oxide, and a mixed barium/strontium oxide [65].

To further understand the phase transformation of BSCF ceramics, Müller et al. investigated the influence of temperature (700–1000 °C) and annealing time (10–1350 h) across large ranges [66]. They found that the fraction of the hexagonal phase increased slowly with increasing annealing time, reaching 10 vol% (normalized with respect to a dense sample) after the sample was annealed at 800 °C for 1350 h; however, the hexagonal phase was not observed at 700 or 900 °C annealing. Unlike the hexagonal phase, platelike regions were observed in the temperature

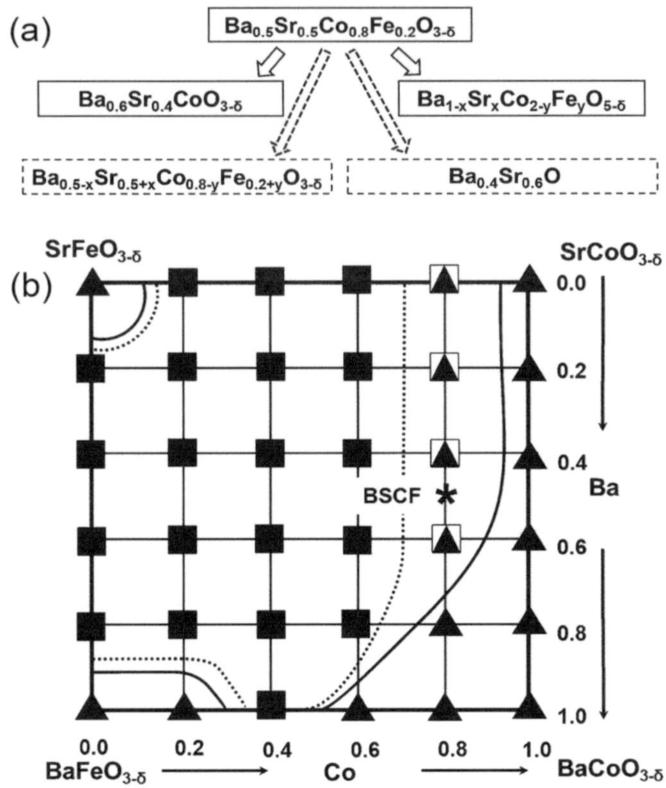

Fig. 6.19 (**a**) Schematic representation of the BSCF decomposition process. (**b**) Schematic phase diagram of the quasi-quaternary Ba–Sr–Co–Fe oxide system. The cubic phases are marked with *square symbols*; *triangle symbols* relate to the multiphase compounds. The *borderlines* each of which is associated with miscibility gap are drawn tentatively as *solid lines* at 1273 K and as dashed lines at temperatures below 1073 K (Reproduced from [65] with permission of the American Chemical Society)

range of 700–900 °C. The fraction of the platelike regions, which was maintained at 3 ± 1.5 vol% throughout the investigated annealing period at 800 °C, was not significantly influenced by the annealing time (from 10 to 1350 h), as shown in Fig. 6.20a. However, the annealing temperature had a significant influence on the content of the platelike regions, reaching a maximum value of ~13 % after annealing for 10 h at 900 °C, as shown in Fig. 6.20b.

An additional study of the platelike regions showed that they were made of three phases, i.e., a cubic perovskite phase, a hexagonal phase, and a $Ba_3Co_{10}O_{17}$ phase [67]. The $Ba_3Co_{10}O_{17}$ phase, a member of the $Ba_{n+1}Co_nO_{3n+3}(Co_8O_8)$ family, was first reported by Sun et al. The phase was also similar to the 15R phase described by Efimov et al. [65]. Figure 6.21 shows the high-resolution HAADF STEM images of the $Ba_3Co_{10}O_{17}$ phase with the corresponding atomic models in two different zone

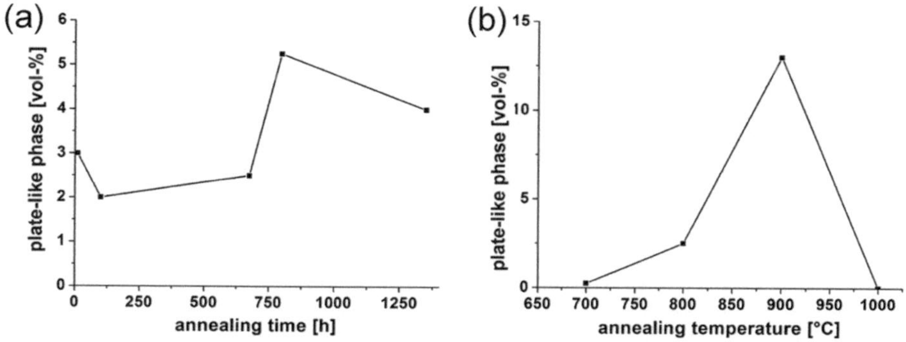

Fig. 6.20 (**a**) Fraction of the platelike phase in vol.% as a function of the annealing time at 800 °C (BSCF800-xx) and (**b**) fraction of the platelike phase as a function of the annealing temperature (BSCFxx-10) (*right*). The fractions are given normalized with respect to a dense sample (Reproduced from [66] with permission of Elsevier)

axes ($\left[2\overline{1}\overline{1}0\right]$ and $\left[10\overline{1}0\right]$) and the projected structure of the $Ba_3Co_{10}O_{17}$ phase described by Sun et al. along the $\left[2\overline{1}\overline{1}0\right]$ direction [68]. The $Ba_3Co_{10}O_{17}$ phase consists of a periodic arrangement of two (111) planes of O octahedra with a cubic stacking order and three planes with a CdI_2-type structure and two interface planes, as shown in Fig. 6.22. Specifically, the two cubic unit cells are composed of $Ba_2Co_2O_6$, leaving a composition of $BaCo_8O_{11}$ for the two interface planes and the CdI_2 plane. The formation of platelike regions was to be found faster at the grain boundaries than at the hexagonal phase in the BSCF bulk. Thus, the faster degradation of the oxygen permeation flux at the initial stage may be related to the formation of the platelike regions. Recently, we found that only a $Ba_3Co_{10}O_{17}$ phase was formed at the grain boundaries of a BSCF membrane after it was operated for oxygen permeation at 600 °C for 500 h [69]. Meanwhile, the reconstructed three-dimensional reciprocal lattice of the new phase, as derived from selected area electron diffraction pattern analysis, indicated a trigonal structure (Fig. 6.23). The reflection conditions determined from these patterns are (i) hkl: $h-k+l=3n$, (ii) h0l: $-h+l=3n$, (iii) hhl: $l=3n$, and (iv) 00l: $l=3n$. According to the International Tables for Crystallography, two space groups, *R*3*m* (No. 160) and *R*-3*m* (No. 166), fulfilled these reflection conditions. The composite diffraction pattern observed along the $[110]_{BSCF}/[100]_T$ direction demonstrated the parallel orientation relationship between BSCF and the trigonal phase, as follows: $[110]_{BSCF}$ $[100]_T$, $(\overline{1}1\overline{1})_{BSCF}(001)_T$, $(32\overline{2})_{BSCF}(01\overline{1})_T$, and $(41\overline{1})_{BSCF}(01\overline{2})_T$ Accordingly, the unit cell parameters were determined from the composite SAED to be $a=b=5.65$ Å and $c=35.76$ Å. Detailed EDS analysis of this new phase suggested an elemental composition of approximately $Ba_{0.45}Sr_{0.07}Co_{1.6}Fe_{0.18}O_{2.8}$. However, because of the direction limitations and the small size of the trigonal phase, it was difficult to determine the space group of the trigonal phase. Sun et al. determined that $Ba_3Co_{10}O_{17}$ has a space group of *R*-3*m* (No. 166) and unit cell parameters of $a=b=5.69$ Å and $c=35.94$ Å. This compound showed a similar chemical

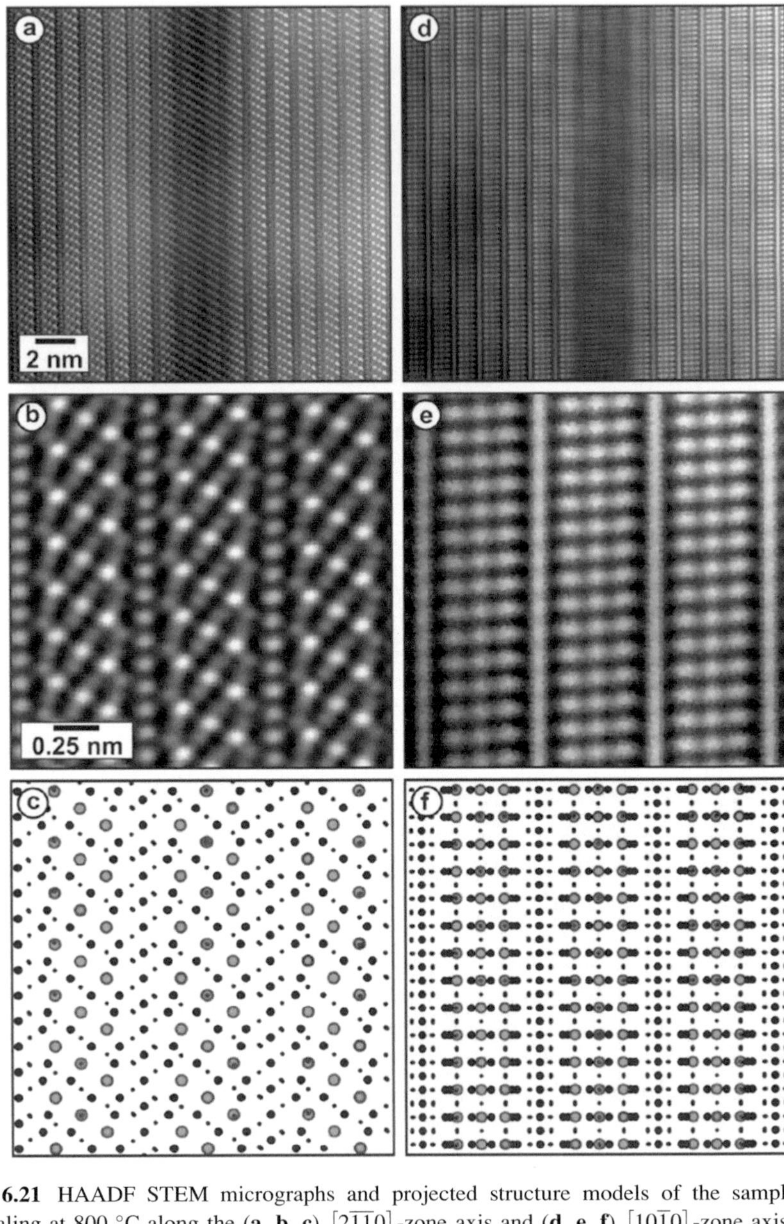

Fig. 6.21 HAADF STEM micrographs and projected structure models of the sample after annealing at 800 °C along the (**a**, **b**, **c**) $[2\bar{1}\bar{1}0]$-zone axis and (**d**, **e**, **f**) $[10\bar{1}0]$-zone axis of the BCO-type phase. A-site cations are marked in *green*, B-site cations in *blue*, and oxygen in *red* in (**c**, **f**) (Reproduced from [67] with permission of the American Chemical Society)

composition and cell parameters as $Ba_{0.45}Sr_{0.07}Co_{1.6}Fe_{0.18}O_{2.8}$, and thus the two compounds may have identical space groups and similar atomic coordinates. The $Ba_3Co_{10}O_{17}$ phase formed at lower temperatures and decomposed into a Co_3O_4 spinel phase and a Co-based perovskite phase at temperatures above 800 °C. That is

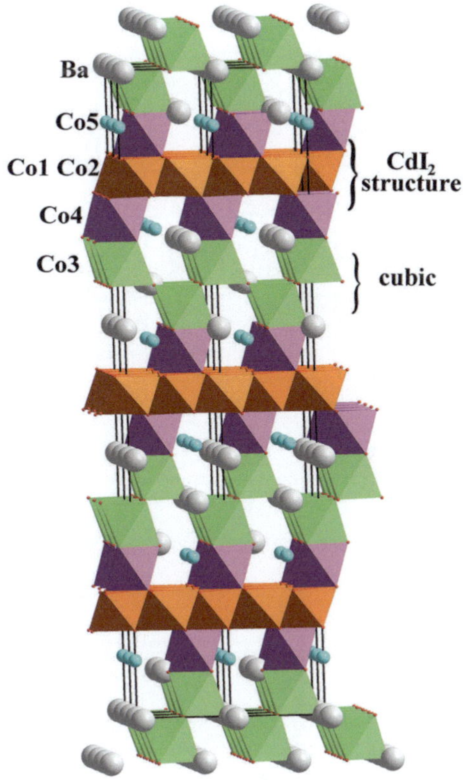

Fig. 6.22 Crystal structure of $Ba_3Co_{10}O_{17}$

why the trigonal phase of $Ba_{0.45}Sr_{0.07}Co_{1.6}Fe_{0.18}O_{2.8}$ was only observed at lower temperatures (<700 °C) and the Co_3O_4 spinel phase and the Co-based perovskite phase were observed at higher temperatures (700–850 °C) for the $Ba_{0.5}Sr_{0.5}Co_{0.8}Fe_{0.2}O_{3-\delta}$ material.

In the above investigations on the phase transformation of BSCF, powders or sintered ceramics treated under static air were used as the samples; however, BSCF membranes work under a certain oxygen chemical potential gradient for oxygen separation. Furthermore, demixing of metal oxides is frequently found under an oxygen chemical potential gradient. Therefore, the phenomena reported in the above works on the phase transformation of BSCF may not differ from that which occurs on the membrane under a certain oxygen chemical potential gradient. After an operation at 950 °C for ~100 h and then at 750 °C for another ~120 h under an oxygen partial pressure gradient of 250/10 kPa, Liang et al. investigated the spent BSCF membrane by electron microscopies [70]. Figure 6.24 shows the backscattered electron channeling contrast images of a BSCF perovskite tube membrane cross section after a long-term permeation experiment. A crack, produced as a result of phase transformation-induced tensile stress, had a length of ~200 μm, and it was observed running into the membrane from the feed side. It is interesting to find that most of the secondary phases crisscrossed the BSCF grains

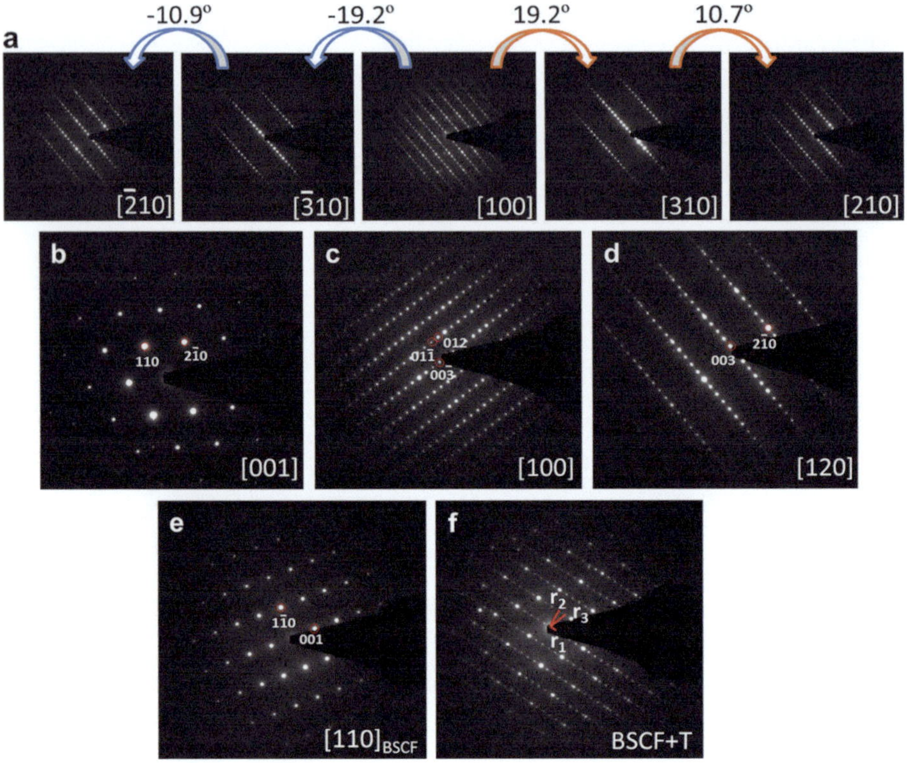

Fig. 6.23 Selected-area electron diffraction (SAED) analysis of the phase formed at BSCF grain boundaries after 500 h operation at 600 °C. (**a**) $[\bar{2}10]$ and [210] and $[\bar{3}10]$ and [310] indicate mirror symmetry perpendicular to the a^*-axis. (**b–d**) Three main SAED patterns indexed: [001], [100], and [120]. (**e**, **f**) The parallel orientation relationship between the BSCF perovskite phase and the trigonal phase, as follows: $[110]_{BSCF} \| [100]_R$, $(\bar{1}1\bar{1})_{BSCF} \| (001)_R$, $(32\bar{2})_{BSCF} \| (01\bar{1})_R$, and $(41\bar{1})_{BSCF} \| (012)_R$ (Reproduced from [69] with permission of the American Chemical Society)

and only a few appeared in the grain boundaries in areas close to the feed side, whereas secondary phases appeared only in the grain boundaries in areas close to the permeation side. This result indicates that the phase transformation mechanism is related to the oxygen chemical potential in the membrane bulk, the oxygen chemical potential gradient, or both. Hexagonal $Ba_{0.6}Sr_{0.4}CoO_{3-\delta}$ and 15R-related $Ba_{1-x}Sr_xCo_{2-y}Fe_yO_{5-\delta}$ (similar to $Ba_3Co_{10}O_{17}$) were found in areas close to the feed and permeation sides, but the content of the hexagonal phase on the feed side was much larger than that of the permeation side [70]. Additionally, hexagonal perovskite arrays with migration channels ([110] direction) of oxygen in cubic BSCF perovskite were found because that direction is the most strongly oxidizing location in ionic conducting perovskite.

Although much effort, as introduced above, has focused on the phase transformation of BSCF, and the possible products of the phase transformation have been clarified, the factors that have a significant influence on the products of the phase

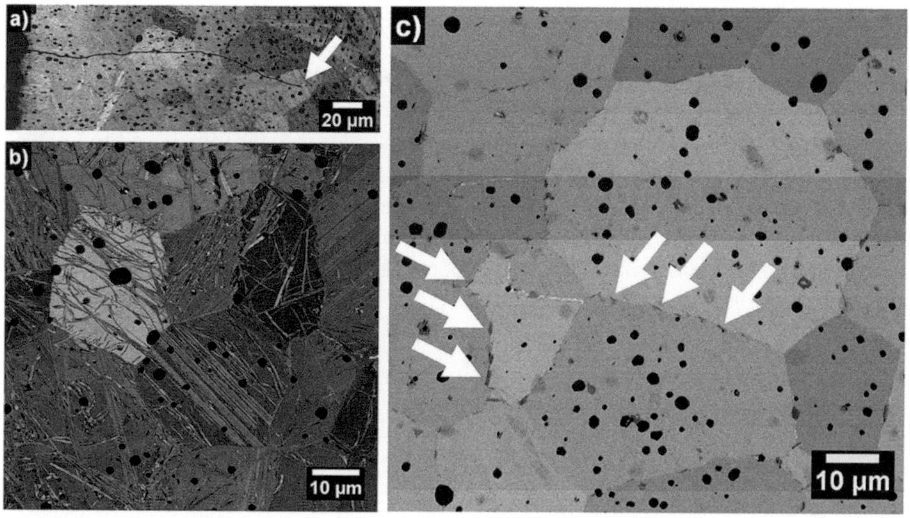

Fig. 6.24 Backscattered electron channeling contrast images of the BSCF perovskite tube membrane cross section after the long-term permeation experiment (100 h at 950 °C, 120 h at 750 °C): (**a**) Overview of shell/feed side showing an approximately 200 µm long crack running into the membrane from the *left* (shell/feed side). The crack tip is marked by an *arrow*. (**b**) Area close to the shell/feed side showing secondary phases crisscrossing BSCF grains. (**c**) Area close to core/permeate side showing secondary phases in the grain boundaries. Some phases are marked by *arrows*. Note that the closed porosity of the ceramic membrane appears as *black dots* (Reproduced from [70] with permission of the American Chemical Society)

transformation are still not clear. Except for the most investigated factor, i.e., temperature, the other factors such as oxygen partial pressure, impurities, and nonstoichiometry of the A and B site ions are still not clear. Müller et al. found that platelike regions are easily generated around CoO impurities [67]. A possible reason is that nucleation of the $Ba_3Co_{10}O_{17}$ phase occurs easily on the surfaces of CoO impurities because of the high cobalt content of the $Ba_3Co_{10}O_{17}$ phase. In our recent research [69], we found that impurities such as $BaCeO_3$ at BSCF grain boundaries can inhibit the phase transformation at low temperatures by restraining the movement of grain boundaries. A detailed discussion on the microstructural effects is given in Chap. 8.

6.5 Co-free Perovskite Membranes

Perovskite oxides with cobalt doped in B sites frequently show high oxygen permeability because the cobalt elements in a perovskite lattice can significantly improve the oxygen ionic diffusion coefficient and interfacial oxygen exchange coefficient. However, weak Co–O bonds decrease the stability of the perovskite structure under a reducing atmosphere and lead to low mechanical strength. These two drawbacks diminish the application of cobalt-based perovskite membrane reactors to the conversion of hydrocarbons. Therefore, some researchers have proposed that cobalt-free membranes with high stability under a reducing atmosphere may be alternatives for the conversion of hydrocarbons in MIEC perovskite membrane reactors. However, the increase in structural stability is at the cost of permeability. Therefore, a trade-off between stability and permeability should be found for cobalt-free perovskite membranes. Much effort has focused on improving the permeability of cobalt-free perovskite membranes. Among the materials, $LaGaO_3$- and $BaFeO_3$-based cobalt-free perovskite oxides have been widely investigated due to their high structural stability and considerable permeability.

6.5.1 $LaGaO_3$ System

Sr- and Mg-co-doped $LaGaO_3$ perovskite oxides ($La_{1-x}Sr_xGa_{1-y}Mg_yO_{3-\delta}$, $0 \leq x \leq 0.2$, $0 \leq y \leq 0.2$) are good oxygen ionic conductors, and their conductivities are as high as those of ceria-based electrolytes (see Chap. 3). Partial substitution of gallium by iron in perovskite B sites can introduce electronic conductivity into the material, thus making it an interesting material for membrane applications. The resultant $La_{1-x}Sr_xGa_{1-y}Fe_yO_{3-\delta}$ ($0 \leq x \leq 0.85$, $0 \leq y \leq 1$) series were reported by Ishihara and coworkers [71]. They found that Fe-doped $La_{0.8}Sr_{0.2}Ga_{1-y}Fe_yO_{3-\delta}$ was stable under a reducing atmosphere, although the oxygen permeability was lower than that of their Co- and Ni-doped counterparts. The oxygen permeability of the materials increased with increasing Fe content, reaching a maximum at $y = 0.4$, i.e., $La_{0.8}Sr_{0.2}Ga_{0.6}Fe_{0.4}O_{3-\delta}$. An oxygen permeation flux up to 0.83 mL cm^{-2} min^{-1} was achieved on a 0.5-mm-thick membrane at 900 °C. In a US patent, Mackay et al. reported the fabrication of membranes by doping Mg into a $La_{1-x}Sr_xGa_{1-y}Fe_yO_{3-\delta}$ lattice, i.e., $La_{0.15}Sr_{0.85}Ga_{0.3}Fe_{0.6}Mg_{0.1}O_{3-\delta}$. A high oxygen permeation flux of 0.94 mL cm^{-2} min^{-1} at 900 °C was reported on a 1-mm-thick membrane, which was higher than that of $La_{0.15}Sr_{0.85}Ga_{0.3}Fe_{0.7}O_{3-\delta}$, (0.75 mL cm^{-2} min^{-1}) under the same experimental conditions [72]. The improvement of permeability was ascribed to the fact that the Mg–O bond was weaker than the Fe–O bond and that the lower oxidation state of Mg ions created more oxygen vacancies in the lattice. The permeation fluxes of $LaGaO_3$-based membranes reported by Ishihara and Mackay were higher than those of other membranes. One important reason is that the catalysts were coated on the membrane surfaces to speed up the interfacial oxygen exchange kinetics. Reichmann et al. found that approximately 17 and 75 % of the driving force were consumed by the oxygen exchange on the gas–solid interface of the feed and permeation sides, respectively, and that only 8 % of the driving force was consumed by the bulk diffusion [73]. As a result, a low oxygen permeation flux of ~0.07 mL cm^{-2} min^{-1} was obtained on a 1-mm-thick $La_{0.5}Sr_{0.5}Fe_{0.7}Ga_{0.3}O_{3-\delta}$ membrane at 900 °C under an air/Ar gradient. Lee et al. prepared a $La_{0.7}Sr_{0.3}Fe_{0.4}Ga_{0.6}O_{3-\delta}$ membrane with a thickness of 1.7 mm and found that its flux was only ~0.03 mL cm^{-2} min^{-1} at 900 °C, but it increased to

0.4 mL cm^{-2} min^{-1} after both sides of the membrane were coated by $La_{0.6}Sr_{0.4}CoO_{3-\delta}$ porous layers to activate the oxygen reduction and evolution reactions [74].

In summary, $LaGaO_3$-based perovskite membranes have considerable ionic conductivity (for instance, for $La_{0.5}Sr_{0.5}Fe_{0.7}Ga_{0.3}O_{3-\delta}$, $\sigma_i = 0.22$ S cm^{-1} at 900 °C) but slow interfacial oxygen exchange kinetics, especially on the permeation side. To improve their permeation flux, one simple and effective method is to coat oxygen activation catalysts on both sides of the membranes. To further increase the permeation flux, researchers can fabricate the membranes into hollow fibers and asymmetric membranes with a thin dense layer and a thick porous support. Thus, in terms of permeability and stability, $La_{1-x}Sr_xGa_{1-y}Fe_yO_{3-\delta}$ is a competitive candidate in the fields of oxygen separation and catalytic membrane reactors.

6.5.2 *$BaFeO_{3-\delta}$ System*

$BaFeO_{3-\delta}$ is a hexagonal perovskite oxide at room temperature, and phase transformation from hexagonal to cubic, which is accompanied by significant increases in both electronic and ionic conductivity, takes place at 800 °C in air. The high-temperature cubic phase exhibits high permeability. Proper doping in the A sites, B sites, or both to stabilize the cubic phase to room temperature is required to use the $BaFeO_{3-\delta}$-based materials as MIEC membranes for oxygen separation. The formation of the hexagonal structure for $BaFeO_{3-\delta}$ perovskite is driven by its large tolerance factor (>1), thus it needs to stabilize the cubic structure by doping. Larger metallic ions for B sites, smaller metallic ions for A sites, or both are good doping choices.

Ce^{4+} (87 pm) was first used to dope $BaFeO_{3-\delta}$ to obtain a $BaCe_xFe_{1-x}O_{3-\delta}$ ($0 \leq x \leq 0.85$) series of oxides. Single-phase cubic perovskites were prepared in the range of $0.05 \leq x \leq 0.15$ and $x \geq 0.85$, and two cubic perovskite phase mixtures were produced in the range $0.15 < x < 0.85$ [75–79]. It was found that the permeation flux of these membranes decreased with increasing cerium doping amount at $x \geq 0.1$ at 900 °C and increase with cerium doping amounts of $0 \leq x \leq 0.1$, as shown in Fig. 6.25a. However, $BaCe_{0.05}Fe_{0.95}O_{3-\delta}$ showed the highest oxygen permeation flux when the operational temperature was below 750 °C. Unlike the $BaFeO_{3-\delta}$ membrane, the $BaCe_{0.05}Fe_{0.95}O_{3-\delta}$ membrane showed high structural stability during the cooling/heating cycles, as indicated by the disappearance of the hysteresis loop during the temperature cycles, as shown in Fig. 6.25b. The Ce-doped membrane materials retained their cubic perovskite structure, even when reduced by a 5 % H_2/Ar mixed gas at 900 °C for 1 h. In contrast, the perovskite structure of $La_{0.5}Sr_{0.5}Ga_{0.3}Fe_{0.7}O_{3-\delta}$ was partially destroyed, and that of $Ba_{0.5}Sr_{0.5}Co_{0.8}Fe_{0.2}O_{3-\delta}$ completely disappeared after reduction under the same conditions.

Similarly, researchers have reported the use of other metallic ions with larger ionic radii to stabilize $BaFeO_{3-\delta}$ perovskite oxide, such as Zr^{4+}, Y^{3+}, La^{3+}, and Nb^{5+}

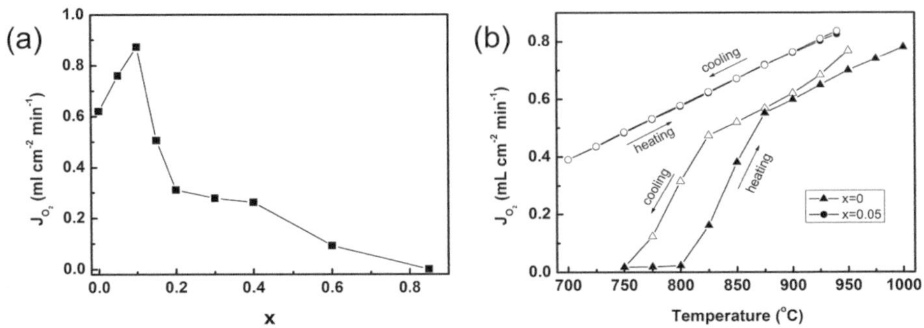

Fig. 6.25 (**a**) The relationship of the 1.0 mm $BaCe_xFe_{1-x}O_{3-\delta}$ membranes' oxygen permeation flux with the cerium doping amount at 900 °C. The oxygen permeation fluxes of $x = 0.3$, 0.4 samples were obtained by adjusting the 1.5 mm membrane to 1.0 mm. (**b**) Effects of cooling–heating circle on the oxygen permeation fluxes of the $x = 0$ and 0.05 samples. Thickness: 1.0 mm (Reproduced from [78] with permission of Elsevier)

[80–84]. Five percent doping in the iron sites was enough to stabilize the cubic perovskite structure to room temperature, and these membranes showed similar oxygen permeation fluxes and stabilities to those of the Ce-doped one. Five percent Na^+, Rb^+, Ca^{2+}, Y^{3+}, and La^{3+} were substituted in the A sites of $BaFeO_{3-\delta}$, and only Ca^{2+}, Y^{3+}, and La^{3+}, which have smaller ionic radii than Ba^{2+}, succeeded in stabilizing $BaFeO_{3-\delta}$ to the cubic phase [81]. Among the three membranes, $Ba_{0.95}La_{0.05}FeO_{3-\delta}$ showed the highest oxygen permeability in the temperature range of 600–930 °C, but the oxygen permeability was comparable to those of $BaFeO_{3-\delta}$ doped by Ce^{4+}, Zr^{4+}, Y^{3+} and La^{3+} in the B sites.

6.6 Perovskite-Related MIEC Membranes

6.6.1 Ruddlesden–Popper Series Materials

Ruddlesden–Popper (R–P) series materials, $A_{n+1}B_nO_{3n+1}$ (A = lanthanide or alkaline earth and $n = 1, 2, 3\ldots$), have perovskite-related intergrowth structures in which a number of perovskite blocks (n is the number of perovskite blocks) with corner-shared BO_6 octahedra alternate with AO rock salt layers along the c axis. The $La_2NiO_{4+\delta}$ type is the first member ($n = 1$) of this series, whereas perovskite is the last member ($n = \infty$). If the A site is occupied by alkaline earth metal ions in the R–P series materials, a larger number of oxygen vacancies are produced along the c axis, such as $Sr_3FeCoO_{7-\delta}$; as a result, the materials are extremely sensitive to atmospheric moisture [85]. However, if the A site is occupied by lanthanide (Ln) metallic ions, such as La^{3+} and Pr^{3+}, interstitial oxygen is generated in the AO rock salt layers along the c axis, such as $La_2NiO_{4+\delta}$; this type of material is not sensitive to atmospheric moisture. Therefore, most investigations on the R–P series have focused on $La_2NiO_{4+\delta}$-type materials.

Oxygen permeation through $La_2NiO_{4+\delta}$-type membranes is via both oxygen vacancies, as is the case with perovskite membranes, and interstitial oxygen ion migration in the rock salt layers. As compared to its perovskite counterparts, this type of oxide has many advantages, such as low activation energy for oxygen ion diffusion [86] and a low thermal expansion coefficient [87]. Therefore, they are deemed competitive materials for oxygen separation membranes. The oxygen permeability, conductivity, stability, and catalytic activity of $La_{2-x}Sr_xNi_{1-y}(Fe,Co,Cu)_yO_{4+\delta}$ were investigated for methane partial oxidation to syngas in the membrane reactors [87–90]. It was found that the migration of interstitial oxygen ions in the rock salt layers contributed more to the ionic conductivity of the $La_2NiO_{4+\delta}$-type membrane than oxygen ion diffusion through oxygen vacancies. The contribution of interstitial oxygen ions to ionic conductivity decreased with increasing oxygen partial pressure. However, because the membranes were used for oxygen permeation and catalytic oxidation of hydrocarbons, the oxygen partial pressure of the permeation side was low. As a net result, the oxygen permeability of $La_2NiO_{4+\delta}$-type membranes was lower than those of $La_{1-x}Sr_xCo_{1-y}Fe_yO_{3-\delta}$ membranes.

Recently, Ishihara et al. reported a $La_2NiO_{4+\delta}$-type membrane using Pr^{3+} to replace La^{3+} in the A sites and using copper and iron ions to partially replace nickel in the B sites [91–94]. After both sides of the membranes were coated by a $La_{0.6}Sr_{0.4}CoO_{3-\delta}$ porous layer for oxygen activation, a high oxygen permeation flux of ~1.0 mL cm^{-2} min^{-1} was achieved through a 0.5-mm-thick $Pr_2Ni_{0.75}Cu_{0.2}Fe_{0.05}O_{4+\delta}$ membrane at 900 °C. The improvement in oxygen permeation flux was due to the increase of interstitial oxygen by doping 5 % iron in the nickel sites. However, much more iron doping would produce secondary phases of Pr_6O_{11} and $PrFeO_3$ and lower the oxygen permeation flux. Subsequently, Yashima and Ishihara optimized the composition of the material, and a much higher oxygen permeation flux of ~2.0 mL cm^{-2} min^{-1} was achieved at 900 °C through a 0.5-mm-thick $(Pr_{0.9}La_{0.1})_2Ni_{0.74}Cu_{0.21}Ga_{0.05}O_{4+\delta}$ membrane with both sides coated by cobalt-based perovskite for oxygen activation [94]. Through in situ neutron powder diffraction of the material by Rietveld analysis, the maximum entropy method (MEM) [95], and MEM-based pattern fitting to neutron and synchrotron diffraction from 25 to 1015.6 °C, they found that bulk oxygen ionic diffusion occurred through the interstitial oxygen and anisotropic apical oxygen sites, as shown in Fig. 6.26. On the oxygen diffusion path, the oxygen probability density at the bottleneck increased with increasing temperature, resulting in an increase of oxygen permeation flux. However, the activation energy of the probability density or interstitial oxygen diffusion activation energy decreased with temperature, reaching 12 kJ mol^{-1} at approximately 900 °C; this low value indicates possible higher bulk oxygen permeation rates in quality single crystals and epitaxial thin films [95].

Wei and Wang fabricated $(Pr_{0.9}La_{0.1})_2Ni_{0.74}Cu_{0.21}Ga_{0.05}O_{4+\delta}$ into hollow fibers by using a phase-inversion spinning process for oxygen separation and combined it with an oxyfuel process for CO_2 capture [96, 97]. The oxygen permeation flux was

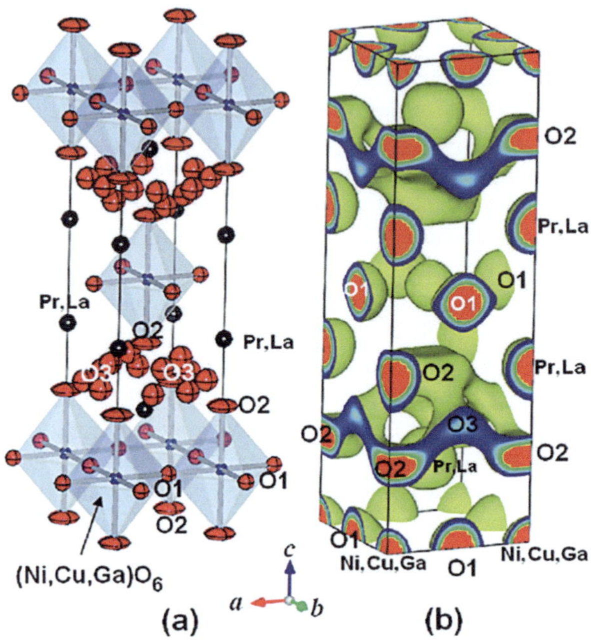

Fig. 6.26 (**a**) Refined crystal structure and (**b**) isosurface of nuclear density at 0.05 fm Å^{-3} of the mixed oxide-ionic and electronic conductor $(Pr_{0.9}La_{0.1})_2(Ni_{0.74}Cu_{0.21}Ga_{0.05})O_{4+\delta}$ determined in situ at 1015.6 °C. Unit cell: tetragonal I4/mmm, $a = 3.875(3)$ Å, and $c = 12.738(9)$ Å (Reproduced from [95] with permission of the American Chemical Society)

0.47 mL cm^{-2} min^{-1} at 900 °C through 0.23-mm-thick hollow fiber membranes. After the feed side surface was etched by HCl and then coated by $La_{0.8}Sr_{0.2}CoO_{3\text{-}\delta}$ porous layers for oxygen activation, the oxygen permeation flux improved only slightly to 0.52 mL cm^{-2} min^{-1} at 900 °C. However, a significant improvement from 0.06 to 0.21 mL cm^{-2} min^{-1} at 800 °C was recorded. The much lower oxygen permeation flux in Wei's report than that reported by Yashima may have been caused by two factors. One is that only the feed side surface was coated with a catalyst for oxygen activation in Wei's experiment; the other factor is that sulfate impurities in the hollow fiber, which were introduced through the use of polyethersulfone in the fabrication process of the hollow fibers, may have blocked ionic transport in the bulk membrane and interfacial oxygen exchange.

6.6.2 *Other Types*

Double-layered perovskite oxides, $LnBaCo_2O_{5+\delta}$ (Ln = lanthanide elements), showing good catalytic activity toward oxygen reduction, have recently been applied as cathodes for solid oxide fuel cells. However, efforts to make them into membranes with high oxygen permeabilities have been frustrated. As compared to their perovskite counterparts, the lower oxygen permeation flux and regular structural stability of these materials have not attracted much attention. Similarly, garnet-type oxides such as $Ln_3Fe_5O_{12}$ have been fabricated into membranes for oxygen permeation [98, 99]. Although this type of material has good stability, its permeation flux is far lower than those of its perovskite counterparts.

Aurivillius-type $(Bi_2O_2)(A_{n-1}B_nO_x)$ oxides have been found to be good oxygen ionic conductors with an alternate $(Bi_2O_2)^{2+}$ layer/perovskite layer structure. $Bi_4V_2O_{11}$ is considered the first member of this series, but only the high-temperature γ-$Bi_4V_2O_{11}$ phase shows a high ionic conductivity up to 1 S cm^{-1} at temperatures higher than 570 °C (see Chap. 3). In the crystalline phase, the average V–O polyhedron appears as a squashed octahedron, and the different oxygen-deficient octahedra are connected by sharing corners. Copper doping in the vanadium site stabilizes the γ phase to room temperature, and it only slightly decreases the ionic conductivity of the material. However, electronic conduction should be introduced in the material to form a mixed conductor before it is fabricated into membranes for oxygen permeation. Although its stability under a reducing environment is not good, its high ionic conductivity makes it possible to prepare membranes for oxygen separation at low temperatures (300–600 °C).

References

1. Kruidhof H, Bouwmeester HJM, van Doorn RHE, Burggraaf AJ (1993) Influence of order–disorder transitions on oxygen permeability through selected nonstoichiometric perovskite-type oxides. Solid State Ionics 63–65:816–822
2. Shao ZP, Yang WS, Cong Y, Dong H, Tong JH, Xiong GX (2000) Investigation of the permeation behavior and stability of a $Ba_{0.5}Sr_{0.5}Co_{0.8}Fe_{0.2}O_{3-\delta}$ oxygen membrane. J Membr Sci 172:177–188
3. Teraoka Y, Zhang HM, Furukawa S, Yamazoe N (1985) Oxygen permeation through perovskite-type oxides. Chem Lett 11:1743–1746
4. Teraoka Y, Nobunaga T, Yamazoe N (1988) Effect of cation substitution on the oxygen semipermeability of perovskite-type oxides. Chem Lett 3:503–506
5. Kharton VV, Naumovich EN, Kovalevsky AV, Viskup AP, Figueiredo FM, Bashmakov IA, Marques FMB (2000) Mixed electronic and ionic conductivity of $LaCo(M)O_3$ (M = Ga, Cr, Fe or Ni) IV. Effect of preparation method on oxygen transport in $LaCoO_{3-\delta}$. Solid State Ionics 138:135–148
6. Kharton VV, Kovalevsky AV, Tikhonovich VN, Naumovich EN, Viskup AP (1998) Mixed electronic and ionic conductivity of $LaCo(M)O_3$ (M = Ga, Cr, Fe or Ni) II. Oxygen permeation through Cr- and Ni-substituted $LaCoO_3$. Solid State Ionics 110:53–60
7. Yaremchenko AA, Kharton VV, Viskup AP, Naumovich EN, Tikhonovich VN, Lapchuk NM (1999) Mixed electronic and ionic conductivity of $LaCo(M)O_3$ (M = Ga, Cr, Fe or Ni). V. Oxygen permeability of Mg-doped La(Ga, Co)$O_{3-\delta}$ perovskites. Solid State Ionics 120:65–74
8. Cherry M, Islam MS, Catlow CRA (1995) Oxygen ion migration in perovskite-type oxides. J Solid State Chem 118:125–132
9. Islam MS, Davies RA (2004) Atomistic study of dopant site-selectivity and defect association in the lanthanum gallate perovskite. J Mater Chem 14:86–93

10. Islam MS (2002) Computer modelling of defects and transport in perovskite oxides. Solid State Ionics 154:75–85
11. Tsai CY, Dixon AG, Ma YH, Moser WR, Pascucci MR (1998) Dense Perovskite, $La_{1-x}A'_xFe_{1-y}Co_yO_{3-\delta}$ (A′ = Ba, Sr, Ca), membrane synthesis, applications, and characterization. J Am Ceram Soc 81:1437–1444
12. Stevenson JW, Armstrong TR, Carneim RD, Pederson LR, Weber WJ (1996) Electrochemical properties of mixed conducting perovskites $La_{1-x}M_xCo_{1-y}Fe_yO_{3-\delta}$ (M = Sr, Ba, Ca). J Electrochem Soc 143:2722–2729
13. Hayashi H, Inaba H, Matsuyama M, Lan NG, Dokiya M, Tagawa H (1999) Structural consideration on the ionic conductivity of perovskite-type oxides. Solid State Ionics 122:1–15
14. Li SG, Jin WQ, Huang P, Xu N, Shi J (1999) Comparison of oxygen permeability and stability of perovskite type $La_{0.2}A_{0.8}Co_{0.2}Fe_{0.8}O_{3-\delta}$ (A = Sr, Ba, Ca) membranes. Ind Eng Chem Res 38:2963–2972
15. Kovalevsky AV, Kharton VV, Tikhonovich VN, Naumovich EN, Tonoyan AA, Reut OP, Boginsky LS (1998) Oxygen permeation through $Sr(Ln)CoO_{3-\delta}$ (Ln = La, Nd, Sm, Gd) ceramic membranes. Mater Sci Eng B-Solid State Mater Adv Technol 52:105–116
16. Lane JA, Benson SJ, Waller D, Kilner JA (1999) Oxygen transport in $La_{0.6}Sr_{0.4}Co_{0.2}Fe_{0.8}O_{3-\delta}$. Solid State Ionics 121:201–208
17. Bouwmeester HJM, Den Otter MW, Boukamp BA (2004) Oxygen transport in $La_{0.6}Sr_{0.4}Co_{1-y}Fe_yO_{3-\delta}$. J Solid State Electrochem 8:599–605
18. Xu SJ, Thomson WJ (1999) Oxygen permeation rates through ion-conducting perovskite membranes. Chem Eng Sci 54:3839–3850
19. Tai LW, Nasrallah MM, Anderson HU (1995) Thermochemical stability, electrical conductivity, and seebeck coefficient of Sr-doped $LaCo_{0.2}Fe_{0.8}O_{3-\delta}$. J Solid State Chem 118:117–124
20. Liu Y, Zhu XF, Yang WS (2016) Stability of sulfate doped $SrCoO_{3-\delta}$ MIEC membrane. J Membr Sci 501:53–59
21. Nagai T, Ito W, Sakon T (2007) Relationship between cation substitution and stability of perovskite structure in $SrCoO_{3-\delta}$-based mixed conductors. Solid State Ionics 177:3433–3444
22. Zeng PY, Shao ZP, Liu SM, Xu ZP (2009) Influence of M cations on structural, thermal and electrical properties of new oxygen selective membranes based on $SrCo_{0.95}M_{0.05}O_{3-\delta}$ perovskite. Sep Purif Technol 67:304–311
23. Chen XZ, Liu H, Wei YY (2011) Tantalum stabilized $SrCoO_{3-\delta}$ perovskite membrane for oxygen separation. J Membr Sci 368:159–164
24. Kharton VV, Li SB, Kovalevsky AV, Naumovich EN (1997) Oxygen permeability of perovskites in the system $SrCoO_{3-\delta}$-$SrTiO_3$. Solid State Ionics 96:141–151
25. Zhang K, Ran R, Ge L, Shao ZP, Jin WQ, Xu NP (2009) Double-site yttria-doped $Sr_{1-x}Y_xCo_{1-y}Y_yO_{3-\delta}$ perovskite oxides as oxygen semi-permeable membranes. J Alloys Compd 474:477–483
26. Zhang K, Ran R, Ge L, Shao ZP, Jin WQ, Xu NP (2008) Systematic investigation on new $SrCo_{1-y}Nb_yO_{3-\delta}$ ceramic membranes with high oxygen semi-permeability. J Membr Sci 323:436–443
27. McIntosh S, Vente JF, Haije WG, Blank DHA, Bouwmeester HJM (2006) Structure and oxygen stoichiometry of $SrCo_{0.8}Fe_{0.2}O_{3-\delta}$ and $Ba_{0.5}Sr_{0.5}Co_{0.8}Fe_{0.2}O_{3-\delta}$. Solid State Ionics 177:1737–1742
28. Wang ZF, Zhao HL, Xu NS, Shen YN, Ding WZ, Lu XG, Li FS (2011) Electrical conductivity and structural stability of $SrCo_{1-x}Fe_xO_{3-\delta}$. J Phys Chem Solids 72:50–55
29. Shin MJ, Yu JH (2012) Oxygen transport of A-site deficient $Sr_{1-x}Fe_{0.5}Co_{0.5}O_{3-\delta}$ (x = 0–0.3) membranes. J Membr Sci 401:40–47
30. He YF, Zhu XF, Guo ZF, Yang WS (2010) Phase transitions in $Sr_{1+x}Co_{0.8}Fe_{0.2}O_{3-\delta}$ oxides. Mater Lett 64:1618–1621
31. Liu HY, Zhu XF, Cong Y, Zhang TY, Yang WS (2012) Remarkable dependence of electrochemical performance of $SrCo_{0.8}Fe_{0.2}O_{3-\delta}$ on A-site nonstoichiometry. Phys Chem Chem Phys 14:7234–7239

32. Qiu L, Lee TH, Liu LM, Yang YL, Jacobson AJ (1995) Oxygen permeation studies of $SrCo_{0.8}Fe_{0.2}O_{3-\delta}$. Solid State Ionics 76:321–329
33. Lee TH, Yang YL, Jacobsona AJ, Abelesa B, Zhou M (1997) Oxygen permeation in dense $SrCo_{0.8}Fe_{0.2}O_{3-\delta}$ membranes: surface exchange kinetics versus bulk diffusion. Solid State Ionics 100:77–85
34. Yi JX, Feng SJ, Zeng Q, Chen CS (2015) Structure, electrical and oxygen transport properties of Fe-doped $SrCoO_{3-\delta}$ perovskites. Chin J Chem Phys 28:189–192
35. Yang L, Tan L, Gu XH, Jin WQ, Zhang LX, Xu NP (2003) A new series of Sr(Co, Fe, Zr) O3−δ perovskite-type membrane materials for oxygen permeation. Ind Eng Chem Res 42:2299–2305
36. Lu H, Son SH, Kim JP, Park JH (2011) A Fe/Nb co-doped $Sr(Co_{0.8}Fe_{0.1}Nb_{0.1})O_{3-\delta}$ perovskite oxide for air separation: structural, sintering and oxygen permeating properties. Mater Lett 65:702–704
37. Zhao HL, Cheng YF, Xu NS, Li Y, Li FS, Ding WZ, Lu XG (2010) Oxygen permeability of A-site nonstoichiometric $BaxCo_{0.7}Fe_{0.2}Nb_{0.1}O_{3-\delta}$ perovskite oxides. Solid State Ionics 181:354–358
38. Luo HX, Tian BB, Wei YY, Wang HH, Jiang HQ, Caro J (2010) Oxygen permeability and structural stability of a novel Tantalum-doped perovskite $BaCo_{0.7}Fe_{0.2}Ta_{0.1}O_{3-\delta}$. AIChE J 56:604–610
39. Shao ZP, Xiong GX, Cong Y, Yang WS (2000) Synthesis and oxygen permeation study of novel perovskite-type $BaBi_xCo_{0.2}Fe_{0.8-x}O_{3-\delta}$ ceramic membranes. J Membr Sci 164:167–174
40. Tong JH (2001) Highly stable oxygen permeation membrane material and its application in the membrane reactors of the partial oxidation of methane to synthesis gas. Dissertation, Dalian Institute of Chemical Physics, Chinese Academy of Sciences
41. Shao ZP (2000) Mixed oxygen ionic and electronic conducting membrane and its application in the partial oxidation of methane to synthesis gas process. Dissertation, Dalian Institute of Chemical Physics, Chinese Academy of Sciences
42. Luo HX, Wei YY, Jiang HQ, Yuan WH, Lv YX, Caro J, Wang HH (2010) Performance of a ceramic membrane reactor with high oxygen flux Ta-containing perovskite for the partial oxidation of methane to syngas. J Membr Sci 350:154–160
43. Li QM, Zhu XF, Yang WS (2010) Investigation of structure and oxygen permeability of Ba–Ce–Co–Fe–O system. Mater Res Bull 45:1112–1117
44. Cheng HW, Lu XG, Hu DH, Zhang YW, Ding WZ, Zhao HL (2011) Hydrogen production by catalytic partial oxidation of coke oven gas in $BaCo_{0.7}Fe_{0.2}Nb_{0.1}O_{3-\delta}$ membranes with surface modification. Int J Hydrog Energy 36:528–538
45. Yang ZB, Ding WZ, Zhang YY, Lu XG, Zhang YW, Shen PJ (2010) Catalytic partial oxidation of coke oven gas to syngas in an oxygen permeation membrane reactor combined with NiO/MgO catalyst. Int J Hydrog Energy 35:6239–6247
46. Tong JH, Yang WS, Cai R, Zhu BC, Lin LW (2002) Novel and ideal zirconium-based dense membrane reactors for partial oxidation of methane to syngas. Catal Lett 78:129–137
47. Shao ZP, Haile SM (2004) A high-performance cathode for the next generation of solid-oxide fuel cells. Nature 431:170–173
48. Bucher E, Egger A, Ried P, Sitte W, Holtappels P (2008) Oxygen nonstoichiometry and exchange kinetics of $Ba_{0.5}Sr_{0.5}Co_{0.8}Fe_{0.2}O_{3-\delta}$. Solid State Ionics 179:1032–1035
49. Wang HH, Cong Y, Yang WS (2002) Oxygen permeation study in a tubular $Ba_{0.5}Sr_{0.5}Co_{0.8}Fe_{0.2}O_{3-\delta}$ oxygen permeable membrane. J Membr Sci 210:259–271
50. Zhu XF, Liu HY, Cong Y, Yang WS (2012) Permeation model and experimental investigation of mixed conducting membranes. AIChE J 58:1744–1754
51. Chen DJ, Shao ZP (2001) Surface exchange and bulk diffusion properties of $Ba_{0.5}Sr_{0.5}Co_{0.8}Fe_{0.2}O_{3-\delta}$ mixed conductor. Int J Hydrog Energy 36:6948–6956
52. Girdauskaite E, Ullmann H, Vashook VV, Guth U, Caraman GB, Bucher E, Sitte W (2008) Oxygen transport properties of $Ba_{0.5}Sr_{0.5}Co_{0.8}Fe_{0.2}O_{3-\delta}$ and $Ca_{0.5}Sr_{0.5}Mn_{0.8}Fe_{0.2}O_{3-\delta}$

obtained from permeation and conductivity relaxation experiments. Solid State Ionics 179:385–392
53. Hong WK, Choi GM (2010) Oxygen permeation of BSCF membrane with varying thickness and surface coating. J Membr Sci 346:353–360
54. Haworth PF, Smart S, Serra JM, Diniz da Costa JC (2012) Combined investigation of bulk diffusion and surface exchange parameters of silver catalyst coated yttrium-doped BSCF membranes. Phys Chem Chem Phys 14:9104–9111
55. Fu YP, Subardi A, Hsieh MY, Chang WK (2016) Electrochemical properties of $La_{0.5}Sr_{0.5}Co_{0.8}M_{0.2}O_{3-\delta}$ (M = Mn, Fe, Ni, Cu) perovskite cathodes for IT-SOFCs. J Am Ceram Soc 99:1345–1352
56. Katsukia M, Wang SR, Dokiya M, Hashimoto T (2003) High temperature properties of $La_{0.6}Sr_{0.4}Co_{0.8}Fe_{0.2}O_{3-\delta}$ oxygen nonstoichiometry and chemical diffusion constant. Solid State Ionics 156:453–461
57. Subardi A, Cheng MH, Fu YP (2014) Chemical bulk diffusion and electrochemical properties of $SmBa_{0.6}Sr_{0.4}Co_2O_{5+\delta}$ cathode for intermediate solid oxide fuel cells. Int J Hydrog Energy 39:20783–20790
58. Cox-Galhotra RA, McIntosh S (2012) Electrical conductivity relaxation of polycrystalline $PrBaCo_2O_{5+\delta}$ thin films. Solid State Ionics 228:14–18
59. Zhu XF, Sun SM, Cong Y, Yang WS (2009) Operation of perovskite membrane under vacuum and elevated pressures for high-purity oxygen production. J Membr Sci 345:47–52
60. Leo A, Smart S, Liu SM, Diniz da Costa JC (2011) High performance perovskite hollow fibres for oxygen separation. J Membr Sci 368:64–68
61. Baumann S, Serra JM, Lobera MP, Escolástico S, Schulze-Küppers F, Meulenberga WA (2011) Ultrahigh oxygen permeation flux through supported $Ba_{0.5}Sr_{0.5}Co_{0.8}Fe_{0.2}O_{3-\delta}$ membranes. J Membr Sci 377:198–205
62. Švarcová S, Wiik K, Tolchard J, Bouwmeester HJM, Grande T (2008) Structural instability of cubic perovskite $Ba_xSr_{1-x}Co_{1-y}Fe_yO_{3-\delta}$. Solid State Ionics 178:1787–1791
63. Arnold M, Gesing TM, Martynczuk J, Feldhoff A (2008) Correlation of the formation and the decomposition process of the BSCF perovskite at intermediate temperatures. Chem Mater 20:5851–5858
64. Mueller DN, De Souza RA, Weirich TE, Roehrens D, Mayer J, Martin M (2010) A kinetic study of the decomposition of the cubic perovskite-type oxide $Ba_xSr_{1-x}Co_{0.8}Fe_{0.2}O_{3-\delta}$ (BSCF) (x = 0.1 and 0.5). Phys Chem Chem Phys 12:10320–10328
65. Efimov K, Xu Q, Feldhoff A (2010) Transmission electron microscopy study of $Ba_{0.5}Sr_{0.5}Co_{0.8}Fe_{0.2}O_{3-\delta}$ perovskite decomposition at intermediate temperatures. Chem Mater 22:5866–5875
66. Müller P, Störmer H, Dieterle L, Niedrig C, Ivers-Tiffée E, Gerthsen D (2012) Decomposition pathway of cubic $Ba_{0.5}Sr_{0.5}Co_{0.8}Fe_{0.2}O_{3-\delta}$ between 700 °C and 1000 °C analyzed by electron microscopic techniques. Solid State Ionics 206:57–66
67. Müller P, Störmer H, Meffert M, Dieterle L, Niedrig C, Wagner SF, Ivers-Tiffée E, Gerthsen D (2013) Secondary phase formation in $Ba_{0.5}Sr_{0.5}Co_{0.8}Fe_{0.2}O_{3-\delta}$ studied by electron microscopy. Chem Mater 25:564–573
68. Sun J, Yang M, Li G, Yang T, Liao F, Wang Y, Xiong M, Lin J (2006) New barium cobaltite series $Ba_{n+1}Co_nO_{3n+3}(Co_8O_8)$: intergrowth structure containing perovskite and CdI_2-type layers. Inorg Chem 45:9151–9153
69. Liu Y, Zhu XF, Li MR, O'Hayre RP, Yang WS (2015) Nanoparticles at grain boundaries inhibit the phase transformation of perovskite membrane. Nano Lett 15:7678–7683
70. Liang FY, Jiang HQ, Luo HX, Caro J, Feldhoff A (2011) Phase stability and permeation behavior of a dead-end $Ba_{0.5}Sr_{0.5}Co_{0.8}Fe_{0.2}O_{3-\delta}$ tube membrane in high-purity oxygen production. Chem Mater 23:4765–4772
71. Ishihara T, Yamada T, Arikawa H, Nishiguchi H, Takita Y (2000) Mixed electronic–oxide ionic conductivity and oxygen permeating property of Fe-, Co- or Ni-doped $LaGaO_3$ perovskite oxide. Solid State Ionics 135:631–636

72. MacKay R, Schwartz M, Sammells AF (2003) Materials and methods for the preparation of oxygen from air. US Patent 6,592,782, 15 July 2003
73. Geffroy PM, Fouletier J, Richet N, Chartier T, Reichmann M (2014) Effect of cation substitution in the A site on the oxygen semi-permeation flux in $La_{0.5}A_{0.5}Fe_{0.7}Ga_{0.3}O_{3-\delta}$ and $La_{0.5}A_{0.5}Fe_{0.7}Co_{0.3}O_{3-\delta}$ dense perovskite membranes with A = Ca, Sr and Ba (part I). J Power Sources 261:175–183
74. Lee KS, Lee S, Kim JW, Woo SK (2002) Enhancement of oxygen permeation by $La_{0.6}Sr_{0.4}CoO_{3-\delta}$ coating in $La_{0.7}Sr_{0.3}Ga_{0.6}Fe_{0.4}O_{3-\delta}$ membrane. Desalination 147:439–444
75. Zhu XF, Wang HH, Yang WS (2004) Novel cobalt-free oxygen permeable membrane. Chem Commun 9:1130–1132
76. Zhu XF, Wang HH, Cong Y, Yang WS (2006) Oxygen permeability and structural stability of $BaCe_{0.15}Fe_{0.85}O_{3-\delta}$ membranes. J Membr Sci 283:38–44
77. Zhu XF, Cong Y, Yang WS (2006) Effects of synthesis methods on the oxygen permeable $BaCe_{0.15}Fe_{0.85}O_{3-\delta}$ ceramic membranes. J Membr Sci 283:158–163
78. Zhu XF, Wang HH, Cong Y, Yang WS (2006) Structural stability and oxygen permeation of cerium light doped $BaFeO_{3-\delta}$ ceramic membranes. Solid State Ionics 117:2917–2921
79. Zhu XF, Wang HH, Cong Y, Yang WS (2006) Partial oxidization of methane to syngas in $BaCe_{0.15}Fe_{0.85}O_{3-\delta}$ membrane reactor. Catal Lett 111:179–185
80. Watanabe K, Takauchi D, Yuasa M, Kida T, Shimanoe K, Teraoka Y, Yamazoe N (2009) Oxygen permeation properties of Co-free perovskite-type oxide membranes based on $BaFe_{1-y}Zr_yO_{3-\delta}$. J Electrochem Soc 156:E81–E85
81. Kida T, Takauchi D, Watanabe K, Yuasa M, Shimanoe K, Teraoka Y, Yamazoe N (2009) Oxygen permeation properties of partially A-Site substituted $BaFeO_{3-\delta}$ perovskites. J Electrochem Soc 156:E187–E191
82. Liu XT, Zhao HL, Yang JY, Li Y, Chen T, Lu XG, Ding WZ, Li FS (2011) Lattice characteristics, structure stability and oxygen permeability of $BaFe_{1-x}Y_xO_{3-\delta}$ ceramic membranes. J Membr Sci 383:235–240
83. Watenabe K, Yuasa M, Kida T, Teraoka Y, Yamazoe N, Shimanoe K (2010) High-performance oxygen-permeable membranes with an asymmetric structure using $Ba_{0.95}La_{0.05}FeO_{3-\delta}$ perovskite-type oxide. Adv Mater 22:2367–2370
84. Dong FF, Chen YB, Ran R, Chen DJ, Tadé MO, Liu SM, Shao ZP (2013) $BaNb_{0.05}Fe_{0.95}O_{3-\delta}$ as a new oxygen reduction electrocatalyst for intermediate temperature solid oxide fuel cells. J Mater Chem A 1:9781–9791
85. Bréard Y, Michel C, Hervieu M, Studer F, Maignan A, Raveau BB (2002) Large oxygen deficiency in a n = 2 member of the RP series: Sr_3FeCoO_{7-x} (x □ 1.55). Chem Mater 14:3128–3135
86. Daroukh MA, Vashook VV, Ullmann H, Tietz F, Raj IA (2003) Oxides of the AMO_3 and A_2MO_4 -type: structural stability, electrical conductivity and thermal expansion. Solid State Ionics 158:141–150
87. Bochkov DM, Khatton VV, Kovalevsky AV, Viskup AP, Naumovich EN (1999) Oxygen permeability of $La_2Cu(Co)O_{4+\delta}$ solid solutions. Solid State Ionics 120:281–288
88. Yaremchenko AA, Kharton VV, Patrakeev MV, Frade JR (2003) p-type electronic conductivity, oxygen permeability and stability of $La_2Ni_{0.9}Co_{0.1}O_{4+\delta}$. J Mater Chem 13:1136–1144
89. Kharton VV, Yaremchenko AA, Valente AA, Sobyanin VA, Belyaev VD, Semin GL, Veniaminov SA, Tsipis EV, Shaula AL, Frade JR (2005) Methane oxidation over Fe-, Co-, Ni- and V-containing mixed conductors. Solid State Ionics 176:781–791
90. Kovalevsky AV, Kharton VV, Yaremchenko AA, Pivak YV, Tsipis EV, Yakovlev SO, Markov AA, Naumovich EN, Frade JR (2007) Oxygen permeability, stability and electrochemical behavior of $Pr_2NiO_{4+\delta}$-based materials. J Electroceram 18:205–218
91. Ishihara T, Nakashima K, Okada S, Enoki M, Matsumoto H (2008) Defect chemistry and oxygen permeation property of $Pr_2Ni_{0.75}Cu_{0.25}O_4$ oxide doped with Ga. Solid State Ionics 179:1367–1371

92. Miyoshi S, Furuno T, Matsumoto H, Ishihara T (2006) Conductivity and oxygen permeability of a novel oxide $Pr_2Ni_{0.8-x}Cu_{0.2}Fe_xO_4$ and its application to partial oxidation of CH_4. Solid State Ionics 177:2269–2273
93. Yashima M, Yamada H, Nuansaeng S, Ishihara T (2012) Role of Ga^{3+} and Cu^{2+} in the high interstitial oxide-ion diffusivity of Pr_2NiO_4-based oxides: design concept of interstitial ion conductors through the higher-valence d10 dopant and Jahn – Teller effect. Chem Mater 24:4100–4113
94. Yashima M, Yamada H, Sirikanda N, Ishihara T (2010) Crystal structure, diffusion path, and oxygen permeability of a Pr_2NiO_4-based mixed conductor $(Pr_{0.9}La_{0.1})_2(Ni_{0.74}Cu_{0.21}Ga_{0.05})O_{4+\delta}$. J Am Chem Soc 132:2385–2392
95. Yashima M, Enoki M, Wakita T, Ali R, Matsushita Y, Izumi F, Ishihara T (2008) Structure disorder and diffusional pathway of oxide ions in a doped Pr_2NiO_4-based mixed conductor. J Am Chem Soc 130:2762–2763
96. Wei YY, Ravkina O, Klande T, Wang HH, Feldhoff A (2013) Effect of CO_2 and SO_2 on oxygen permeation and microstructure of $(Pr_{0.9}La_{0.1})_2(Ni_{0.74}Cu_{0.21}Ga_{0.05})O_{4+\delta}$ membranes. J Membr Sci 429:147–154
97. Wei YY, Liao Q, Li Z, Wang HH (2013) Enhancement of oxygen permeation through U-shaped K_2NiF_4-type oxide hollow fiber membranes by surface modifications. Sep Purif Technol 110:74–80
98. Kharton VV, Shaula AL, Naumovich EN, Vyshatko NP, Marozau IP, Viskup AP, Marques FMB (2003) Ionic transport in $Gd_3Fe_5O_{12}$- and $Y_3Fe_5O_{12}$-based garnets. J Electrochem Soc 150:J33–J42
99. Shaula AL, Kharton VV, Marques FMB (2004) Mixed conductivity of garnet phases based on gadolinium ferrite. J Eur Ceram Soc 24:1309–1312

Chapter 7
Dual-Phase MIEC Membranes

Abstract In this chapter, the critical factors affecting oxygen permeation through dual-phase membranes are discussed with a view to improving the permeability of dual-phase membranes. Based on the current knowledge in solid-state electrochemistry, two series of dual-phase membranes, made of ceria-based ionic conductors and Fe-based perovskite mixed conductors, are suggested as promising dual-phase materials. Oxygen exchange on the membrane surface, preparation methods, sintering temperature, and phase composition are all critical factors to affect the membrane performance and discussed with experiment demonstration. Other factors, such as the elemental composition of the phases, grain size, grain size ratio between the two phases, the conductivity of the mixed conducting phase, the microstructure, and thickness, all affect the oxygen transport process and deserve thorough investigation.

Keywords Dual-phase membranes • Fluorite • Perovskite • Microstructure effect • Interfacial oxygen exchange • Oxygen permeation • Stability and permeability

7.1 Introduction

When one talks about MIEC membranes for oxygen permeation, a single-phase perovskite (ABO_3) is usually regarded as the typical structure for MIEC membranes. The name "perovskite membranes" has almost become synonymous with MIEC membranes. The overall stability of perovskite oxides includes their thermodynamic and kinetic stabilities. The thermodynamic stability depends mainly on the average bond energy (ABE) of the A- and B-site ions with oxygen. High ABE values lead to high stability, but the motion of oxygen ions in the lattice becomes difficult at high ABE values. The kinetic stability of perovskite-type membranes depends on the self-diffusion coefficients of the metal cations. When a chemical potential gradient generated by oxygen is applied across a membrane, both oxygen ions and metal cations are driven from one side to the other. The kinetic demixing rate largely depends on the self-diffusion coefficients of the cations, the temperature, atmosphere, membrane microstructure, oxygen chemical potential gradient across the membrane, and the oxygen exchange rates at the surfaces. Obviously, the ABE values of a perovskite material have a great influence on the self-diffusion

X. Zhu, W. Yang, *Mixed Conducting Ceramic Membranes*, Green Chemistry and Sustainable Technology, DOI 10.1007/978-3-662-53534-9_7

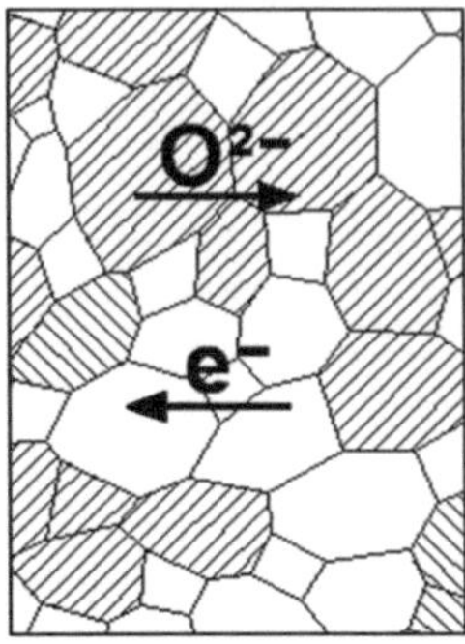

Fig. 7.1 Traditional dual-phase membranes

coefficients of the cations in the lattice. As a result, it is difficult to obtain a membrane with excellent permeability and stability simultaneously.

However, another important structure, i.e., dual-phase membranes, is attracting more and more attention. The "dual-phase" concept was developed only a few years after the single-phase perovskite membranes. In a dual-phase membrane, there is one phase for the diffusion of oxygen ions and another phase for the transport of electrons, so the transport paths of ions and electrons are separated in the membrane bulk. This kind of membrane is made of oxygen ionic conductors and electronic conductors. Figure 7.1 gives a schematic diagram of dual-phase membranes for oxygen separation. The motivation for developing dual-phase membranes is to design membranes with both high oxygen permeability and stability, since researchers have found that stability and permeability are mutually restricted for membranes with a single-phase perovskite structure, as illustrated in Chap. 6.

7.2 Traditional Dual-Phase MIEC Membranes

Initial research was focused on the cermet dual-phase membranes made of noble metals (such as Au, Pd, Pt, Ag, etc.) and oxygen ionic conductors (such as Bi_2O_3-, ZrO_2-, and CeO_2-based solid electrolytes) [1–3]. The dual-phase concept was first proposed by Mazanec and coworkers in 1992 in the investigation of membrane reactors for partial oxidation of methane to syngas [4]. They employed yttria-stabilized zirconia (YSZ) as the oxygen ionic conductor and used Pd, Pt, $In_{0.9}Pr_{0.1}O_{1.5+\delta}$, and $In_{0.95}Pr_{0.025}Zr_{0.025}O_{1.5+\delta}$ as the electronic conductor, with a volume ratio of around 1:1. A high permeation flux of up to 7.8 mL cm^{-2} min^{-1} was achieved through a 0.3-mm membrane at 1100 °C when air and hydrogen were fed to the two sides of the $In_{0.95}Pr_{0.025}Zr_{0.025}O_{1.5+\delta}$–YSZ dual-phase membrane, respectively. To prove the dual-phase concept for an oxygen separation process, Chen et al. prepared several dual-phase membranes made of YSZ and palladium [2], erbia-stabilized bismuth oxide (BE), and noble metals (silver, gold) [1, 5, 6]. A method for mechanically mixing the two kinds of powders was used to prepare the composite powders. An oxygen flux of 0.21 mL cm^{-2} min^{-1} was achieved through a BE–Ag dual-phase membrane with a thickness of 0.23 mm at 750 °C under an oxygen partial pressure gradient of 0.21 atm/0.056 atm [6]. The examination of the effect of membrane thickness on permeability revealed that the oxygen transport kinetics were controlled by the diffusion of oxygen ions in the membrane bulk as the thickness was larger than 1.0 mm. A high oxygen permeation flux of 0.78 mL cm^{-2} min^{-1} was achieved through a $Bi_{1.5}Y_{0.3}Sm_{0.2}O_3$ (BYS)–Ag dual-phase membrane with thickness of 1.3 mm at 850 °C under an oxygen partial pressure gradient of 0.21 atm/0.009 atm [7]. However, the permeation fluxes of the YSZ-based dual-phase membranes are more than one order lower than those of the bismuth oxide–silver (or other noble metals) membranes. Although an asymmetric dual-phase membrane of YSZ–Pt was prepared with a dense layer of less than 10 μm by a chemical vapor deposition (CVD) technique, only a low permeation flux of around 0.06 mL cm^{-2} min^{-1} was achieved at 1050 °C [8]. A high volume fraction of up to 40 vol.% is usually required to make the metal phase form a continuous conduction network across the bulk of the membrane. As a result, the high cost makes this kind of dual-phase membrane far from commercially viable. Furthermore, their much lower permeability compared with that of perovskite membranes does not attract much attention from researchers. Therefore, less than twenty papers on the cermet dual-phase MIEC membranes have been published up until now.

To reduce the material costs of the dual-phase membranes, electronic conducting oxides have been suggested as replacements for the noble metals. This kind of ceramic dual-phase membrane was first proposed by Mazanec and coworkers in 1992 [4], but it was not concerned until 8 years later that a detailed investigation on this type of membrane (like $Ce_{0.8}Gd_{0.2}O_{1.9}$–$La_{0.7}Sr_{0.3}MnO_3$) was reported by Kharton and coworkers [9]. Soon after, the dual-phase membranes $Ce_{0.8}Sm_{0.2}O_{1.9}$–$La_{0.8}Sr_{0.2}Co_{0.2}Fe_{0.8}O_3$, $Ce_{0.8}Gd_{0.2}O_{1.9}$–$Gd_{0.7}Ca_{0.3}CoO_3$, $La_{0.8}Sr_{0.2}Ga_{0.8}Mg_{0.2}O_3$–$La_{0.8}Sr_{0.2}Co_{0.2}Fe_{0.8}O_3$, $Ce_{0.8}Sm_{0.2}O_{1.9}$–$La_{0.8}Sr_{0.2}CrO_3$, and $Zr_{0.6}Y_{0.4}O_{1.8}$–$La_{0.7}Sr_{0.3}MnO_3$, among others, were reported by several groups [10–15]. Most of these membranes have high stability but all show poor permeability. In addition, the oxygen permeation process is unstable over prolonged periods. For $Ce_{0.8}Gd_{0.2}O_{1.9}$–$La_{0.7}Sr_{0.3}MnO_3$, the oxygen flux decreases with time and does not reach a steady state [9]. However, the oxygen fluxes of $Ce_{0.8}Sm_{0.2}O_{1.9}$–$La_{0.8}Sr_{0.2}Co_{0.2}Fe_{0.8}O_3$ and $Ce_{0.8}Sm_{0.2}O_{1.9}$–$La_{0.8}Sr_{0.2}CrO_3$ initially increase and then achieve steady states [10, 14]. In addition to perovskite-type oxides, spinel-type oxides, such as $NiFe_2O_4$, $NiCo_2O_4$, $MnFe_2O_4$, $MnCo_2O_4$, etc., have been employed as the electronic conductor in dual-phase systems [16–22]. Ceria-based oxygen ionic conductors are the most commonly used ionic conductors in dual-phase membranes because of their high ionic conductivity and good chemical compatibility with perovskite and spinel oxides. There has been no detailed investigation of these dual-phase membranes because their permeation fluxes are far lower than that required for practical purposes. The poor permeability of this type of dual-phase membrane is related to an effect blocking the electronic conducting phase transporting oxygen ions across membrane bulk. Figure 7.1 gives a schematic diagram of oxygen ionic and electronic transport through the bulk of the membrane. The ionic conductivity of the perovskite and spinel oxides is far lower than that of the ionic conductors. Therefore, the electronic conducting phase in this type of dual-phase membrane not only blocks the ionic diffusion in the membrane bulk but also significantly reduces the volume fraction of the ionic conducting phase. As a result, the ionic transport paths across the dual-phase membranes are seriously limited by the electronic conducting phase.

7.3 New Type of Dual-Phase MIEC Membranes

Recently, a new type of dual-phase membranes comprising an oxygen ionic conductor and a mixed conductor, and showing high permeability and stability, was reported by our group [23–36]. In this configuration, the ionic conducting phase is employed for the transport of oxygen ions as usual, while the mixed conducting phase is employed for the transport of both oxygen ions and electrons, rather than only electrons as with a noble metal phase or a pure electronic conducting ceramic phase. The mixed conducting phase is the key to improving the permeability, because the ionic diffusion in the bulk membrane is not blocked by the mixed conducting phase, and the volume fraction of the ionic conductor is always equal to unity because of the high ionic conductivity of the mixed conducting phase, as illustrated by Fig. 7.2. Therefore, it is expected that the new dual-phase membranes, consisting of an oxygen ionic conductor and a mixed conductor, will have a high oxygen permeation flux.

The separation of the ionic and electronic transport paths allows the respective design and selection of the chemical composition of the two phases. However, various interactions between the two phases, including chemical, thermal, and mechanical, among others, introduce further complications compared with single-phase membranes. Therefore, there are many new factors that may influence the performance of dual-phase membranes in addition to those frequently encountered

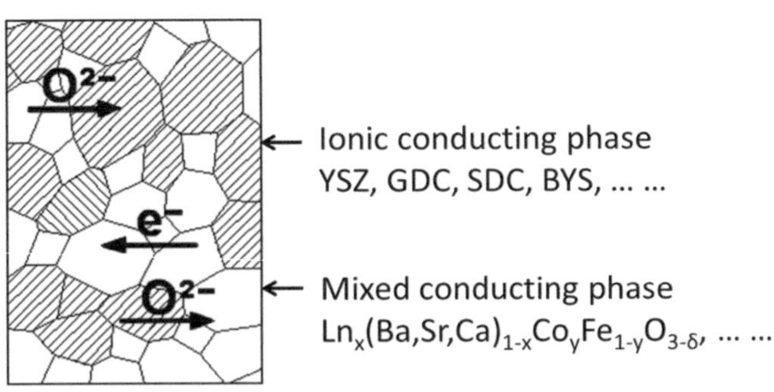

Fig. 7.2 New type of dual-phase membranes

in single-phase perovskite membranes. Furthermore, the performance behaved and rules applied may be very different from those occurring in single-phase perovskite membranes even for the same factors acting on the dual-phase membranes. Therefore, a summary of the achievements on dual-phase membranes by our group and other researchers is presented to identify the critical factors determining oxygen permeation through dual-phase ceramic membranes and also to throw light on the development of new dual-phase membranes.

7.3.1 Design of Dual-Phase Membranes with High Stability and Permeability

For dual-phase membranes, the transport of oxygen ions and electrons takes place separately in the two phases. This facilitates tailoring of the stability and permeability of the membranes by choosing the ionic and electronic conducting phases, respectively. However, the design of dual-phase membranes must follow the basic principles of solid-state chemistry. MIEC membranes can be used in many fields, for instance, as membrane reactors for the conversion of natural gas to syngas, selective oxidation of alkanes to alkenes, and production of high-purity hydrogen from water splitting and as membrane separators for high-purity oxygen production and for CO_2 capture by integration with an oxyfuel process. Different applications have different requirements for the membrane materials. High permeability may be more important for some applications than stability and vice versa. Suitable design of the dual-phase membranes can quickly generate high-performance membrane materials. Three basic requirements must be considered in the design of dual-phase membranes, i.e.:

1. High stability under working conditions
2. Considerable ionic conductivity under working conditions
3. Good chemical compatibility of the two phases under working conditions

Here, we give two examples, presented in Figs. 7.3 and 7.4, for the design of dual-phase membrane materials for two potential applications: one as a membrane reactor for the conversion of natural gas to syngas and the other as a membrane separator for CO_2 capture integrated with an oxyfuel process.

When a dual-phase membrane is used as a membrane reactor for natural gas conversion to syngas, the retentate side of the membrane is exposed to an oxidizing atmosphere, while the reaction side is exposed to a strong reducing atmosphere containing hydrogen and carbon monoxide. Therefore, the membrane materials are required to keep their crystal structure under both oxidizing and reducing environments. First of all, an oxygen ionic conductor with high stability must be selected from those currently available for dual-phase membranes. Although bismuth-based fluorite or Aurivillius oxides have the highest ionic conductivities among the known oxygen ionic conductors, they are all easily reduced to metals under a low oxygen partial pressure. Their stability is nowhere near that required for a

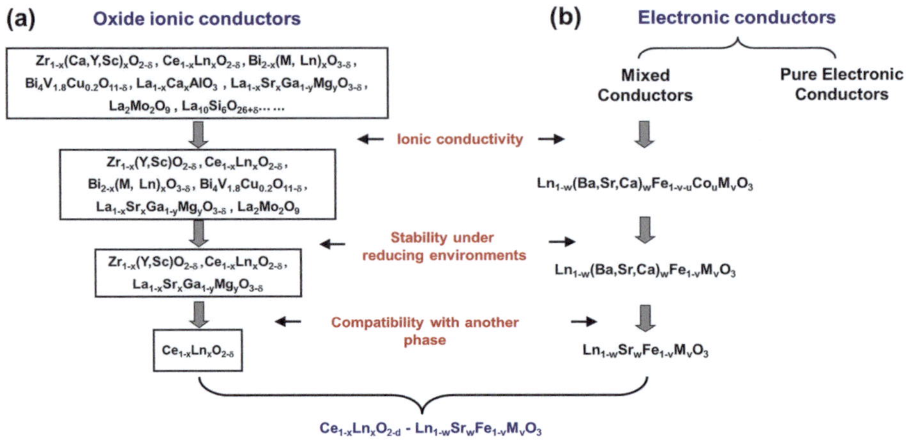

Fig. 7.3 The design of dual-phase membrane materials for the conversion of natural gas to syngas

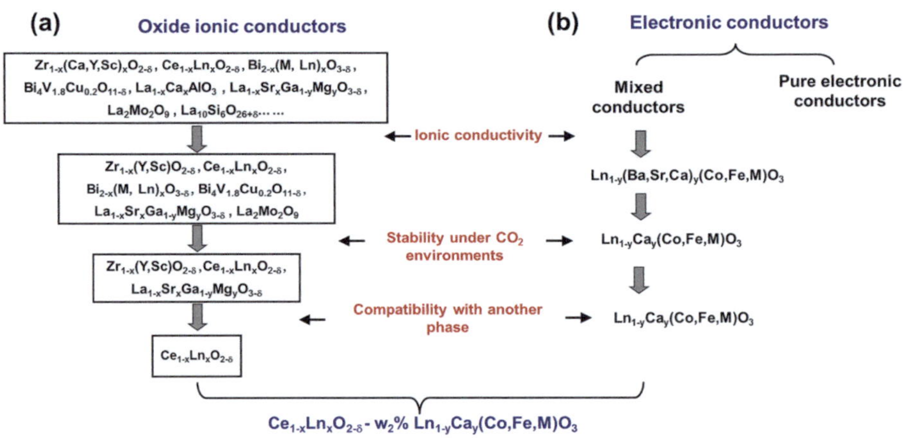

Fig. 7.4 The design of dual-phase membrane materials as membrane separator for CO_2 capture by integrating with oxyfuel process

membrane reactor for syngas generation. $La_2Mo_2O_9$ and its substitutes are also unstable in reducing environments at high temperatures [37]. In addition, the volatilization of MoO_3 during sintering and high-temperature operation limits their applications. The newly invented $NaBi_{0.5}Ti_{0.5}O_3$ is unstable under a strong reducing atmosphere at elevated temperature or in an atmosphere containing CO_2; therefore, it is also ruled out in the first round of selection. The second criterion for selecting an ionic conducting phase for a dual-phase membrane is high ionic conductivity. High oxygen ionic conductivity is a prerequisite for obtaining high oxygen permeability. Among the ionic conductors left after the first round selection, Y- or Sc-stabilized zirconia, doped ceria, and $La_{1-x}Sr_xGa_{1-y}Mg_yO_{3-\delta}$ ($0 \leq x \leq 0.2$, $0 \leq y \leq 0.2$) possess high ionic conductivity and are suitable as the ionic conducting phase in a dual-phase membrane. The third criterion for the selection of the ionic conducting phase is chemical compatibility with the potential electronic conducting phase. Perovskite-type or spinel oxides (such as $MnFe_2O_4$, $MnCo_2O_4$, $NiCo_2O_4$) are good electronic conductors at high temperatures. Considering that a series of high-temperature (usually 1000–1500 °C) steps are needed to prepare dual-phase membranes, and they are operated at temperatures higher than 800 °C, the chemical compatibility between the ionic conducting phase and electronic conducting phase is critical to obtain a dual-phase membrane with high permeability and permeation stability. Therefore, when the chemical compatibility between the ionic and electronic conducting phases is considered, only doped ceria remains. Zirconia-based materials easily react with perovskite-type oxides to produce zirconates and pyrochlores when sintered at elevated temperatures, and the resulting new compounds have poor ionic and electronic conductivity. The reactions between Y/Sc-stabilized zirconia and spinel oxides are not as severe as those between perovskite-type oxides, but the cubic/tetragonal structures with high ionic conductivity are degraded to monoclinic structures with poor ionic conductivity after elevated-temperature sintering. The elements in electronic conducting perovskite oxides easily diffuse into the $La_{1-x}Sr_xGa_{1-y}Mg_yO_{3-\delta}$ lattice and produce a new perovskite oxide at high temperatures, and the new materials produced usually have normal stability and permeability. In the investigation of $La_{0.8}Sr_{0.2}Ga_{0.8}Mg_{0.2}O_{3-\delta}$–$La_{0.8}Sr_{0.2}Co_{0.2}Fe_{0.8}O_{3-\delta}$, Kharton et al. found that the composite material becomes almost a pure perovskite phase after sintering at high temperature [13]. The similar crystal structures of the two perovskite oxides above facilitate the diffusion of metal ions from one perovskite oxide to the other. Similarly, almost all the elements making up spinel oxides can be substituted by Ga $^{3+}$ and Mg^{2+} to produce new composite systems such as $La_{0.8}Sr_{0.2}$(Ga, Mg, Mn, Fe, Co)O_3 + (Ga, Mg, Fe, Co)O_4 or even more complex phases. Therefore, these two kinds of ionic conductors (zirconia-based or $LaGaO_3$-based perovskite oxides) are unsuited as oxygen ionic conducting phases in dual-phase membranes except compatible electronic conducting oxides are developed.

In selecting the electronic conducting phase for dual-phase membranes, if the stability under a reducing environment is considered (i.e., for the purpose of converting natural gas to syngas in a membrane reactor), perovskite or spinel oxides doped with cobalt or other reducible metal ions are excluded. Electronic conducting oxides can be classified into two groups: (1) pure electronic conductors (or those where the ionic conductivity is extremely low), such as the perovskite-type $La_{1-x}Sr_xMn(Cr)O_3$ and spinels $(M,M')_3O_4$), and (2) mixed conductors (such as $Ln_x(Ba,Sr,Ca)_{1-x}Co_yFe_{1-y}O_{3-\delta}$). If a pure electronic conductor, or a mixed conductor with extremely low ionic conductivity, such as $La_{0.8}Sr_{0.2}Co_{0.2}Fe_{0.8}O_3$, is used as the second phase in the dual-phase membrane, it blocks the transport of oxygen ions between ionic conductor grains in the bulk membrane. For example, when $La_{0.8}Sr_{0.2}Co_{0.2}Fe_{0.8}O_{3-\delta}$, a mixed conductor with very low ionic conductivity, is used, the permeation flux is as low as 0.05 mL cm^{-2} min^{-1} at 950 °C for the $Ce_{0.8}Gd_{0.2}O_{1.9}$–$La_{0.8}Sr_{0.2}Co_{0.2}Fe_{0.8}O_{3-\delta}$ dual-phase membrane with thickness of 1 mm [10], which is even lower than that of the $Ce_{0.8}Gd_{0.2}O_{1.9}$–$La_{0.7}Sr_{0.3}MnO_3$

dual-phase membrane under the same conditions [9]. However, when a mixed conductor with significant ionic conductivity is used as the second phase, it will contribute to, rather than block, the transport of oxide ions, as illustrated in Fig. 7.2. It is expected that dual-phase membranes made up of an ionic conductor and a mixed conductor would have higher oxygen permeation fluxes than those made of a pure electronic conductor or a mixed conductor with poor ionic conductivity. Therefore, only the iron-based perovskite with high ionic conductivity can be employed as the electronic conducting phase. If the compatibility between doped ceria and the $Ln_x(Ba,Sr,Ca)_{1-x}FeO_{3-\delta}$ perovskite is considered, barium and calcium may be excluded because barium-doped perovskite oxides easily react with ceria to produce $BaCeO_3$ [38], and calcium can easily diffuse into the ceria lattice and decrease the ionic conductivity. Of course, strontium-doped perovskite oxides may also react with ceria to produce $SrCeO_3$. However, the low tolerance factor of 0.87 gives Sr^{2+} a higher chemical potential in $SrCeO_3$ than in $Ln_{1-x}Sr_xFeO_3$, so $SrCeO_3$ does not appear in the dual-phase membranes. Ba^{2+} and Ca^{2+} are not completely excluded from being doped into the perovskite lattice, as much depends on the particular dual-phase system. In general, Sr^{2+}-doped $LnFeO_3$ has better compatibility with the ceria-based ionic conductors than Ba^{2+} and Ca^{2+} counterparts. Therefore, one can obtain dual-phase membranes with the composition $Ce_{1-x}Ln_xO_{2-\delta}$–$Ln_{1-y}Sr_yFeO_{3-\delta}$. Among possible doped ceria, Gd^{3+} and Sm^{3+} doping produces the highest conductivities, so the matched perovskite oxides need to be $Gd_{1-y}Sr_yFeO_{3-\delta}$ and $Sm_{1-y}Sr_yFeO_{3-\delta}$. The perovskite can be improved by doping an appropriate amount of Al^{3+}, Ga^{3+}, $Cr^{3+/4+}$, $Mn^{3+/4+}$, Ti^{4+}, Zr^{4+}, etc., into the B site to improve their stability under reducing conditions. However, the diffusion of the Zr^{4+} ions from the perovskite phase to the ceria fluorite phase must be considered. Because the Zr^{4+} ion is easily incorporated into the ceria lattice, it produces a solid solution with a lower ionic conductivity. Here, the selection of the materials is based on current knowledge, and more reasonable designs for the dual-phase membranes are expected in the future.

In the other example, of the design of a dual-phase membrane for CO_2 capture integrated with an oxyfuel process, high stability under a CO_2 environment is required of the membranes. As a result, the perovskite-type ionic conductors and mixed conductors doped with Ba^{2+} or Sr^{2+} cannot be used to construct dual-phase membranes. Ceria, zirconia, bismuth oxides, and $La_2Mo_2O_9$ are good ionic conductors coupled with high stability under a CO_2 atmosphere and are good choices for the ionic conducting phase. To make an MIEC perovskite oxide with high oxygen ionic conductivity, alkaline earth metal ions are doped into the A site. Among the alkaline earth metal ions doped into perovskite membranes, Ca^{2+}-doped ones demonstrate the highest CO_2 tolerance due to the lower decomposition temperature of $CaCO_3$ (~800 °C at 1 atm CO_2 partial pressure). Therefore, Ca-doped perovskite may be a good electronic conducting phase for the construction of the dual-phase membranes. However, the Ca-doped perovskite oxides usually have much lower ionic conductivity than those of the Ba^{2+}- or Sr^{2+}-doped ones, so they are not the only alternative to the CO_2-tolerant dual-phase membranes. Other perovskite oxides without Ca doping and spinel oxides should be considered even though their ionic conductivity is extremely low, because the stability should be the primary consideration in the application of the dual-phase membranes to CO_2 capture. Bismuth oxide-based ionic conductors have much higher conductivity than others; however, it has been difficult to find an electronic conducting oxide having good chemical compatibility with bismuth oxide-based ionic conductors up till now. However, once a proper electronic conducting oxide is found, the bismuth oxide-based dual-phase membranes will show high permeability and stability under a CO_2 atmosphere. Among the ceria, zirconia, and $La_2Mo_2O_9$ ionic conductors, the ceria-based conductors show the highest conductivity and best chemical compatibility with perovskite and spinel oxides. Therefore, most of the CO_2-tolerant dual-phase membranes are now made of ceria-based ionic conductors and perovskite- or spinel-based electronic conductors. Although the above arrangements are the best for the CO_2-tolerant dual-phase membranes, their permeability is usually lower than those of the dual-phase membranes designed for syngas production under an oxygen partial pressure gradient of air/inert gas. The lower permeability is related to the blockage effect of the electronic conducting phase in the bulk membrane due to their extremely low ionic conductivity. The key to improving the permeability of the CO_2-tolerant dual-phase membranes is to find an electronic conducting phase with both high ionic conductivity and good CO_2 durability.

7.3.2 Comparison Between the Traditional and New Dual-Phase Membranes

In the above section, a mixed conducting phase was deemed better than a pure electronic conducting phase for the construction of dual-phase membranes, because mixed conductors with significant ionic conductivity would contribute to, rather than block, the transport of oxide ions. In the following discussion, a comparative experimental investigation of the two types of dual-phase membranes was performed in our laboratory and some interesting results were obtained. The following two dual-phase membranes were selected to illustrate the difference experimentally. The material 75 wt.% $Ce_{0.85}Sm_{0.15}O_{1.925}$–25 wt.% $Sm_{0.6}Sr_{0.4}FeO_{3-\delta}$ is designated as SDC75–SSF25 and 75 wt.% $Ce_{0.85}Sm_{0.15}O_{1.925}$–25 wt.% $Sm_{0.6}Sr_{0.4}CrO_{3-\delta}$ as SDC75–SSFCr25. Only one element is different in the perovskite phase for the two dual-phase membranes. The difference means that the former dual-phase membrane comprises a perovskite-type mixed conductor and a ceria-based ionic conductor, and the latter comprises a pure electronic conductor and a ceria-based ionic conductor. The two composite powders were prepared via the same procedure, i.e., a one-pot method which will be discussed in detail in Sect. 7.3.4.1, followed by pressing and sintering at the same pressure and temperature. All the metal ions required are mixed in one beaker, so that every element has the same chemical potential in each phase after synthesis and

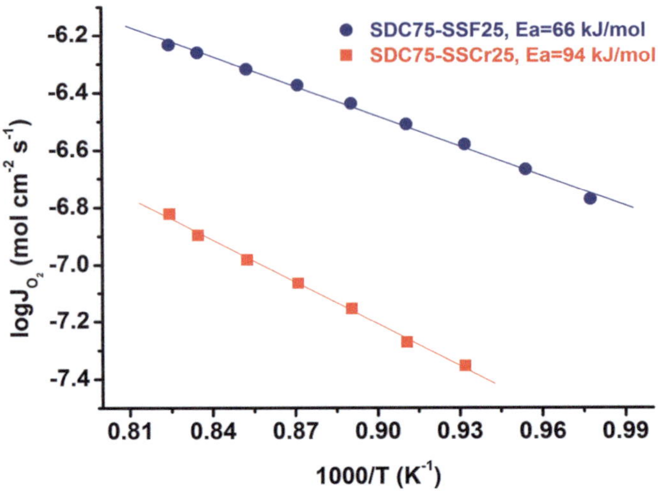

Fig. 7.5 Temperature dependence of oxygen permeation fluxes of dual-phase membranes with both sides coated with $La_{0.6}Sr_{0.4}CoO_3$ porous layers. Thickness: 0.5 mm for SDC75–SSF25 and 0.38 mm for SDC75–SSCr25 (Reproduced from [31] with permission of the Elsevier)

sintering. Therefore, the interdiffusion of metallic elements between the two phases is completed during the preparation process of the dual-phase membranes. XRD data from the as-synthesized powders reveal that no new phase appears, such as $SrCeO_3$, for the two dual-phase membranes. This result demonstrates that the perovskite phases have good compatibility with the fluorite phase even though the powders were prepared by the one-pot method. The compatibility arises because the two phases contain the same rare earth element, samarium, and because strontium has a lower dissolution energy in the $SmFe(Cr)O_{3-\delta}$ perovskite lattice than in the fluorite or $SrCeO_3$ lattices [39, 40].

The oxygen permeation flux for SDC75–SSF25 is more than four times that of SDC75–SSCr25 under the same testing conditions, as shown in Fig. 7.5. The oxygen permeation activation energy of SDC75–SSF25 (66 ± 1 kJ mol^{-1}) is smaller than that of SDC75–SSCr25 (94 ± 2 kJ mol^{-1}) and close to the reported oxygen ionic conduction activation energy of SDC [41]. This reveals that the presence of a mixed conducting phase, SSF, in the dual-phase membrane system has little influence on the ionic transport through the SDC grains. However, the appearance of a pure electronic conductor in the dual-phase system will block the ionic transport and leads to a higher permeation activation energy and lower permeation flux. Table 7.1 lists the oxygen permeation performance of the dual-phase membranes with a pure electronic conducting phase or with a mixed conducting phase. As indicated by the table, all the dual-phase membranes containing a mixed conducting phase exhibit much higher oxygen permeation fluxes than those of dual-phase membranes containing a pure electronic conducting phase.

Table 7.1 The oxygen permeation performance of the typical dual-phase membranes with a pure electronic conducting phase or a mixed conducting phase

Membrane materials	Membrane configurations/ thickness (mm)	T (°C)	J_{O_2} (mL cm^{-2} min^{-1})	Oxygen gradients	Catalysts[b]	References
40 vol.% Pd–YSZ	Disk/2	1100	0.06	Air(0.209 atm)/He(0.014 atm)	N	[2]
40 vol.% Au–BE	Disk/1.03	800	0.046	Air(0.209 atm)/He(0.0015 atm)	N	[5]
40 vol.% Ag–$(Bi_2O_3)_{0.75}(Er_2O_3)_{0.25}$	Disk/1.5	852.5	0.41	Air(0.21 atm)/He(0.046 atm)	N	[1]
40 vol.% Ag–$Bi_{1.5}Y_{0.3}Sm_{0.2}O_3$	Disk/1.3	850	0.78	Air(0.21 atm)/He(0.09 atm)	N	[7]
40 vol.% Ag–$BaBi_8O_{13}$	Disk/1.5	650	0.1	Air(0.21 atm)/He(0.012 atm)	N	[3]
40 vol.% Ag–$(Bi_2O_3)_{0.74}(SrO)_{0.26}$	Disk/1	700	0.067	Air(0.209 atm)/He(0.0024 atm)	N	[42]
60 wt.% $Ce_{0.9}Gd_{0.1}O_{2-\delta}$–40 wt.% Fe_2O_3	Disk/0.5	950	0.1	Air (150 mL min^{-1})/He (29 mL min^{-1}) + Ne (1 mL min^{-1})	N	[16]
60 wt.% $Ce_{0.9}Gd_{0.1}O_{2-\delta}$–40 wt.% $NiFe_2O_4$	Disk/0.6	950	0.18	50 % O_2 (150 mL min^{-1})/He (29 mL min^{-1}) + Ne (1 mL min^{-1})	N	[17]
60 vol.% $Ce_{0.8}Tb_{0.2}O_{2-\delta}$–40 vol.% $NiFe_2O_4$	Disk/0.77	975	0.2	Air (100 mL min^{-1})/Ar (300 mL min^{-1})	N	[18]
50 wt.% $Ce_{0.8}Tb_{0.2}O_{2-\delta}$–50 wt.% $NiFe_2O_4$	Disk/0.59	900	0.95	Air (100 mL min^{-1})/Ar (300 mL min^{-1})	Y	[18]
40 wt.% $Ce_{0.8}Tb_{0.2}O_{2-\delta}$–60 wt.% $NiFe_2O_4$	Disk/0.68	950	0.9	Air (100 mL min^{-1})/Ar (300 mL min^{-1})	Y	[18]
60 vol.% $Ce_{0.8}Sm_{0.2}O_{2-\delta}$–40 vol.% $La_{0.8}Sr_{0.2}CrO_{3-\delta}$	Disk/1	900	0.09	Air/He	Y	[14]
YSZ-$La_{0.8}Sr_{0.2}MnO_{3-\delta}$	Tubular/0.16	950	0.28	Air/He(30 mL min^{-1})	N	[43]
75 wt.% $Ce_{0.8}Sm_{0.2}O_{2-\delta}$–25 wt.% $SmMn_{0.5}Co_{0.5}O_{3-\delta}$	Disk/0.5	900	0.37	Air(0.21 atm)/He(0.005 atm)	Y	[30]
75 wt.% $Ce_{0.8}Sm_{0.2}O_{2-\delta}$–25 wt.% $Sm_{0.8}Ca_{0.2}Mn_{0.5}Co_{0.5}O_{3-\delta}$	Disk/0.5	900	0.4	Air(0.21 atm)/He(0.005 atm)	Y	[30]
75 wt.% $Ce_{0.8}Sm_{0.2}O_{1.9}$–25 wt.% $Sm_{0.6}Ca_{0.4}FeO_{3-\delta}$	Disk/0.5	950	0.45	Air(0.21 atm)/He(0.006)	Y	[34]

(continued)

Table 7.1 (continued)

Membrane materials	Membrane configurations/ thickness (mm)	T (°C)	J_{O_2} (mL cm^{-2} min^{-1})	Oxygen gradients	Catalysts[b]	References
67 wt.% $Ce_{0.8}Sm_{0.2}O_{1.9}$–33 wt.% $Sm_{0.6}Ca_{0.4}CoO_{3-\delta}$	Disk/0.5	950	0.55	Air(0.21 atm)/He(0.0066)	Y	[34]
60 wt.% $Ce_{0.8}Gd_{0.2}O_{2-\delta}$–40 wt.% $Ga_{0.2}Sr_{0.8}FeO_{3-\delta}$	Disk/0.5	950	0.8	Air(100 mL min^{-1})/He(30 mL min^{-1})	Y	[23]
75 wt.% $Ce_{0.85}Sm_{0.15}O_{2-\delta}$–25 wt.% $Sm_{0.6}Sr_{0.4}FeO_{3-\delta}$	Disk/0.5	940	0.5	Air(100 mL min^{-1})/He(30 mL min^{-1})	Y	[27]
75 wt.% $Ce_{0.85}Sm_{0.15}O_{2-\delta}$–25 wt.% $Sm_{0.6}Sr_{0.4}Al_{0.3}Fe_{0.7}O_{3-\delta}$	Disk/0.5	950	0.7	Air(100 mL min^{-1})/He(30 mL min^{-1})	Y	[32]
75 wt.% $Ce_{0.85}Sm_{0.15}O_{2-\delta}$–25 wt.% $Sm_{0.6}Sr_{0.4}Cr_{0.3}Fe_{0.7}O_{3-\delta}$	Disk/0.5	950	0.53	Air(100 mL min^{-1})/He(30 mL min^{-1})	Y	[36]
$Zr_{0.8}Y_{0.2}O_{2-\delta}$–$La_{0.8}Sr_{0.2}Cr_{0.5}Fe_{0.5}O_{3-\delta}$	Asymmetric disk/0.06	900	0.041	Air/Ar	N	[44]
$Zr_{0.86}Y_{0.14}O_{1.92}$–$La_{0.8}Sr_{0.2}Cr_{0.5}Fe_{0.5}O_{3-\delta}$	Hollow fiber/0.08[a]	950	0.16	Air/He	N	[45]
80 wt.% $Ce_{0.85}Gd_{0.1}Cu_{0.05}O_{2-\delta}$–20 wt.% $SrFeO_{3-\delta}$	Disk/0.5	900	0.56	Air(100 mL min^{-1})/He(35 mL min^{-1})	Y	[46]
70 vol.% $Ce_{0.8}Sm_{0.2}O_{2-\delta}$–30 vol.% $La_{0.9}Sr_{0.1}FeO_{3-\delta}$	Disk/1.1	900	0.15	Air/He(40 mL min^{-1})	Y	[47]
60 wt.% $Ce_{0.9}Nd_{0.1}O_{2-\delta}$–40 wt.% $Nd_{0.6}Sr_{0.4}FeO_{3-\delta}$	Disk/0.6	950	0.26	Air/He	N	[48]
60 wt.% $Ce_{0.9}Pr_{0.1}O_{2-\delta}$–40 wt.% $Pr_{0.6}Sr_{0.4}FeO_{3-\delta}$	Disk/0.6	950	0.27	Air/He	Y	[49]
50 vol.% $Ce_{0.9}Gd_{0.1}O_{2-\delta}$–50 vol.% $La_{0.6}Sr_{0.4}Co_{0.2}Fe_{0.8}O_{3-\delta}$	Disk/1	900	0.15	Air/He	N	[50]
50 vol.% $Ce_{0.8}Gd_{0.2}O_{2-\delta}$–50 vol.% $La_{0.8}Sr_{0.2}Co_{0.2}Fe_{0.8}O_{3-\delta}$	Disk/1	950	0.08	Air/He(0.03 atm)	N	[10]
75 wt.% $Ce_{0.85}Gd_{0.1}Cu_{0.05}O_{2-\delta}$–25 wt.% $La_{0.6}Ca_{0.4}FeO_{3-\delta}$	Disk/0.5	900	0.61	Air/He	Y	[51]

60 wt.% $Ce_{0.8}Sm_{0.2}O_{2-\delta}$–40 wt.% $Ba_{0.95}La_{0.05}FeO_{3-\delta}$	Disk/0.6	925	0.54	Air(300 mL min^{-1})/He (100 mL min^{-1})	N	[52]
60 wt.% $Ce_{0.8}Sm_{0.2}O_{2-\delta}$–40 wt.% $Ba_{0.95}La_{0.05}Fe_{0.85}Zr_{0.15}O_{3-\delta}$	Disk/1.0	925	0.3	Air(300 mL min^{-1})/He (100 mL min^{-1})	N	[52]
60 wt.% $Ce_{0.8}Sm_{0.2}O_{2-\delta}$–40 wt.% $Ba_{0.95}La_{0.05}Fe_{0.95}Zr_{0.05}O_{3-\delta}$	Disk/1.0	925	0.33	Air(300 mL min^{-1})/He (100 mL min^{-1})	N	[52]
60 wt.% $Ce_{0.9}Gd_{0.1}O_{2-\delta}$–40 wt.% $Ba_{0.5}Sr_{0.5}Co_{0.8}Fe_{0.2}O_{3-\delta}$	Disk/0.5	950	1.79	Air(150 mL min^{-1})/He (100 mL min^{-1})	Y	[53]
60 wt.% $Ce_{0.8}Sm_{0.2}O_{2-\delta}$–40 wt.% $LaBaCo_2O_{5+-\delta}$	Disk/1.0	900	0.28	Air/He	N	[54]
60 wt.% $Ce_{0.8}Sm_{0.2}O_{2-\delta}$–40 wt.% $NdBaCo_2O_{5+-\delta}$	Disk/1.0	900	0.17	Air/He	N	[54]
60 wt.% $Ce_{0.8}Sm_{0.2}O_{2-\delta}$–40 wt.% $SmBaCo_2O_{5+-\delta}$	Disk/1.0	900	0.09	Air/He	N	[54]
60 wt.% $Ce_{0.8}Sm_{0.2}O_{2-\delta}$–40 wt.% $YBaCo_2O_{5+-\delta}$	Disk/1.0	900	0.08	Air/He	N	[54]
80 vol.% $Ce_{0.8}Sm_{0.2}O_{2-\delta}$−20 vol.% $PrBaCo_2O_{5+-\delta}$	Disk/1.0	940	0.51	Air(150 mL min^{-1})/He (65.8 mL min^{-1})	N	[55]

[a] The thickness of dense layer is not given directly in the literatures but is observable in the SEM images. The wall thickness of the hollow fiber is 0.34 mm, and the thickness of porous layer is about 0.26 mm

[b] *Y* and *N* mean the membrane coated with or without catalyst, respectively

7.3.3 *Interfacial Oxygen Exchange*

Oxygen exchange on both gas–solid interfaces (or membrane surfaces) is important in the process of oxygen permeation across MIEC membranes. It includes the oxygen reduction reactions (ORR) on the surface of the high oxygen partial pressure side and the oxygen evolution reactions (OER) on the surface of the side at low oxygen partial pressure. In the investigation of perovskite membranes, it was found that besides the material composition, temperature, and oxygen partial pressure, the surface microstructure of the membrane is a critical factor influencing both ORR and OER. For a dual-phase membrane, the surface microstructure has a great influence not only on ORR and OER but also on the permeation stability.

As mentioned above, the impedance spectra of the SDC75–SSF25 and SDC75–SSCr25 membranes at elevated temperatures show only a single arc, as shown in Fig. 7.6, which is related to the interfacial exchange processes. The first two intersection points between the origin and the arcs along the abscissa represent the bulk resistance including contributions from the grains and grain boundaries. The values between the first and second intersection points can be used to define the resistance of the membrane surface [28]. Compared with the SDC75–SSF25 membrane, the SDC75–SSCr25 membrane shows much bigger surface resistance and

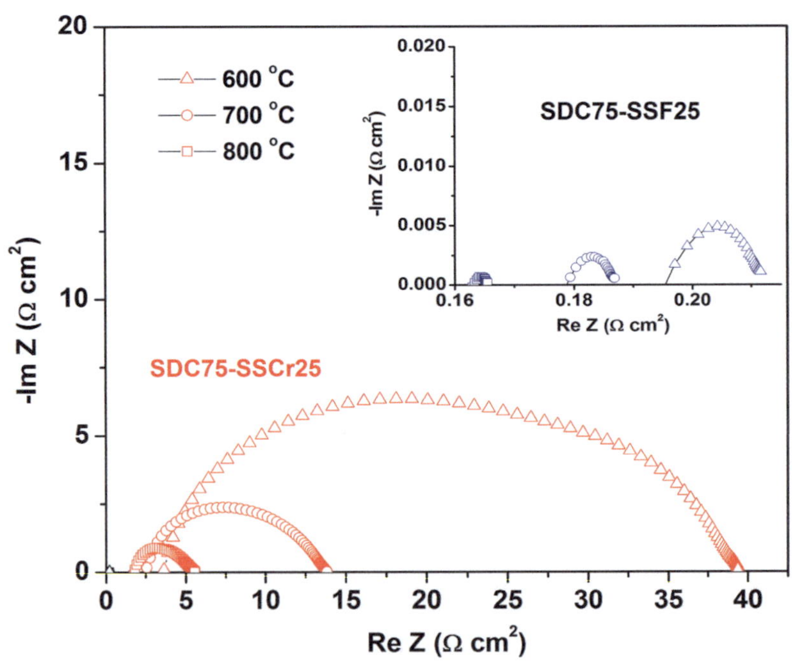

Fig. 7.6 Impedance spectroscopy of SDC75–SSF25 and SDC75–SSCr25 membranes at different temperatures (Reproduced from [31] with permission of the Elsevier)

Table 7.2 Situations of SDC–SSF composite membrane with or without LSC porous layers, oxygen permeation energy (Ea), and permeation unsteady period

	LSC porous layers			
Samples	He side	Air side	Unsteady period (h)	Ea (kJ/mol)
SDC–SSF-I	–	–	42	117 ± 3
SDC–SSF-II	–	+	18	91.4 ± 0.9
SDC–SSF-III	+	–	8	106 ± 3
SDC–SSF-IV	+	+	0	62.1 ± 0.8

Reproduced from [29] with permission of Elsevier
"+, –" represent the membrane coated with or without LSC porous layers, respectively

bulk resistance in the temperature range investigated. The higher surface resistance reveals slow surface exchange kinetics, because the exchange reactions can only take place at the triphase boundaries for SDC75–SSCr25. However, the lower surface resistance of SDC75–SSF25 results from the fast exchange reactions that can take place at the triphase boundaries as well as on the SSF grains. Therefore, with dual-phase membranes made of pure electronic conductors and ionic conductors, the surface exchange kinetics will inevitably be slow except where porous catalyst layers are coated on both surfaces.

Different groups have reported variable oxygen permeation through a number of dual-phase membranes in the initial stages. Usually, it takes tens or hundreds of hours to reach a permeation steady state [14, 17, 23, 25, 27]. The oxygen permeation flux at steady state is often twice the initial value. This unsteady phenomenon is occasionally observed on perovskite membranes, which has been related to the readjustment of the lattice structure from the as-prepared state to an invariant state under permeation conditions or related to the reduction of surface oxygen. Here, the permeation flux at the final steady state is usually lower than the initial value, which is different from that usually observed on dual-phase membranes.

A $La_{0.6}Sr_{0.4}CoO_3$ (LSC) porous layer with a thickness of ~20 μm was coated on one or both sides of the dual-phase membrane to speed up the oxygen exchange on the membrane surfaces, as shown in Table 7.2. The permeation results revealed that when both sides of the membrane were coated by LSC porous layers, the membrane reached the permeation steady state immediately, and showed the highest oxygen flux, higher than those with only one side coated or naked membranes, as shown in Fig. 7.7 [29]. It takes longer for the naked membranes to reach steady state than the membranes coated with LSC layers on the feed or sweep side. For example, the naked membrane needs about 42 h to reach steady state, but the membrane with its sweep side coated with an LSC porous layer needs only 8 h. This unsteady period for the SDC75–SSF25 dual-phase membranes is irreversible. When the membranes reached their permeation steady state, the helium was replaced by air on the permeation side, and after about 50 h, the helium flow was switched back. However, the oxygen flux maintained its steady value. From the above results, we deduced that the unsteady permeation of dual-

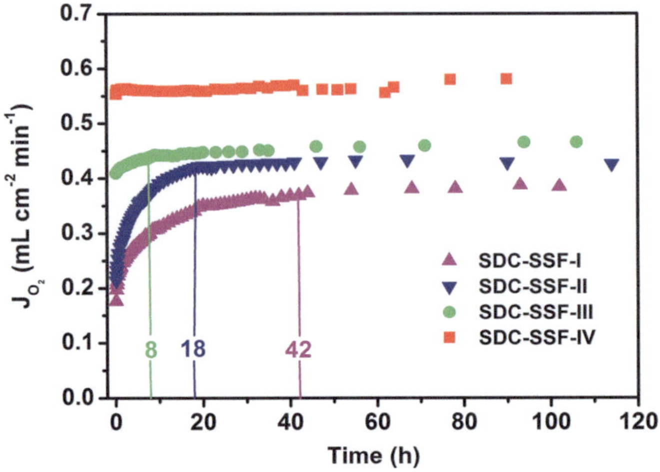

Fig. 7.7 Dependence of oxygen permeation fluxes on time at 940 °C of SDC–SSF dual-phase membranes with or without LSC porous layers on surfaces. Air flow rate, 100 ml min^{-1}; He flow rate, 30 ml min^{-1} (Reproduced from [29] with permission of Elsevier)

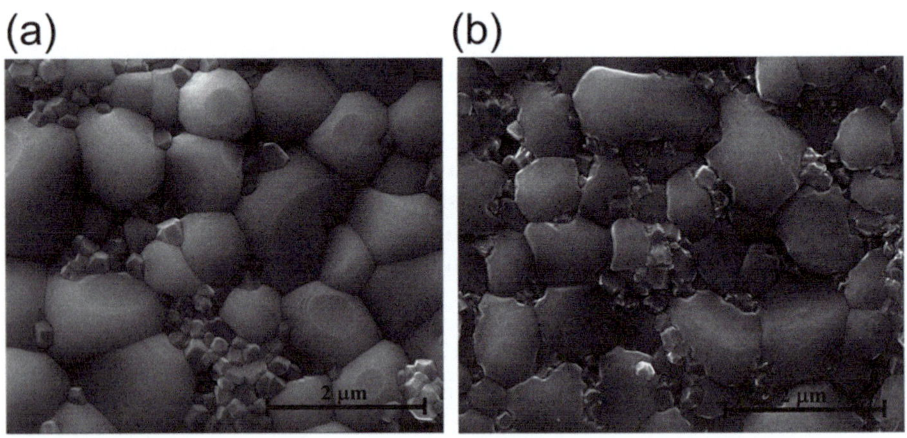

Fig. 7.8 SEM pictures of the fresh SDC–SSF dual-phase membrane and the membranes after reaching the permeation steady state. The LSC porous layers were removed by ultrasonic cleaning before observing them on SEM. (**a**) Fresh membrane; (**b**) naked surface of air side after oxygen permeation (Reproduced from [29] with permission of Elsevier)

phase membranes is related to the surface oxygen exchange on the membrane surface [29]. A microstructure change was observed on the dual-phase membrane surfaces without an LSC porous coat, as shown in Fig. 7.8. Furthermore, the unsteady permeation can be turned on only at higher temperatures up to 940 °C, as shown in Fig. 7.9. For the SDC–SSF dual-phase membrane with the air side coated with a porous LSC layer, the oxygen permeation flux rapidly decreases

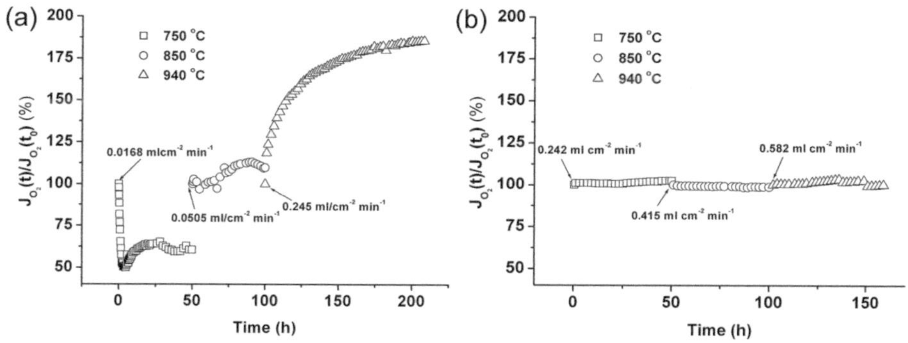

Fig. 7.9 Time dependence of relative oxygen permeation fluxes at different temperatures of (**a**) SDC–SSF-II and (**b**) SDC–SSF-IV dual-phase membranes coated with LSC porous layer on air side and on both sides, respectively. Air flow rate, 100 ml min^{-1}; He flow rate, 30 ml min^{-1} (Reproduced from [29] with permission of Elsevier)

with time during the first 7 h at 750 °C and then increases slightly. The decrease in flux at the initial stage is due to a decrease in the initial rate of surface oxygen desorption [29]. When the operating temperature was increased from 750 to 850 °C, the initial flux increased from 0.0168 to 0.0505 mL cm^{-2} min^{-1}, which is still much lower than the value obtained after reaching a steady state at the same temperature. During the 50 h operation at 850 °C, the permeation flux increases by only ~12 %; however, when the operating temperature was increased to 940 °C, the initial flux increased to 0.245 mL cm^{-2} min^{-1}, and the flux increased by ~90 % during the 100 h of operation and reached a value similar to that shown in Fig. 7.8. This reveals that an increase in operating temperature can accelerate the changes in the surface structure. However, for the membrane with both sides coated with porous LSC layers, the oxygen permeation is constant even at the low temperature in the initial stage.

From the above experimental results, one can conclude that the unsteady permeation is related to oxygen exchange reactions at the gas–solid interface. If there is no LSC porous layer, the gas–solid interface is exactly the same as the membrane surface; if the membrane surface is coated with an LSC porous layer, the gas–solid interface is located in the porous layer. For the dual-phase membranes, their interface microstructures are changed by the asymmetric oxygen partial pressure gradient across the membranes from the as-prepared state to a new state under oxygen permeation conditions. Therefore, when there is no LSC porous layer, the adjustment of the microstructure is on the membrane surface; if the membrane surface is coated by an LSC porous layer, the adjustment is on the porous layer. However, on the porous structure, the changes in morphologies are not notable. It can be anticipated that the adjustment of the microstructure on the LSC porous layer would be much faster than that on the dense dual-phase membrane surface due to the large specific area and the fine oxide grains. Consequently, there is no unsteady period observed for the membrane with LSC coated on both sides.

7.3.4 Microstructure Effects

It is well known that the ceramic microstructure has a great influence on the performance of the membrane material. Grain sizes, grain size distribution, and grain boundaries, as well as the surface morphology, are important ceramic microstructure facets. Compared with the structure of a single phase, the dual-phase arrangement shows many new and unique microstructural characteristics, for instance, the relative grain sizes of the two phases, the grain boundaries between the two phases, and the distribution of the grains of the two phases between the bulk and surfaces of the membranes. These microstructural characteristics may affect the ionic and electronic transport in the bulk, and the oxygen activation on the membrane surface, while they are easily affected by many factors, such as the preparation methods of the composite powders, the elemental composition of the two phases, and the sintering temperature of the dual-phase membranes. In this section, the effects of these microstructure factors on membrane performance are discussed to elucidate the unique characteristics of the dual-phase membrane materials.

7.3.4.1 Preparation Methods of the Composite Powders

Significant differences in the reported permeability have always been found in literatures, even for the same oxygen-permeable materials. One important factor influencing the reproducibility is the synthesis method of the composite powders. There are many methods available for preparing perovskite powders, such as the solid-state reaction (SSR) method, the EDTA–citric acid (EC) process, the glycine–nitrate combustion process (GNP), and the chemical co-precipitation method, among others. A detailed introduction to these methods is given in Chap. 4. Powders synthesized via different processes have different particle sizes, sintering activity, and chemical composition. All these factors will lead to differences in the microstructure after the ceramic membranes have been sintered at elevated temperatures. Qi et al. prepared $La_{0.8}Sr_{0.2}Co_{0.6}Fe_{0.4}O_{3-\delta}$ dense membranes by four methods, a sol–gel method with citrate as the complexing agent, the solid-state reaction method, the spray pyrolysis method, and a co-precipitation method [56]. They found that the powders prepared by the first three methods had compositions close to the desired stoichiometry, whereas the powder synthesized by the co-precipitation method had a large strontium deficiency due to the filtration and aqueous washing steps involved in this method. For the three membranes with similar composition, the activation energy of oxygen permeation decreases with increasing grain size. The membrane microstructure and composition exert important influences on the oxygen vacancy concentration depending on the oxygen pressure. Usually, perovskite membranes prepared by the solid-state reaction method have been reported to show higher oxygen permeability than those prepared by the sol–gel method. For example, the oxygen permeability of $Ba_{0.5}Sr_{0.5}Co_{0.8}Fe_{0.2}O_{3-\delta}$ and $BaCe_{0.15}Fe_{0.85}O_{3-\delta}$ membranes derived from the solid-state reaction method was higher than that of the membranes synthesized by the sol–gel process [57, 58]. One possible reason is that membranes derived from solid-state reaction method usually have more defects at the grain boundaries and in the lattice, and these cationic/anionic defects can enhance oxygen ion conductivity and improve oxygen exchange. For example, $LaCoO_{3-\delta}$ ceramic membranes can be prepared by several methods, but the membrane prepared using the solid-state reaction method shows the highest oxygen permeability. This is believed to be caused by a low grain boundary resistance to the transport of ionic oxygen [59]. Furthermore, it was reported that mixed conducting oxides synthesized by different methods had different phase compositions. For example, the phase composition of $SrFeCo_{0.5}O_y$ is largely determined by the synthetic method. Materials synthesized by the solid-state reaction method contain a perovskite phase, $SrFe_{1-x}Co_xO_{3-\delta}$, as the major phase and an $Sr_4Fe_{6-x}Co_xO_{13+\delta}$ intergrowth phase and a spinel phase, $Co_{3-x}Fe_xO_4$, as the minor phases; however, the same material synthesized through the sol–gel process usually contains the intergrowth phase, $Sr_4Fe_{6-x}Co_xO_{13+\delta}$, as the major phase [60].

The composite powders for the preparation of dual-phase membranes are usually prepared by simple mixing of the two oxide powders. This method was widely used for the preparation of traditional dual-phase membranes. The effects of the powder preparation methods on the microstructures, chemical composition, defects, conductivity, and oxygen permeability of perovskite materials have been extensively investigated. However, the influences on those corresponding facets of dual-phase membranes had still not been well addressed until several papers were published by our group [25, 28, 32]. In this section, the preparation methods of the composite powders for dual-phase membranes are discussed.

To understand the influence of the distribution of the two phases in a dual-phase membrane on the oxygen permeability, four methods were used to prepare the powders, namely, mechanical mixing of the two powders (method 1), mixing the perovskite phase powder with a solution for the preparation of the fluorite phase (method 2), mixing the fluorite phase powder with a solution for the preparation of the perovskite phase (method 3), and a one-pot method (method 4). The dual-phase membrane with a composition of 75 wt.% GDC–25 wt.% GSF (corresponding to a volume ratio of 71:29) was chosen as an example [25]. In detail, in method 1, the required fluorite and perovskite oxides were prepared by the EDTA–citric acid procedure and then mixed at a weight ratio of 75:25; in method 2, the GDC powder was mixed with a complex solution containing Gd^{3+}, Sr^{2+}, and Fe^{3+}; similarly, in method 3, a complex solution containing Gd^{3+} and Ce^{3+} was mixed with the GSF powder; in method 4, all the required metal nitrates were mixed in one beaker and then the EDTA–citric acid procedure was followed, so this method is also called a one-pot method. After evaporation, all the resultant gels (in methods 2–4) were burnt on an electrical furnace at 600–800 °C to remove most of the organics and then calcined in a muffle at 900 °C for 5 h. The prepared dual-phase membranes are labeled GDC–GSF-1–4, respectively.

The surface morphologies of the as-prepared dual-phase membranes were observed using an SEM, as shown in Fig. 7.10. The white grains are fluorite

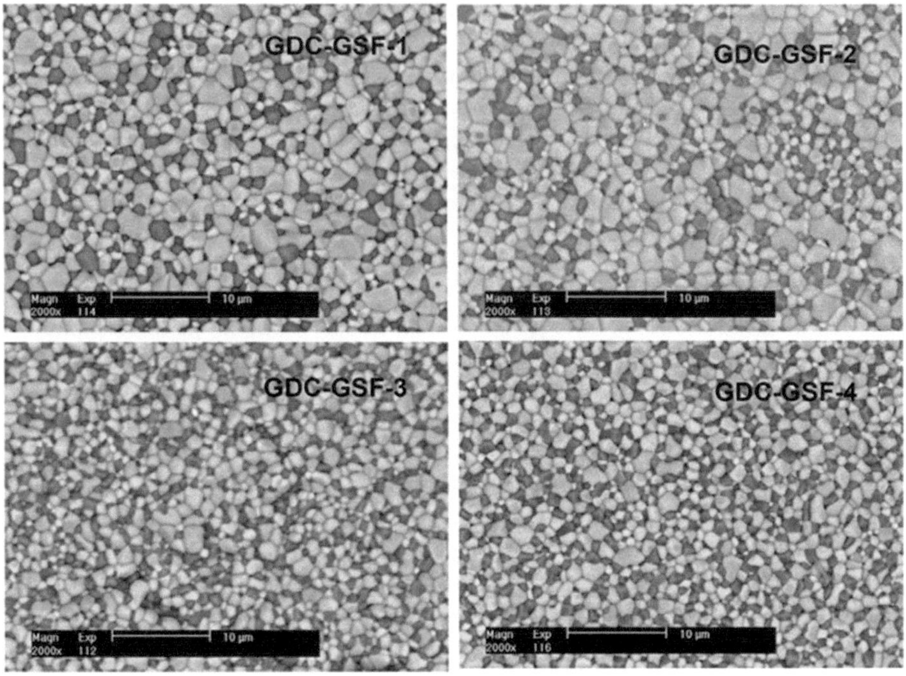

Fig. 7.10 Backscattered electron images of the composite membranes prepared by different methods (Reproduced from [25] with permission of Elsevier)

oxide and the black grains are perovskite oxide. It is easy to see that the uniformity of composition and the grain sizes of the two phases are different for the four membranes produced from the different methods. The uniformity of composition in GDC–GSF-4 is the highest among the four samples. The grain sizes of these membranes are mainly in the range 0.5–3 μm, and the grain size of the perovskite phase is usually smaller than that of the fluorite phase for all membranes. GDC–GSF-4 has a relatively narrow grain size distribution, around 1 μm for the perovskite phase and 1–2 μm for the fluorite phase. Oxygen transport through the membranes is controlled by the ambipolar conductivity of the oxide ions and electrons. In inhomogeneous membranes, like GDC–GSF-2, some perovskite grains were enwrapped by fluorite grains, forming islands. It is certain that the islands make little contribution to the electronic transfer because those are blocked by the surrounding fluorite grains. However, the influence of this blockage is less for homogeneous membranes than inhomogeneous ones. Therefore, the ambipolar conductivity of the homogeneous sample is higher than that of the inhomogeneous sample. Other researchers also found that the more homogeneous the distribution of the two phases in dual-phase membranes, the higher the oxygen permeation flux in investigations of ceria–spinel-type membranes [16, 17, 48]. They also found that the one-pot method gave the most homogeneous mixture of the two phases, as was observed in our investigation. The preparation methods have little influence on the

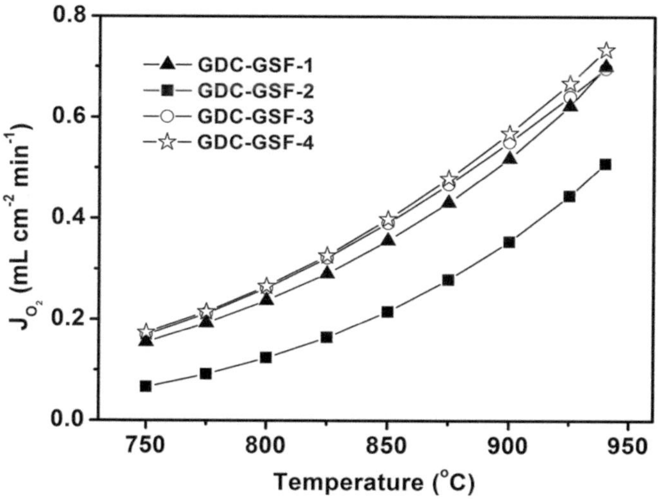

Fig. 7.11 Influence of temperatures on the oxygen permeation fluxes through the composite membranes. Membrane thickness, 0.5 mm; He flow rate, 30 ml min^{-1} (Reproduced from [25] with permission of Elsevier)

permeation stability but a significant effect on the oxygen permeability. Figure 7.11 shows the temperature and time dependence of the oxygen permeation fluxes of the four membranes. The unsteady permeation is observed on all four membranes prepared by different methods. As discussed above, the unsteady permeation is related to oxygen exchange on the membrane surface [29]. However, another important factor that affects the oxygen permeation flux of dual-phase membranes is the oxygen exchange on the membrane surface. For ionic conductor–mixed conductor dual-phase membranes, the oxygen exchange reactions take place at both the perovskite areas and at the triphase boundaries (TPB). As other external conditions cannot be changed, prolonging the lengths of the TPB is a way of improving the homogeneity of the two phases. Therefore, it can be expected that the most homogeneous mixture of the two phases corresponds to the one with longest TPB and will result in the highest oxygen permeability. Experimental results showed a good agreement between the structural homogeneity and the oxygen permeability based on the concept of TPB. Due to the exchange reaction occurring on the TPB, the surface exchange will play a more important role in the oxygen permeation if the dual-phase membrane is a more inhomogeneous mixture of the two phases. Therefore, it can be inferred that the GDC–GSF-2 dual-phase membrane gives the slowest surface exchange rate due to the lowest uniformity of the two phases. On the other hand, the homogeneity can also influence the activation energy for the oxygen permeation, i.e., the activation energy decreases with the homogeneity of the two phases in dual-phase membranes. Greater homogeneous mixing of the two phases leads to higher oxygen permeability and lower permeation activation energy.

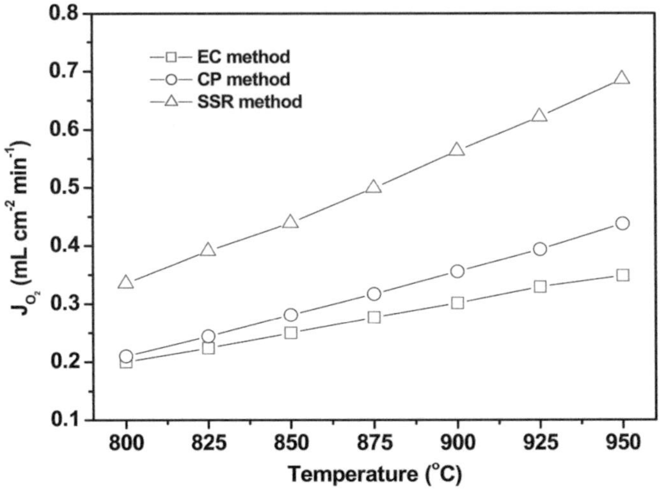

Fig. 7.12 Temperature dependence of the permeation fluxes of the dual-phase membranes derived from the EC, CP, and SSR method. Both sides of the membranes coated with $Sm_{0.5}Sr_{0.5}CoO_{3-\delta}$ porous layers for oxygen activation. Air flow rate, 100 ml min^{-1}; He flow rate, 30 ml min^{-1} (Reproduced from [32] with permission of Elsevier)

The one-pot method is a good and simple method for preparing composite powders with high homogeneity in terms of the distribution of the two phases in the derived dual-phase membranes. As mentioned at the beginning of this section, there are many methods for the preparation of composite powders. Therefore, combining the one-pot synthesis with the traditional methods for powder preparation can reveal some new aspects concerning the effect of membrane microstructure on the performance of dual-phase membranes. The solid-state reaction method is a standard process for preparing ceramic powders. It is very commonly used in preparing mixed conducting ceramic membranes. Usually, membranes derived from the solid-state reaction method have higher oxygen permeability than those derived from liquid-phase synthesis [57–60]. Changes in the processing route may affect the surface exchange processes, and these include the surface concentration of active adsorption centers. Here, both sides of the membranes were coated with SSC porous layers to eliminate oxygen exchange limitations, providing a direct comparison of oxygen ionic conductivity of these dual-phase membranes. Figure 7.12 shows the temperature dependence of the oxygen permeation flux of 75 wt.% $Ce_{0.85}Sm_{0.15}O_{1.925}$–25 wt.% $Sm_{0.6}Sr_{0.4}Al_{0.3}Fe_{0.7}O_{3-\delta}$ (SDC75–SSAF25) dual-phase membranes prepared via different methods, namely, the EDTA–citric acid method (EC, a type of sol–gel method), the solid-state reaction method (SSR), and the co-precipitation method (CP) [32]. All the three methods are examples of one-pot processes, i.e., all the required starting nitrate precursors were mixed together, followed by the individually relevant steps. The SSR-derived membrane shows the highest permeation flux of the three membranes, being about twice that of

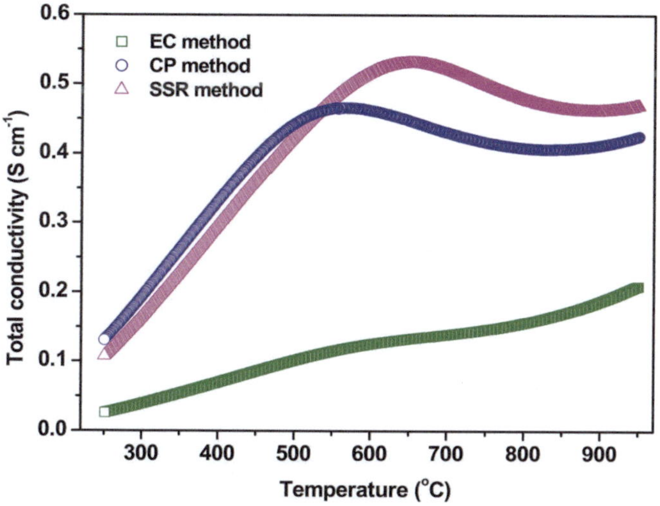

Fig. 7.13 Total conductivity of the dual-phase membranes derived from EC, CP, and SSR methods in air from 950 °C to 250 °C (Reproduced from [32] with permission of Elsevier)

the EC membrane, though there is no difference in phase structure among the three dual-phase membranes. In other words, the microstructure of the membrane has a great influence on the oxygen transport process.

The conductivities of the samples derived from the CP and SSR powders are more than twice that of the sample derived from the EC powder over the temperature range investigated, as shown in Fig. 7.13. The results reveal that the electronic conduction predominated in the CP- and SSR-derived membranes, and the conduction characteristics of the membranes show the behavior of an Fe-based perovskite in the high-temperature region [61]. However, the total conductivity of the EC membrane increases monotonically with an increase in temperature. Considering the ionic conductivity of SDC is ~0.01 S/cm at 600 °C and ~0.1 S/cm at 800 °C, for the EDTA-derived composite the electronic conduction is prevalent when the temperature is lower than 600 °C, and the electronic conductivity is comparable to the ionic conductivity when the temperature is higher than 800 °C [62]. That is to say, the electrical conduction behavior is controlled by the balance of electronic and ionic transport at > 800 °C for the EC membrane, and it shows a conduction behavior that is a composite of those in perovskite and fluorite oxides. For a dual-phase membrane, the amounts of each phase need to be high enough to form continuous phases in the bulk and surfaces for oxygen permeation. In other words, both phases need to exceed the percolation threshold, so the volume fraction of the minor phase should normally be no less than 30 % [63]. However, for the dual-phase materials investigated here, the weight ratio of SDC and SSAF is 75:25, which corresponds to a volume ratio of 74:26. Consequently, the volume fraction of the perovskite phase is smaller than the required percolation threshold. Under this condition, the distribution of the two phases in the membrane bulk is important and determines whether the mixed conducting phase, SSAF, can form a continuous conduction network through the bulk of the membrane. Percolation theory shows that, for a homogeneous composite material, once the volume ratio of the conducting phase is smaller than the percolation threshold, it is difficult to achieve a continuous conduction network through the material [63]. However, for an inhomogeneous composite system, it is still possible to build a continuous conduction network across the material [64]. In other words, once a continuous conducting network is built in a composite system with the volume fraction of conducting phase less than the percolation threshold value, the distribution of the conducting phase in the composite system must be inhomogeneous. In this case, for the membrane derived from the EC method, its low total conductivity is due to the homogeneous distribution of SSAF grains among the SDC grains. For the membrane derived from the CP or SSR method, both the grains are not as uniformly mixed as those in the EC membrane. The inhomogeneous distribution of the grains means the mixed conducting phase SSAF has a chance to form a continuous conduction network through the membrane bulk. This is also verified by a recently published study on a dual-phase membrane containing only 20 vol.% of the electronic conducting phase [55], as shown in Fig. 7.14. In this paper, the researchers developed a coating strategy to fabricate a dual-phase membrane with 20 vol.% $PrBaCo_2O_{5+\delta}$ (PBCO) dispersed in an 80 vol.% $Ce_{0.8}Sm_{0.2}O_{2-\delta}$ (SDC) matrix. In the dual-phase membrane, the fiber-shaped PBCO forms a percolative network and thus provides sufficient electronic conductivity. The oxygen permeation flux through a 1-mm-thick SDC–PBCO membrane was high at up to 0.51 mL cm^{-2} min^{-1} at 940 °C. They attributed the high oxygen permeability to the high volume percentage of the SDC phase in the dual-phase membrane. In their previous work, the PBCO electronic conducting phase in the dual-phase membrane, prepared by a conventional ceramic route, could not form percolative networks until it was at least 40 vol.% [65]. The low SDC content thus leads to low oxygen permeability.

In contrast to the conventional perovskite mixed conducting membranes, here the electronic conduction is the main limiting step for the SDC–SSAF membrane derived from the EC process. The volume fraction of the perovskite phase is smaller than the required percolation threshold, so the distribution of the two phases in the membrane bulk determines the electronic conduction across the membranes, as discussed above. As discussed in the earlier part of this section, the permeation flux of the GDC–GSF dual-phase membrane increases with increasing homogeneity of the two phases [25]. The volume ratio of the two phases is 71:29, very close to the percolation threshold, so the electronic conductivity is high enough for oxygen permeation. This was verified by the investigation of a 75 wt.% $Ce_{0.85}Sm_{0.15}O_{1.925}$–25 wt.% $Sm_{0.6}Sr_{0.4}FeO_{3-\delta}$ (volume ratio of 72:28) dual-phase membrane, where the total conductivity was 0.5–0.6 S cm^{-1} in the temperature range of 600–950 °C [31]. Therefore, improving the homogeneity of the two phases aids the oxygen permeation across dual-phase membranes, but only if electronic transport is not the limiting step.

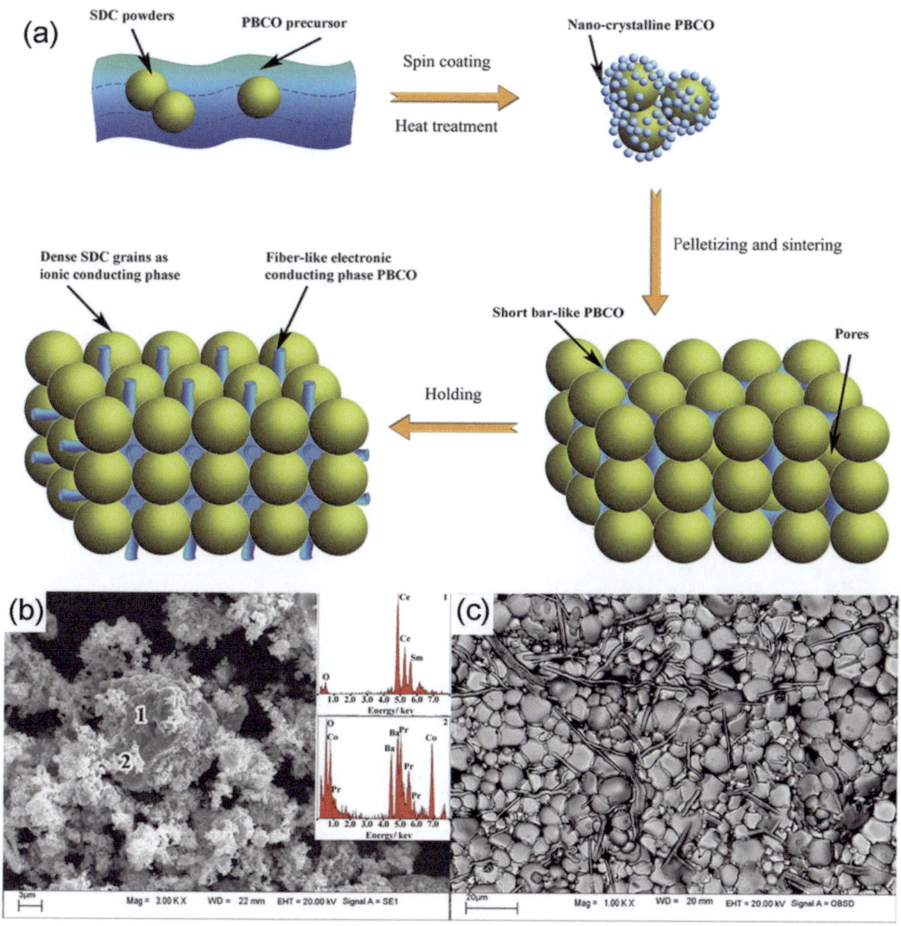

Fig. 7.14 (**a**) Schematic illustration of the novel strategy for the synthesis of SDC–PBCO membrane with fiber-like PBCO phase distributing in the matrix of SDC phase; (**b**) PBCO-coated SDC powders and the insert is the EDS spectrum of the corresponding part in the coated particle; (**c**) backscattered electron micrograph of the prepared SDC–PBCO (8/2) membrane. The fiber-like parts correspond to PBCO phase (Reproduced from [55] with permission of Elsevier)

7.3.4.2 Elemental Composition

Many investigations have revealed that Fe- and Co-based mixed conducting perovskite membranes show great differences in permeation performance, electrical conduction properties, stability, and microstructure [66–68]. However, these differences are not evident when Fe- and Co-based perovskite oxides are employed in dual-phase membranes. Dual-phase membranes, comprising Ca^{2+}-doped perovskite oxides of $Sm_{0.6}Ca_{0.4}CoO_{3-\delta}$ (SCC) or $Sm_{0.6}Ca_{0.4}FeO_{3-\delta}$ (SCF) as the mixed conducting phase and fluorite oxides of $Ce_{0.8}Sm_{0.2}O_{1.9}$ (SDC) as the ionic

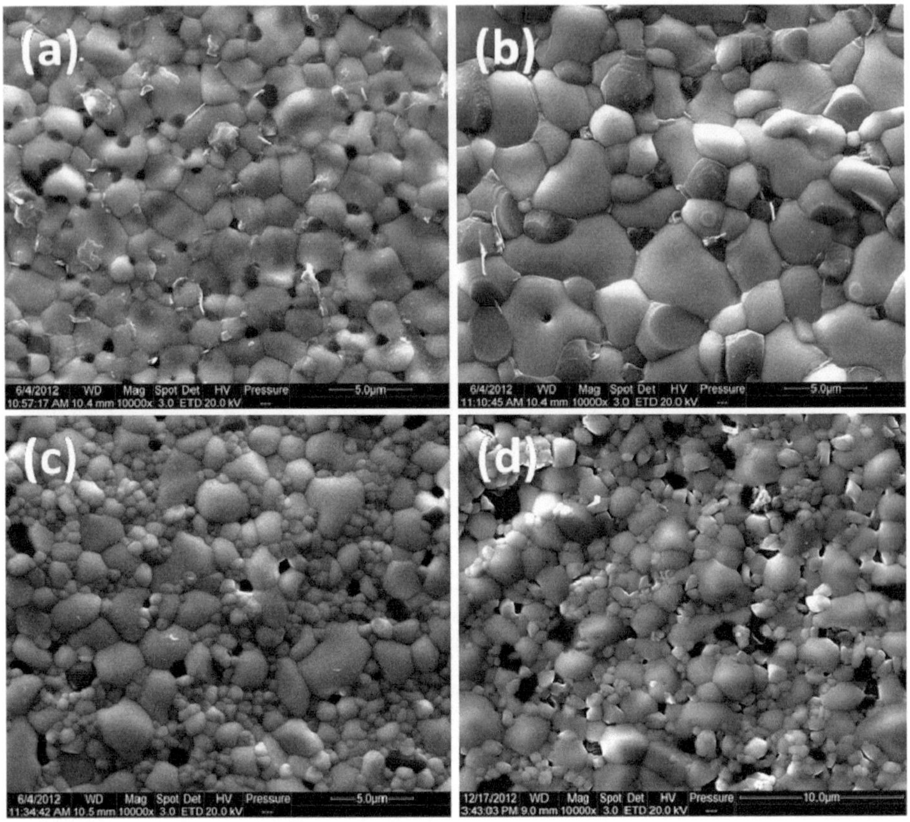

Fig. 7.15 SEM pictures of the dual-phase membranes. Top views: (**a**) SDC75–SCC25, (**b**) SDC67–SCC33, (**c**) SDC75–SCF25, and (**d**) SDC67–SCF33. Top and cross-views of the typical porous coating on membranes (Reproduced from [34] with permission of Elsevier)

conducting phase, were prepared and compared to find out what influences the B-site metal ions of the perovskite phases have on the properties of the dual-phase membranes [34]. The weight ratio of the fluorite phase and the perovskite phase used was 3:1 for the preparation of the dual-phase membranes. All the dual-phase composite oxides contain a fluorite phase and a perovskite phase, and no impurities were detected. That is to say, the two phases have good structural compatibility with each other. The sizes of fluorite phase grain are in the range of 2–5 μm for the two dual-phase membranes; however, the perovskite grains are in the range of 0.5–1 μm for SDC75–SCC25, and 0.3–0.5 μm for SDC75–SCF25, as shown in Fig. 7.15. A much smaller perovskite grain size was achieved in the iron-containing dual-phase membrane though it was sintered at a higher temperature (1370 °C) than that used with the cobalt-containing sample (1200 °C). For dual-phase membranes, the sintering temperature mainly depends on the phase which has the lower sintering temperature. The sintering temperature of the $Ln_{1-x}(Sr,Ca)_xCoO_{3-\delta}$

perovskite is around 1200 °C [66, 69], while that of SDC is high at 1600 °C [70]. Therefore, the SCC perovskite acts as a sintering aid here and helps the densification of SDC. However, the $Ln_{1-x}(Sr,Ca)_xFeO_{3-\delta}$ perovskite has a higher sintering temperature of 1350 °C [71], and as a result the corresponding dual-phase membranes need a higher sintering temperature than the cobalt-containing dual-phase membranes.

For single-phase perovskite membrane materials, their electronic conductivities are almost equal to their total conductivities, because the oxygen ionic transfer numbers are usually of the order of 10^{-2}–10^{-4}. The total conductivities are 0.12, 630, 22, 0.17, and 1.0 S cm^{-1} at 950 °C for SDC, SCC, SCF, SDC75–SCC25, and SDC75–SCF25, respectively. It can be estimated that the electronic conductivities of SDC75–SCF25 are high enough compared with the ionic conductivity of SDC. However, that of SDC75–SCC25 is insufficient for oxygen permeation. It is interesting to find that the total conductivity of the iron-containing dual-phase membrane is higher than that of the cobalt-containing one though the total conductivity of SCC is ~30 times higher than that of SCF at 950 °C. The weight ratio of 75:25 of the two phases corresponds to a volume ratio of 71:29. In general, the volume fraction of one phase should be larger than 30 % in a dual-phase system to exceed the percolation threshold. Here, the perovskite volume fractions of SDC75–SCC25 and SDC75–SCF25 are very close to the percolation threshold. However, only the latter demonstrates the presence of a continuous electronic conducting network throughout the membrane. The difference results from the ceramic microstructure developed during the sintering step [28, 32]. In other words, the smaller perovskite phase grains increase the possibility of percolation across the membranes. As a result, the permeation flux of the SDC75–SCF25 dual-phase membrane is much higher than that of SDC75–SCC25 under the same testing conditions, as shown in Fig. 7.16.

Another possible microstructure effect caused by the elements in the perovskite phase is the segregation of impurities at the grain boundaries [31]. The HRTEM investigation of the SDC75–SSCr25 dual-phase membrane shows that impurities appear at the grains boundaries, which block ionic transport between fluorite grains and lead to the low oxygen permeation flux, as shown in Fig. 7.17. However, the SDC75–SSF25 membranes show clear grain boundaries whether between perovskite grains and fluorite grains or between fluorite grains. Mn element segregation was found on the grain boundaries of the GDC–LSM dual-phase membranes and results in a significant degradation of the permeation flux during long-term operation [9].

7.3.4.3 Sintering Temperature-Induced Microstructural Effects

The fact that should be noted is that the microstructure and texture of the ceramic membranes is greatly influenced by the sintering temperature, with effects on both oxygen permeability and structural stability. This is despite the fact that the oxygen permeability of a given material can vary in a wide range due to differences in

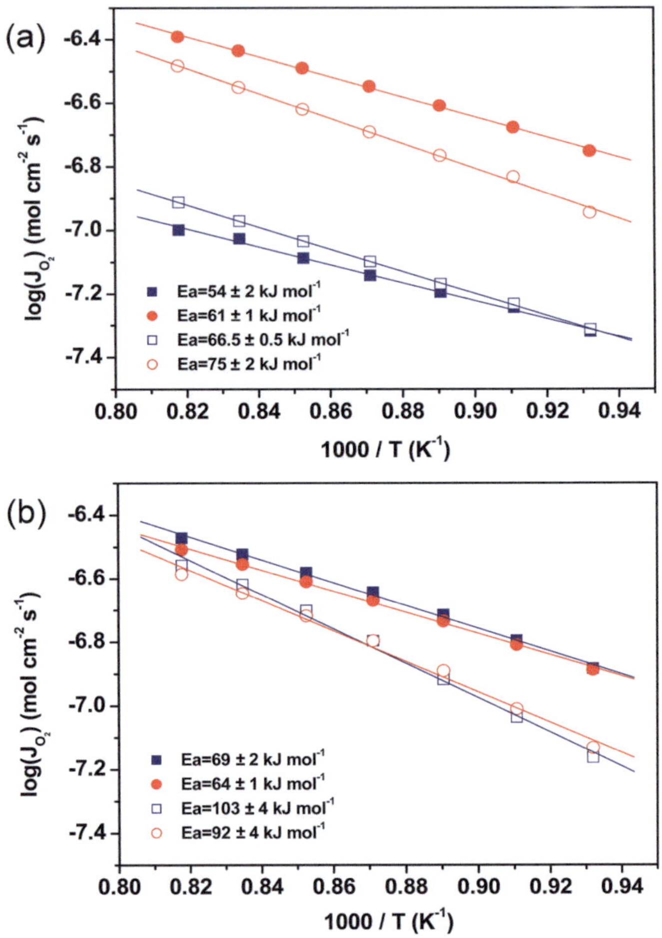

Fig. 7.16 Arrhenius plots of oxygen permeation performance of the four dual-phase membranes swept by helium and CO_2, respectively. Solid symbols, swept by helium; open symbols, swept by CO_2; squares, SDC75–SCC25 or SDC75–SCF25; circles, SDC67–SCC33 or SDC67–SCF33 (Reproduced from [34] with permission of Elsevier)

microstructure [72–75]. As an example, Wang et al. investigated the effects of sintering temperature on $Ba_{0.5}Sr_{0.5}Co_{0.8}Fe_{0.2}O_{3-\delta}$ membranes and found that the oxygen permeability increases with grain size [75]. Usually, for perovskite-type mixed conducting ceramics, the grain boundary resistance plays an important role in ionic and electronic transport and decreases with an increase in grain size (or sintering temperature). However, the above investigations on the sintering temperature were all focused on single-phase membrane materials. The composition of dual-phase membranes is more complex, so the influence of sintering temperature on the microstructure and permeability of dual-phase membranes is also more complex.

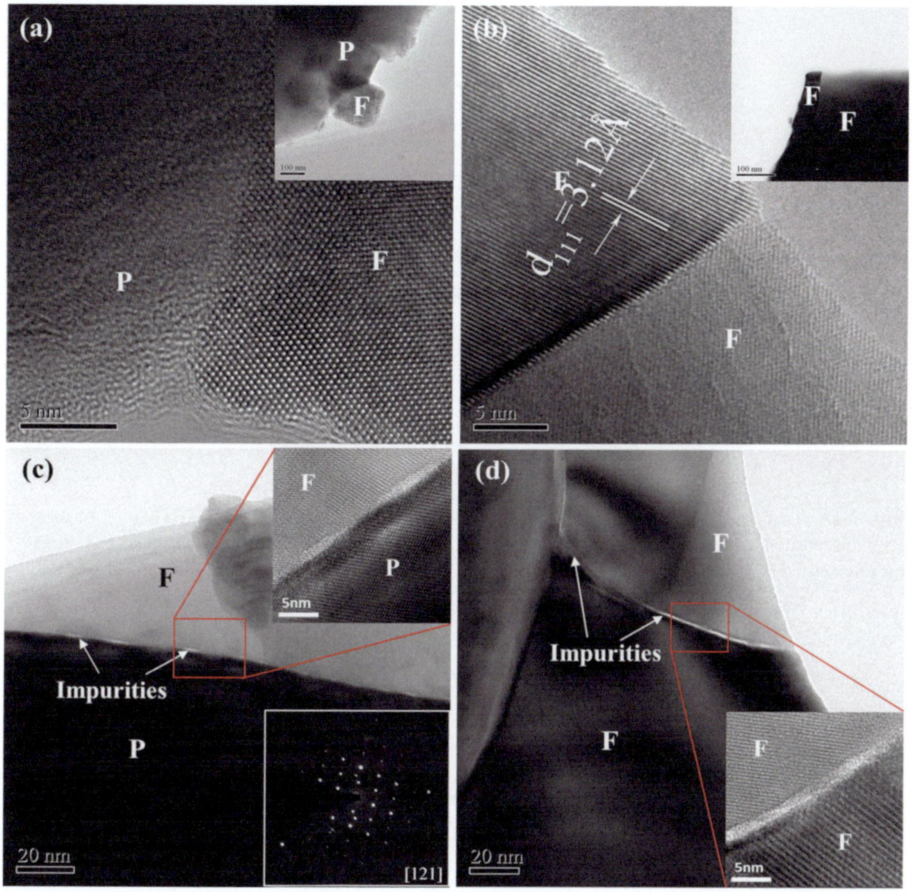

Fig. 7.17 HR-TEM pictures and SAED of the grain boundaries of (*a* and *b*) SDC75–SSF25 and (*c* and *d*) SDC75–SSCr25. (**a**) The grain boundary between the perovskite grain and the fluorite grain along the [110] direction of fluorite grain; (**b**) the grain boundary between the fluorite grains along [110] directions. (**c**) The grain boundary between the perovskite grain and the fluorite grain along the [121] direction of perovskite grain; (**d**) the grain boundary between fluorite grains. "F" and "P" denote the fluorite grain and perovskite grain, respectively. The SAED was obtained in the red square areas (Reproduced from [31] with permission of Elsevier)

Figure 7.18 shows the oxygen permeation flux of the SDC75–SSF25 dual-phase membranes sintered at different temperatures [28]. All the membranes were prepared by the one-pot method. It can be seen that the oxygen permeation flux of the membrane disks at a given operating temperature varies with the sintering temperature. The membrane sintered at 1425 °C exhibits the highest oxygen permeation flux (950 °C, 0.42 mL cm^{-2} min^{-1}), and the membrane sintered at 1525 °C has the lowest oxygen permeation flux (950 °C, 0.21 mL cm^{-2} min^{-1}). As shown in Fig. 7.19, the membrane surface is composed of fluorite and perovskite grains after sintering at 1400 °C and 1425 °C, but only the fluorite

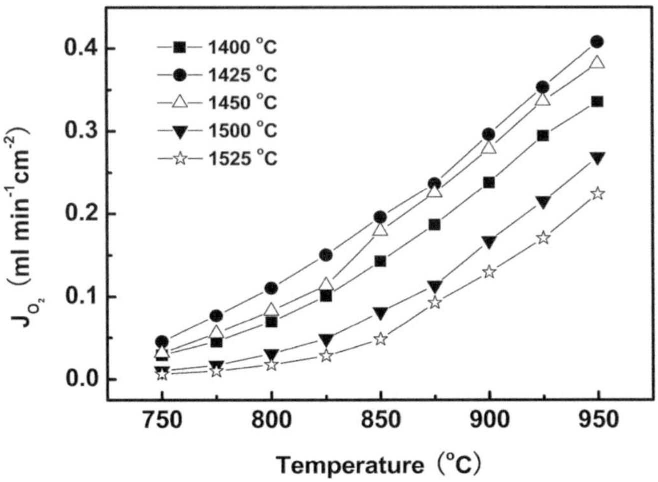

Fig. 7.18 Temperature dependence of oxygen permeation flux through dual-phase membranes with different sintering temperatures. He flow rate, 40 ml min^{-1}; air flow rate, 100 ml min^{-1}; membrane thickness, 0.6 mm (Reproduced from [28] with permission of Elsevier)

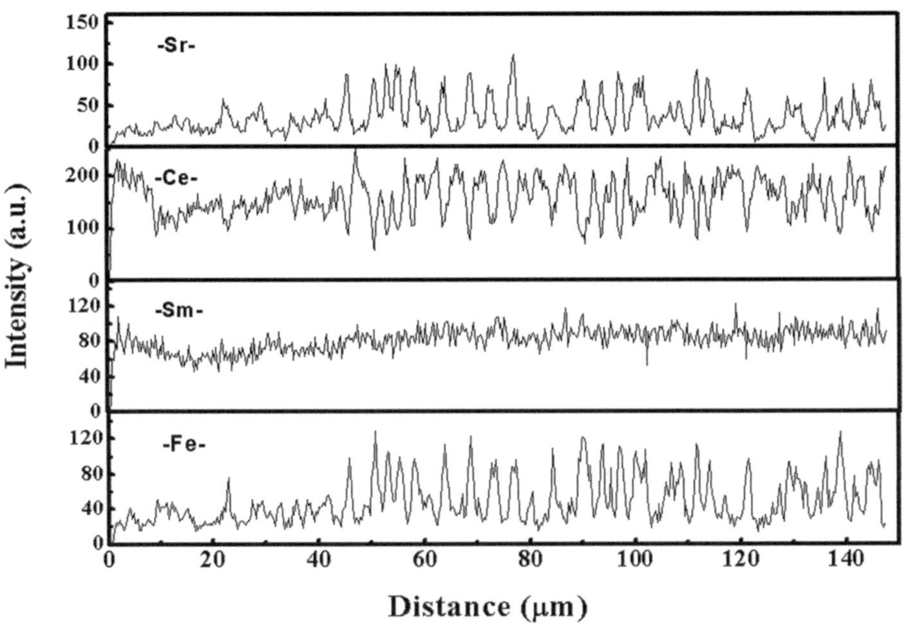

Fig. 7.19 The relative element distribution vs. the distance of the EDX analysis on the membrane sintered at 1500 °C for 5 h. EDX linear scan analysis was performed from surface to bulk over a 145 μm range (Reproduced from [28] with permission of Elsevier)

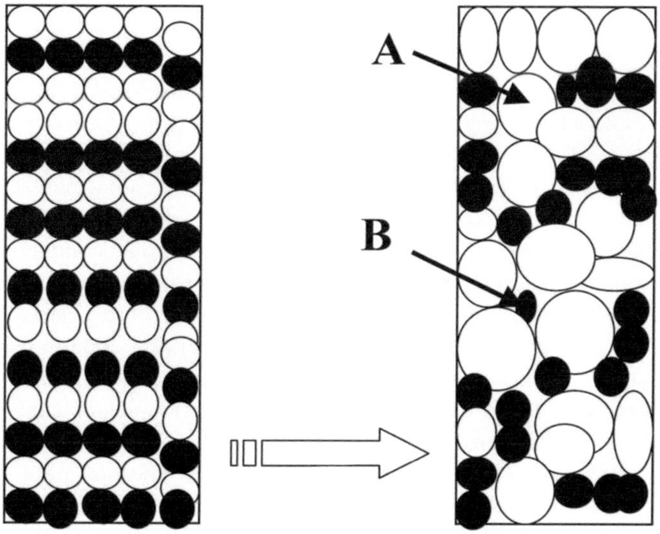

Fig. 7.20 The sketch of dual-phase membrane grains under high sintering temperature. *A* denotes that SDC blocked the connectivity of SSF; *B* denotes that small SSF grains were encapsulated by large SDC grains (Reproduced from [28] with permission of Elsevier)

phase was detected when the sintering temperature was raised to 1450 °C. After a top layer with thickness of ~200 μm was removed by polishing, all the membranes, irrespective of sintering temperature, showed both the fluorite and perovskite phases. The EDX analysis shows that the increase in sintering temperature leads to a difference in elemental composition between the membrane surface and membrane bulk, which impairs the homogenous distribution of SSF and SDC grains in the membranes. In the dual-phase membranes, the oxygen permeation process depends on the cooperation between the SDC and SSF phases. It should be noted that a homogenous distribution of the dual-phase grains improves the oxygen permeability of the dual-phase membranes through ionic transport and surface exchange. However, increasing the sintering temperature leads to the growth of fluorite and perovskite grains. The primary continuous electronic conducting network is broken by the growth of perovskite grains; as a result, the oxygen permeation flux decreases. Figure 7.20 gives a schematic of the break effect in the continuous electronic conducting network as the perovskite grains grow [28]. This phenomenon can also be found in Fig. 7.15, in which the dual-phase membrane has larger perovskite grains than the Fe-containing one [34]. Consequentially, the conductivity of the former is much lower than the latter. This grain size-dependent phenomenon is significantly different from that in the single-phase perovskite membranes because their crystal grains can transport oxygen ions and electrons simultaneously.

7.3.5 Ratio Between the Two Phases

Besides the aforementioned effects on microstructure, the volume ratio between the two phases obviously has a great influence on the electronic transport across dual-phase membranes. Ceria-based fluorite ionic conductors have high conductivity and stability at elevated temperatures, so it was hoped that dual-phase membranes containing more fluorite oxides would give good performance as long as the mixed conducting phase grains could form a continuous conducting network through the membranes. The lowest volume fraction for the perovskite phase was investigated that would give dual-phase membranes of high performance, especially when the membranes are employed in harsh working conditions.

For comparison purposes, both sides of the dual-phase membranes were coated with LSC porous layers to eliminate the disturbance in surface exchange on the evaluation of bulk transport. Dual-phase membranes with perovskite contents of 15 wt%, 25 wt%, and 35 wt% were prepared by the one-pot method [27]. The volume fractions of the perovskite phase were 17 %, 28 %, and 39 %, respectively. The oxygen permeation flux within the investigated temperature range decreases in the sequence SDC65–SSF35 $\approx$ SDC75–SSF25 $>$ SDC85–SSF15. The ionic conductivity of SDC at 950 °C is about 0.1 S cm^{-1}, and that of SSF is also about 0.1 S cm^{-1} at this temperature; so the ionic conductivities of the SDC–SSF dual-phase membranes are close to 0.1 S cm^{-1} no matter what the volume ratio of the two phases is, if the slight elemental diffusion between the two phases is negligible. From the total conductivity data and keeping the Wagner equation in mind, it was found that the electronic conductivities were usually high enough for SDC65–SSF35 ($\sim$1.6 S cm^{-1}) and SDC75–SSF25 ($\sim$0.5 S cm^{-1}) but insufficient for SDC85–SSF15 ($\sim$0.17 S cm^{-1}). The former two materials with similar ionic conductivities have comparable oxygen permeation fluxes and even activation energies. A value for the oxygen permeation flux as high as 0.8 mL cm^{-2} min^{-1} was achieved on 0.5-mm-thick SDC65–SSF35 and SDC75–SSF25 membranes. The optimal ratio was around 75:25 for the SDC–SSF dual-phase membrane system but may be different for others.

The perovskite phase content can be further reduced by optimizing the preparation method of the membranes. Chen et al. reported a coating strategy to fabricate a dual-phase membrane with 20 vol.% $PrBaCo_2O_{5+\delta}$ (PBCO) dispersed in an 80 vol.% $Ce_{0.8}Sm_{0.2}O_{2-\delta}$ (SDC) matrix, where the fiber-shaped PBCO forms a percolative network and provides sufficient electronic conductivity [55]. The total conductivity of the composite materials reached 6 S cm^{-1} at 900 °C. A tunable segmented configuration was reported by Joo et al. [76]. The new type of membrane consisted of a robust fluorite oxide matrix and an electron-conducting perovskite oxide segment, as shown in Fig. 7.21. Mixed electron–ion conduction in the membrane can be optimized by controlling the number of electron-conducting segments. One of the advantages of the new configuration is that the ionic and electronic conductivities of the membrane can be adjusted by controlling the areas of the ionic and electronic phases.

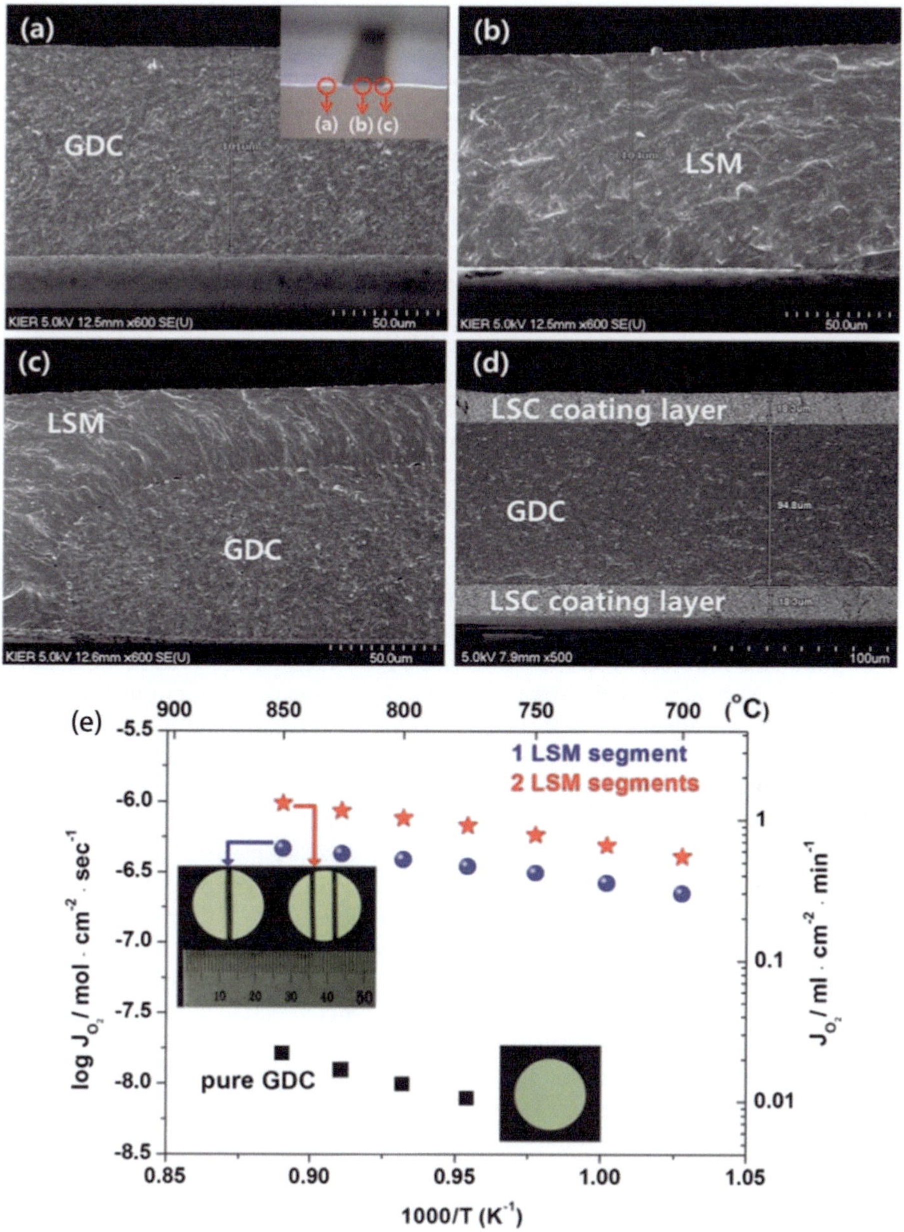

Fig. 7.21 SEM micrographs of the membrane, images of the cross-section of sintered (**a**) GDC, (**b**) LSM, (**c**) GDC–LSM interface, and (**d**) the SEM image of the GDC coated by LSC layers; (**e**) temperature dependence of the oxygen fluxes of the segmented and pure GDC membranes (membrane thickness ~100 mm) (Reproduced from [76] with permission of the Royal Society of Chemistry)

7.3.6 Other Potential Factors

Besides the factors discussed above, there are many other factors that may influence the oxygen transport through dual-phase membranes, such as the grain size ratio between the two phases, the conductivity of the mixed conducting phase, and the ionic conducting phase. Some factors are closely correlated to each other, so it is difficult to separately investigate the influence of each factor. Elemental diffusion between the two phases is inevitable during synthesis and high-temperature operation. However, the correct choice of elemental composition for dual-phase membranes can prevent the variable permeation rate with time caused by the elemental interdiffusion under operating conditions. The grain size of the dual-phase membranes is dependent on the sintering temperature, preparation method, and the volume ratio between the two phases, as well as the grain size ratio between the two phases. An appropriate grain size ratio between the two phases is needed to provide a percolation path when the minimum amounts of perovskite phases are used. For the SDC75–SSF25 dual-phase membranes, the increase of sintering temperature leads to a growth of the grains and a simultaneous decrease of the grain size ratio. Usually, 30 vol.% of the electronic conducting phase is needed to exceed the percolation threshold for a dual-phase system where the grain sizes of the two phases are comparable. A low percolation threshold can be achieved for a high grain size ratio between the two phases. New methods need to be developed for the construction of such a microstructure. The defects in structures, including lattice defects and boundary defects, provide other unquantifiable factors affecting oxygen permeation through dual-phase membranes; these depend mainly on the preparation procedure for a given composition. Defects sometimes favor the ionic and electronic transport in the bulk membrane and oxygen exchange on the membrane surfaces, but sometimes it results in variable permeation due to their instability at elevated temperatures. The defects along the grain boundaries of the two phases are still not clearly understood, and their effects on membrane performance should be investigated in the future. An investigation of these problems using high-resolution electron microscopy may reveal some important information and help in the development and optimization of mixed conducting ceramic materials for high-temperature electrochemical applications. Note that, if the limitation of oxygen exchange can be totally removed by coating catalyst layers onto the membrane surfaces, the oxygen permeation flux of a single-phase membrane increases linearly with the reciprocal of its thickness. However, this is not suitable for dual-phase membranes. The percolation threshold will decrease quickly with the reduction in membrane thickness when the membrane thickness is comparable to the grain size. In addition, this leads to a rapid increase in both electronic and ionic conductivity, especially for the dual-phase membranes containing a small volume fraction of the electronic conducting phase or made of pure electronic and ionic conductors. Under these circumstances, a small difference in the oxygen permeability will be found between the two kinds of dual-phase membranes, such as SDC–SSF and SDC–SSCr. There may be other important factors not discussed in this section since the investigation on dual-phase membranes is in its infancy. As new materials and techniques are developed to improve the permeation flux, new ideas about the selection of dual-phase materials will also be needed to make further progress.

7.4 Outside/Inside Short Circuit

For the traditional noble metal-ionic conductor dual-phase membranes, the high material costs cannot be avoided because of the large amount of noble metals used to form a continuous electronic conducting phase. Any mismatch of thermal expansion between the ceramic and metal would lead to the membrane breaking after several thermal cycles. Additionally, a much lower permeation flux than those of the perovskite membranes means the cermet membranes cannot be applied on a large scale. To overcome the difficulties occurring in cermet membranes, Zhang and Liu suggested outside/inside short-circuited membranes for oxygen permeation [77, 78]. Their structures are just like those of solid oxide fuel cells but with an outside/inside short circuit. Figure 7.22 shows the working principles of the membranes with an outside/inside short circuit for oxygen separation. Of course, this type of structure can also be applied to protons or carbonate ions in electrolyte membranes for hydrogen or CO_2 permeation. Usually, the typical fluorite- and perovskite-type ionic conductors are converted into electrolyte membranes, and noble metal wires (Au, Ag, Pt, etc.) are used for electronic transport. Compared with the traditional dual-phase membranes made of noble metals and ionic conductors, the amount of the noble metals is reduced remarkably, but the permeability is enhanced by more than ten times as there is no pure conducting phase to block the diffusion of oxygen ions in membrane bulk. Figure 7.23 illustrates the dependence

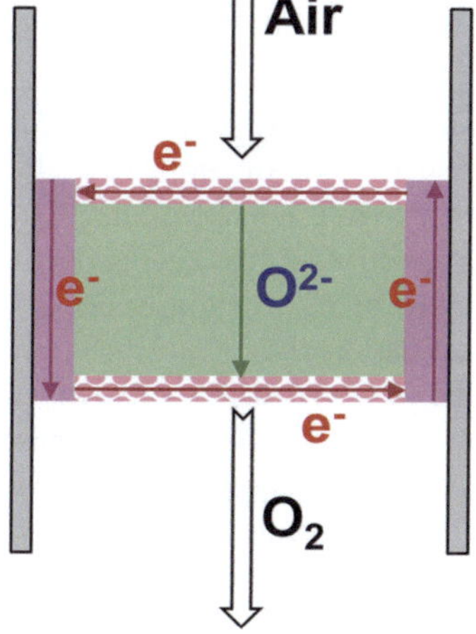

Fig. 7.22 Diagrams of novel oxygen ionic conducting ceramic membranes with a metal coating layer and silver paste for oxygen separation (Reproduced from [77] with permission of the Royal Society of Chemistry)

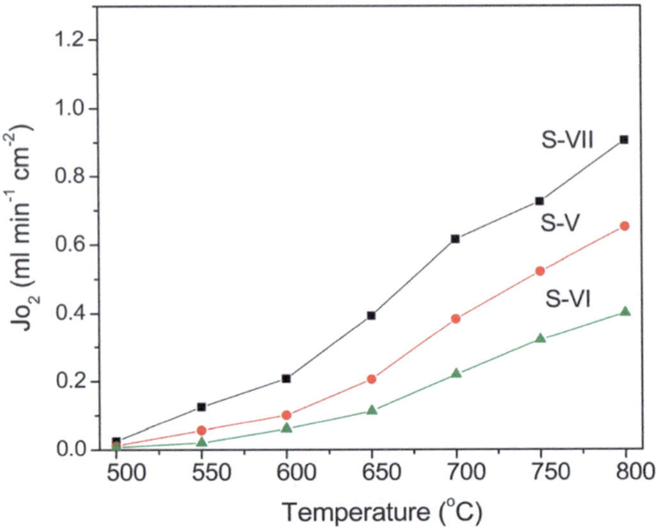

Fig. 7.23 Dependence of oxygen permeation fluxes (±5 %) on the thickness of the SDC membrane (S-V: 1 mm, S-VI: 1.5 mm, S-VII: 0.4 mm) (Reproduced from [77] with permission of the Royal Society of Chemistry)

of oxygen permeation flux on the membrane thickness for outside short-circuited SDC membranes [77]. The two surfaces of the SDC membranes were covered with porous Pt and sealed by an Ag paste. A high permeation flux of up to 0.6 mL cm^{-2} min^{-1} across a 0.4 mm membrane at 700 °C has been achieved. However, this flux is about twice the theoretical flux based on the oxygen ion conductivity. These membranes have high stability in the working environment containing CO_2, H_2O, etc., because they do not contain alkaline earth elements in the lattice. Although the noble metal content is significantly decreased, the costs of this type of membrane are still too high to attract wide interest.

One possible and feasible method is to utilize the electronic conducting oxides to replace the noble metals. Wang and Yang et al. reported a type of outside short-circuited ionic conducting membrane with surfaces coated by MIEC oxides and sealed by Au or Ag [79]. In this work, $La_{0.8}Sr_{0.2}Ga_{0.8}Mg_{0.115}Co_{0.085}O_{3-\delta}$ (LSGMC) and $Ce_{0.8}Sm_{0.2}O_{2-\delta}$ (SDC) were used as the electrolyte membranes, respectively; $La_{0.8}Sr_{0.2}CoO_{3-\delta}$ (LSC113) and $LaSrCoO_{4\pm\delta}$ (LSC214) were coated by using a pulsed laser deposition technique on the membrane surfaces for oxygen activation and electronic conduction (from the permeation side to the air side) with the Ag or Au sealant. However, a permeation flux of ~0.05 mL cm^{-2} min^{-1} across a 1 mm LSGMC- or SDC-based outside short-circuited membrane was achieved at 700 °C. Such a permeation flux is slightly lower than the maximum flux estimated from oxygen ion conductivity, i.e., ~0.32 and ~0.09 mL cm^{-2} min^{-1} for LSGMC and SDC, respectively. The slow oxygen reduction reaction kinetics at the feed side are considered the limiting factor for oxygen permeation. As illustrated by the above results, there is no remarkable difference in terms of oxygen permeability between the outside/inside short-circuited membranes and the dual-phase membranes made of ionic conductors and mixed conductors.

7.5 Asymmetric Dual-Phase Membranes

Though dual-phase membranes have high stability and provide potential applications for air separation and membrane reactors for light hydrocarbon conversion, their permeability remains far lower than those of most perovskite-type membranes and the requirements for the commercialization of the oxygen membrane technology. Therefore, many researchers have focused their research on the preparation of asymmetric dual-phase membranes with a thin dense layer for oxygen permeation and a thick porous layer as the support. The porous support provides the mechanical strength and allows the diffusion of gases. In previous work, asymmetric membranes were prepared by various methods, such as tape casting [80, 81], freeze casting [82], acid leaching [26], and phase inversion [83, 84]. Among these, tape casting is one of the most promising methods for the preparation of asymmetric membranes on a large scale and is discussed in detail in Chap. 4. In our laboratory, we chose 75 wt.% $Ce_{0.85}Sm_{0.15}O_{2-\delta}$–25 wt.% $Sm_{0.6}Sr_{0.4}Al_{0.3}Fe_{0.7}O_{3-\delta}$ (SDC–SSAF) as an example to develop the tape-casting preparation technique for dual-phase membranes. We found that SDC–SSAF shows excellent performance in syngas generation and oxygen separation. The success of a 1100-h syngas generation experiment using the SDC–SSAF membrane reactor verified the excellent stability of the membrane under both strongly oxidizing and reducing conditions [24].

As shown in Fig. 7.24, the surface of the asymmetric membrane is dense without any pinholes or cracks. After being sintered at 1450 °C, the membrane had a total thickness of ~500 μm and a dense layer thickness of ~40 μm [35]. It can be seen that the dense layer is closely bonded with the porous support, and there is no delamination and cracks. Pores with a size of 5–10 μm were created by burning out pore formers during the sintering process. An SSC layer with a thickness of ~5 μm and pore size of ~2–5 μm was coated on the dense layer side of the asymmetric membrane, which was thus named as cat-Asym. To speed up the oxygen exchange reactions of the porous support side, a solution of SSC nitrates was dropped into the porous support to form SSC nanoparticles on the porous support skeleton. Both sides of the asymmetric dual-phase membrane were coated with SSC catalysts, and this was denoted as cat-Asym-cat. Figure 7.25 shows the oxygen permeation fluxes of the membranes in the temperature range of 750–950 °C under an oxygen partial pressure gradient of 21 kPa/0.5 kPa. The oxygen permeation flux reached as high as 3.9 mL cm^{-2} min^{-1} at 950 °C for the cat-Asym-cat. It is worthwhile noting that as the thickness of the membrane was reduced from 400 to 40 μm, the oxygen permeation flux only increased from 0.9 to 3.9 mL cm^{-2} min^{-1} at 950 °C under the same oxygen partial pressure gradient.

Fig. 7.24 SEM images of asymmetric membranes prepared via tape casting and co-lamination. (**a**) Top view of the dense layer, (**b**) cross-view of the asymmetric membrane, (**c**) cross-view of the dense layer of the asymmetric membrane, (**d**) cross-view of the porous support, (**e**) SSC catalyst layer on the surface of the dense layer side, (**f**) SSC catalyst particles impregnated into the porous support and some of the particles marked by *arrows* (Reproduced from [35] with permission of Elsevier)

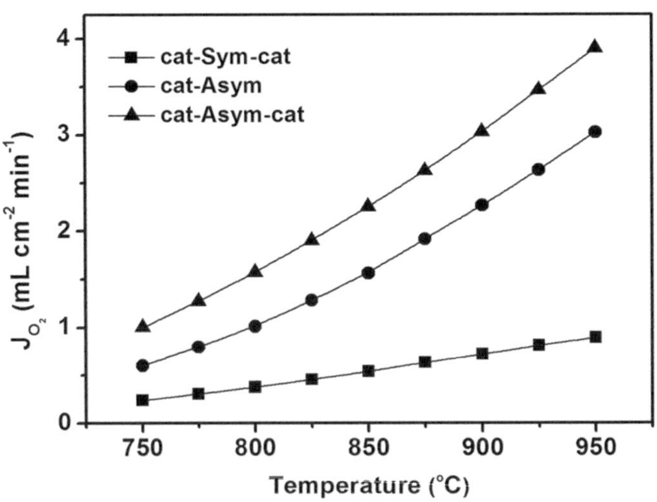

Fig. 7.25 Relationship between oxygen permeation flux and temperature for the symmetric and asymmetric membranes. Temperature-dependent oxygen permeation fluxes. Oxygen partial pressure gradient, 21 kPa/0.5 kPa; air flow rate, 100 mL min^{-1} (Reproduced from [35] with permission of Elsevier)

This high value is still far lower than the predicted value of 9.0 mL cm^{-2} min^{-1} according to the Wagner equation. Two factors can explain this phenomenon. One is that the surface exchange resistances cannot be completely eliminated in the oxygen permeation process even though both sides were modified by the SSC catalyst. The other is the concentration polarization involved in the gas diffusion across the porous support. It should be pointed out that the permeation flux reaches 1.0 mL cm^{-2} min^{-1} at 750 °C for the cat-Asym-cat membrane, which is close to the lower limit given by Steele [85].

References

1. Chen CS, Kruidhof H, Bouwmeester HJM (1997) Thickness dependence of oxygen permeation through stabilized bismuth oxide-silver composite. Solid State Ionics 99:215–219
2. Chen CS, Boukamp BA, Bouwmeester HJM, Cao GZ, Kruidhof H, Winnubst AJA, Burggraaf AJ (1995) Microstructural development, electrical properties and oxygen permeation of zirconia-palladium composites. Solid State Ionics 76:23–28
3. Lee TH, Yang YL, Jacobson AJ (2000) Electrical conductivity and oxygen permeation of $Ag/BaBi_8O_{13}$ composites. Solid State Ionics 134:331–339
4. Mazanec TJ, Cable TL, Frye JG (1992) Electrocatalytic cells for chemical reaction. Solid State Ionics 53:111–118
5. Chen CS, Kruidhof H, Bouwmeester HJM, Verweij H, Burggraaf AJ (1996) Oxygen permeation through oxygen ion oxide-noble metal dual phase composites. Solid State Ionics 86–88:569–572
6. Chen CS, Burggraaf AJ (1999) Stabilized bismuth oxide noble metal mixed conducting composites as high temperature oxygen separation membranes. J Appl Electrochem 29:355–360
7. Kim J, Lin YS (2000) Synthesis and oxygen permeation properties of ceramic-metal dual-phase membranes. J Membr Sci 167:123–133
8. Kim J, Lin YS (2000) Synthesis and oxygen-permeation properties of thin YSZ/Pd composite membranes. AIChE J 46:1521–1529
9. Kharton VV, Kovalevsky AV, Viskup AP, Figueiredo FM, Yaremchenko AA, Naumovich EN, Marques FMB (2000) Oxygen permeability of $Ce_{0.8}Gd_{0.2}O_{2-\delta}$-$La_{0.7}Sr_{0.3}MnO_{3-\delta}$ composite membranes. J Electrochem Soc 147:2814–2821
10. Kharton VV, Kovalevsky AV, Viskup AP, Shaula AL, Figueiredo FM, Naumovich EN, Marques FMB (2003) Oxygen transport in $Ce_{0.8}Gd_{0.2}O_{2-\delta}$-based composite membranes. Solid State Ionics 160:247–258
11. Sirman JD, Chen JC (2003) Ceramic membrane structure and oxygen separation method. US Patent 6,514,314
12. Nigge U, Wiemhofer HD, Romer EWJ, Bouwmeester HJM, Schulte TR (2002) Composites of $Ce_{0.8}Gd_{0.2}O_{1.9}$ and $Gd_{0.7}Ca_{0.3}CoO_{3-delta}$ as oxygen permeable membranes for exhaust gas sensors. Solid State Ionics 146:163–174
13. Shaula AL, Kharton VV, Marques FMB (2004) Phase interaction and oxygen transport in $La_{0.8}Sr_{0.2}Fe_{0.8}Co_{0.2}O_3$-$(La_{0.9}Sr_{0.1})_{0.98}Ga_{0.8}Mg_{0.2}O_3$ composites. J Eur Ceram Soc 24:2631–2639
14. Yi JX, Zuo YB, Liu W, Winnubst L, Chen CS (2006) Oxygen permeation through a $Ce_{0.8}Sm_{0.2}O_{2-\delta}$-$La_{0.8}Sr_{0.2}CrO_{3-\delta}$ dual-phase composite membrane. J Membr Sci 280:849–855
15. Wang B, Zhan MC, Zhu DC, Liu W, Chen CS (2006) Oxygen permeation and stability of $Zr_{0.8}Y_{0.2}O_{1.9}$-$La_{0.8}Sr_{0.2}CrO_{3-\delta}$ dual-phase composite. J Solid State Electrochem 10:625–628

16. Luo HX, Jiang HQ, Efimov K, Liang FY, Wang HH, Caro J (2011) CO_2-tolerant oxygen-permeable Fe_2O_3-$Ce_{0.9}Gd_{0.1}O_{2-\delta}$ dual phase membranes. Ind Eng Chem Res 50:13508–13517
17. Luo HX, Jiang HQ, Efimov K, Wang HH, Caro J (2011) Influence of the preparation methods on the microstructure and oxygen permeability of a CO_2-stable dual phase membrane. AIChE J 57:2738–2745
18. Garcia-Fayos J, Balaguer M, Serra JM (2015) Dual-phase oxygen transport membranes for stable operation in environments containing carbon dioxide and sulfur dioxide. ChemSusChem 8:4242–4249
19. Takamura H, Okumura K, Koshino Y, Kamegawa A, Okada M (2004) Oxygen permeation properties of ceria-ferrite-based composites. J Electroceram 13:613–618
20. Takamura H, Kobayashi T, Kasahara T, Kamegawa A, Okada M (2006) Oxygen permeation and methane reforming properties of ceria-based composite membranes. J Alloy Compd 408:1084–1089
21. Takamura H, Sugai H, Watanabe M, Kasahara T, Kamegawa A, Okada M (2006) Oxygen permeation properties and surface modification of acceptor-doped CeO_2/$MnFe_2O_4$ composites. J Electroceram 17:741–748
22. Takamura H, Ogawa M, Suehiro K, Takahashi H, Okada M (2008) Fabrication and characteristics of planar-type methane reformer using ceria-based oxygen permeable membrane. Solid State Ionics 179:1354–1359
23. Zhu XF, Yang WS (2008) Composite membrane based on ionic conductor and mixed conductor for oxygen permeation. AIChE J 54:665–672
24. Zhu XF, Li QM, Cong Y, Yang WS (2008) Syngas generation in a membrane reactor with a highly stable ceramic composite membrane. Catal Commun 10:309–312
25. Zhu XF, Wang HH, Yang WS (2008) Relationship between homogeneity and oxygen permeability of composite membranes. J Membr Sci 39:120–127
26. Li QM, Zhu XF, Yang WS (2008) Single-step fabrication of asymmetric dual-phase composite membranes for oxygen separation. J Membr Sci 325:11–15
27. Zhu XF, Li QM, He YF, Cong Y, Yang WS (2010) Oxygen permeation and partial oxidation of methane in dual-phase membrane reactors. J Membr Sci 360:454–460
28. Li QM, Zhu XF, He YF, Cong Y, Yang WS (2011) Effects of sintering temperature on properties of dual-phase oxygen permeation membranes. J Membr Sci 367:134–140
29. Zhu XF, Liu HY, Li QM, Cong Y, Yang WS (2011) Unsteady-state permeation and surface exchange of dual-phase membranes. Solid State Ionics 185:27–31
30. Zhu XF, Liu HY, Cong Y, Yang WS (2012) Novel dual-phase membranes for CO_2 capture via an oxyfuel route. Chem Commun 48:251–253
31. Zhu XF, Li MR, Liu HY, Zhang TY, Cong Y, Yang WS (2012) Design and experimental investigation of oxide ceramic dual-phase membranes. J Membr Sci 394–395:120–130
32. Zhu XF, Liu Y, Cong Y, Yang WS (2013) $Ce_{0.85}Sm_{0.15}O_{1.925}$–$Sm_{0.6}Sr_{0.4}Al_{0.3}Fe_{0.7}O_3$ dual-phase membranes: One-pot synthesis and stability in a CO_2 atmosphere. Solid State Ionics 253:57–63
33. Li HB, Liu Y, Zhu XF, Cong Y, Xu SP, Xu WQ, Yang WS (2013) Oxygen permeation through Ca-contained dual-phase membranes for oxyfuel CO_2 capture. Sep Purif Technol 114:31–37
34. Li HB, Zhu XF, Liu Y, Wang WP, Yang WS (2014) Comparative investigation of dual-phase membranes containing cobalt and iron-based mixed conducting perovskite for oxygen permeation. J Membr Sci 462:170–177
35. Cao ZW, Zhu XF, Li WP, Xu B, Yang LN, Yang WS (2015) Asymmetric dual-phase membranes prepared via tape-casting and co-lamination for oxygen permeation. Mater Lett 147:88–91
36. Cai LL, Li WP, Cao ZW, Zhu XF, Yang WS (2016) Improving oxygen permeation of MIEC membrane reactor by enhancing the electronic conductivity under intermediate-low oxygen partial pressures. J Membr Sci. 520:607–615

37. Li D, Wang XP, Zhuang Z, Wang JX, Li C, Fang QF (2009) Reducibility study of oxide-ion conductors $La_{2-x}Ba_xMo_{2-y}A_yO_{9-\delta}$ (A = W, Al, Ga) assessed by impedance spectroscopy. Mater Res Bull 44:446–450
38. Sun M, Chen XW, Hong L (2013) Influence of the interfacial phase on the structural integrity and oxygen permeability of a dual-phase membrane. ACS Appl Mater Interfaces 5:9067–9074
39. Read MSD, Islam MS, Watson GW, King F, Hancock FE (2000) Defect chemistry and surface properties of $LaCoO_3$. J Mater Chem 10:2298–2305
40. Islam MS, Davies RA (2004) Atomistic study of dopant site-selectivity and defect association in the lanthanum gallate perovskite. J Mater Chem 14:86–93
41. Peng RR, Xia CR, Fu QX, Meng GY, Peng DK (2002) Sintering and electrical properties of $(CeO_2)_{0.8}(Sm_2O_3)_{0.1}$ powders prepared by glycine-nitrate process. Mater Lett 56:1043–1047
42. Wu K, Xie S, Jiang GS, Liu W, Chen CS (2001) Oxygen permeation through $(Bi_2O_3)_{0.74}(SrO)_{0.26}$-Ag (40% v/o) composite. J Membr Sci 188:189–193
43. Li W, Liu JJ, Chen CS (2009) Hollow fiber membrane of yttrium-stabilized zirconia and strontium-doped lanthanum manganite dual-phase composite for oxygen separation. J Membr Sci 340:266–271
44. Fang W, Gao JF, Chen CS (2013) $La_{0.8}Sr_{0.2}Cr_{0.5}Fe_{0.5}O_{3-\delta}$ (LSCF)-$Zr_{0.8}Y_{0.2}O_{2-\delta}$ (YSZ) based multilayer membrane for CO_2 decomposition. Ceram Int 39:7269–7272
45. Liu JJ, Liu T, Wang WD, Gao JF, Chen CS (2012) $Zr_{0.84}Y_{0.16}O_{1.92}$-$La_{0.8}Sr_{0.2}Cr_{0.5}Fe_{0.5}O_{3-\delta}$ dual-phase composite hollow fiber membrane targeting chemical reactor applications. J Membr Sci 389:435–440
46. Fang W, Steinbach F, Chen CS, Feldhoff A (2015) An approach to enhance the CO_2 tolerance of fluorite-perovskite dual-phase oxygen-transporting membrane. Chem Mater 27:7820–7826
47. Wang ZT, Sun WP, Zhu ZW, Liu T, Liu W (2013) A novel cobalt-free, CO_2-stable, and reduction-tolerant dual-phase oxygen-permeable membrane. ACS Appl Mater Interfaces 5:11038–11043
48. Luo HX, Efimov K, Jiang HQ, Feldhoff A, Wang HH, Caro J (2011) CO_2-stable and cobalt-free dual-phase membrane for oxygen separation. Angew Chem Int Ed 50:759–763
49. Luo HX, Jiang HQ, Klande T, Cao ZW, Liang FY, Wang HH, Caro J (2012) Novel cobalt-free, noble metal-free oxygen-permeable $40Pr_{0.6}Sr_{0.4}FeO_{3-\delta}$-$60Ce_{0.9}Pr_{0.1}O_{2-\delta}$ dual-phase membrane. Chem Mater 24:2148–2154
50. Joo JH, Park GS, Yoo C-Y, Yu JH (2013) Contribution of the surface exchange kinetics to the oxygen transport properties in $Gd_{0.1}Ce_{0.9}O_{2-\delta}$-$La_{0.6}Sr_{0.4}Co_{0.2}Fe_{0.8}O_{3-\delta}$ dual-phase membrane. Solid State Ionics 253:64–69
51. Fang W, Liang FY, Cao ZW, Steinbach F, Feldhoff A (2015) A mixed ionic and electronic conducting dual-phase membrane with high oxygen permeability. Angew Chem Int Ed 54:4847–4850
52. Cheng HW, Luo LF, Yao WL, Lu XG, Zou XL, Zhou ZF (2015) Novel cobalt-free CO_2-tolerant dual-phase membranes of $Ce_{0.8}Sm_{0.2}O_{2-\delta}$-$Ba_{0.95}La_{0.05}Fe_{1-x}Zr_xO_{3-\delta}$ for oxygen separation. J Membr Sci 492:220–229
53. Xue J, Liao Q, Wei YY, Li Z, Wang HH (2013) A CO_2-tolerance oxygen permeable $60Ce_{0.9}Gd_{0.1}O_{2-\delta}$-$40Ba_{0.5}Sr_{0.5}Co_{0.8}Fe_{0.2}O_{3-\delta}$ dual phase membrane. J Membr Sci 443:124–130
54. Chen T, Zhao HL, Xie ZX, Xu NS, Lu Y (2015) Oxygen permeability of $Ce_{0.8}Sm_{0.2}O_{2-\delta}$-$LnBaCo_2O_{5+\delta}$ (Ln = La, Nd, Sm, and Y) dual-phase ceramic membranes. Ionics 21:1683–1692
55. Chen T, Zhao HL, Xie ZX, Wang J, Lu Y, Xu NS (2013) $Ce_{0.8}Sm_{0.2}O_{2-\delta}$-$PrBaCo_2O_{5+\delta}$ dual-phase membrane: novel preparation and improved oxygen permeability. J Power Sources 223:289–292
56. Qi XW, Lin YS, Swartz SL (2000) Electric transport and oxygen permeation properties of lanthanum cobaltite membranes synthesized by different methods. Ind Eng Chem Res 39:646–653

57. Tan L, Gu XH, Yang L, Jin WQ, Zhang LX, Xu NP (2003) Influence of powder synthesis methods on microstructure and oxygen permeation performance of $Ba_{0.5}Sr_{0.5}Co_{0.8}Fe_{0.2}O_{3-\delta}$ perovskite-type membranes. J Membr Sci 212:157–165
58. Zhu XF, Cong Y, Yang WS (2006) Effects of synthesis methods on the oxygen permeable $BaCe_{0.15}Fe_{0.85}O_{3-\delta}$ ceramic membranes. J Membr Sci 283:158–163
59. Kharton VV, Marques FMB (2002) Mixed ionic-electronic conductors: effects of ceramic microstructure on transport properties. Curr Opin Solid State Mater Sci 6:261–269
60. Deng ZQ, Liu W, Peng DK, Chen CS, Yang WS (2004) Combustion synthesis, annealing, and oxygen permeation properties of $SrFeCo_{0.5}Oy$ membranes. Mater Res Bull 39:963–969
61. Tai L-W, Nasrallah MM, Anderson HU, Sparlin DM, Sehlin SR (1995) Structure and electrical properties of $La_{1-x}Sr_xCo_{1-y}Fe_yO_3$. Part 1. The system $La_{0.8}Sr_{0.2}Co_{1-y}Fe_yO_3$. Solid State Ionics 76:259–271
62. Yahiro H, Eguchi Y, Eguchi K, Arai H (1988) Oxygen ion conductivity of the ceria-samarium oxide system with fluorite structure. J Appl Electrochem 18:527–531
63. Hunt AG (2005) Percolation theory for flow in porous media. Lecture notes in physics, vol 674. Springer, Berlin
64. Feng ZT, Wang JF, Zhao D (2011) Effect of rock nonhomogeneous characteristics on threshold permeability. Adv Mater Res 243–249:3217–3222
65. Chen T, Zhao HL, Xie ZX, Feng LC, Lu XG, Ding WZ, Li FS (2012) Electrical conductivity and oxygen permeability of $Ce_{0.8}Sm_{0.2}O_{2-\delta}$-$PrBaCo_2O_{5+\delta}$ dual-phase composites. Int J Hydrog Energy 37:5277–5285
66. Stevenson JW, Armstrong TR, Carneim RD, Pederson LR, Weber WJ (1996) Electrochemical properties of mixed conducting perovskites $La_{1-x}M_xCo_{1-y}Fe_yO_{3-\delta}$ (M = Sr, Ba, Ca). J Electrochem Soc 143:2722–2729
67. Ishihara T, Yamada T, Arikawa H, Nishiguchi H, Takita Y (2000) Mixed electronic-oxide ionic conductivity and oxygen permeating property of Fe-, Co- or Ni doped $LaGaO_3$ perovskite oxide. Solid State Ionics 135:631–636
68. Yaremchenko AA, Kharton VV, Viskup AP, Naumovich EN, Tikhonovich VN, Lapchuk NM (1999) Mixed electronic and ionic conductivity of LaCo(M)O3$-\delta$ (M = Ga, Cr, Fe or Ni).: V. Oxygen permeability of Mg-doped La(Ga, Co)O3$-\delta$ perovskites. Solid State Ionics 120:65–74
69. Li SG, Jin WQ, Huang P, Xu NP, Shi J, Lin YS, Hu MZ-C, Payzant A (1999) Comparison of oxygen permeability and stability of perovskite type $La_{0.2}A_{0.8}Co_{0.2}Fe_{0.8}O_{3-\delta}$ (A = Ba, Sr, Ca) membranes. Ind Chem Eng Res 38:2963–2972
70. Yoshida H, Miura K, Fujita J-I, Inagaki T (1999) Effect of gallia addition on the sintering behavior of samaria-doped ceria. J Am Ceram Soc 82:219–221
71. Efimov K, Klande T, Juditzki N, Feldhoff A (2012) Ca-containing CO_2-tolerant perovskite materials for oxygen separation. J Membr Sci 389:205–215
72. Zeng PY, Ran R, Chen ZH, Gu HX, Shao ZP, da Costa JCD, Liu SM (2007) Significant effects of sintering temperature on the performance of $La_{0.6}Sr_{0.4}Co_{0.2}Fe_{0.8}O_{3-\delta\delta}$ oxygen selective membranes. J Membr Sci 302:171–179
73. Kharton VV, Tikhonovich VN, Li SB, Naumovich EN, Kovalevsky AV, Viskup AP, Bashmakov IA, Yaremchenko AA (1998) Ceramic microstructure and oxygen permeability of SrCo(Fe, M)O3$-\delta$ (M = Cu or Cr) perovskite membranes. J Electrochem Soc 145:1363–1373
74. Tan L, Gu XH, Yang L, Zhang LX, Wang CQ, Xu NP (2003) Influence of sintering condition on crystal structure, microstructure, and oxygen permeability of perovskite-related type $Ba_{0.8}Sr_{0.2}Co_{0.8}Fe_{0.2}O_{3-\delta\delta}$membranes. Sep Purif Technol 32:307–312
75. Wang HH, Tablet C, Feldhoff A, Caro H (2005) Investigation of phase structure, sintering, and permeability of perovskite-type $Ba_{0.5}Sr_{0.5}Co_{0.8}Fe_{0.2}O_{3-\delta\delta}$ membranes. J Membr Sci 262:20–26
76. Joo JH, Yun KS, Yoo C-Y, Yu JH (2014) Novel oxygen transport membranes with tunable segmented structures. J Mater Chem A 2:8174–8178

77. Zhang K, Shao ZP, Li CZ, Liu SM (2012) Novel CO_2-tolerant ion-transporting ceramic membranes with an external short circuit for oxygen separation at intermediate temperatures. Energy Environ Sci 5:5257–5264
78. Zhang K, Liu LH, Shao ZP, Xu R, da Costa JCD, Wang SB, Liu SM (2013) Robust ion-transporting ceramic membrane with an internal short circuit for oxygen production. J Mater Chem A 1:9150–9156
79. Wang L, Imashuku S, Grimaud A, Lee D, Mezghani K, Habib MA, Shao-Horna Y (2013) Enhancing oxygen permeation of electronically short-circuited oxygen-ion conductors by decorating with mixed ionic-electronic conducting oxides. ECS Electrochem Lett 2:F77–F81
80. Weirich M, Gurauskis J, Gil V, Wiik K, Einarsrud M-A (2012) Preparation of lanthanum tungstate membranes by tape casting technique. Int J Hydrog Energy 37:8056–8061
81. Julian A, Juste E, Geffroy PM, Coudert V, Degot S, Del Gallo P, Richet N, Chartier T (2009) Elaboration of $La_{0.8}Sr_{0.2}Fe_{0.7}Ga_{0.3}O_{3-\delta}$/$La_{0.8}M_{0.2}FeO_{3-\delta}$ (M = Ca, Sr and Ba) asymmetric membranes by tape-casting and co-firing. J Membr Sci 333:132–140
82. Koh YH, Jun IK, Sun JJ, Kim HE (2006) In situ fabrication of a dense/porous Bi-layered ceramic composite using freeze casting of a ceramic-camphene slurry. J Am Ceram Soc 89:763–766
83. Huang H, Cheng SY, Gao JF, Chen CS, Yi JX (2014) Phase-inversion tape-casting preparation and significant performance enhancement of $Ce_{0.9}Gd_{0.1}O_{1.95}$-$La_{0.6}Sr_{0.4}Co_{0.2}Fe_{0.8}O_{3-\delta}$ dual-phase asymmetric membrane for oxygen separation. Mater Lett 137:245–248
84. He W, Huang H, Gao JF, Winnubst L, Chen CS (2014) Phase-inversion tape casting and oxygen permeation properties of supported ceramic membranes. J Membr Sci 452:294–299
85. Steele BCH (1992) Oxygen ion conductors and their technological applications. Mater Sci Eng B 13:79–87

Chapter 8
Oxygen Permeation at Intermediate–Low Temperatures

Abstract Although the intermediate–low-temperature operation has many advantages in practical applications, the operation temperature of MIEC membranes is limited in the range of 700–1000 °C in literatures, because membranes have low oxygen permeation flux and low permeation stability when the temperature is lower than 700 °C. With the development of preparation techniques of MIEC membranes, the low oxygen permeation flux can be easily solved; however, it is still a challenge to overcome the degradation of oxygen permeation in the intermediate–low temperature range until we published several papers to disclose the degradation mechanism and the stabilization methods recently. Thus, in chapter, an introduction of our achievements is presented on the stabilization of oxygen permeation of MIEC membranes in the intermediate–low temperature range.

Keywords Oxygen permeation • Intermediate–low temperature • Degradation mechanism • Stabilization mechanism • Impurities migration • Permeation stability • Phase transformation • Nanoparticles • Grain boundary • $Ba_{0.5}Sr_{0.5}Co_{0.8}Fe_{0.2}O_{3-\delta}$

8.1 Introduction

As mentioned in Chaps. 6 and 7, the perovskite-type and dual-phase membranes are all operated at high temperatures, with the typical temperature range being 700–950 °C. Although high-temperature operation would lead to a high permeation flux for an MIEC membrane, the drawbacks related to the high temperature cannot be ignored when the membranes are used for oxygen separation on a large scale. The creep rate of an MIEC membrane increases quickly with temperature since the creep activation energy is usually in the range of 300–500 kJ mol^{-1} for perovskite-type MIEC membranes. Additionally, few perovskite-type MIEC membranes have a creep rate as low as 10^{-10} s^{-1} over the temperature range of 700–950 °C and at a strain of 10 MPa, which is regarded as the bottom boundary for engineering ceramics [1]. A report from Air Products & Chemicals Inc. showed that creep due to the pressure load on a wafer-type membrane had resulted in a compressive creep strain of 3.4 % on the entire stack of wafers after 15,000 h of operation at an air pressure of 1.7 MPa. The creep deformation resulted in macroscopic wafer

X. Zhu, W. Yang, *Mixed Conducting Ceramic Membranes*, Green Chemistry and Sustainable Technology, DOI 10.1007/978-3-662-53534-9_8

displacements and a decrease in the cross-sectional area of the internal flow channels of the membrane wafers at the central manifold (see Chap. 10) [2]. Although the operation temperature of the membrane was not shown in the report, it would be in the temperature range of 850–900 °C estimated from their other reports. Besides creep, the high-temperature operation of the MIEC membranes needs more energy to heat the air and more expensive stainless steel in the construction of the separation unit. These would increase the operation and investment costs of the MIEC membrane-based air separation unit. Furthermore, the sealing of ceramics at elevated temperatures is a great challenge in MIEC membrane technology, especially under high-pressure gradients (up to 2 MPa), long operation periods (5–10 years), and high temperatures.

8.2 Difficulties Related to Oxygen Permeation at Intermediate–Low Temperatures

The problems encountered at high temperature as discussed above are easily solved by decreasing the operating temperature to the intermediate–low range (350–700 °C), most typically the range of 500–650 °C. However, there are two new problems to be solved with this temperature range. One is the low permeation flux, i.e., the oxygen exchange on the membrane surfaces and oxygen transport in the membrane bulk are heavily temperature dependent; another is the permeation stability, i.e., no MIEC perovskite membranes have been reported showing a stable oxygen permeation flux at a temperature lower than 650 °C [3–6]. The first difficulty can be overcome by preparing asymmetric MIEC membranes with a thick porous support layer and a thin dense layer for oxygen separation [7, 8]. There are many methods that can be used to prepare these asymmetric MIEC membranes, as discussed in Chap. 4. The most popular methods are the tape-casting–lamination method for plate membrane production and the phase-inversion method for hollow fiber membrane production. Oxygen exchange on membrane surfaces can be accelerated by depositing a nano-catalyst in the porous support or coating the porous catalyst onto the surfaces of the MIEC membranes. Although the permeation flux of current MIEC membranes at intermediate–low temperatures is still lower than the commonly cited threshold (>1 mL cm^{-2} min^{-1}) required for commercial viability [9], this threshold value is likely to be reached in the near future given that gradual progress is being made on membrane materials and the preparation techniques for membranes and catalyst materials. For example, Cao et al. reported an investigation on asymmetric dual-phase MIEC membranes prepared by the tape-casting−lamination method [8]. The membrane had a thin dense layer of 40 μm and a porous support layer of 460 μm. The $Sm_{0.5}Sr_{0.5}CoO_{3-\delta}$ catalyst was deposited in the porous support layer and also coated on the dense layer. The oxygen permeation flux of the dual-phase membrane of 75wt%$Ce_{0.85}Sm_{0.15}O_{2-\delta}$–25 wt%$Sm_{0.6}Sr_{0.4}Al_{0.3}Fe_{0.7}O_{3-\delta}$ was improved by five times at 750 °C compared with the 400-μm-thick self-supported control one, i.e., from 0.2 to 1.0 mL cm^{-2} min^{-1}. Further improvement on the permeation flux is possible by reducing the thickness of the dense layer down to about 10 μm and applying more active catalysts. Watenabe et al. prepared an asymmetric $Ba_{0.95}La_{0.05}FeO_{3-\delta}$ (BLF) membrane by dropping BLF-ethanol slurry onto the green porous support and co-sintering at 1175 °C [10]. The resultant asymmetric membrane had a dense layer 46 μm in thickness and an oxygen permeation flux as high as ~2.5 mL cm^{-2} min^{-1} at 660 °C, which is about four times that of the 1-mm-thick BLF membrane.

8.3 Degradation Mechanisms

There had been no progress on the permeation stability of MIEC membranes at intermediate–low temperatures in the past 30 years until recently when we published a series of papers [11–14]. In general, the degradation processes of the materials become more severe with increasing temperature, although this can happen with MIEC membranes at lower temperatures. The fact that MIEC membrane degradation becomes more severe at lower temperatures stands in stark contrast to the conventional wisdom about membranes.

Three main mechanisms have been proposed to account for the intermediate–low-temperature degradation in perovskite MIEC membranes. The first is that oxygen vacancy ordering (association of substitutional defects and oxygen vacancies, such as $(Sr_{La}V_OSr_{La})^{\times}$; see Chap. 2) occurs in these materials during sustained operation at an intermediate–low temperature, thus severely decreasing the ionic conductivity. Kruidhof et al. suggested that although no new phase was detected by XRD analysis in the $La_{0.6}Sr_{0.4}CoO_{3-\delta}$ sample treated at 700 °C for 100 h, the small microdomains (5–20 nm) with ordered oxygen vacancies could be detected by HRTEM [5]. They obtained evidence for an ordered defect structure at low temperatures for $La_{0.6}Sr_{0.4}CoO_{3-\delta}$ samples provided by data from Differential Scanning Calorimeter (DSC) and oxygen permeability. DSC measurements revealed an endothermic (disordering) peak with a heat effect, $\Delta H = 9.0$ J g^{-1}, at about 910 °C upon heating a disk sample in air. Thus, the permeation flux in the $La_{0.6}Sr_{0.4}CoO_{3-\delta}$ membrane decreasing by two orders after 400 h on-stream operation at 750 °C was attributed to the great decrease in the oxygen vacancy ordering, as shown in Fig. 8.1 [5].

The second mechanism is one of the phase transformations from a cubic perovskite phase to a phase structure with low ionic conductivity as the operation temperature is lowered. This phenomenon has been observed in many perovskite materials, such as $SrCoO_{3-\delta}$, $Sr_{0.9}Ca_{0.1}Co_{0.89}Fe_{0.11}O_{3-\delta}$, $SrCo_{0.8}Fe_{0.2}O_{3-\delta}$, $Ba_{0.5}Sr_{0.5}Co_{0.8}Fe_{0.2}O_{3-\delta}$, and $BaFeO_{3-\delta}$ [5, 15–18]. For $SrCoO_{3-\delta}$ and $BaFeO_{3-\delta}$ perovskite oxides, the cubic phase transforms to a 2-H hexagonal phase at a temperature around 850–900 °C. Figure 8.2 shows the temperature-dependent

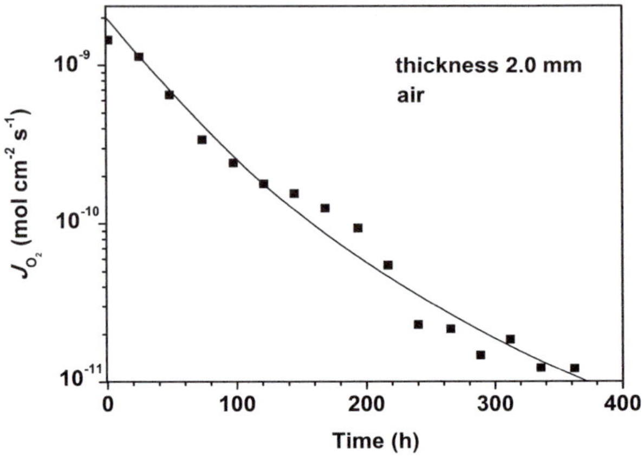

Fig. 8.1 Transient toward steady-state oxygen permeation observed for $La_{0.6}Sr_{0.4}CoO_{3-\delta}$ at 750 °C (Reproduced from [5] with permission of Elsevier)

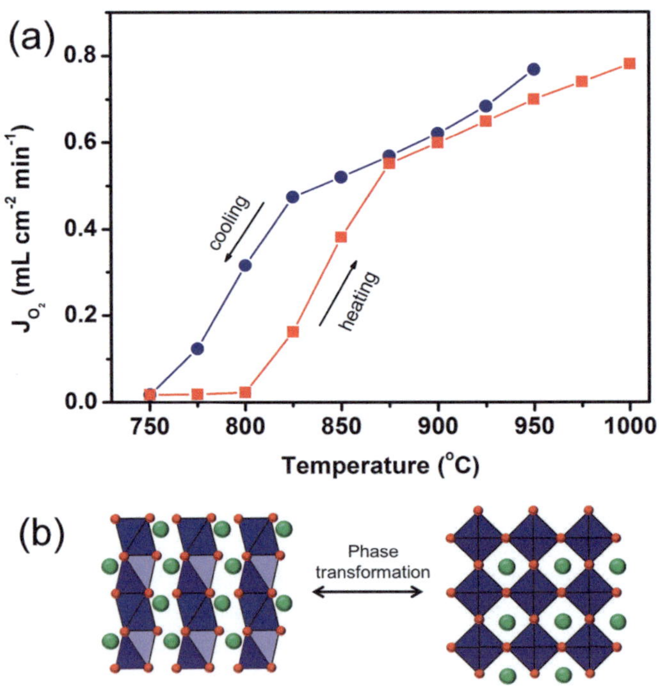

Fig. 8.2 The temperature-dependent oxygen permeation flux of $BaFeO_{3-\delta}$ perovskite membrane (**a**); phase transformation from a hexagonal phase to a cubic phase (**b**) (Reproduced from [18] with permission of Elsevier)

oxygen permeation flux of $BaFeO_{3-\delta}$ perovskite membranes. The oxygen permeation flux decreases suddenly when the operation temperature is lower than the phase transformation temperature. Cooling and heating circles of membranes in the oxygen permeation experiment revealed a reversible phase transition process

[18]. An obvious hysteresis effect appears during the heating step for the two membranes, which is typical for a phase transformation process of an MIEC material. For $SrCo_{0.8}Fe_{0.2}O_{3-\delta}$, the cubic perovskite phase transforms to a brownmillerite phase when the temperature is lower than 790 °C [5]. The phase transformation of $Ba_{0.5}Sr_{0.5}Co_{0.8}Fe_{0.2}O_{3-\delta}$ is complicated, and instead of a perovskite-type hexagonal phase, a lamellar trigonal phase appears depending on the temperature and oxygen partial pressure. The detailed discussion on the phase transformation of $Ba_{0.5}Sr_{0.5}Co_{0.8}Fe_{0.2}O_{3-\delta}$ appears in Chap. 6.

The third mechanism is kinetic demixing. When an MIEC membrane is operated under an oxygen partial pressure gradient, thermodynamic driving forces favor the kinetic demixing of the membrane, and then new phases with low ionic conductivity are produced, or undesirable changes occur in the surface elemental composition and morphology [19]. However, there is no report which shows sufficient evidence to support the kinetic demixing mechanism at intermediate–low temperatures. It should be noted that the oxygen permeation flux of a membrane degrades at low temperatures but remains stable at high temperatures. However, kinetic demixing occurs at both low and high temperatures but is faster at high temperatures due to the thermally activated diffusion of cations. For example, a $BaZr_{0.2}Co_{0.4}Fe_{0.4}O_{3-\delta}$ membrane exhibits stable oxygen permeation at 850 °C for more than 1100 h [20, 21], but the oxygen permeation flux decreases quickly at 500 °C [4]. Therefore, the third mechanism can be ruled out in explaining the degradation of perovskite MIEC membranes at intermediate–low temperatures.

8.4 Degradation and Stabilization Mechanisms of Phase-Stable Membranes

8.4.1 Sulfur-Containing Membranes

Oxygen vacancy ordering is the most frequently used to interpret the degradation of oxygen permeation flux at intermediate–low temperatures. However, in most cases, researchers have not done a thorough investigation or characterization on their membranes before invoking this mechanism to interpret the experimental results. To explore the degradation mechanism of membranes with a stable phase structure (i.e., those membrane materials that do not undergo phase transformation under the conditions used for oxygen permeation), we selected three perovskite membranes, $BaCe_{0.1}Co_{0.4}Fe_{0.5}O_{3-\delta}$ (BCCF) [22], $BaZr_{0.2}Co_{0.4}Fe_{0.4}O_{3-\delta}$ (BZCF) [20], and $BaCo_{0.7}Fe_{0.22}Nb_{0.08}O_{3-\delta}$ (BCFN) [23], which were recently developed by different groups and showed stable phase structures under the conditions for oxygen permeation and operating temperatures higher than 750 °C. However, for all three membrane materials derived from high-purity precursors, their permeation fluxes decreased quickly over time at 600 °C, as shown in Fig. 8.3. After about 480 h on-stream at 600 °C, the permeation flux of the BCCF membrane decreased by

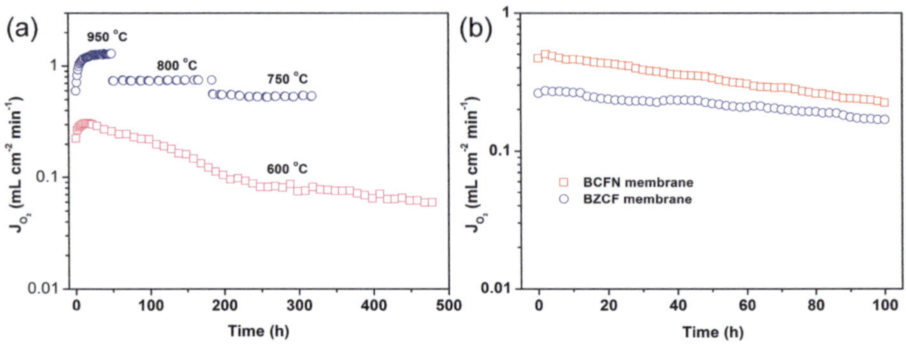

Fig. 8.3 Time-dependent oxygen permeation fluxes of (**a**) BCCF membranes at 950, 800, 750 (thickness 1.0 mm), and 600 °C (thickness 0.5 mm), (**b**) $BaZr_{0.2}Co_{0.4}Fe_{0.4}O_{3-\delta}$ (*blue open circles*) and $BaCo_{0.7}Fe_{0.22}Nb_{0.08}O_{3-\delta}$ (*red open squares*) membranes at 600 °C (thickness 0.5 mm) (Reproduced from [11] with permission of Wiley)

about 80 %, while after just 100 h of operation at 600 °C, the permeation fluxes of BZCF and BCFN decreased by 38.5 % and 55.7 %, respectively [11]. XRD analysis showed that no change in the phase structure had occurred, but $BaSO_4$ was detected on all membrane surfaces. For all three membranes, the intensity of the $BaSO_4$ peak was significantly higher on the permeation side compared to the feed side. For BZCF, both $BaSO_4$ and $Zr(SO_4)_2$ impurity phases were detected on the permeation-side surface. HRTEM analysis confirmed that there was no phase change for the perovskite membranes (Fig. 8.4). Thus, oxygen vacancy ordering is not the correct explanation for the degradation of oxygen permeation fluxes for the three membranes. Based on these initial findings, additional studies were carried out to investigate the $BaSO_4$ surface impurities as a potential new explanation for permeation degradation in these systems at intermediate–low temperatures.

There are two potential sources of sulfur impurities; one is from the high-purity gases (N_2, O_2, and He) used in the experiments, and the other is from the chemical raw materials (such as various nitrates, deionized water, citric acid, EDTA, etc.) used in the preparation of the perovskite materials. All the high-purity gases were passed through cylinders containing KOH and 4A zeolites to remove potential impurities, such as CO_2, H_2O, H_2S, SO_2, etc. Among all the chemical raw materials used in the experiments, only $Co(NO_3)_2{\cdot}6H_2O$ is documented as containing sulfur impurities, at less than 17 ppm, by Sigma-Aldrich. The starting sulfur impurity content in the BCCF membrane made from Sigma-Aldrich regents was determined to be less than 20 ppm. It was difficult for us, at first, to accept that this small background sulfur impurity content was sufficient to cause such a large and rapid surface enrichment of $BaSO_4$ and a commensurately large and rapid degradation of the membrane permeation flux during low-temperature operation.

To quickly verify that sulfur was indeed responsible for this behavior, BCCF membranes with a higher sulfur content (135 ppm) were prepared and named as BCCF-s. Figure 8.5 shows that the oxygen permeation flux of the BCCF-s membrane decreased by 70 % in 100 h at 650 °C (disk 1), while it remained stable at an

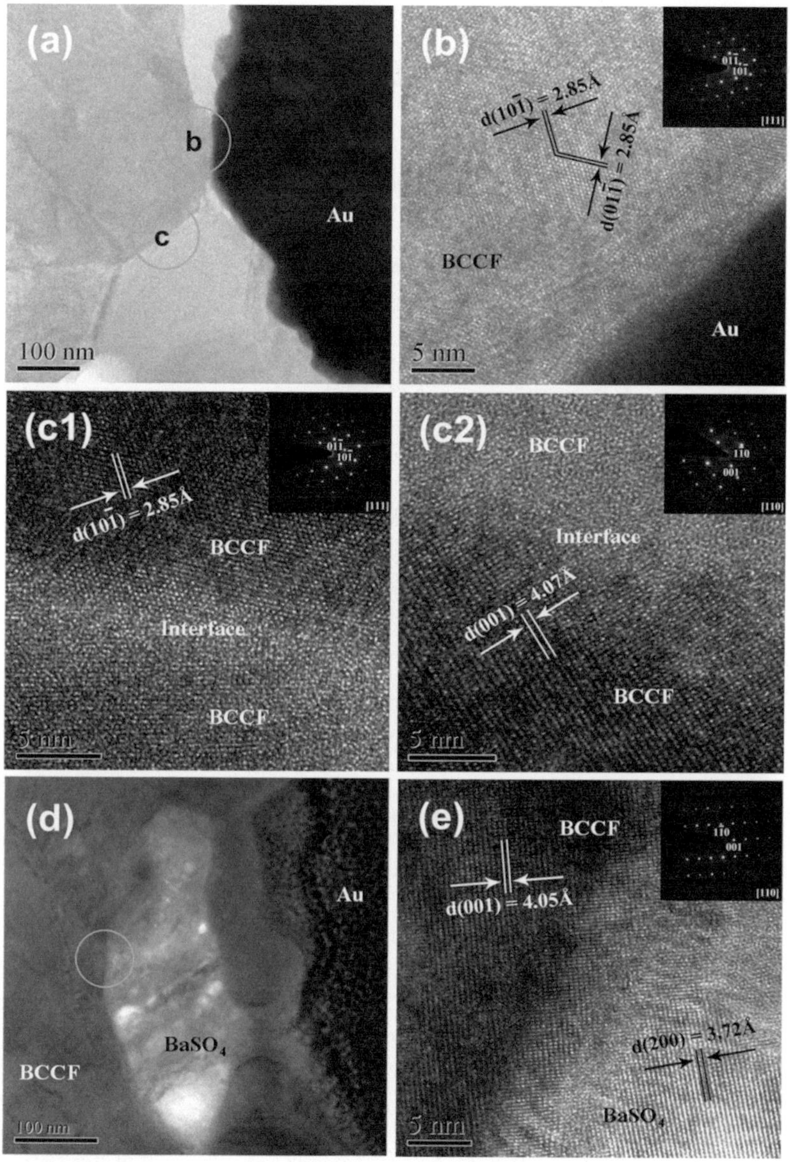

Fig. 8.4 HRTEM analysis of the perovskite membrane after long-term operation at 600 °C. The *circles* highlighted in the large-scale images (**a**, **d**) correspond to the areas shown at higher resolution in (**b**, **c**, and **e**). (**a**, **b**) "Clean" area of the BCCF surface not covered by $BaSO_4$. (**c1**, **c2**) High-resolution close-up of a perovskite grain boundaries viewed along the [111] (**c1**) and [110] (**c2**) directions. (**d**, **e**) Grain boundary between the perovskite and $BaSO_4$ viewed along the [110] direction of the perovskite (Reproduced from [11] with permission of Wiley)

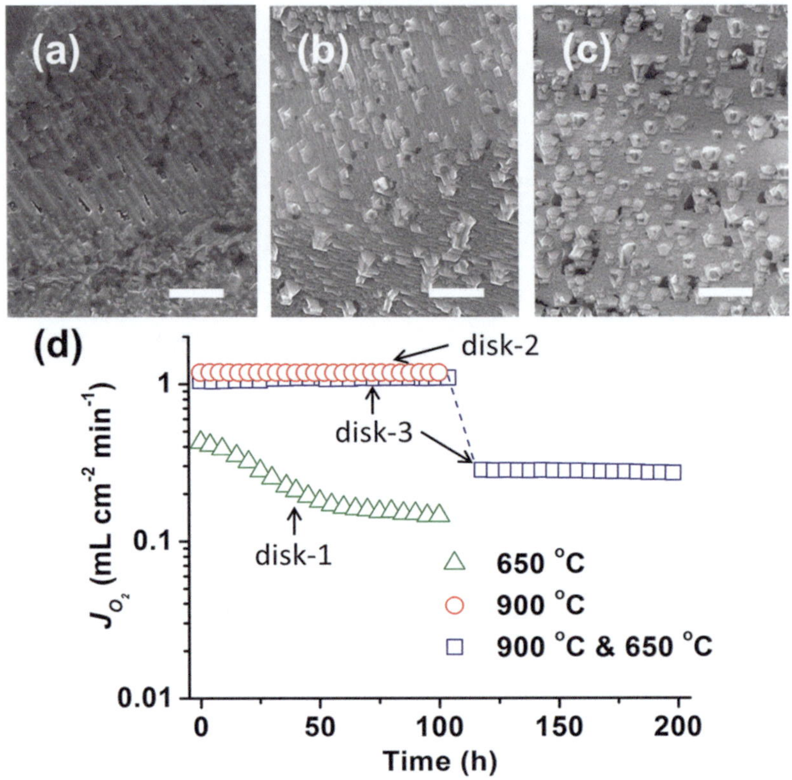

Fig. 8.5 Analysis of the high sulfur-containing BCCF-s membranes operated under different conditions. SEM pictures of the spent membrane surfaces (permeate side) after oxygen permeation testing at (**a**) 650 °C for 100 h, (**b**) 900 °C for 100 h, (**c**) 900 °C for 100 h, and then 650 °C for 100 h. (**d**) Oxygen permeation flux versus time behavior of the three membranes (1 mm in thickness). Scale bar: 2 μm (Reproduced from [11] with permission of Wiley)

operating temperature of 900 °C (disk 2) [11]. The excellent high-temperature permeation stability is consistent with observations from other perovskite-type MIEC membrane materials, which typically show good permeation stability above 850 °C. If the membrane is operated first at 900 °C for 100 h, and then at 650 °C for another 100 h, the oxygen permeation flux stays stable at 900 °C but decreases slowly at 650 °C (disk 3). SEM/EDS analysis shows that the permeation surface of disk 1 was covered by $BaSO_4$ with a sulfur content of 7.5 at%, while the surface of disk 2 had a lower sulfur content of 1.6 at%. The surface sulfur content for disk 3 was higher than that of disk 2 but lower than that of disk 1.

Thermodynamic calculations reveal that $BaSO_4$ can gradually decompose at elevated temperatures and the equilibrium constant for the decomposition reaction increases exponentially with temperature. A simple calculation shows that if the bulk sulfur diffusion and surface $BaSO_4$ decomposition kinetics are fast enough to maintain the SO_2 partial pressure in equilibrium with the flowing gas phase, it

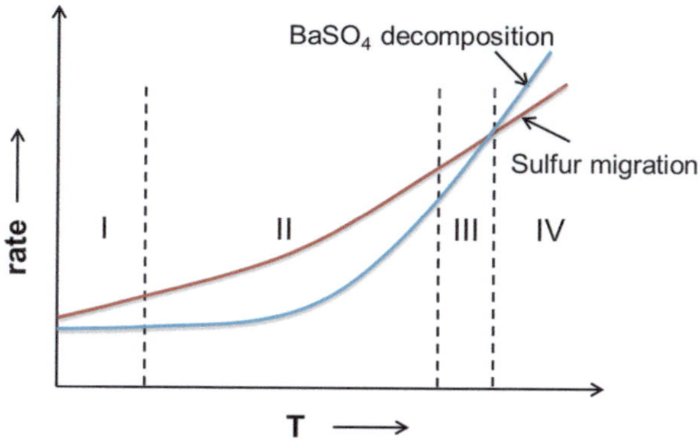

Fig. 8.6 Schematic of the temperature-dependent rates of $BaSO_4$ decomposition and sulfur diffusion. Zone I, the rates of $BaSO_4$ decomposition and sulfur migration are both slow; zone II, sulfur migration is faster than $BaSO_4$ decomposition; zone III, sulfur migration is only a little faster than $BaSO_4$ decomposition; zone IV, sulfur migration is slower than $BaSO_4$ decomposition

would take more than 4000 days at 600 °C to remove all sulfur from a 0.5-mm-thick BCCF membrane (containing 20 ppm S). In contrast, at 900 °C it would take less than 1 day to accomplish the same thing. At low operating temperatures, the rate of enrichment of $BaSO_4$ on the membrane surface is faster than $BaSO_4$ decomposition; as a result, $BaSO_4$ grains aggregate on the membrane surface leading to the degradation of the permeation flux. At higher temperatures, even though sulfur segregation from the bulk to the surface should also be accelerated, this is more than offset by significantly faster $BaSO_4$ decomposition, and as a result, only a few $BaSO_4$ grains can be found, and no degradation of the permeation flux is observed. This mechanism is diagrammatically presented in Fig. 8.6. This mechanism also helps to explain why disk 3 shows a slow degradation rate at 650 °C as compared to disk 1. Much of the sulfur was likely already removed from this membrane during the first 100 h of operation at 900 °C. This mechanism is not just limited to barium-containing materials. The permeation stability of other membranes containing alkaline earth and rare earth elements will be likely degraded by sulfur impurities since their sulfates are generally stable at low temperatures.

8.4.2 Silicon-Containing Membranes

The negative effect of sulfur impurities on the permeation stability of MIEC membranes at low temperatures inspired us to explore the potential effect of silicon impurities in these membranes [13]. Because silicon is one of the most universal contaminants (it exists widely in dust, water, refractory materials used in high-temperature furnaces, glass-based sealants, silicon-based greases, and lubricants, as well as original chemicals) [24, 25], it is impossible to obtain silicon-free ceramic materials. The BCCF membranes prepared using another batch of chemicals

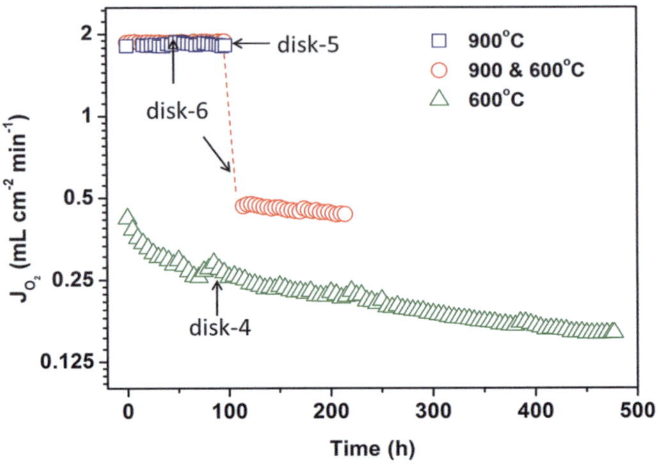

Fig. 8.7 Time-dependent oxygen permeation fluxes of three BCCF membranes. *Disk 4*, 600 °C for 477 h; *disk 5*, 900 °C for 100 h; *disk 6*, 900 °C for 96 h and then 600 °C for 100 h (Reproduced from [13] with permission of Elsevier)

purchased from Sigma-Aldrich showed high silicon content up to 140 ppm, even though super-high-purity water was used in the preparation process. Three disks were employed for oxygen permeation at 600 °C for 477 h (disk 4), 900 °C for 100 h (disk 5), 900 °C for 96 h, and then 600 °C for 100 h (disk 6), as shown in Fig. 8.7. The oxygen permeation flux of disk 4 decreased by 62 % (gradually from 0.42 to 0.16 mL cm^{-2} min^{-1}) during the 477 h on-stream at 600 °C, while that of disk 5 remained constant at 1.8 mL cm^{-2} min^{-1} during the 100 h on-stream at 900 °C. It is worth noting that the disk 6 exhibited a higher oxygen permeation flux and a slower degradation rate at 600 °C than did disk 4, which was initially operated at the same temperature [13].

The phenomenon of the time-dependent permeation stability at high and low temperatures occurring on silicon-containing BCCF membranes is similar to that happening on sulfur-containing ones (see Fig. 8.5). Therefore, the degradation mechanism for silicon-containing membranes should be similar to those containing sulfur. However, unlike sulfates, the enriched silicates on membrane surfaces cannot be removed by decomposition, so why do the silicon-containing membranes still keep a stable permeation flux at high temperature? Figure 8.8 shows the SEM pictures of both surfaces of the three spent membranes above [13]. For the membrane operated at 600 °C for 477 h, there was no morphology change for the feed side, but there was a great change on the permeation side. EDS analysis showed that the surface of the permeation side had a silicon content reaching 4.7 at%, while that of the feed side was close to zero. The extremely low silicon content on the feed-side surface will be discussed below. For the membrane operated at 900 °C for 100 h, larger white particles were formed on the feed and permeation sides of disk 5. EDS analysis showed that these white particles were silicates and the silicon content for the feed- and permeation-side surfaces were 0.2 and 0.5 at%,

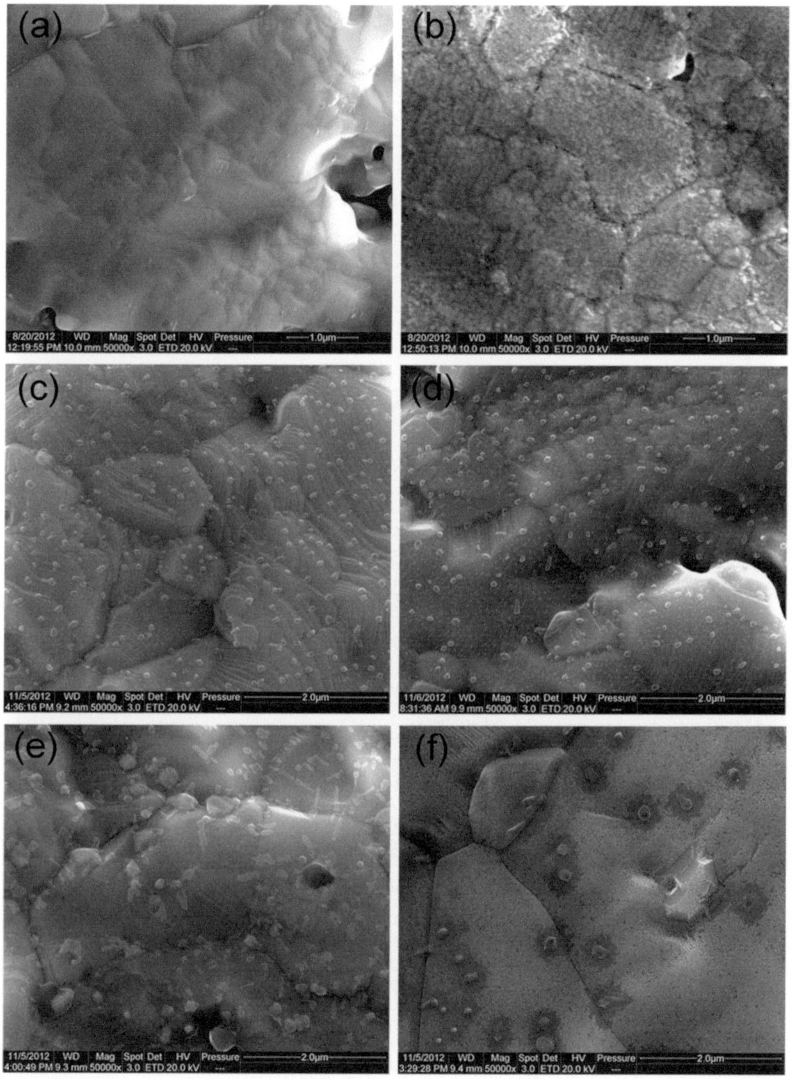

Fig. 8.8 SEM images of the spent BCCF membranes. (**a**) and (**b**) are feed and sweep sides of disk 4 operated at 600 °C for 477 h, respectively; (**c**) and (**d**) are feed and sweep sides of disk 5 operated at 900 °C for 100 h, respectively; (**e**) and (**f**) are feed and sweep sides of disk 6 operated at 900 °C for 96 h and then 600 °C for 100 h, respectively (Reproduced from [13] with permission of Elsevier)

respectively. The silicon content of disk 6 was higher than in disk 5 but still lower than that on the permeation-side surface of disk 4. The silicon-containing particles grow large at an operation temperature of 900 °C but stay small at a lower temperature of 600 °C. Therefore, the surface coverage of the silicon impurity is lower for disk 5 than it is on the permeation-side surface of disk 4. If small particles appeared around big particles, they would be "swallowed" by the big particles. As a result, a specific morphology was observed on the membrane surface (Fig. 8.8f). The coverage of the silicon impurity on the permeation-side surface was lower for disk 6 than for the disk 4. These observations can well explain the time-dependent permeation flux shown in Fig. 8.7.

A detailed analysis is shown in Fig. 8.9 using HRTEM/EDS mapping of the cross section of the permeation side of disk 4 [13]. The segregation of barium on the membrane surface was induced by silicon enrichment leading to the formation of amorphous barium silicates. Meanwhile, the BCCF perovskite was decomposed into cobalt and iron oxides which can be identified in Fig. 8.9a–f, i.e., a layer with

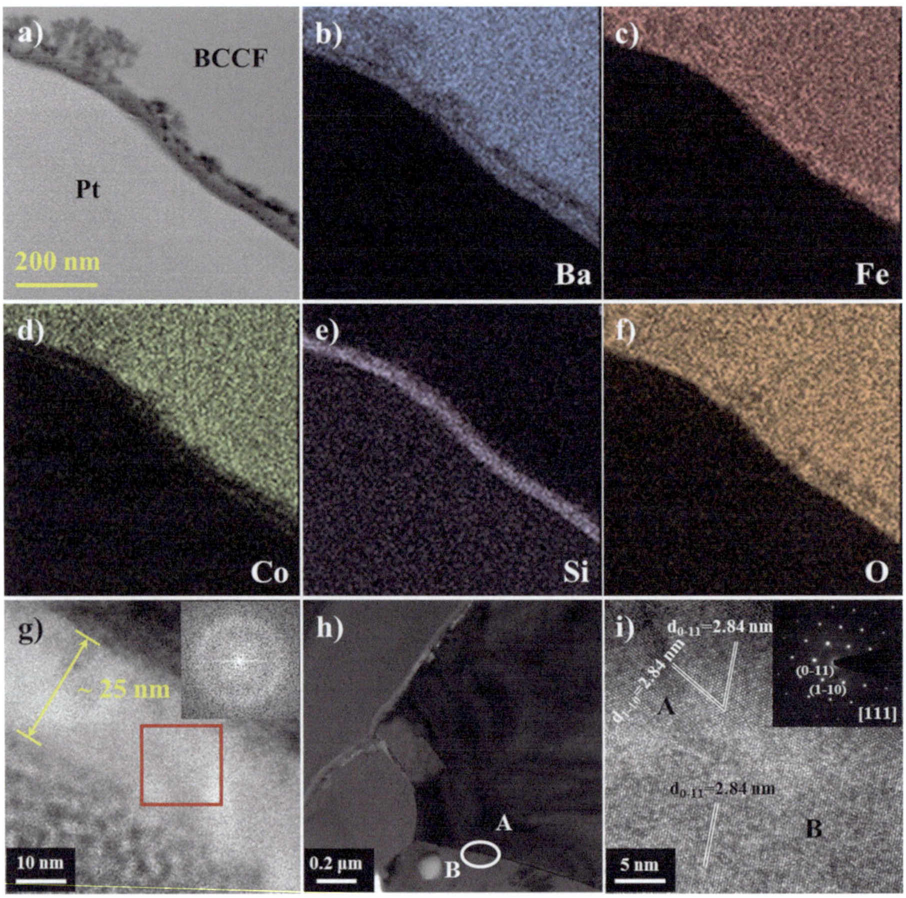

Fig. 8.9 Microscopic analysis of the sweep side surface of the perovskite membrane after 477 h on-stream at 600 °C. (**a**) STEM image of the BCCF slice embedded in platinum, (**b**) ~ (**f**) EDS mapping of the impurity layer on sweep side surface, (**g**) silicon impurity layer, (**h**) grain boundary of BCCF highlighted by a *circle*, (**i**) HRTEM image of the *circled* zone in (**h**) viewed along the [111] direction of the cubic BCCF crystal (Reproduced from [13] with permission of Elsevier)

thicknesses of 20–30 nm showing stronger signals for Fe and Co elements can be observed beneath the barium silicate layer. The outermost layer of the BCCF slice contains Si, Ba, and O as the major elements and Co as the minor element and is a ~25-nm-thick amorphous layer, indicated by the halo ring in the inset Fig. 8.9g derived from the fast Fourier transform (FFT) pattern of the squared area. As shown in Fig. 8.9h, there is no secondary phase in the grain boundaries, and the BCCF membrane retains the cubic perovskite structure after long-term operation. Therefore, the reduction in oxygen permeation does not result from a phase transformation at low temperature (600 °C). It is clear that the change in composition of the BCCF membrane in this study is due to silicon enrichment-induced decomposition of the perovskite phase. Therefore, there was no clear evidence to support the assertion that the degradation was induced by the kinetic demixing mechanism. From the above discussion, it was inferred that the enrichment of silicon on BCCF membrane surfaces is the reason for the degradation of oxygen permeation at low temperature [13].

8.4.3 Mechanism of Sulfur and Silicon Migration to the Membrane Surface

Slater and coworkers found that sulfate and silicate groups can be doped into the perovskite lattice with content up to 15 at% [26–29]. High temperature, for instance, 1100 °C, is usually needed for incorporation of the sulfate and silicate groups into the lattice. In their works, they successfully incorporated oxyanions, including borate, silicate, phosphate, and sulfate, into a perovskite lattice in order to stabilize the cubic perovskite structure of $SrCoO_{3-\delta}$ and $SrMnO_{3-\delta}$ at room temperature and also incorporated borate and phosphate into cathode materials to improve the structural stability of the cathodes at intermediate temperatures [29]. Li et al. investigated a phosphate-doped perovskite, $SrCo_{0.85}Fe_{0.1}P_{0.05}O_{3-\delta}$, and observed that the incorporation of phosphate can prevent oxygen vacancy ordering at temperatures higher than 600 °C both in air and atmospheres with lower oxygen partial pressures [30]. However, in the investigation into the degradation of the oxygen permeation flux in MIEC membranes at low temperatures (above), we found that trace sulfur or silicon impurities coming from raw chemicals would migrate from the bulk of the membrane to the membrane surfaces and deposit on the permeation-side surface as sulfates or silicates, respectively [11, 13, 14]. The enrichment of sulfur or silicon impurities on the membrane surfaces then blocks the oxygen exchange reactions and leads to the degradation of oxygen permeation flux under long-term operation.

Three membranes containing sulfur and silicon impurities were prepared, namely, BCCF-S (sulfur content of 140 ppm), BCCF-Si (silicon content of

340 ppm), and $SrCo_{0.97}S_{0.03}O_{3-\delta}$ (SC-S), and all the membranes were treated in flowing air and helium at 650 °C for BCCF-S, 600 °C for BCCF-Si, and 850 °C for SC-S, respectively. All the membranes showed greater enrichment with sulfates or silicates on the membrane surfaces when the membranes were treated in helium. This result indicates that sulfur and silicon can migrate outside of the perovskite lattice during the aging treatment. Further, it reveals that the chemical potentials of sulfur and silicon in the perovskite lattices are higher than those at the gas–solid interfaces. Combined with the results from Slater and coworkers, it was deduced that sulfate and silicate groups can incorporate into the perovskite lattice at elevated temperatures (such as 1100 °C) and that most of these groups stay in the perovskite lattice during the cooling step due to the slow migration rates of these groups and fast cooling rate (usually, 2 °C/min). These groups have high chemical potentials at low temperature in the perovskite lattice; thus, the high chemical potentials are driving forces for sulfur and silicon migration from the membrane bulk to the surfaces when the perovskite materials are treated at low temperatures. However, the above discussion cannot explain why the permeation-side surfaces always have higher contents of sulfur or silicon impurities than those on the feed side.

We found that sulfur or silicon migration is indirectly related to the oxygen chemical potential gradient, i.e., at the gas–solid interface, the chemical potential of sulfur or silicon is related to the oxygen partial pressure. Here, the BCCF membrane containing silicon impurities is discussed as an example to illustrate the mechanism by which the oxygen partial pressure affects the migration of the sulfur and silicon impurities, as shown in Fig. 8.10. When the membrane sintered at 1150 °C is then treated in air at 600 °C, the chemical potential of silicon in the perovskite lattice ($\mu_{\mathbf{Si}}^{\mathrm{Latt}}$) is higher than that at the air–solid interface ($\mu_{\mathbf{Si},\,\mathrm{Air}}^{\mathrm{Interf}}$), so the silicon migrates from the membrane bulk to the surfaces, as illustrated by Fig. 8.10a. When the membrane is treated in helium, similarly, the chemical potential of silicon in the perovskite lattice ($\mu_{\mathbf{Si}}^{\mathrm{Latt}}$) is much higher than that at the helium−solid interface ($\mu_{\mathbf{Si},\,\mathrm{He}}^{\mathrm{Interf}}$), so the silicon again migrates from the bulk of the membrane to the surfaces, as illustrated in Fig. 8.10b. Additionally, greater silicon enrichment was found on the surface of the membrane treated in helium (2.5 at%) than on that treated in air

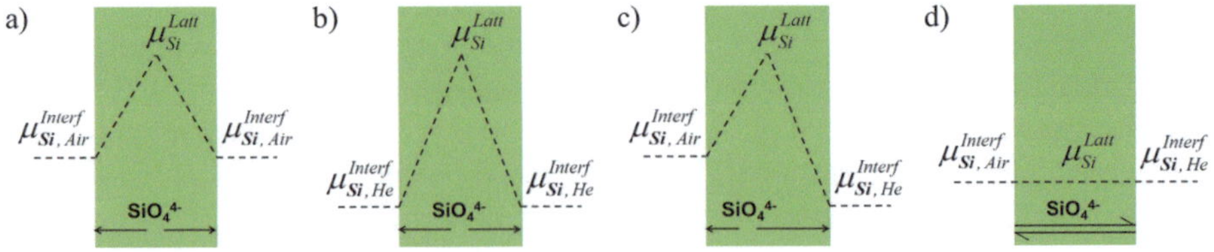

Fig. 8.10 Schematic of the chemical potentials of silicon element and the migration of silicate group across the membranes under different atmospheres. (**a**) Membrane treated in air, (**b**) membrane treated in helium, (**c**) membrane operated for oxygen permeation in the initial stage, (**d**) membrane operated for oxygen permeation reaching an equilibrium state of silicon element migration (Reproduced from [13] with permission of Elsevier)

Fig. 8.11 Feed-side surface (**a**) and permeation-side surface (**b**) of the spent BCCF membrane after 480 h operation at 600 °C all the protuberant grains on the permeation-side surface are $BaSO_4$ (Reproduced from [11] with permission of Wiley)

(1.2 at%), which indicates that $\mu_{\mathbf{Si},\,\mathrm{Air}}^{\mathrm{Interf}}$ is higher than $\mu_{\mathbf{Si},\,\mathrm{He}}^{\mathrm{Interf}}$ at 600 °C. If the membrane is operated for oxygen permeation, silicon migrates from the membrane bulk to both the surfaces in the initial stage, as illustrated by Fig. 8.10c. Therefore, silicon impurities were observed on both surfaces, as shown in Fig. 8.8e, f, but that on the helium side (i.e., sweep side) surface was higher than that on the air side (i.e., feed side) surface. This situation can be found also on disk 5 and disk 6. However, $\mu_{\mathbf{Si}}^{\mathrm{Latt}}$ decreases with time but both $\mu_{\mathbf{Si},\,\mathrm{Air}}^{\mathrm{Interf}}$ and $\mu_{\mathbf{Si},\,\mathrm{He}}^{\mathrm{Interf}}$ increase with time. Thus after a certain time, $\mu_{\mathbf{Si}}^{\mathrm{Latt}}$ will be lower than $\mu_{\mathbf{Si},\,\mathrm{Air}}^{\mathrm{Interf}}$ but higher than $\mu_{\mathbf{Si},\,\mathrm{He}}^{\mathrm{Interf}}$. As a result, there will be a net migration of silicon from the air side surface to the helium side surface. The silicon impurity content on the air side surface will be decreased by further prolonged operation time; meanwhile, it will have accumulated on the helium side surface. In the final equilibrium state, the silicon chemical potentials at the air–solid interface, in the perovskite lattice, and at the helium–solid interface will be equal to each other, as illustrated by Fig. 8.10d. This final equilibrium state may be achieved after the BCCF membrane (disk 4) was operated for 477 h at 600 °C, since there was almost no silicon impurity detected by EDS on the air side surface but much more on the helium side surface. Similarly, the final equilibrium state was almost achieved after the sulfur-containing BCCF membrane was operated for 480 h at 600 °C, as shown in Fig. 8.11, because there were few barium sulfate particles on the air side surface but many more on the helium side.

The above discussion indicates that other anions such as BO_3^{2-} and PO_4^{3-}, which are common impurities in precursor salts and oxides and can be incorporated into a perovskite lattice at high temperatures, may exhibit similar migration behaviors when there is an oxygen chemical potential gradient across the MIEC membranes. Moreover, if an inert and stable phase is produced due to the enrichment of anions such as sulfates and silicates on membrane surfaces, then the oxygen permeation fluxes will decrease with time, especially at low temperatures.

8.4.4 *Stabilization of the Phase-Stable Membranes at Low Temperature*

Based on the above analysis of the degradation mechanism of phase-stable perovskite MIEC membranes, one option to combat the sulfur- or silicon-induced degradation phenomenon at intermediate–low temperatures is to synthesize perovskite membranes using ultrahigh-purity sulfur and silicon-free reagents. However, recall that even the 20-ppm sulfur impurity level introduced by the high-purity reagents used to synthesize the membranes tested in Fig. 8.3 was still sufficient to cause significant degradation; so this method is infeasible for large-scale preparation of MIEC membranes. Additionally, although the same chemicals were purchased from Sigma-Aldrich, the contents of sulfur and silicon impurities changed with batches. Obtaining higher-purity reagents, especially at an industrially relevant scale, is not an easy or cost-effective approach. As an alternative, we examined the possibility of deploying a high-surface-area buffer/catalyst layer with high oxygen activity on both the feed and permeation surfaces of the membrane to mitigate the effects of the surface contamination from sulfates and silicates. Porous $Sm_{0.5}Sr_{0.5}CoO_{3-\delta}$ (SSC) was used for this purpose.

SSC is a perovskite oxide that shows high oxygen catalytic activity and is commonly used as an oxygen exchange catalyst in MIEC membranes and as a cathode material for low-temperature (~600 °C) solid oxide fuel cells [31–34]. A porous SSC, composed of particles 0.5–1 μm calcined at 950 °C, was applied to both sides of a BCCF membrane to create porous surface buffer layers of ~ 10 μm thickness. Figure 8.12 compares the permeation stability of BCCF, BZCF, and BCFN membranes with or without SSC porous layer coatings under identical

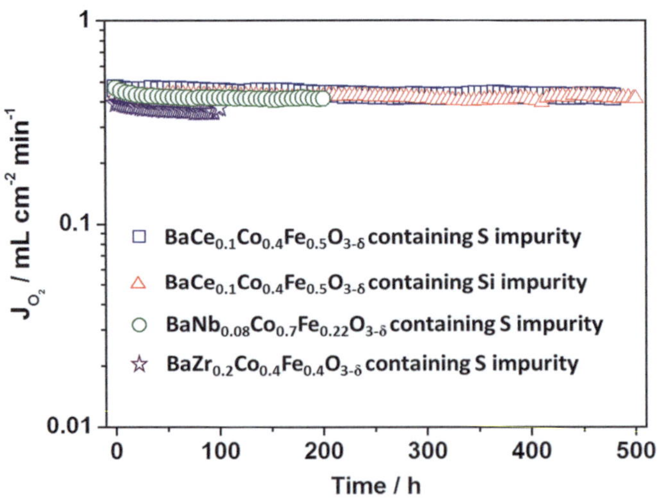

Fig. 8.12 Time-dependent oxygen permeation of BCCF, BZCF, and BCFN membranes with both surfaces coated with SSC porous catalyst layers. Temperature, 600 °C; membrane thickness, 0.5 mm (Ref. [11,13])

testing conditions at 600 °C [11,13], and the changes in oxygen permeation flux are compared in Table 8.1. The flux of the SSC-coated BCCF membrane decreased by only ~5 % in the first 200 h and then maintained stable performance. Similarly, stabilized permeation performances at low temperature were recorded on BCFN and BZCF membranes coated with SSC porous layers. For the SSC-coated BCFN membrane, an 8 % decline in flux was observed in the first 40 h of operation followed by no further decreases in flux over the next 160 h. For the SSC-coated BZCF membrane, a very stable performance was observed with only small fluctuations of ±4 % in the flux during 100 h of operation. Postmortem SEM and EDX analysis was conducted on the spent SSC-coated BCCF membrane after 480 h of operation, as shown in Fig. 8.13. This analysis indicated that the $BaSO_4$ grains

Table 8.1 The permeation stability of BCCF, BZCF, and BCFN membranes coated with or without SSC porous layers under identical testing conditions at 600 °C

	$BaCe_{0.1}Co_{0.4}Fe_{0.5}O_{3-\delta}$	$BaZr_{0.2}Co_{0.4}Fe_{0.4}O_{3-\delta}$	$BaCo_{0.7}Fe_{0.22}Nb_{0.08}O_{3-\delta}$
No catalyst	−80 %, 480 h	−38 %, 100 h	−56 %, 100 h
With catalyst	−5 %, 480 h	±4 %, 100 h	−8 %, 200 h

Ref. [11]

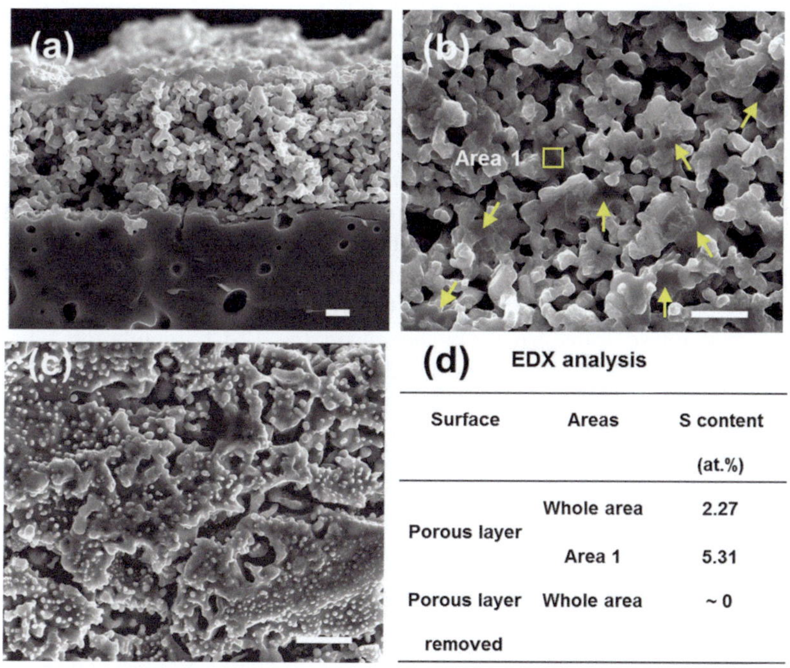

Fig. 8.13 Post-treatment and analysis of the SSC-coated perovskite membrane after 480 h of oxygen permeation testing at 600 °C. (**a**) Cross-section view of the porous SSC catalyst layer. (**b**) Top view of the porous SSC catalyst layer. The *arrows mark* $BaSO_4$ grains. (**c**) Top view of the membrane surface after the porous SSC layer was removed by ultrasonic agitation. (**d**) EDX analysis of the sulfur content of various different areas. Scale bar: 2 mm (Reproduced from [11] with permission of Wiley)

diffused to the outer skin of the porous SSC layer and, as a consequence, there are no $BaSO_4$ grains remaining on the BCCF membrane surface itself [11]. Similarly, XPS analysis on the spent membranes showed that the silicon migrated from the membrane surfaces to the SSC porous layers.

All these findings further verify that the migration of sulfur or silicon is related to the oxygen chemical potential. The SSC coating can effectively act as a "buffer" or a sponge to remove the sulfate and silicate grains from the membrane surface. Furthermore, thanks to the high surface area and high activity of this porous SSC layer, the enrichment of sulfate and silicate on the SSC surface has little effect on the oxygen exchange rate as the sulfate and silicate particles are distributed over a much larger area and cannot completely block the oxygen exchange process. The success in achieving flux stabilization in the perovskite MIEC membrane materials suggests that this stabilization technique may be widely used to overcome the degradation of permeation flux at low temperature for many MIEC perovskite membrane material families. However, the enrichment of silicon or sulfur impurities in the porous layer would decrease the catalytic activity of the SSC toward oxygen exchange reactions to some extent. Therefore, a slight decrease (−5 %) in oxygen permeation is inevitable [11, 13].

8.5 Degradation and Stabilization Mechanisms of Phase-Unstable Membranes

As illustrated in Chap. 6, $Ba_{0.5}Sr_{0.5}Co_{0.8}Fe_{0.2}O_{3-\delta}$ (BSCF), which was developed by Yang's group, exhibits high oxygen permeability and has received significant attention in the last decade. Its high permeability is attributed to its high ionic conductivity and catalytic activity toward both the oxygen reduction reactions (ORR) and oxygen evolution reactions (OER). However, many researchers have reported that phase transformations occur on the BSCF membrane when the operation temperature is lower than 850 °C [17, 35–37]. In this section, BSCF is selected as an example to show its degradation and stabilization mechanisms at intermediate–low temperatures.

8.5.1 *Degradation Mechanism of $Ba_{0.5}Sr_{0.5}Co_{0.8}Fe_{0.2}O_{3-\delta}$ at Intermediate–Low Temperatures*

Hexagonal and lamellar phases were observed after the powder or membranes were treated in air at temperatures below 850 °C for tens to hundreds of hours [17, 35–37]. The new phases have much lower ionic conductivities than the cubic phase; therefore, a degradation of the oxygen permeation flux with time is frequently observed when the BSCF membranes are operated at intermediate–low temperatures. A detailed discussion of the phase transformation of BSCF is shown in Chap. 6. All the studies attributed the degradation of oxygen permeation through the BSCF membranes at intermediate temperatures to these hexagonal and lamellar phases of extremely low ionic conductivity.

For perovskite-type MIEC membranes, the oxygen permeation process is usually controlled by three steps, i.e., oxygen exchange on the gas–solid interface of the feed side, bulk phase diffusion of oxygen ions (the electronic conductivity is typically several orders of magnitude higher than the oxygen ionic conductivity), and oxygen exchange on the gas–solid interface of the permeation side (see Chaps. 1 and 5). Each step corresponds to a permeation resistance. An increase in any one of the three permeation resistances will lead to a decrease in permeation flux. The investigation of the performance degradation of phase-stable membranes inspired us to consider the influence of sulfur and silicon impurities on the permeation stability of the BSCF membrane. As mentioned above, the impurities degrade the permeation flux through the enrichment of sulfate and silicate on the membrane surfaces to retard the oxygen exchange at the gas–solid interfaces. Therefore, one question to be asked is as follows: is the degradation of oxygen permeation flux for BSCF membranes completely caused by the bulk phase transformation at intermediate–low temperature? If the answer is no, how much does the enrichment of sulfur or silicon impurities contribute to the degradation? To correctly answer the two questions, a permeation model which can distinguish the three permeation resistances must be utilized to monitor the time-dependent changes in them. We developed a simple permeation model, shown in Chap. 5, with clear physical meanings for every parameter that has been successfully applied to the acquisition of data on the three resistances to the oxygen permeation through perovskite and dual-phase MIEC membranes [38–40]. We found that this model succeeded in explaining the time-dependent changes of the permeation resistances of a BSCF membrane.

The sulfur content in the as-prepared BSCF powder was determined to be 240 ppm even though high-purity regents were used as the chemical precursors. Figure 8.14 shows the time-dependent oxygen permeation fluxes of the BSCF membranes and the membranes coated with an SSC porous layer, at 700, 650, and 600 °C [12]. If the enrichment of sulfate on the membrane surface has a great influence on the oxygen exchange at the gas–solid interfaces, the degradation rate will be slowed down on the membranes coated with the SSC porous layers in accordance with the result shown in the Sect. 8.4.4 of this chapter. The oxygen permeation fluxes of the membranes, uncoated and coated with SSC, all decreased with time during the 100 h on-stream, but the degradation rate is slow for the membranes coated with SSC porous layers. In the previous investigations, the degradation rate increased with decreasing operating temperature in the range 700–850 °C, while here it was at 650 °C that the SSC-uncoated membranes showed the fastest degradation rate among the three investigated temperatures. However, for the SSC-coated membrane, the degradation rate was slowed down by the decrease in temperature. The difference in the temperature-dependent degradation

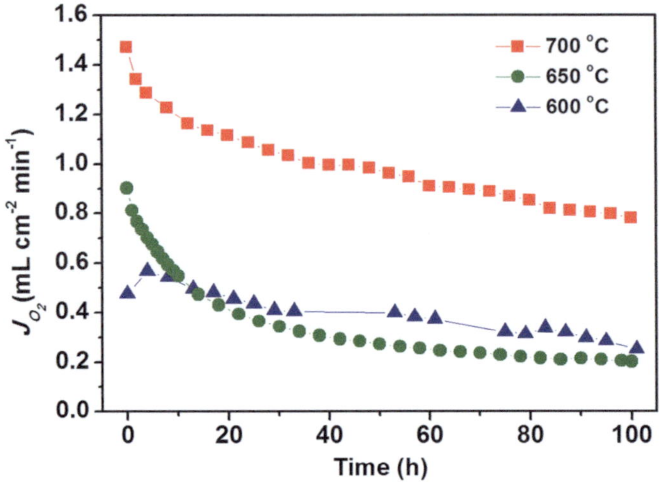

Fig. 8.14 Time-dependent oxygen permeation fluxes of the BSCF membranes and the membrane coated with SSC porous layer at 700, 650, and 600 °C. Air flow rate, 150 ml min^{-1}; He flow rate, 50 ml min^{-1} (Reproduced from [14] with permission of Wiley)

rate indicates that the degradation mechanism may change within the temperature range 600–850 °C for the BSCF uncoated membrane.

Figure 8.15 exhibits the time-dependent changes of the interface exchange resistances of the feed and permeation sides at different temperatures with the SSC-uncoated and SSC-coated BSCF membranes [12]. The oxygen permeation fluxes of the BSCF membranes decreased rapidly with time at intermediate–low temperatures; thus additional errors were introduced for the determination of resistance constants of interface oxygen exchange (r_0' for feed side and r_0'' for permeation side). In fact, the regressed r_0' and r_0'' were the average values over the data collecting periods (2 ~ 3 h). Therefore, the r_0' and r_0'' still effectively reflect the time-dependent changes in the interfacial exchange resistances. However, the additional errors have great effects on the determination of the bulk diffusion resistances because they are much smaller than the interfacial exchange resistances. Although the bulk diffusion resistance could not be reliably acquired by the model under investigation, the increase in the total interfacial exchange resistances $(r' + r'')$ at all temperatures indicates that the degradation of the oxygen permeation flux is related to the interfacial exchange reactions for the SSC-uncoated membranes (Fig. 8.15a–c). Unlike the case of the SSC-uncoated BSCF membrane, as the permeation fluxes with the membranes coated with SSC degraded slowly, it was possible to acquire r^b values with relatively small errors. In addition, the interfacial oxygen exchange resistances on the feed side (interface I), r', and the permeation side (interface II), r'', only exhibited limited fluctuation during the ~ 100 h on-stream at each temperature (Fig. 8.15d–f). This indicates that the SSC porous layers were effective at eliminating the negative influence of the enrichment of the

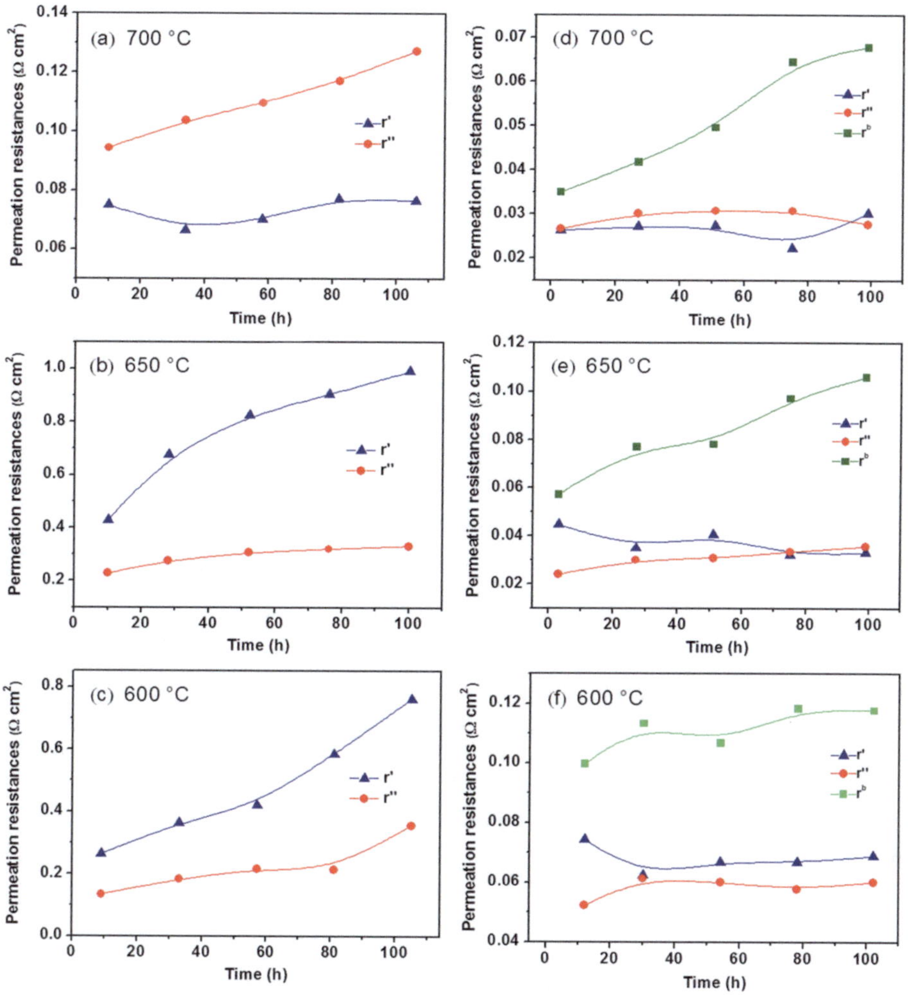

Fig. 8.15 Time-dependent permeation resistances for BSCF membranes (**a**, **b**, **c**) and for BSCF membranes with both surfaces coated with SSC porous layers (**d**, **e**, **f**) at 700, 650, and 600 °C. Air flow rate, 150 ml min^{-1}; membrane thickness, 0.5 mm (Reproduced from [14] with permission of Wiley)

sulfur impurities on the membrane surfaces. The SSC porous layers have two functions: one is to catalyze the interfacial oxygen exchange reactions due to their high oxygen catalytic activity; the other is to accommodate sulfur impurities through their large surface area. Thus, the total interfacial exchange resistances were significantly lower after the membranes were coated with SSC porous layers compared to the SSC-uncoated counterparts. However, the bulk diffusion

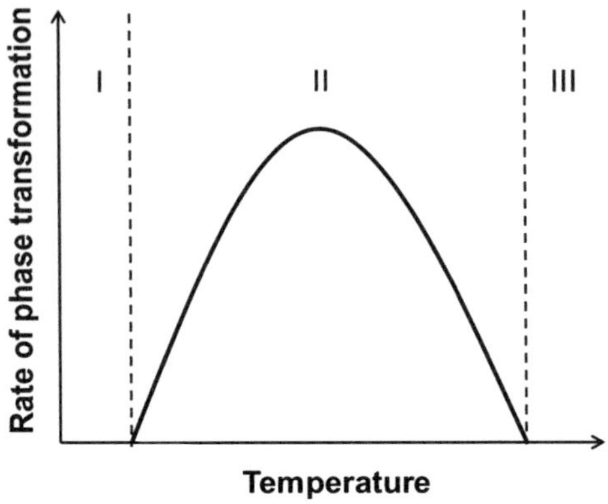

Fig. 8.16 Temperature-dependent phase transformation rate of BSCF. Zone I, > 850 °C; zone II, 600–850 °C; zone III, < 600 °C. The maximum rate appears at around 700 °C

resistances (r^b) increased gradually with time at every operating temperature. According to the previous investigation on the phase transformation of BSCF at intermediate temperatures, the increase in bulk diffusion resistances is the result of the appearance of hexagonal and lamellar phases in the membrane bulk. After ~ 100 h on-stream, r^b increased by 94 %, 83 %, and 18 % for the membranes operated at temperatures of 700, 650, and 600 °C, respectively, which reveals that the phase transformation happens faster at higher temperatures. Combined with other researchers' findings on the phase transformation rate, the above result led to the conclusion that the plot of the phase transformation rate against temperature is a volcano-type curve and the maximum should appear at a temperature around 700 °C, as shown in Fig. 8.16.

If the SSC-uncoated BSCF membranes have the same bulk diffusion resistances as those of the SSC-coated BSCF membranes, the interfacial oxygen exchange resistances $\left(r' + r''\right)$ contribute substantially to the total resistance at each temperature for the SSC-uncoated BSCF membranes, reaching 77–86 % at 700 °C, ~92 % at 650 °C, and 87–91 % at 600 °C during the 100 h on-stream. Therefore, the interfacial exchange kinetics are the rate-limiting steps in the investigated temperature range. As a result, changes in the interfacial exchange kinetics have a significant influence on the permeation flux. However, for the SSC-coated BSCF membranes, the total interfacial oxygen exchange resistances were only slightly higher than the bulk diffusion resistances in the initial stage of the permeation experiments, but this trend gradually reversed with time as the bulk diffusion resistance gradually increased. Therefore, the degradation of the interfacial exchange kinetics, induced by sulfur enrichment on the BSCF membrane surfaces, is the dominant factor that leads to the degradation of the oxygen permeation flux at intermediate temperatures, and bulk phase transformation to hexagonal and lamellar phases has a minor contribution to the degradation.

8.5.2 Stabilization Mechanism of $Ba_{0.5}Sr_{0.5}Co_{0.8}Fe_{0.2}O_{3-\delta}$ at Low Temperatures

It is clear from the above discussion that, to stabilize the oxygen permeation flux of the BSCF membrane at low temperature, two things must be done. One is to eliminate the negative influence of the surface enrichment of sulfur and silicon impurities, and the other is to inhibit the phase transformation from cubic to hexagonal and lamellar phases. The first one can be accomplished by coating both surfaces with a porous MIEC perovskite catalyst (such as SSC) [11, 13]. The second problem can be solved by heavily doping cations with stable valence states, such as Zr^{4+} and Nb^{5+}, into the BSCF lattice. However, this traditional strategy incurs a significant loss of permeability. Recently, we identified a new stabilization mechanism that involves inhibiting the undesirable phase transformation at low temperature by deliberately introducing nanoparticles at grain boundaries. Using this novel approach, a stable oxygen permeation flux (over 500 h) through a nanoparticle-decorated BSCF perovskite membrane was demonstrated with 10–1000 times higher permeation flux than previous studies at 550–650 °C [12].

8.5.2.1 Phase Transformation of BSCF at Low Temperature

The phase transformation of BSCF is different from that of most perovskite materials (see Chap. 6). For $SrCoO_{3-\delta}$, a martensitic transformation from a high-temperature cubic phase to a 2-H hexagonal phase takes place at 900 °C in air [15]. The transformation between these two structures requires very little thermal activation energy because it is a diffusionless transformation, which results in the subtle but rapid rearrangement of atomic positions. Thus, the transformation rate of this type of phase transformation is extremely fast. However, the phase transformation of BSCF is a time-dependent process; long-distance atomic diffusion in the material bulk is needed to reach active sites for the formation of crystal nuclei and the growth of the crystal nuclei to the secondary phase. This type of phase transformation is called diffusion-controlled transformation, i.e., the phase transformation rate depends on the diffusion rate of the atoms. Grain boundaries are the active sites for the formation and growth of such crystal nuclei because the high density of defects and the loose arrangement of atoms lead to facile generation of crystal nuclei. Figure 8.17 shows the new phase growth at a grain boundary in BSCF at 800 °C. The time-dependent phase transformation process originates at the grain boundaries of the parent material. This process can be monitored in real time by determining the time-dependent oxygen permeation flux through the membrane, because the permeation takes place via oxygen ionic diffusion through grains and grain boundaries, as shown in Fig. 8.18. The resultant new phase has extremely low ionic conductivity. As the new phase is a thin sheet, a small volume fraction of the new phase would significantly block the transport of oxygen ions in the

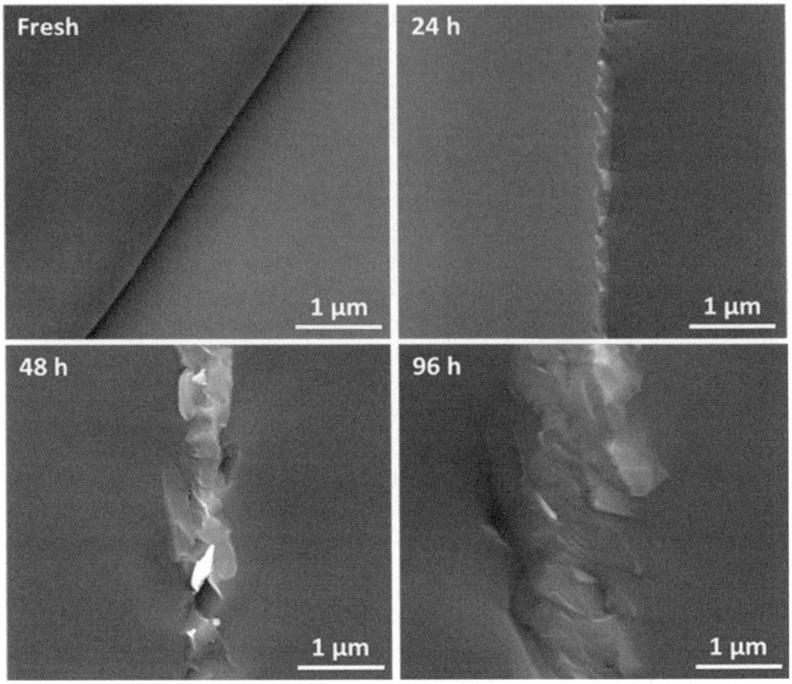

Fig. 8.17 Time-dependent phase transformation at BSCF grain boundaries at 800 °C. The fresh BSCF membrane has clear and smooth grain boundaries; however after the membrane is treated in air for 24 h, a new phase with width of ~200 nm appears at the grain boundaries. The new phase grows gradually bigger with time, reaching, for example, ~1200 nm in width after 96 h of membrane operation (Reproduced from [12] with permission of the American Chemical Society)

membrane bulk. As the hexagonal and/or lamellar phases appear at the grain boundaries, a remarkable decrease in oxygen permeation flux can be detected due to the much lower oxygen ionic conductivity of these newly nucleated grain boundary phases compared to the parent phase.

The BSCF membrane was coated with an $Sm_{0.5}Sr_{0.5}CoO_{3-\delta}$ (SSC) catalyst on both surfaces before permeation testing to eliminate the negative effects of impurity enrichment on the membrane surfaces. The permeation flux declines by up to 30 % during 500 h on-stream oxygen permeation testing at 600 °C. Similar to that shown in Sect. 8.5.1, the result of the model investigation indicates that the bulk resistance of the BSCF membrane increased by ~ 20 % over ~ 110 h, whereas the oxygen exchange resistance for both surfaces remained nearly constant. The time-dependent increase in the oxygen ion diffusion resistance through the BSCF membrane bulk includes the resistances through grains and grain boundaries. Postmortem HRTEM analysis of the spent BSCF membrane shows that a new phase with a width of ~ 200 nm was observed at the grain boundary. This new phase was enriched with cobalt but depleted of strontium, as revealed by EDS analysis in Fig. 8.19. However, the BSCF parent phase maintains the cubic

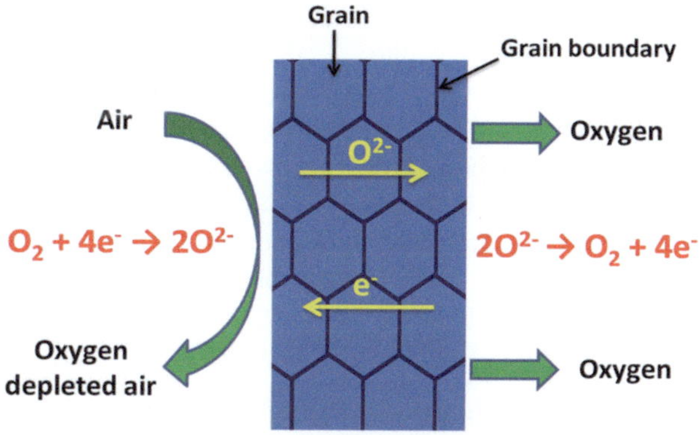

Fig. 8.18 Oxygen ions walk across the grain boundaries in the process of oxygen permeation through an MIEC membrane

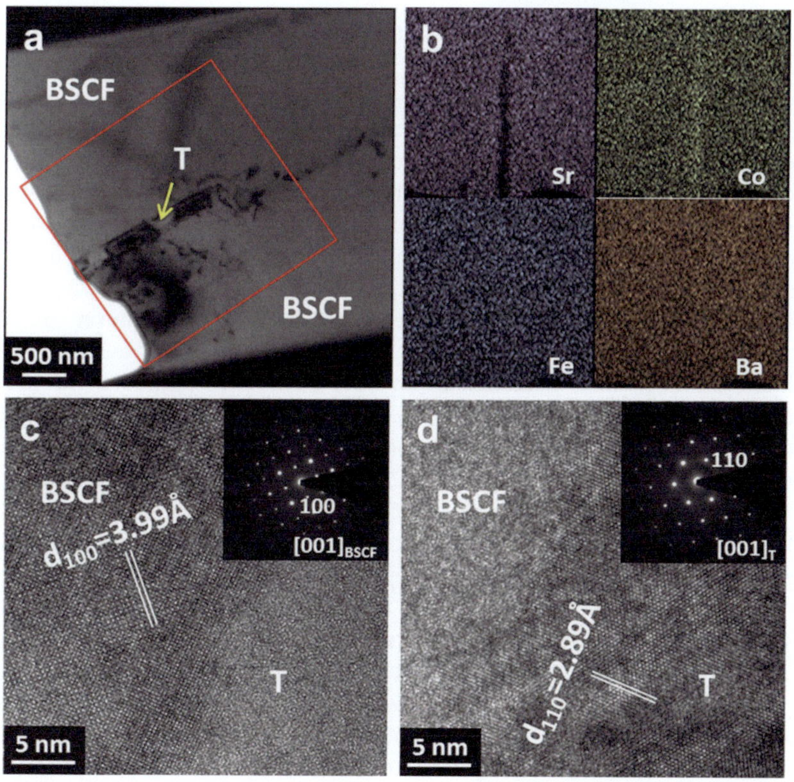

Fig. 8.19 Post-analysis of the BSCF membrane after 500 h on-stream at 600 °C. (**a**) Scanning transmission electron microscopy (STEM) image. (**b**) EDS map analysis of the area marked in the *red frame* in (**a**). (**c**) and (**d**) HRTEM images for the interfacial area between the parent phase (BSCF) and the trigonal phase viewed along their [001] directions, respectively. "T" denotes the trigonal phase precipitated at the grain boundary (Reproduced from [12] with permission of the American Chemical Society)

perovskite structure with a $d_{100} = 0.399$ nm. Detailed EDS analysis of this new phase suggests an elemental composition of approximately $Ba_{0.45}Sr_{0.07}Co_{1.6}Fe_{0.18}O_{2.8}$. Multiple examinations of the transformation to the second phase with a trigonal structure revealed initiation was at the grain boundaries; the grain boundaries act as activation sites for the new phase generation. Additionally, no homogeneous nucleation of the trigonal phase in the grain bulk was observed. Therefore, it is deduced that the phase transformation is controlled by heterogeneous nucleation at the grain boundaries.

8.5.2.2 The Attempts to Inhibit Phase Transformation by Light Doping in B Site

To improve the stability of the perovskite oxide phase, metal ions with high and stable valence states were doped heavily (≥10 mol%) at the B site of the perovskite lattice because of their tendency to form strong metal–oxygen bonds. However, the improved stability was accompanied by a significant loss in permeability, and the oxygen permeation fluxes for the stabilized membranes were usually less than half that of the BSCF membrane. To find the balance point between stability and permeability, light doping (3 mol%) of cations in the B site was attempted. Specifically, Y^{3+}, Zr^{4+}, and Nb^{5+}, a series of dopants that increase in valence charge and decrease in ionic size, were examined. The oxygen permeation fluxes of the membranes doped with Zr and Nb decreased by 20 % in 432 h on-stream and 24 % in 288 h on-stream, respectively, at 600 °C (Fig. 8.20), even though the membrane surfaces were coated with an SSC catalyst to eliminate the negative effects of sulfur and silicon impurities. The low-level doping does not reduce the oxygen permeability of the BSCF membrane, but it failed to stabilize the oxygen permeation fluxes at lower temperatures. The Zr-doped sample was analyzed by HRTEM after long-term operation at 600 °C. The same cobalt-enriched and strontium-deficient trigonal phase was observed at grain boundaries with a width of 20–30 nm. Therefore, the low-level doping cannot inhibit the phase transformation of BSCF. In contrast to the Zr- and Nb-doped samples, the Y-doped sample showed a remarkable improvement in permeation stability, indicated by only a 9 % drop in permeation flux during a 500 h period of testing at 600 °C. The result of SEM analysis on the spent membrane showed that there were some nanoparticles appearing along the grain boundaries, which implied that the Y-doped sample cannot form a uniform, single-phase material. However, no nanoparticles are observed on the grain boundaries of the Zr^{4+}- and Nb^{5+}-doped samples; thus, these two dopants can completely incorporate into the BSCF lattice to form solid solutions. Another interesting finding was that the trigonal phase could be observed at the grain boundaries where no nanoparticles existed in the Y-doped sample. Therefore, we deduced that the trigonal phase transformation of BSCF can be inhibited by introducing nanoparticles in grain boundaries.

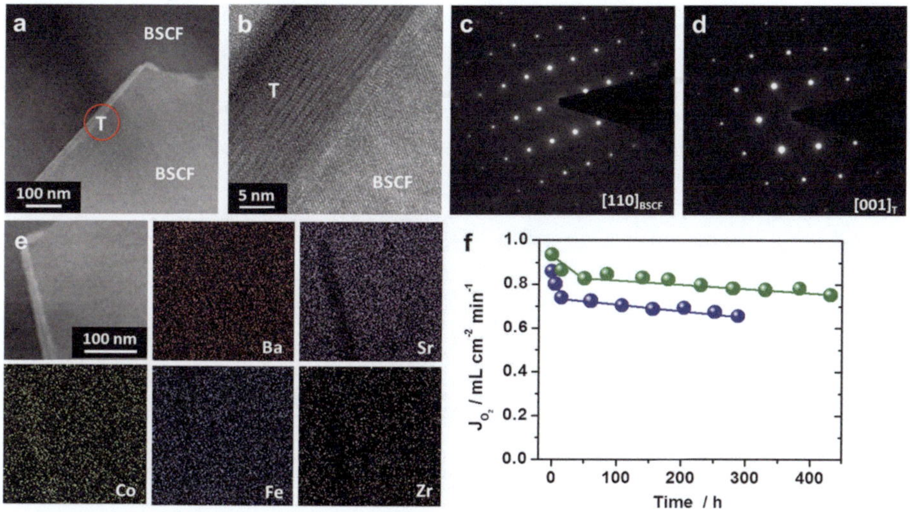

Fig. 8.20 Phase transformation and oxygen permeation of Zr- and Nb-doped BSCF membranes. (**a**) STEM image of the Zr-doped BSCF membrane after 432 h on-stream testing at 600 °C. Trigonal phase appeared at the BSCF grain boundary. (**b**) HRTEM image of the interfacial area marked by the *red circle* in (**a**). (**c**) and (**d**) SAED patterns viewed along the [110] and [001] directions for the BSCF and trigonal phases, respectively. (**e**) EDS map analysis of the zirconium-doped sample after the 432 h LT operation. (**f**) Time-dependent oxygen permeation fluxes through 0.5-mm-thick Zr- (*green*) and Nb- (*blue*) doped BSCF membranes at 600 °C (Reproduced from [12] with permission of the American Chemical Society)

8.5.2.3 Nanoparticles Inhibiting Phase Transformation

Ce^{4+}, with a larger ionic radius and high valence state, shows poor solubility in Co- and Fe-based perovskite oxides and is thus an ideal dopant to produce nanoparticles in the grain boundaries of a BSCF polycrystalline membrane. A one-pot method with EDTA and citrate as complexing agents was used to prepare the powder for the targeted nanoparticle-decorating membranes. $BaCeO_3$ nanoparticles with a size of ~25 nm formed on the BSCF particles (5–10 μm). After being shaped at 200 MPa and sintered at elevated temperature, the nanoparticles grew bigger and were located at the grain boundaries between BSCF grains. 3 mol% Ce-doping in BSCF (chemical composition: $Ba_{0.5}Sr_{0.5}(Co_{0.8}Fe_{0.2})Ce_{0.03}O_{3-\delta}$) resulted in the formation of a relatively widespread coverage of nanoparticles at the grain boundaries (Fig. 8.21). The $BaCeO_3$ nanoparticles were observed with sizes of 50–100 nm and 100–200 nm at the grain boundaries and crystal edges, respectively, and at an average interval of less than 1 μm. The Ce-doped BSCF membrane yields an extremely stable oxygen permeation flux during 500 h of on-stream testing. HRTEM analysis revealed that no trigonal phase was present at the BSCF grain boundaries where the $BaCeO_3$ nanoparticles were present even after 500 h on-stream at 600 °C (Fig. 8.21). The investigation of oxygen permeation resistances using Zhu's model [38] revealed that not only the interfacial exchange resistances

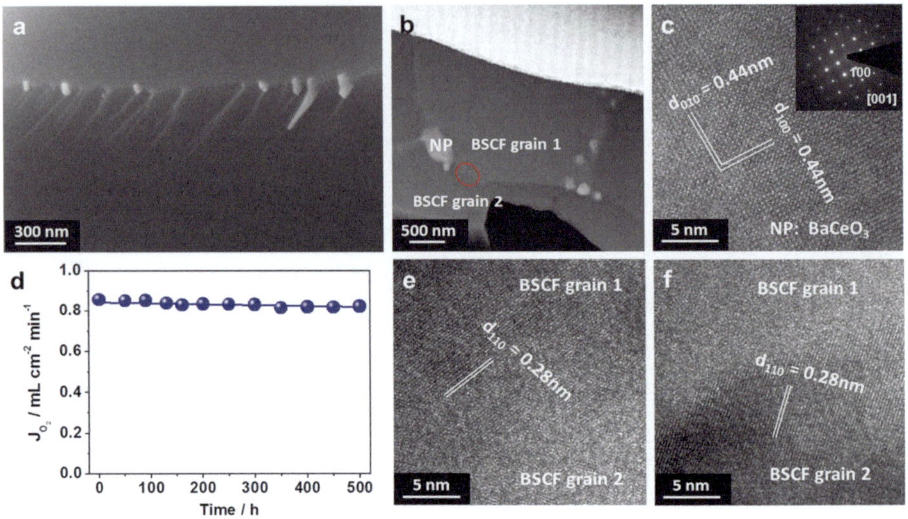

Fig. 8.21 Nanoparticle inhibition of the phase transformation of Ce-doped BSCF perovskite. (**a**) Typical SEM image of grain boundaries for the as-prepared Ce-doped BSCF membrane. (**b**) STEM image of the grain boundaries of the Ce-doped BSCF membrane after running for 500 h at 600 °C. "NP" denotes the nanoparticle. (**c**) HRTEM image and SAED pattern for the "NP" particle viewed along the [001] direction. (**d**) Time-dependent oxygen permeation of the 0.5-mm-thick Ce-doped BSCF membrane at 600 °C. (**e**) and (**f**) HRTEM images of the "BSCF grain 1" and "BSCF grain 2" particles viewed along the axes closest to the [311] and [111] directions, respectively (Reproduced from [12] with permission of the American Chemical Society)

but also the bulk diffusion resistance remained constant during a long period of operation at 600 °C for the BSCF membrane with $BaCeO_3$ nanoparticles at the grain boundaries. To understand the stabilization mechanism of the nanoparticles at the grain boundaries, a thermodynamic analysis and a kinetic analysis were undertaken.

8.5.2.4 Possible Mechanism

A Thermodynamic Analysis

The phase transformation from cubic to trigonal is accompanied by significant changes in the elemental composition (Figs. 8.19 and 8.20) and indicates a diffusion transformation mechanism in the solid phase, where the long-distance migration of metal ions is required to form and grow the nuclei. The free energy change of the system with heterogeneous nucleation at the grain boundaries in a polycrystalline material is given by

$$\Delta G = V\Delta G_V + S\gamma + V\omega + \Delta G_{gb} \tag{8.1}$$

where V is the molar volume of the material, ΔG_V is the free energy difference per unit volume between the parent phase and the new phase, S is the nuclei surface area, γ is the grain boundary energy per unit area, ω is the elastic strain energy per unit volume, and ΔG_{gb} is the grain boundary energy. The first and last terms, which are negative, represent the driving forces for heterogeneous nucleation. The middle two terms are positive and represent forces that restrain the heterogeneous nucleation. Thus, it is possible to inhibit the phase transformation by decreasing the negative values of the grain boundary energy from the viewpoint of thermodynamics. The inherent grain boundary energy of a polycrystalline material is associated with its boundary area and grain boundary energy per unit area. If a spherical grain is considered for a simplified calculation, a grain with radius R_g has a volume of $\frac{4}{3}\pi R_g^3$ and a surface area of $4\pi R_g^2$. Thus, the corresponding grain boundary energy is $-2\pi R_g^2\gamma$. Accounting for the grain boundary being shared between two grains, the grain boundary energy per unit volume is $-3\gamma/2R_g$. Therefore, the inherent grain boundary energy of a polycrystalline material can be written as

$$\Delta G_{gb,\,in} = -V\frac{3\gamma}{2R_g} = -VP_{gb,\,in} \tag{8.2}$$

where $P_{gb,\,in}$ is the inherent grain boundary pressure. Thus, to inhibit the phase transformation at grain boundaries, another pressure should be introduced into the grain boundaries to counteract the inherent grain boundary pressure.

According to the Zener pinning theory [41], to move away from a particle, the grain boundary requires the creation of a new surface because the insoluble and discontinuous secondary-phase particles appear at the grain boundaries of polycrystalline materials. Thus, a net drag force is produced on the grain boundary:

$$F = 2\pi r_p \gamma \sin\theta \cos\theta \tag{8.3}$$

where r_p is the particle radius and θ is the angle between the grain boundary surface and the surface at the point where the grain joins the particle ($0 \le \theta \le \pi/2$). The number of particles per unit volume (N_V) for a volume fraction f_V of a randomly distributed spherical particle is given by

$$N_V = 3f_V/4\pi r_p^3 \tag{8.4}$$

The number of particles intersecting the boundary unit area is

$$N_S = 2r_p N_V = 3f_V/2\pi r_p^2 \tag{8.5}$$

The Zener pining pressure is thus expressed as

$$P_Z = FN_S = \frac{3\gamma f_V}{r_p}\sin\theta\cos\theta = \frac{3\gamma f_V}{2r_p}\sin 2\theta \tag{8.6}$$

Therefore, the maximum pressure for a particle intersecting a grain boundary occurs at $\theta = \pi/4$, i.e.,

$$P_{Z,\max} = \frac{3\gamma f_V}{2r_p} \tag{8.7}$$

The free energy change induced at grain boundaries for a polycrystalline material by introducing particles at the grain boundaries can be written as

$$\Delta G_Z = VP_Z = \frac{3V\gamma f_V}{2r_p}\sin 2\theta \tag{8.8}$$

As nanoparticles appear at the polycrystalline grain boundaries, the grain boundary energy becomes

$$\Delta G_{gb,\text{total}} = \Delta G_{gb,in} + \Delta G_Z = -\frac{3}{2}\gamma V\left(\frac{1}{R_g} - \frac{f_V}{r_p}\sin 2\theta\right) \tag{8.9}$$

As indicated by Eq. (8.9), the grain boundary energy can be negative, zero, or positive at a given temperature (γ is the function of temperature). To make sure that grain boundaries are not the active sites for the phase transformation, the grain boundary energy should be no smaller than zero, i.e.,

$$\frac{1}{R_g} - \frac{f_V}{r_p}\sin 2\theta \leq 0 \tag{8.10}$$

or

$$\frac{f_V R_g}{r_p}\sin 2\theta \geq 1 \tag{8.11}$$

As long as $f_V R_g/r_p$ is greater than one at the maximum Zener pining pressure, the total grain boundary energy will be zero at equilibrium. Therefore, according to Eq. (8.1), the free energy change of the system with heterogeneous nucleation at the grain boundaries can be significantly reduced if secondary-phase particles are introduced into grain boundaries. To fulfill the equation, $f_V R_g/r_p \geq 1$, a large volume fraction of particles with smaller size is needed at the grain boundaries. However, a larger volume fraction of particles in the material leads to the particles growing bigger during the preparation of the material.

A Kinetic Analysis

A stepwise growth mechanism is widely accepted for the motion of grain boundaries. According to this mechanism, the grain boundary velocity is given by

$$u = \frac{h}{k} v \tag{8.12}$$

where h, k, and v are the height of the step, the distance between the two steps, and the motion velocity of steps, respectively. According to kinetic laws, if a driving pressure (P), which is sufficiently small relative to the thermal activation energy per atom, is applied to a grain boundary, it can generate a grain boundary velocity linearly proportional to the driving pressure [42, 43]:

$$u = mP \tag{8.13}$$

where m is the boundary migration mobility. The stepwise motion velocity is derived by combining Eqs. (8.12) and (8.13):

$$v = \frac{km}{h} P \tag{8.14}$$

Grain boundaries with large free volumes and high defect concentrations lead to facile atomic diffusion. The stepwise motion velocity is directly related to the net diffusion flux of atoms at the grain boundaries. For a polycrystalline material, the driving pressure is the inherent grain boundary pressure ($\frac{3\gamma}{2R_g}$); thus, the atoms at the boundaries of small grains have higher activity than those of big grains. This deduction can be verified by the fact that materials with small grains have higher sintering activity than those with big grains.

As secondary-phase particles are introduced into grain boundaries, a Zener pinning pressure is produced by the particles. Thus, the driving pressure for the motion of grain boundaries is the sum of the inherent grain boundary pressure and the Zener pinning pressure, so Eq. (8.14) can be rewritten as

$$v = \frac{3\gamma km}{2h} \left(\frac{1}{R_g} - \frac{f_V}{r_p} \sin 2\theta \right) \tag{8.15}$$

As illustrated by Eq. (8.15), the introduction of particles into grain boundaries can inhibit the stepwise motion; in other words, the net diffusion flux of atoms at the grain boundaries is inhibited by the particles at the grain boundaries. Similar to the previous explanation, as long as the value of $f_V R_g / r_p$ is greater than 1, the stepwise motion velocity can be zero (i.e., the net diffusion flux of the atoms at grain boundaries can be reduced to zero by the particles).

According to the heterogeneous nucleation mechanism, a net cationic diffusion flux along the grain boundaries of the primary material is required for the initial nucleation and the growth of the new phase. However, the appearance of secondary-phase particles at grain boundaries inhibits the net cationic diffusion flux along the grain boundaries, and the potential phase transformation is stopped by the particles.

For the Ce-doped BSCF membrane, the $BaCeO_3$ nanoparticles formed have a volume fraction of 4 % and an average particle size of 100 nm, and the BSCF grains have an average size of 15 μm. Therefore, the value of $f_V R_g / r_p$ is 6, which is much larger than the critical value of 1. In other words, it indicates, theoretically, that the driving force for heterogeneous nucleation and growth of the trigonal phase via stepwise motion has been reduced to zero. The nanoparticles are successfully acting as "roadblocks" and have reduced the net diffusion flux of cations along the nanoparticle-containing grain boundaries to zero. The inhibition of the stepwise motion mechanism is also supported by the zigzag nature of the grain boundaries seen in the Ce-doped material. Since grain boundary mobility in the vicinity of a nanoparticle is greatly reduced compared to grain boundary mobility distal to a nanoparticle, the grain boundary develops a zigzag structure during the high-temperature sintering of the nanoparticle-containing BSCF membrane. In contrast, straight and smooth grain boundaries are observed in undoped, Zr-doped, and Nb-doped BSCF.

8.5.2.5 High Permeation Flux at Low Temperatures

Other than stability, the permeation flux is another important performance index to be considered in the practical applications of MIEC membranes. The value of 1 mL cm^{-2} min^{-1} is deemed to be the lowest value for practical applications in oxygen separation. For a given MIEC membrane, the oxygen permeation flux can be improved by reducing the thickness of the separation layer (or dense layer), as predicted by the Wagner equation. An asymmetric Ce-doped membrane prepared by dry pressing and co-sintering has a 90-μm-thick dense layer and a 700-μm-thick porous layer to support the dense layer, as shown in Fig. 8.22. A porous 30-μm-thick SSC layer coated on the dense layer helps to activate oxygen and eliminate the potential negative influence of sulfur and silicon impurities. The oxygen permeation flux reaches 3.0 mL cm^{-2} min^{-1} when the feed-side oxygen partial pressure is 0.21 atm and 5.2 mL cm^{-2} min^{-1} when the oxygen partial pressure is 1.0 atm at 600 °C. As the temperature is decreased to 550 °C, the oxygen permeation fluxes fall to 1.4 and 2.5 mL cm^{-2} min^{-1} when the oxygen partial pressures of feed side are 0.21 and 1.0 atm, respectively. The asymmetric membrane fluxes in the temperature range 550–650 °C are 10–1000 times higher than those reported for other membranes and exceed the 1 mL cm^{-2} min^{-1} threshold required for commercial viability.

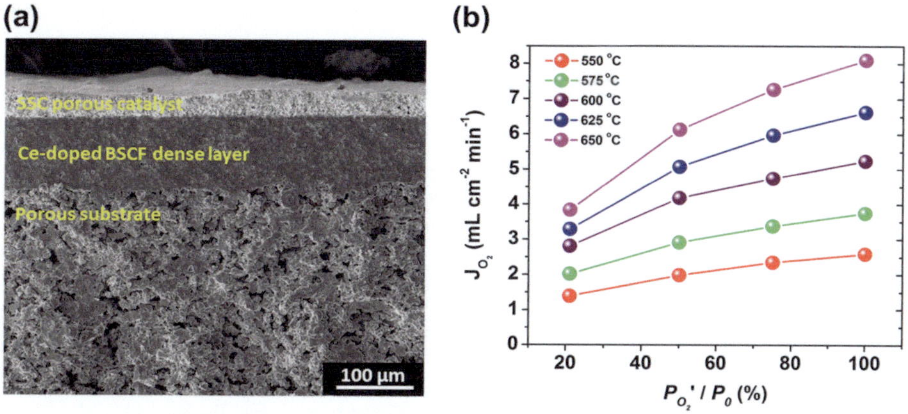

Fig. 8.22 Asymmetric Ce-doped BSCF membrane and oxygen permeation. (**a**) Cross-sectional SEM image of an asymmetric Ce-doped BSCF membrane. (**b**) Influence of feed-side oxygen partial pressure and temperature on the permeation flux (Reproduced from [12] with permission of the American Chemical Society)

References

1. Rutkowski B, Malzbender J, Beck T, Steinbrech RW, Singheiser L (2010) Creep behaviour of tubular $Ba_{0.5}Sr_{0.5}Co_{0.8}Fe_{0.2}O_{3-\delta}$ gas separation membranes. J Eur Ceram Soc 31:493–499
2. Anderson LL, Armstrong PA, Broekhuis RR, Carolan MF, Chen J, Hutcheon MD, Lewinsohn CA, Miller CF, Repasky JM, Taylor DM, Woods CM (2016) Advances in ion transport membrane technology for oxygen and syngas production. Solid State Ionics 288:331–337
3. van Veen AC, Rebeilleau M, Farrusseng D, Mirodatos C (2003) Studies on the performance stability of mixed conducting BSCFO membranes in medium temperature oxygen permeation. Chem Commun 9:32–33
4. Wang HH, Tablet C, Caro J (2008) Oxygen production at low temperature using dense perovskite hollow fiber membranes. J Membr Sci 322:214–217
5. Kruidhof H, Bouwmeester HJM, Doorn RHEV, Burggraaf AJ (1992) Influence of order-disorder transitions on oxygen permeability through selected nonstoichiometric perovskite-type oxides. Solid State Ionics 63–65:816–822
6. Shao ZP, Yang WS, Cong Y, Dong H, Tong JH, Xiong GX (2000) Investigation of the permeation behavior and stability of a $Ba_{0.5}Sr_{0.5}Co_{0.8}Fe_{0.2}O_{3-\delta}$ oxygen membrane. J Membr Sci 172:177–188
7. Baumann S, Serra JM, Lobera MP, Escolástico S, Schulze-Küppers F, Meulenberg WA (2011) Ultrahigh oxygen permeation flux through supported $Ba_{0.5}Sr_{0.5}Co_{0.8}Fe_{0.2}O_{3-\delta}$ membranes. J Membr Sci 377:198–205
8. Cao ZW, Zhu XF, Li WP, Xu B, Yang LN, Yang WS (2015) Asymmetric dual-phase membranes prepared via tape-casting and co-lamination for oxygen permeation. Mater Lett 147:88–91
9. Sunarso J, Baumann S, Serra JM, Meulenberg WA, Liu S, Lin YS, Diniz da Costa JC (2008) Mixed ionic-electronic conducting (MIEC) ceramic-based membranes for oxygen separation. J Membr Sci 320:13–41
10. Watenabe K, Yuasa M, Kida T, Teraoka Y, Yamazoe N, Shimanoe K (2010) High-performance oxygen-permeable membranes with an asymmetric structure using $Ba_{0.95}La_{0.05}FeO_{3-\delta}$ perovskite type oxide. Adv Mater 22:2367–2370

11. Liu Y, Zhu XF, Li MR, Yang WS (2013) Stabilization of low-temperature degradation in mixed ionic and electronic conducting perovskite oxygen permeation membranes. Angew Chem Int Ed 52:3232–3236
12. Liu Y, Zhu XF, Li MR, O'Hayre RP, Yang WS (2015) Nanoparticles at grain boundaries inhibit the phase transformation of perovskite membrane. Nano Lett 15:7678–7683
13. Liu Y, Zhu XF, Li MR, Li WP, Yang WS (2015) Degradation and stabilization of perovskite membranes containing silicon impurity at low temperature. J Membr Sci 492:17–180
14. Liu Y, Zhu XF, Li MR, O'Hayre RP, Yang WS (2015) Degradation mechanism analysis of $Ba_{0.5}Sr_{0.5}Co_{0.8}Fe_{0.2}O_{3-\delta}$ membranes at intermediate-low temperatures. AIChE J 61:3879–3888
15. Liu Y, Zhu X, Yang W (2016) Stability of sulfate doped $SrCoO_{3-\delta}$ MIEC membrane. J Membr Sci 501:53–59
16. Deng ZQ, Yang WS, Liu W, Chen CS (2006) Relationship between transport properties and phase transformations in mixed-conducting oxides. J Solid State Chem 179:362–369
17. Mueller DN, De Souza RA, Weirich TE, Roethrens D, Mayer J, Martin M (2010) A kinetic study of the decomposition of the cubic perovskite-type oxide $Ba_xSr_{1-x}Co_{0.8}Fe_{0.2}O_{3-\delta}$ (BSCF) (x = 0.1 and 0.5). Phys Chem Chem Phys 12:10320–10328
18. Zhu XF, Wang HH, Yang WS (2006) Structural stability and oxygen permeability of cerium lightly doped $BaFeO_{3-\delta}$ ceramic membranes. Solid State Ionics 177:2917–2921
19. Hong JO, Teller O, Martin M, Yoo HI (1999) Demixing of mixed (A, B)O in an oxygen potential gradient: numerical solution the time evolution of the demixing process. Solid State Ionics 123:75–85
20. Tong JH, Yang WS, Zhu BC, Cai R (2002) Investigation of ideal zirconium-doped perovskite-type ceramic membrane materials for oxygen separation. J Membr Sci 203:175–189
21. Tong JH, Yang WS, Cai R, Zhu BC, Lin LW (2002) Novel and ideal zirconium-based dense membrane reactors for partial oxidation of methane to syngas. Catal Lett 78:129–137
22. Li QM, Zhu XF, He YF, Yang WS (2010) Oxygen permeability and stability of $BaCe_{0.1}Co_{0.4}Fe_{0.5}O_{3-\delta}$ oxygen permeable membrane. Sep Purif Technol 73:38–43
23. Cheng YF, Zhao HL, Teng DQ, Li FS, Lu XG, Ding WZ (2008) Investigation of Ba fully occupied A-site $BaCo_{0.7}Fe_{0.3-x}Nb_xO_{3-\delta}$ perovskite stabilized by low concentration of Nb for oxygen permeation membrane. J Membr Sci 322:484–490
24. Bucher E, Gspan C, Hofer F, Sitte W (2013) Post-test analysis of silicon poisoning and phase decomposition in the SOFC cathode material $La_{0.58}Sr_{0.4}Co_{0.2}Fe_{0.8}O_{3-\delta}$ by transmission electron microscopy. Solid State Ionics 230:7–11
25. Viitanena MM, Welzenis RGv, Brongersma HH, van Berkel FPF (2002) Silica poisoning of oxygen membranes. Solid State Ionics 150:223–228
26. Hancock CA, Slade RCT, Varcoe JR, Slater PR (2011) Synthesis, structure and conductivity of sulfate and phosphate doped $SrCoO_3$. J Solid State Chem 184:2972–2977
27. Porras-Vazquez JM, Slater PR (2012) Synthesis and characterization of oxyanion-doped cobalt containing perovskites. Fuel Cells 12:1056–1063
28. Hancock CA, Slater PR (2011) Synthesis of silicon doped $SrMO_3$ (M = Mn, Co): stabilization of the cubic perovskite and enhancement in conductivity. Dalton Trans 40:5599–5603
29. Porras-Vazquez JM, Slater PR (2012) Synthesis of oxyanion-doped barium strontium cobalt ferrites: stabilization of the cubic perovskite and enhancement in conductivity. J Power Sources 209:180–183
30. Li MR, Zhou W, Xu XY, Zhu ZH (2013) $SrCo_{0.85}Fe_{0.1}P_{0.05}O_{3-\delta}$ perovskite as a cathode for intermediate-temperature solid oxide fuel cells. J Mater Chem A 1:13632–13639
31. Ishihara T, Honda M, Shibayama T, Minami H, Nishiguchi H, Takita Y (1998) Intermediate temperature solid oxide fuel cells using a new $LaGaO_3$ based oxide ion conductor. J Electrochem Soc 145:3177–3183
32. Tu HY, Takeda Y, Imanishi N, Yamamoto O (1999) $Ln_{0.4}Sr_{0.6}Co_{0.8}Fe_{0.2}O_{3-\delta}$ (Ln = La, Pr, Nd, Sm, Gd) for the electrode in solid oxide fuel cells. Solid State Ionics 117:277–281

33. Zhang HZ, Yang WS (2007) High efficient electrocatalysts for oxygen reduction reaction. Chem Commun 41:4215–4217
34. Zhu XF, Li QM, Cong Y, Yang WS (2011) Unsteady-state permeation and surface exchange of dual-phase membranes. Solid State Ionics 185:27–31
35. Efimov K, Xu Q, Feldhoff A (2010) Transmission electron microscopy study of $Ba_{0.5}Sr_{0.5}Co_{0.8}Fe_{0.2}O_{3-\delta}$ perovskite decomposition at intermediate temperatures. Chem Mater 22:5866–5875
36. Müller P, Störmer H, Dieterle L, Niedrig C, Ivers-Tiffée E, Gerthsen D (2012) Decomposition pathway of cubic $Ba_{0.5}Sr_{0.5}Co_{0.8}Fe_{0.2}O_{3-\delta}$ between 700 °C and 1000 °C analyzed by electron microscopic techniques. Solid State Ionics 206:57–66
37. Müller P, StÖrmer H, Meffert M, Dieterle L, Niedrig C, Wagner SF, Ivers-Tiffée E, Gerthsen D (2013) Secondary phase formation in $Ba_{0.5}Sr_{0.5}Co_{0.8}Fe_{0.2}O_{3-\delta}$ studied by electron microscopy. Chem Mater 25:564–573
38. Zhu XF, Liu HY, Cong Y, Yang WS (2012) Permeation model and experimental investigation of mixed conducting membranes. AIChE J 58:1744–1754
39. Zhu XF, Liu HY, Cong Y, Yang WS (2012) Novel dual-phase membranes for CO_2 capture via an oxyfuel route. Chem Commun 48:251–253
40. Liu Y, Zhu Y, Li MR, Zhu XF, Yang WS (2016) Oxygen transport kinetics of MIEC membranes coated with different catalysts. AIChE J 62:2803–2812
41. Zener C, (quoted by Smith CS) (1948) Grains, phases, and interfaces-an interpretation of microsture. Trans AIME 175:15–51
42. Sun B, Suo Z, Yang W (1997) A finite element method for simulating interface motion-1. Migration of phase and grain boundaries. Acta Mater 45:1907–1915
43. Toda-Caraballo I, Capdevila C, Pimentel G, De Andrés CG (2012) Drag effects on grain growth dynamics. Comput Mater Sci 68:95–106

Chapter 9
Catalytic Reactions in MIEC Membrane Reactors

Abstract Besides oxygen separation from air, another important application of MIEC membranes is to act as membrane reactors to unite reactions and separation. The integration of reactions and separation has many advantages, thus has attracted much attention of the industrial and academic communities. The important reactions in MIEC membrane reactors include partial oxidation of natural gas to syngas, oxidation of coupling of methane, oxidation dehydrogenation of ethane to ethylene, water splitting for hydrogen production, etc. To safely operate the membrane reactors for oxidation reactions, the membrane materials need to be stable in both oxidative and reductive atmospheres. Thus, in this chapter, the most concerned topic, i.e., membrane materials, is introduced, and the factors influencing the stability of membrane materials are discussed. Additionally, the progresses on the membrane reactors for various oxidation reactions are briefly introduced, and the reaction–separation–reaction membrane reactors developed in the recent years are presented.

Keywords Membrane reactors • Partial oxidation of methane • Syngas ($CO + H_2$) • Stability • Hydrogen production • Oxidation coupling of methane • Oxidation dehydrogenation • Water splitting • Reaction–separation–reaction coupling • Ammonia synthesis gas

9.1 Introduction of Catalytic Membrane Reactors

With the development of the society, improvement of the environmental quality, enhancement of the utilization efficiency of fossil resources, and ultimately the realization of a sustainable development are urgently required. All this prompts us to find new methods and processes to intensify and refresh the traditional industrial technologies. Catalytic membrane reactors (CMRs), a concept firstly introduced in 1963 by Gryaznov and coworkers [1], have since then received great attention from both academic and industrial communities. In CMRs, the catalytic reactions are coupled with the separation processes. Process intensification is easily achieved with the improvement of energy efficiency, simplification of technological processes, and reduction of investment and operational costs, as synergistic effects between reaction and separation are frequently found in CMRs. For example, for a

X. Zhu, W. Yang, *Mixed Conducting Ceramic Membranes*, Green Chemistry and Sustainable Technology, DOI 10.1007/978-3-662-53534-9_9

reaction limited by thermodynamic equilibrium, such as water–gas shifting, the conversion can be raised to 100 % in a hydrogen-permeable Pd-based membrane reactor. Except for reaction–separation coupling in CMRs, reaction–separation–reaction coupling can also fulfill several complex reactions and separation processes in one CMR, detailed examples of which will be shown in the coming sections. The quick development of inorganic membranes in the last two decades led to a considerable progress on the investigation of CMRs, greatly expanding the application of membrane technology. The International Conference on Catalysis in Membrane Reactors (ICCMR) is held every 2 years since 1994 in Villeurbanne, France, to communicate the new progress in this fast growing field.

MIEC membranes are typical inorganic membranes used in CMRs for coupling oxygen-related reactions and oxygen separation. For MIEC CMRs, membranes are the key element of the reactors. Different reactions have various requirements concerning the performance of MIEC membranes, such as oxygen permeation flux and permeation stability under reaction conditions, structural stability under reducing atmosphere, tolerance toward CO_2, etc. Therefore, developing MIEC membrane materials that meet the requirement of catalytic reactions is one of the important investigation tasks in the field of MIEC CMRs, and it will be introduced in this chapter. The interaction between MIEC membranes and catalysts and the one between oxygen permeation and catalytic reactions are newly emerging problems in CMRs, which are remarkably complex in different membrane–catalyst systems; the investigation on these interactions will be introduced in this chapter.

9.2 Types of Membrane Reactors

According to the functions of MIEC membranes in CMRs, and independently from their configuration (disk, tube, hollow fiber, etc.), MIEC membrane reactors can be classified into two types, namely, "membrane reactors with controllable oxygen input" and "membrane reactors with controllable oxygen output," as shown in Fig. 9.1.

Membrane reactors with controllable oxygen input (Fig. 9.1a) can be used for selective oxidation reactions. For this kind of reactions, the reaction free energies are negative and large, so the formation of target products is not limited by the thermodynamic equilibrium but by the reaction kinetics. Low selectivity and complex products are common problems occurring in this type of reactions. Additionally, it is difficult to overcome the explosion risk derived by co-feeding oxygen and reductive reactants in traditional reactors, such as in the case of the selective oxidation of ethane to ethylene. However, when the reactions are conducted in an MIEC membrane reactor, the input of oxygen into the reactor can be controlled by the MIEC membrane. As a consequence, the deep oxidation of the target products is avoidable due to the low oxygen concentration in the membrane reactor, and thus the selectivity is significantly improved [2]. Furthermore, the explosion risk can be completely overcome since oxygen and reductive reactants are separately

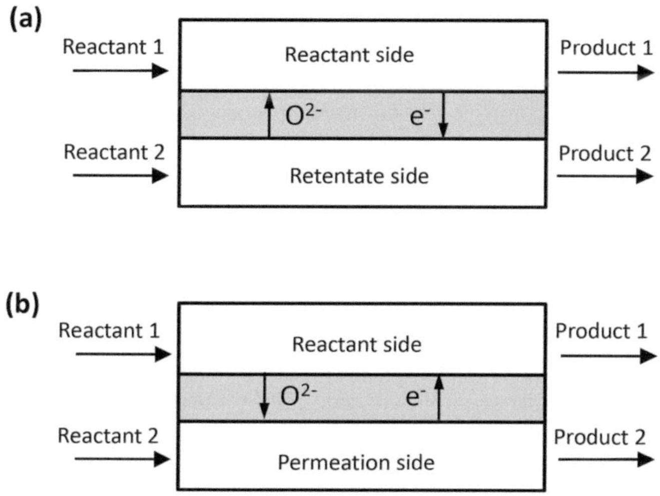

Fig. 9.1 Types of MIEC membrane reactors. (**a**) Membrane reactors with controllable oxygen input; (**b**) membrane reactors with controllable oxygen output

introduced into the two sides of the membrane reactor. If the membrane has high catalytic activity and selectivity toward the selective oxidation reaction, the membrane has two functions, i.e., controlling the input of oxygen and catalyzing the reaction. If the membrane has low catalytic activity and selectivity toward the target reactions, the addition of a catalyst in the membrane reactor is needed. In this instance, the membrane has one function, i.e., controlling the input of oxygen.

On the other hand, a membrane reactor with controllable oxygen output (Fig. 9.1b) can be used to improve the conversion of reduction reactions, because oxygen can be removed from the reaction system, thus shifting the reaction equilibrium. For this kind of reactions, the reaction free energies are positive, so they are limited by the thermodynamic equilibrium, as for instance the water-splitting reaction. When reactions are conducted in MIEC membrane reactors, oxygen can be selectively removed by the MIEC membrane, so that the decomposition reactions are not limited by the thermodynamic equilibrium anymore, and the conversion can be significantly improved by breaking the equilibrium limitation [3].

9.3 Partial Oxidation of Hydrocarbons for Syngas or Hydrogen Production

During the past decades, extensive efforts have been focused on using MIEC membranes to improve the process efficiency of the catalytic oxidative conversion of alkanes. Of particular interest is the partial oxidation of methane to syngas ($CO + H_2$) in MIEC membrane reactors. Synthesis gas is an important intermediate product in the gas-to-liquid conversion (natural gas-to-liquid fuels) via existing processes, such as Fischer–Tropsch (F–T) and methanol synthesis. There are two routes for the conversion of natural gas to syngas, i.e., steam reforming and partial oxidation. Steam reforming is the dominant process for the production of syngas from methane and has been commercialized for tens of years. However, this process is strongly endothermic:

$$CH_4 + H_2O \rightleftharpoons CO + 3H_2 \qquad \Delta H(25\ ^\circ\mathrm{C}) = 206.16\ \mathrm{kJ\ mol^{-1}} \tag{9.1}$$

and the space velocity of the reaction process is low (~3000 h^{-1}) even under elevated temperature (850–1000 °C) and pressure (1.5–3.0 MPa). Great equipment investment and huge energy input are required by this technology. Another disadvantage of this process is that the produced syngas has a high H_2/CO ratio up to 3, which needs to be adjusted to 2 for the subsequent F–T and methanol synthesis.

In the last 20 years, the partial oxidation process has attracted the attention of both academic and industrial communities. This reaction is weakly exothermal:

$$CH_4 + 0.5O_2 \rightleftharpoons CO + 2H_2 \qquad \Delta H(25\ ^\circ\mathrm{C}) = -35.67\ \mathrm{kJ\ mol^{-1}} \tag{9.2}$$

and can be operated at high space velocity up to 70,000 h^{-1}. Furthermore, the produced H_2/CO ratio is 2, thus suitable for the subsequent F–T and methanol synthesis. All these assets make it a promising technology for methane conversion. However, three issues occurring in traditional reactors have severely inhibited the industrialization process of this technology:

1. Hot spot formation is inevitable; thus, the reactor is at the risk of temperature runaway.
2. High explosion risk due to the co-feeding of oxygen and methane in the traditional membrane reactor.
3. An air separation unit is needed to produce pure oxygen for this process because the downstream processes cannot tolerate nitrogen; thus, the economic benefit of this technology is severely degraded.

The investment cost in a cryogenic oxygen plant may constitute up to 45 % of the total investment cost of a GTL factory [4]. Therefore, there is a great incentive to find an economical way to supply oxygen to the partial oxidation route.

9.3.1 MIEC Membrane Reactors for Methane Conversion to Syngas

The advent of MIEC membranes encouraged researchers to combine the use of such membranes with chemical reactions. This membrane technology was deemed a

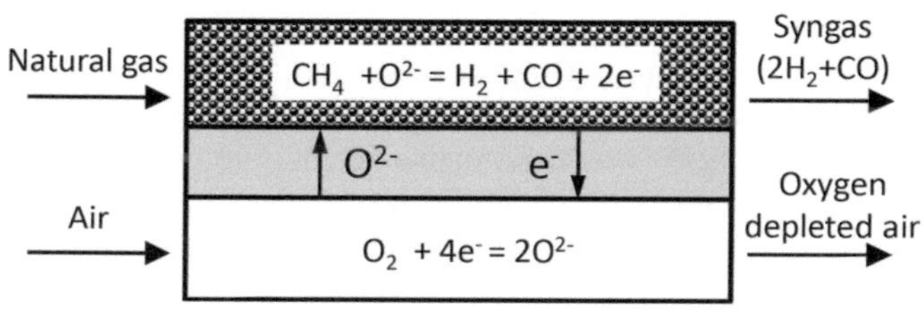

Fig. 9.2 Catalytic partial oxidation of natural gas in an MIEC membrane reactor for syngas production

very promising route to unite the catalytic partial oxidation of methane and the MIEC membrane for oxygen separation into a CMR, as shown in Fig. 9.2. In the CMR, one side of the membrane is filled with a catalyst for the partial oxidation of methane, and another side is exposed to flowing air. As methane is introduced into the membrane reactor, it reacts with the permeated oxygen from the air side, and the oxygen partial pressure at the methane side is reduced to an extremely low value ($\sim 10^{-21}$ atm). In contrast, the decrease of oxygen partial pressure at the permeation side leads to an increase of the driving force for oxygen permeation across the MIEC membrane. At a steady state, the oxygen permeation flux of the membrane is usually increased by 5 ~ 10 times for the partial oxidation of methane in MIEC membrane reactors compared with that achieved under an air/inert gas gradient for oxygen permeation testing. In CMRs, the dense MIEC membranes isolate air from methane by the two-chamber structure; thus, the premixing of oxygen and natural gas in CMRs is avoidable. As a result, there is no explosion risk for CMRs, while there is a high risk for traditional co-feed reactors. The reaction rate of the partial oxidation of methane is controlled by the oxygen permeation flux of the MIEC membrane; thus, the operation process is controllable in CMRs. Therefore, the uncontrollable temperature runaway induced by the formation of hot spots does not occur in MIEC membrane reactors. The catalytic reaction can improve the oxygen permeation flux by enhancing the driving force across the membrane and thus increase the oxygen separation efficiency; meanwhile, the membrane can separate the reactor into two chambers to avoid the premixing of oxygen and methane and thus improve the safety of the reaction process. Therefore, it can be clearly seen that the combination of an MIEC membrane and the partial oxidation of methane reaction is a win-win situation. Due to the potential advantages and economic benefits offered by this technology, large enterprises, like Air Products & Chemicals, Inc., have launched some big programs to solve the scale-up problems.

Although the combination of an MIEC membrane with the partial oxidation of methane can overcome the aforementioned drawbacks, several challenges remain to be faced for the practical application of MIEC membranes in CMRs. In a CMR, one side of the membrane is exposed to air and another side to a strongly reductive atmosphere containing methane, hydrogen, carbon monoxide, water, and carbon dioxide. To ensure safe operation of the CMR, the membrane should be stable under both oxidative and reductive atmospheres. Therefore, membrane materials are the current focus of concern in the CMR technology.

In 1994, Balachandran and coworkers were the first to propose the concept of combining the partial oxidation of methane with an MIEC membrane for oxygen separation at elevated temperature [5–7]. A non-perovskite composite oxide with composition of $SrCo_{0.5}FeO_x$ was used in the membrane reactor for the methane conversion to syngas. The composite oxide was fabricated into tubular membranes and filled with an Rh-based reforming catalyst. High methane conversion and CO selectivity were achieved on the CMR. However, the oxygen permeation flux decreased from 4 to 2 mL $cm^{-2}min^{-1}$ after 1000 h on-stream [6]. When a perovskite-type oxide with a composition of $SrCo_{0.8}Fe_{0.2}O_{3-\delta}$ was used as membrane, the integrity of the tubular membrane degraded with time, eventually leading to catastrophic failure [8]. Two types of fractures took place; one occurred at the hot reaction zone shortly after the introduction of methane into the membrane reactor and another after a certain reaction period along the axis of the tube. A detailed investigation disclosed that the first fracture was caused by a strain (or lattice mismatch) that was created by the great difference in oxygen partial pressure across the membrane, while the secondary fracture was the consequence of an expansion of the membrane on the reaction side that arose from the decomposition of the membrane material under reductive atmosphere.

9.3.2 Membrane Materials

From the pioneering work of Balachandran and coworkers on MIEC CMRs for methane conversion, it can be seen that the stability of the membrane materials is one of the important factors determining the safe long-term operation of MIEC CMRs. The permeation flux of the membrane is another key factor for syngas production in CMRs, since it determines the membrane area and the costs of the CMRs. Therefore, membranes with higher stability were used in membrane reactors for partial oxidation of methane to syngas. The successful running of the $La_{0.2}Ba_{0.8}Co_{0.2}Fe_{0.8}O_{3-\delta}$ disk-type membrane reactor for 850 h at 850 °C with a stable oxygen permeation flux of 4.2 mL $cm^{-2}min^{-1}$ showed that the concept proposed by Balachandran and coworkers was feasible [9], although the post-analysis showed that there were remarkable changes in the crystal structure and elemental composition of the membrane surface. Shao et al. reported a study on the $Ba_{0.5}Sr_{0.5}Co_{0.8}Fe_{0.2}O_{3-\delta}$ (BSCF) disk-type membrane reactor for methane conversion to syngas with a 10 % Ni loaded LiLaNiO/γ - Al_2O_3 catalyst [10]. The membrane was stable for methane conversion during the investigated 500 h. High methane conversion (98.5 %), CO selectivity (93.0 %), and oxygen permeation flux (11.5 mL $cm^{-2}min^{-1}$) were achieved on a 1.5-mm-thick membrane under a steady state at 850 °C. The membrane reactor was also operated at elevated pressure for methane conversion, and 92 % methane conversion, 90 % CO selectivity, and 10.5 mL $cm^{-2}min^{-1}$ oxygen permeation flux were achieved at 850 °C and 5 atm. H_2–TPR and XRD results showed that the BSCF material was unstable under reductive atmosphere and could even be reduced to metals and alkaline earth oxides by hydrogen at 850 °C. However, its crystal structure could be quickly recovered by the oxidative atmosphere and showed excellent phase reversibility. This feature saved the membrane from being destroyed by the reductive gases during the long-term operation.

9.3.2.1 Co-based Perovskite Membranes with Improved Stability

Industrial-scale chemical processes require stringent safety measures, especially in the case of oxidation reactions. The risk of explosion of a membrane reactor induced by the breakage of an MIEC membrane inside the reactor is not acceptable. Under a reductive atmosphere (like syngas), the membrane material loses part of its lattice oxygen, and the metal ions are reduced to low valence states, while the other side of the membrane is exposed to air, and thus less oxygen vacancies and high oxidation states of metal ions are produced. While the reduction makes the lattice expand, the oxidation makes it shrink; thus, the great tensile force applied by the syngas to the air side would lead to the breakage of the membrane. Although the abovementioned perovskite membrane reactor can safely run with high performance up to 1000 h, the damage of the surface structure down to tens of microns indicates that the membrane materials are not stable enough to avoid breakage during a long-term operation. Thus, extensive efforts have focused on the improvement of the stability of perovskite membranes by doping less reducible metals ions, i.e., Al^{3+}, Ga^{3+}, Ti^{4+}, Zr^{4+}, Ce^{4+}, Nb^{5+}, and Ta^{5+}, in the B site of Co-based perovskite oxides [11–18].

Tong et al. reported a Zr^{4+}-doped Co-based perovskite membrane with composition of $BaZr_{0.2}Co_{0.4}Fe_{0.4}O_{3-\delta}$ (BZCF) [14]. While its stability improved significantly compared to BSCF, its oxygen permeation flux was only half of that of BSCF. The partial oxidation of methane to syngas was finished within 2200 h on-stream on the BZCF membrane reactor, with a stable oxygen permeation flux of 5.6 mL cm^{-2}min^{-1}, high methane conversion, and CO selectivity up to 98 % at 850 °C. However, post-analysis results revealed that the membrane surface at the methane side was destroyed at depths of about 100 microns, although its bulk still maintained the cubic perovskite structure. Li et al. reported a Ce^{4+}-doped Co-based perovskite membrane with composition of $BaCe_{0.1}Co_{0.4}Fe_{0.5}O_{3-\delta}$ (BCCF) [16]. The membrane was operated for syngas production for more than 1000 h at 875 °C without any decline and reached a higher oxygen permeation flux up to 9mL cm^{-2}min^{-1}, and methane conversion and CO selectivity were higher than 97 %, as shown in Fig. 9.3. The results of the membrane post-analysis showed that the surfaces at both sides of the membrane became porous after 1000 h on-stream. However, the porous layer thickness was extremely thinner (< 5 μm) than the membrane thickness (1.0 mm).

Makoto et al. developed an Nb^{5+}-doped Co-based perovskite membrane with composition of $BaCo_{0.7}Fe_{0.2}Nb_{0.1}O_{3-\delta}$ (BCFN) [17]. The membrane showed

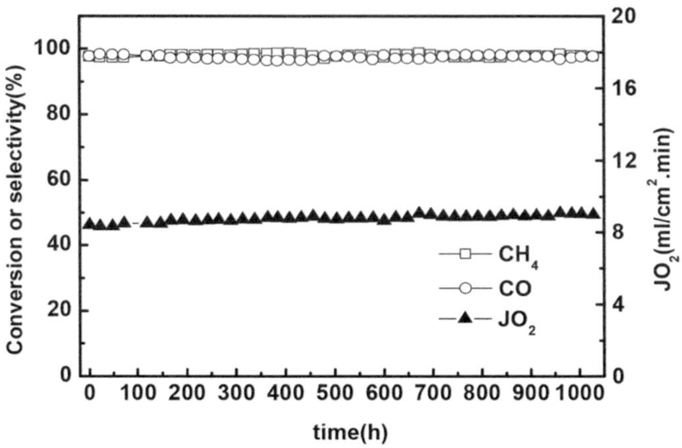

Fig. 9.3 Life experiment of $BaCe_{0.1}Co_{0.4}Fe_{0.5}O_{3-\delta}$ membrane reactor for 1000 h. Air flow rate, 250 ml/min; CH_4 flow rate, 20 ml/min; thickness of the membrane, 1 mm; temperature, 875°C (Reproduced from Ref. [16] with permission of Elsevier)

oxygen permeation flux higher than 18 mL $cm^{-2}min^{-1}$ at 850 °C when it was filled with a Ru(2.0 wt %)Rh(0.2 wt %)/$Co_{0.3}Mg_{0.7}O$ catalyst for the partial oxidation of methane. During the 160 h on-stream, both methane conversion and CO selectivity remained constant at 88 % and 97 %, respectively. A similar material with Nb^{5+} replaced by Ta^{5+} also showed a similar stability and high oxygen permeation flux (~14 mL $cm^{-2}min^{-1}$ at 850 °C) for methane conversion [18]. After 400 h on-stream, the spent membrane was analyzed by SEM/EDS. The results showed that the membrane surface in contact with the syngas was significantly eroded and large amounts of small particles enriched with cobalt were observed. Jin and coworkers found that the addition of Al_2O_3 or yttria-stabilized zirconia (YSZ) powders in Sr(Fe, Co)$O_{3-\delta}$ materials can make the composite membranes stable for the partial oxidation of methane to syngas [11, 19]. For example, a 3 wt% Al_2O_3-doped $SrCo_{0.8}Fe_{0.2}O_{3-\delta}$ membrane was successfully operated for methane conversion to syngas for more than 500 h at 850 °C with an oxygen permeation flux of 2 mL $cm^{-2}min^{-1}$ [11]. The dopants (Al_2O_3 and YSZ) react with $SrCo_{0.8}Fe_{0.2}O_{3-\delta}$ to produce $SrAl_2O_4$ and $SrZrO_3$ particles that locate in the grain boundaries, greatly improving the structural stability through a restricting oxygen-loss mechanism at low oxygen partial pressure. Then, an optimal composition of 5 wt% $SrAl_2O_4$ - 95 wt. % $SrCo_{0.8}Fe_{0.2}O_{3-\delta}$ was used to fabricate membranes for methane conversion to syngas at 850 °C. After more than 1200 h on-stream, the methane conversion, hydrogen selectivity, and oxygen permeation flux remained at about 75 %, 91 %, and 8 mL $cm^{-2}min^{-1}$, respectively [20].

The stability of perovskite membrane materials under reductive atmosphere can also be improved by reducing the doping amount of the cobalt element in the B site. A study on the $Sm_{0.4}Ba_{0.6}Fe_{0.8}Co_{0.2}O_{3-\delta}$ disk-type membrane reactor for methane conversion to syngas showed that the surface microstructure of both sides of the membrane was extensively changed after a period of on-stream [21]. Diethelm et al. prepared $La_{0.6}Ca_{0.4}Fe_{0.75}Co_{0.25}O_{3-\delta}$ tubular membranes with a thickness of 0.5 mm [22]. A 1400 on-stream methane conversion to syngas was performed at 900 °C. The oxygen flux gradually reached 3.2 mL $cm^{-2}min^{-1}$ as the methane content gradually increased from 10 to 30 % but then decreased to a stable value of 2.4 mL $cm^{-2}min^{-1}$ as the methane content increased from 30 to 100 %. The methane conversion and the CO selectivity exceeded 95 % and 90 %, respectively. Post-analysis results revealed that severe kinetic demixing and decomposition occurred in the membrane bulk at depths of hundreds of microns.

Cobalt-based perovskite membranes usually possess high oxygen permeability since the low Co–O band energy and easily changeable valence state result in high oxygen ionic diffusion and fast oxygen exchange kinetics. However, these characteristics are also the source of their low stability under reductive atmosphere. In view of the fact that the stability (or safety) of a membrane reactor is the most important factor concerning the practical applications of CMRs, ideal membranes should have high structural stability under reductive atmosphere and considerable oxygen permeability.

9.3.2.2 Co-free Perovskite Membranes

Cobalt-free perovskite membranes were proposed by researchers to improve the structure stability under reductive environment. Sr and Fe, Co and Ni co-doped $LaGaO_3$ perovskite membranes were reported by Ishihara and coworkers [23]. Among Fe, Co, and Ni, the Fe-doped $La_{0.8}Sr_{0.2}Ga_{1-x}Fe_xO_{3-\delta}$ materials showed higher stability under reductive environment than Co- and Ni-doped materials. The optimal composition $La_{0.8}Sr_{0.2}Ga_{0.6}Fe_{0.4}O_{3-\delta}$ was employed for methane conversion to syngas, and an oxygen permeation flux higher than that of $La_{0.6}Sr_{0.4}Co_{0.2}Fe_{0.8}O_{3-\delta}$ was achieved in the temperature range of 700–1000 °C. A $La_{0.7}Sr_{0.3}Ga_{0.6}Fe_{0.4}O_{3-\delta}$ membrane with a thickness of 0.5 mm coated with $La_{0.6}Sr_{0.4}CoO_{3-\delta}$ at the air side was used to construct a membrane reactor for methane conversion [24]. The modification of the other surface with nickel and ruthenium catalysts led to the best performance. An oxygen permeation rate of 8 mL cm^{-2} min^{-1} at 1000 °C was achieved together with 37 % methane conversion and 100 % CO selectivity. However, no data showing the membrane structure after the reaction. In another research, the stability of $La_{0.9}Sr_{0.1}Ga_{0.1}Fe_{0.9}O_{3-\delta}$ and $La_{0.6}Sr_{0.4}Ga_{0.1}Fe_{0.9}O_{3-\delta}$ was investigated by treating these materials with 3 vol% CH_4–Ar at 900 °C for 24 h [25]. All the $LaGaO_3$-based perovskite oxides were decomposed into several compounds ($SrLaGaO_4$, $SrLaGa_3O_7$, and $La(OH)_3$), and their stability decreased with the increase of Fe content. The $La_{0.5}Sr_{0.5}Ga_{0.3}Fe_{0.7}O_{3-\delta}$ perovskite oxide was found to be unstable under a reductive atmosphere of 5 vol% H_2–Ar at 900 °C. A similar composition of $La_{0.5}Sr_{0.5}Ga_{0.2}Fe_{0.8}O_{3-\delta}$ perovskite oxide was reported to be decomposed at 860 °C under a syngas environment with an oxygen partial pressure of about 10^{-17} atm [26]. The tolerance factors of $LaGaO_3$-based perovskite oxides are all smaller than 1 (0.94 ~ 0.97 for $La_{1-y}Sr_yGa_{1-x}Fe_xO_{3-\delta}$). Under a reductive environment, the reduction of iron from $+4/+3$ to $+3/+2$ could further decrease such tolerance factors, thus resulting in the decomposition of the perovskite phase into $SrLaGaO_4$, $SrLaGa_3O_7$, and FeO_x. It was found that the volatilization of Ga under a reductive atmosphere (such as $H_2 - 1.2\%\ H_2O$ at 900 °C), through formation of Ga_2O gas, was accelerated by the addition of Sr to the La sites [27]. A large amount of Ga depletion caused the formation of secondary phases such as La_2O_3 and $LaSrGaO_4$. Long-term annealing under reductive environment resulted in the whole surface of the $LaGaO_3$-based perovskite oxides being enriched with La_2O_3-related phases. Figure 9.4 gives a schematic view of the change of Ga concentration in the LSGM electrolyte near the surface under a reducing atmosphere at high temperatures.

$BaFeO_{3-\delta}$ perovskite oxide has attracted the attention of researchers because it maintains the perovskite structure at oxygen partial pressures even down to 10^{-17} atm at 827 °C [28]. Doping larger metallic ions, such as Ce^{4+}, Zr^{4+}, Y^{3+}, and La^{3+}, can stabilize the cubic structure down to room temperature [29–32]. $BaCe_xFe_{1-x}O_{3-\delta}$ ($x = 0.05, 0.10, 0.15$) sustained its cubic perovskite structure after treatment

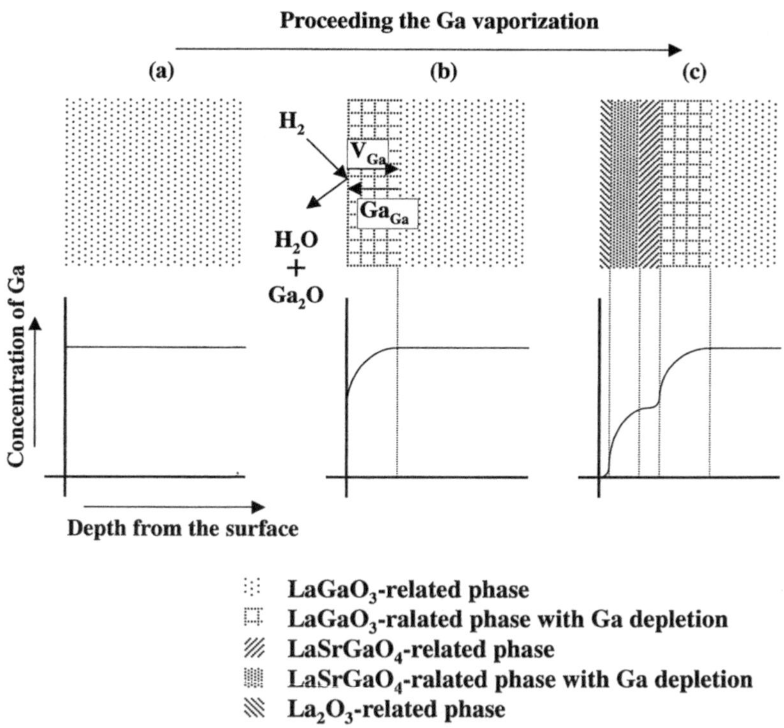

Fig. 9.4 Schematic view of the transition of LSGM electrolyte near the surface with Ga concentration in a high reducing atmosphere at a high temperature (Reproduced from Ref. [27] with permission of Elsevier)

with 10 % H_2–Ar gas at 900 °C for 1 h [29]. The membrane showed high oxygen permeation flux up to 1.0 mL cm^{-2}min^{-1} through a 1-mm-thick disk at 950 °C and under an oxygen partial pressure gradient of air/He. Similar stability and permeability were found for the $BaY_xFe_{1-x}O_{3-\delta}$ ($x = 0.05, 0.10, 0.20$) membranes [31]. As the $BaFeO_{3-\delta}$-based membranes were fabricated into membrane reactors for methane conversion to syngas, an oxygen permeation flux of 4.2 mL cm^{-2}min^{-1} was achieved through a 1.0-mm-thick membrane at 850 °C, accompanied by methane conversion and CO selectivity greater than 96 % [33]. Post-analysis of the spent membrane indicated that both sides of the membrane surface remained dense, in contrast with the porous structure of the cobalt-based perovskite membranes. The changes of the membrane microstructure were deep to several microns on both sides. A La^{3+}- and Zn^{2+}-doped $BaFeO_{3-\delta}$ perovskite membrane (1.0 mm in thickness) with composition of $La_{0.4}Ba_{0.6}Zn_{0.2}Fe_{0.8}O_{3-\delta}$ showed an oxygen permeation flux of 3.8 mL cm^{-2}min^{-1} at 900 °C for methane conversion to syngas with 95–100 % methane conversion and CO selectivity [34]. After a 500 h on-stream, the XRD patterns showed no change of the perovskite structure, although microstructure changes on the membrane surfaces were inevitable.

Al-doped $SrFeO_{3-\delta}$ perovskite oxide, $SrFe_{0.7}Al_{0.3}O_{3-\delta}$, has high structural stability and can retain its perovskite structure even if annealed under a H_2–H_2O–N_2 atmosphere at elevated temperature, as shown in Fig. 9.5a [35]. However, after 215 h on-stream for methane conversion to syngas in the membrane reactor, a significant microstructure change was observed. The corresponding SEM images are shown in Fig. 9.5b. The magnified micrographs of selected zones, marked by numbers, are given below the cross section. The depth of the surface layers with clearly visible microstructural changes extends up to 120–150 μm. This investigation reveals that the high stability of membrane materials under reductive atmosphere is not sufficient to make the membrane reactor safe and that the structural stability under a great oxygen chemical potential gradient is also important.

Table 9.1 lists the performance of typical perovskite membranes reported in the literature for the partial oxidation of methane.

9.3.2.3 Degradation Mechanism of Perovskite Membranes for Methane Conversion

From the above experiments, it is evident that the operation models have great influence on the stability of membrane materials. Upon treatment of a membrane material with a reductive gas, a homogeneous atmosphere surrounds the material, and a new balance between material and atmosphere can build up in a short period. However, when the same material is fabricated into a membrane for methane conversion, the material is under a great oxygen chemical potential gradient, which can continuously drive not only oxygen ions but also cations from one side to the other. Therefore, it is this oxygen chemical potential gradient that is

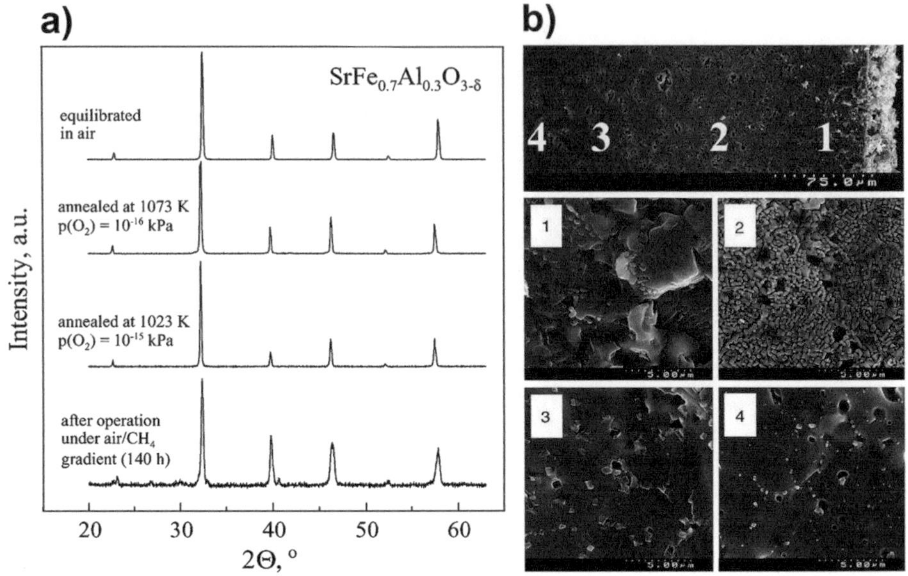

Fig. 9.5 (**a**) XRD patterns of $SrFe_{0.7}Al_{0.3}O_{3-\delta}$ ceramics equilibrated at low temperatures in air and annealed at reduced oxygen pressures and the permeate side surface of $SrFe_{0.7}Al_{0.3}O_{3-\delta}$ membrane after operation under air/CH_4 gradient at 1123–1223 K for 140 h. (**b**) Cross section of $SrFe_{0.7}Al_{0.3}O_{3-\delta}$ membrane after operation under air/CH_4 gradient at 1023–1223 K for 215 h (Reproduced from Ref. [35] with permission of Elsevier)

responsible for the decomposition of perovskite materials. The formation of different microstructures at different depths is a typical reflection of kinetic demixing of composite oxides under a certain oxygen chemical potential gradient. The kinetic demixing phenomenon may lead to a time-dependent degradation of the membrane performance, particularly concerning the integrity of ceramic membranes. van Doorn et al. used the XPS technique to characterize the spent membrane and treated powder and observed that elemental enrichment occurred on the $La_{0.3}Sr_{0.7}CoO_{3-\delta}$ membrane surface as long as there was an oxygen partial pressure gradient across the membrane [51]. However, no elemental enrichment took place on the $La_{0.3}Sr_{0.7}CoO_{3-\delta}$ particles surface whenever the powder was treated by flowing air or nitrogen for several days. This experimental result can well explain the phenomena observed on the $SrFe_{0.7}Al_{0.3}O_{3-\delta}$ and $BaCe_{0.15}Fe_{0.85}O_{3-\delta}$ membranes. In an MIEC membrane reactor for methane conversion, in addition to the damage action applied to the membrane structure by strong reductive gases, i.e., H_2 and CO, the by-products, i.e., CO_2 and H_2O, can also significantly speed up the decomposition of the perovskite structure. The microstructure at the air side of a membrane is always found to be severely destroyed after the methane conversion to syngas experiment. However, since no corrosive gases, like H_2, CO, CO_2, and H_2O, are present at the air side, it can be concluded that the changes observed are not related to the above gases but to the oxygen chemical potential gradient.

Table 9.1 Performance of the typical perovskite membranes for the partial oxidation of methane reaction

Membrane	Reactor configuration and catalyst	Temperature (°C)	X (CH_4)	S(CO)	J_{O2} (mL cm^{-2} min^{-1})	Stability	Ref.
$SrFeCo_{0.5}O_x$	Tube packed with Ru cat.	850	> 99 %	> 98 %	2–4	>1000 h	[7]
$La_{0.6}Sr_{0.4}Co_{0.2}Fe_{0.8}O_{3-\delta}$	Tube packed with Ni/γ-Al_2O_3	825–885	> 96 %	> 97 %	–	3–7 h	[36]
$Ca_{0.8}Sr_{0.2}Ti_{0.7}Fe_{0.3}O_{3-\alpha}$	Disk coated with Ni/$Ca_{0.8}Sr_{0.2}Ti_{0.9}Fe_{0.1}O_{3-\alpha}$ cat.	900	13.7 %	98 %	–	–	[37]
$La_{0.5}Sr_{0.5}Fe_{0.8}Ga_{0.2}O_{3-\delta}$	Tube, Ru cat. packed on shell	850	97 %	100 %	–	–	[26]
$Ba_{0.5}Sr_{0.5}Co_{0.8}Fe_{0.2}O_3$	Disk packed with LiLaNiO/γ-Al_2O_3 cat.	875	98.5 %	93 %	11.5	500 h	[10]
$BaCo_{0.4}Fe_{0.4}Zr_{0.2}O_{3-\delta}$	Disk with catalyst	850	98 %	100 %	5.6	2200 h	[14]
$Ba_{0.5}Sr_{0.5}Co_{0.8}Fe_{0.2}O_{3-\delta}$	Tube packed with LiLaNiO/γ-Al_2O_3	875	94 %	95 %	8.0	500 h	[38]
$Sm_{0.4}Ba_{0.6}Co_{0.2}Fe_{0.8}O_{3-\delta}$	Disk packed with Rh/MgO	900	90%	98%	–	–	[21]
$SrFe_{0.7}Al_{0.3}O_{3-\delta}$	Disk packed with $SrFe_{0.7}Al_{0.3}O_{3-\delta}$ cat.	950	65 %	48 %	–	–	[39]
$Ba_{0.5}Sr_{0.5}Co_{0.8}Fe_{0.2}O_{3-\delta}$	Disk packed with LiLaNiO/γ-Al_2O_3 cat.	850	92 %	90 %	15.5	40 h	[40]
$SrCo_{0.4}Fe_{0.5}Zr_{0.1}O_{3-\alpha}$	Disk packed with NiO/Al_2O_3	950	–	> 90 %	–	–	[41]
$BaCe_{0.15}Fe_{0.85}O_{3-\delta}$	Disk packed with LiLaNiO/γ-Al_2O_3 cat.	850	> 96 %	> 98 %	4.2	140 h	[33]
$La_{0.8}Sr_{0.2}Fe_{0.7}Ga_{0.3}O_{3-\delta}$	Tube coated with $La_{0.8}Sr_{0.2}Fe_{0.7}Ga_{0.3}O_{3-\delta}$ cat.	900	74 %	–	–	142 h	[42]
$BaCo_{0.7}Fe_{0.2}Ta_{0.1}O_{3-\delta}$	Disk packed with Ni cat	900	99 %	–	16.2	400 h	[18]
$BaCe_{0.1}Co_{0.4}Fe_{0.5}O_{3-\delta}$	Disk packed with LiLaNiO/γ-Al_2O_3 cat.	950	99	93	9.5	1000 h	[16]
$La_{0.4}Ba_{0.6}Fe_{0.8}Zn_{0.2}O_{3-\delta}$	Disk packed with Ni cat.	900	~100%	95%	–	500 h	[34]
$BaCo_{0.7}Fe_{0.2}Nb_{0.1}O_{3-\delta}$	Disk packed with Ni cat.	875	92 %	–	15	550 h	[43]
$Ba_{0.5}Sr_{0.5}Co_{0.8}Fe_{0.1}Ni_{0.1}O_{3-\delta}$	Disk packed with Ni/α-Al_2O_3 cat.	850	98 %	97.5 %	11.5	–	[44]
$Ca_{0.8}Sr_{0.2}Ti_{0.7}Fe_{0.3}O_{3-\delta}$	Dense layer painted with $La_{0.6}Sr_{0.4}CoO_{3-\delta}$ and support layer painted with Ni/$Ca_{0.8}Sr_{0.2}Ti_{0.7}Fe_{0.3}O_{3-\delta}$	900	82.3 %	99.8%	3.0	–	[45]
$Ba_{0.9}Co_{0.7}Fe_{0.2}Nb_{0.1}O_{3-\delta}$	Disk packed with NiO/MgO cat.	875	93 %	94.6 %	7.2	400	[46]
$BaCo_{0.7}Fe_{0.2}Nb_{0.1}O_{3-\delta}$	Tube packed with Ni-based cat. on shell	850	94 %	–	11.3	110 h	[47]
$La_{0.6}Sr_{0.4}Co_{0.8}Ga_{0.2}O_{3-\delta}$	Hollow fiber membrane packed with Ni/$LaAlO_3$–Al_2O_3 cat. on shell	750	97 %	91 %	~2.2	–	[48]
$BaCo_{0.7}Fe_{0.2}Ta_{0.1}O_{3-\delta}$	Hollow fiber membrane packed with Ni-based cat.	875	96 %	99 %	20	31 h	[49]
$SrFe_{0.8}Nb_{0.2}O_{3-\delta}$	Multichannel hollow fiber packed with Ni/Al_2O_3 cat. on shell	900	94.6 %	99 %	19.2	120 h	[50]

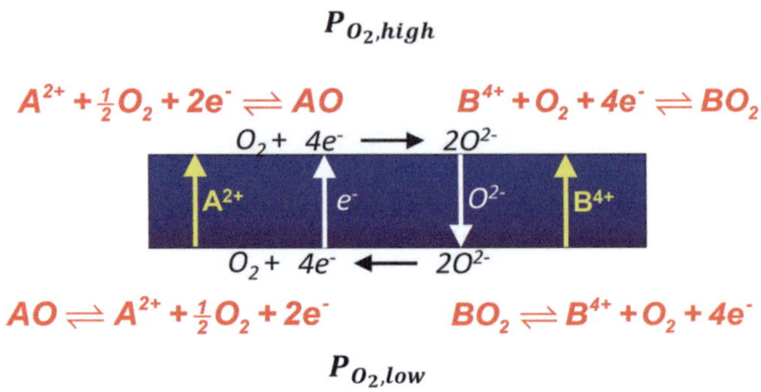

Fig. 9.6 Transport of cations under an oxygen chemical potential

The stability of perovskite structures can be evaluated by the formation free energy of perovskite oxides, which is predominately determined by the stabilization energy of the BO_6 octahedron and slightly affected by the A-site ions. B-site ions with high valence state and small ionic radius usually provide high stabilization energy of the BO_6 octahedron, for instance, the ZrO_6 octahedron. Thus, $BaZrO_3$ perovskite oxide is stable even under CO_2 and strong reductive atmosphere. For ABO_3 perovskite, the A-site ions are usually alkaline earth and/or rare earth elements with larger ionic radii and lower valence states, while the B-site ions are usually transition elements in the fourth period with variable valence states under reductive and oxidative atmospheres. These two kinds of metallic ions have very different physiochemical properties, for instance, metallic ion diffusion kinetics and bonding energy to oxygen. As MIEC membranes are operated under a certain oxygen chemical gradient, except for the diffusion of oxygen ions in the membrane bulk, all metallic ions are transported from the low oxygen partial pressure side to the high oxygen partial pressure side due to a difference in free energy, as illustrated by Fig. 9.6. Of course, the diffusion rate of metallic ions is extremely slow compared to that of oxygen ions. The free energy difference across the membrane for certain cations can be described as

$$\Delta G = \Delta G^{II}_{M^{n+}}(P_{O_2,\ \mathrm{low}}) - \Delta G^{I}_{M^{n+}}(P_{O_2,\ \mathrm{high}}) \tag{9.3}$$

Since different cations have various diffusion rates in the membrane bulk, kinetic demixing inevitably takes place (see Chap. 3). Usually, smaller cations with lower valence states (such as Li^+) have high diffusion coefficients, while bigger cations with higher valence states (such as Zr^{4+}) have low diffusion coefficients. The diffusion coefficients of cations are also determined by the solid crystalline, elemental composition, stoichiometric ratio, microstructure, etc. Figure 9.7 gives the temperature-dependent cationic diffusion coefficients in perovskite materials [52–57]. As shown in the figure, the diffusion coefficients of cations in the grain boundaries is 3 ~ 6 orders higher than those in the lattices. MIEC membrane materials are made of 2 ~ 5 metallic elements; thus, fast cations will be enriched

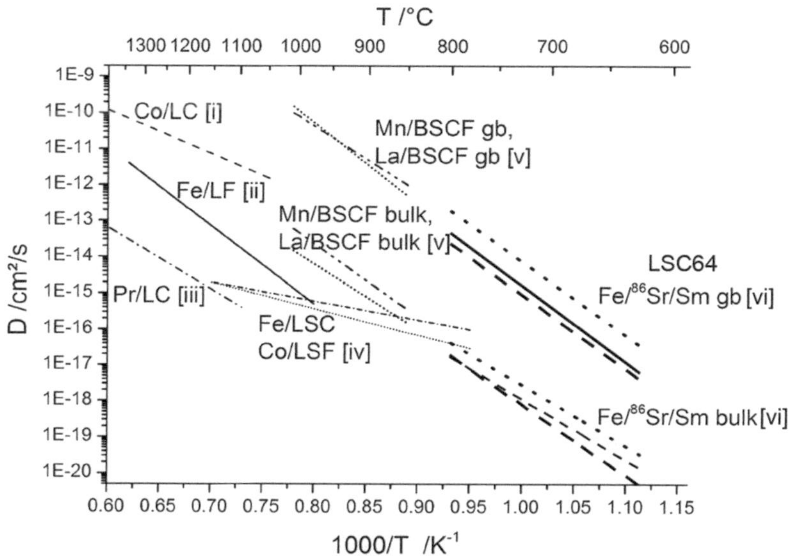

Fig. 9.7 Temperature-dependent cationic diffusion coefficients in perovskite materials. Solid-state reaction, (*i*) Co in $LaCoO_3$, (*ii*) Fe in $LaFeO_3$; impurity diffusion: (*iii*) Pr in $LaCoO_3$, (*iv*) Co in $La_{0.8}Sr_{0.2}FeO_{3-\delta}$ and Fe in $La_{0.8}Sr_{0.2}CoO_{3-\delta}$, (*v*) La and Mn in $Ba_{0.5}Sr_{0.5}Co_{0.8}Fe_{0.2}O_{3-\delta}$ (Reproduced from Ref. [57] – Published by the PCCP Owner Societies)

on the membrane surface exposed to high oxygen partial pressure; meanwhile slow cations will be left on the surface exposed to low oxygen partial pressure. If one type of cation has a significantly high diffusion coefficient, a visible kinetic demixing will be observed.

From the above analysis, it is easy to understand why perovskite oxides that are stable under reductive atmosphere are unstable for methane conversion to syngas. Accordingly, the kinetic demixing of the perovskite-type membranes is inevitable, especially under a high oxygen chemical potential gradient. To suppress the kinetic demixing, five approaches may work:

1. Introducing a small amount of secondary phase to inhibit the movement of grain boundaries
2. Doping low-mobility cations in the perovskite lattice to hamper cation migration
3. Increasing the relative density of ceramic membranes to avoid the fast surface diffusion of cations in the inner surfaces of the membranes
4. Preparing membranes with elemental compositional gradients to compensate the driving force for cations transport under oxygen chemical potential gradients
5. Selecting cations with close diffusion coefficients in the resultant perovskite lattice

Additionally, two approaches could be employed to improve the operational conditions:

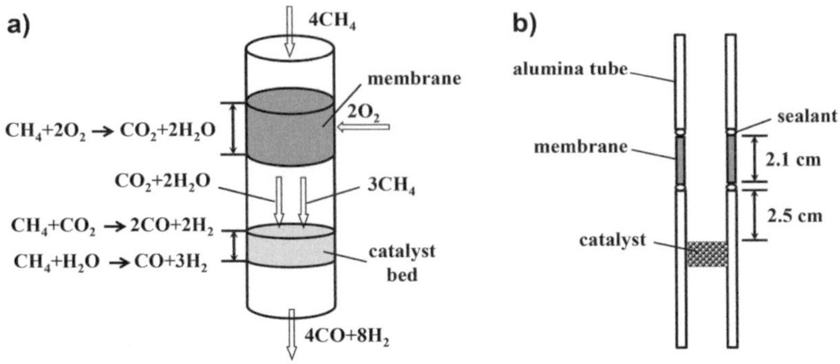

Fig. 9.8 Schematic diagrams of two-stage oxygen-permeable membrane reactor for syngas production. (**a**) The chemical conversions in different areas of the membrane reactor. (**b**) The construction and dimensions of the reactor (Reproduced from Ref. [58] with permission of Wiley)

1. Decreasing the operation temperature of membranes since the diffusion activation energy for cations is much higher than that of oxygen ions
2. Decreasing the oxygen chemical potential gradient of the membrane exposed by improving the design of membrane reactors

Chen et al. reported a two-stage membrane reactor for methane conversion to syngas, as shown in Fig. 9.8 [58]. In the reactor, a part of methane is converted into CO_2 and H_2O by reaction with the oxygen permeated through the membrane from the air; subsequently, the resultant mixture is transferred to a catalyst bed where the remaining methane is reformed to syngas. Unlike the traditional membrane reactors filled with catalysts in the membrane separation zone, the two-stage reactor has less stringent requirements for the stability of membrane materials. The combustion reaction takes place in the membrane zone with a thermodynamic equilibrium at an oxygen partial pressure of 10^{-13}–10^{-14} atm. However, in reality the reaction does not reach the equilibrium state, and a small amount of oxygen can be detected in the effluent. Therefore, the membrane in the two-stage reactor is under much lower oxygen chemical gradient than that in the traditional one.

After a comprehensive survey of the well-known oxygen ionic conducting oxides, it was deduced that fluorite-type ionic conducting oxides, especially YSZ and ceria-based ionic conductors, are good candidates for the preparation of high stable membranes with slow kinetic demixing. Four reasons led to this conclusion:

1. The metallic ions involved in the structure have low diffusion coefficients.
2. These ionic conductors have high ionic conductivity.
3. They are stable under a reductive atmosphere.
4. There is only one type of lattice site for the cations, and these cations have similar diffusion coefficients.

However, these oxides are almost pure ionic conductors in oxidation environments, and the electronic transfer number is far lower than the ionic transfer number under reductive atmosphere. Thus, to turn them into MIEC membranes, a secondary phase with high electronic conductivity should be incorporated to form dual-phase type membranes.

9.3.2.4 Dual-Phase Membranes

Dual-phase membranes are suggested as good alternatives with compositions that can be adjusted according to practical requirements. One phase is usually an oxygen ionic conductor, such as Sm_2O_3-doped CeO_2(SDC), for oxygen ionic transport, and the other is a noble metal or a perovskite oxide for electronic transport (see Chap. 7). Traditional dual-phase membranes are cermet and made of a noble metal (Ag, Au, Pd, Pt, etc.) and an oxygen ionic conductor. This type of dual-phase membrane is not suitable for practical applications because of its high cost and poor permeability. The cost can be reduced by replacing the noble metals with pure electronic conducting perovskite or spinel oxides, such as $La_{1-x}Sr_xMnO_3$, $La_{1-x}(Sr, Ca)_xCrO_3$ and $(Mn_{3-x}Co_x)O_4$, that have high electronic conductivity under typical operational conditions. However, the permeability is far lower than that of perovskite membranes.

Our group proposed a new kind of dual-phase membranes made of mixed electronic conducting perovskite oxides and ionic conducting fluorite oxides. We proposed a design strategy to select each phase of this new type of dual-phase membranes by considering three important factors, i.e., ionic conductivity, stability, and chemical compatibility. The detailed elaboration of this method is shown in Chap. 7. The optimal dual-phase membranes composed of 75 wt.%$Ce_{0.85}Sm_{0.15}$ −25 wt.%$Sm_{0.6}Sr_{0.4}Al_xFe_{1-x}O_{3-\delta}$ (SDC–SSAF) showed an oxygen permeation flux one order higher than that of traditional ones. These dual-phase membranes were applied for methane conversion to syngas. Figure 9.9 shows the typical long-term on-stream data of a dual-phase membrane reactor [59]. After 1100-h operation at 950 °C, the experiment was deliberately stopped. During the whole on-stream, the oxygen permeation flux reached 4.2 mL $cm^{-2}min^{-1}$, methane conversion and CO selectivity were >98 %, and the H_2/CO ratio remained around 2. Post-analysis of the spent membrane indicated that the surface phase structures of the dual-phase membrane were not destroyed under syngas production conditions. Figure 9.10 shows the SEM images of the as-prepared and spent membranes. The dual-phase membranes comprise two types of grains, i.e., the white and big SDC grains and the black and small SSAF grains (Fig. 9.10a). The big SDC grains pack compactly together, while the small SSAF grains fill the interspaces between the SDC grains. This type of structure allows easy percolation of the two phases. Therefore, it is expected that a lower amount of mixed conducting phase would be required to realize an electronic path through dual-phase membranes. As shown in Fig. 9.10c, d, both sides of the membrane retained a dense structure after 1100 h on-stream for syngas production. In addition, no difference in morphology between cross sections (the one close to the air side and the one close to the methane side) of the spent and

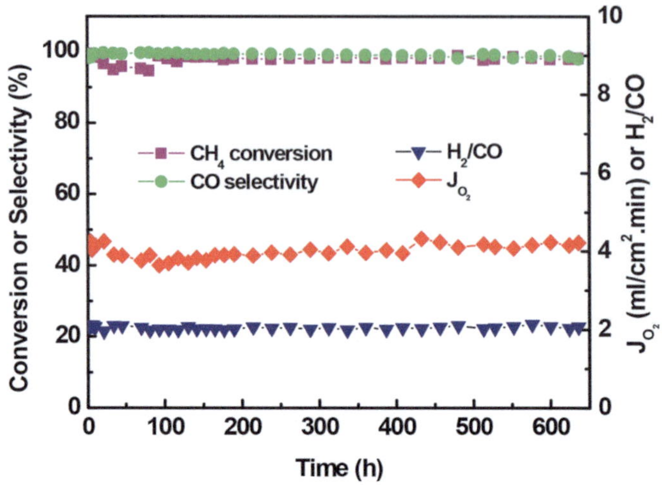

Fig. 9.9 The typical long-term on-stream of the dual-phase membrane reactor for syngas production (Reproduced from Ref. [59] with permission of Elsevier)

as-prepared membranes was observed. No delamination and elemental enrichment were observed on the cross section of each side, as shown in Fig. 9.10c, d. This result confirms the good stability of the SDC–SSAF membrane used as a membrane reactor for syngas application.

As stated above, MIEC perovskite membranes do not usually retain their structure under syngas generation conditions. The atomic ratio between alkaline earth and total metal ions is usually greater than 20 % and sometimes up to 50 %. Thus, carbonates are easily formed on the perovskite membrane surface as the membrane is exposed to an atmosphere containing CO_2. The unstable phase structure under reductive atmosphere can be attributed to the high content of reducible metallic ions in the B site. To obtain high oxygen permeability, mixed conducting perovskite membranes usually have a high atomic ratio of reducible to total metal ions. The value is often greater than 30 % and up to 50 % for materials with high permeability.

However, for SDC–SSAF, the situation is different. This dual-phase membrane material contains low amounts of both alkaline earth (6.7 %) and reducible metal ions (11.7 %). Thus, this composition makes it a highly stable membrane material. Furthermore, the metal elements affect the material stability, especially the choice of B-site ions for the mixed conducting perovskite phase. Iron is optimal for the B site among all the light transition elements. Metal ions such as cobalt, nickel, and copper are easily reduced under reducing environments and thus are not suitable for the B sites. In addition, aluminum substitution of iron in B sites can increase the metal–oxygen bond energy and improve the stability under reductive environments.

The microstructure of dual-phase membranes can easily inhibit the kinetic demixing of the perovskite oxides under great oxygen chemical potential gradients. The membrane comprises two phases, the fluorite phase of SDC and the perovskite phase of SSAF. Decomposition of the two phases is limited by their mutual

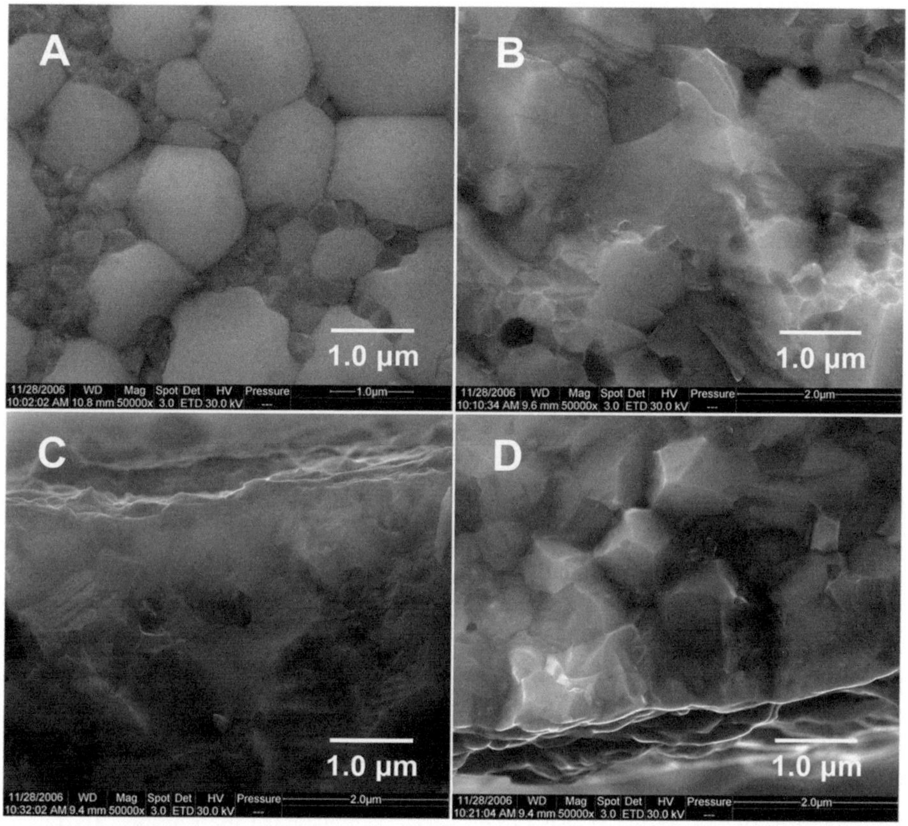

Fig. 9.10 SEM images of the as-prepared (a and b) and used membranes (c and d). Top view of as-prepared membrane (**a**), cross section of as-prepared membrane (**b**), cross section of the reaction side of the used membrane (**c**), and cross section of the air side of the used membrane (**d**) (Reproduced from Ref. [59] with permission of Elsevier)

isolation, so that cationic diffusion in the membrane bulk is difficult. For example, it is not easy for strontium in a perovskite grain to reach another perovskite grain if a fluorite grain is present between the two SSAF grains, since the strontium element has a high solution energy in a ceria lattice when Sr^{2+} passes through SDC grains. Material creep under elevated temperatures is also restrained by the dual-phase structure. Another advantage makes ceria-based dual-phase membranes very suitable for large-scale use., i.e., an accurate ratio between all metal ions is not needed during the preparation of these dual-phase membranes, because a slight excess or deficit of metal ions can be automatically compensated by the system or has little effect on the physical properties (such as permeability, sintering temperature, mechanical strength, thermal expansion, etc.) of the membranes. For example, if there is a little deficiency of Sr^{2+} in SSAF, some of the Sm^{3+} ions can move from SDC into the SSAF; if there is a little excess of Sm^{3+} in SSAF, some of the Sm^{3+}

ions can transfer from SSAF into the SDC. A one-pot method was used to synthesize SDC–SSAF, and therefore the above adjustments occur during the powder preparation process. These membrane materials can successfully be synthesized by a one-pot method, also allowing to verify whether the two phases with different structure are compatible with each other, which also directly supports a long-term stable operation without degradation.

Recently, we found that the oxygen permeation flux of a dual-phase membrane can be further improved for the production of syngas in a membrane reactor by enhancing the electronic conductivity under intermediate–low oxygen partial pressures. The oxygen chemical potential decreases continuously across a mixed ionic–electronic conducting membrane as one side is fed with air and the other with methane for syngas production, as shown in Fig. 9.11. Therefore, at a certain position in the membrane bulk, the oxygen chemical potential is intermediate–low, usually corresponding to an oxygen partial pressure of $10^{-10} - 10^{-15}$ atm. For example, for $La_{1-x}Sr_xFeO_{3-\delta}$ [60–62] and $La_{1-x}Sr_xM_{1-y}Fe_yO_{3-\delta}$ (M = Al, Ti, Cr, etc.) [63–67], the total conductivity of these perovskite materials is predominantly p-type electronic conduction in a high oxygen partial pressure regime (> 10^{-10} atm) and n-type electronic in a very low oxygen partial pressure regime (< 10^{-15} atm). In the intermediate–low oxygen partial pressure range, usually 10^{-10}–10^{-15} atm, a minimum value appears. At this point, oxygen ionic conduction dominates, and the contributions of p-type and n-type electronic transport are low and comparable. Therefore, although an MIEC membrane has high electronic conductivities under both oxidizing and strong reducing atmosphere, its oxygen permeation may still be limited by the low electronic conductivity in zone II, as shown in Fig. 9.11. It is possible to improve the electronic conductivity in zone II by changing the chemical composition of the MIEC phase in ceria-based dual-phase membranes. Cr-based perovskite oxides exhibit high electronic conductivity in the oxygen partial pressure range of $0.21 - 10^{-21}$ atm. Therefore, a combination of the transition metal Cr and Fe on the B site of perovskites may produce considerable ionic conductivity and high electronic conductivity at intermediate–low oxygen partial pressures. A new ceria-based dual-phase membrane composed of 75 wt% $Ce_{0.85}Sm_{0.15}O_{1.925}$ – 25

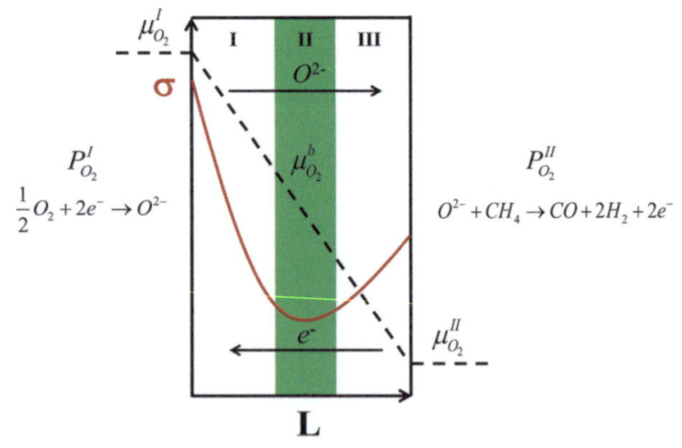

Fig. 9.11 Schematic illustration of oxygen chemical potential drop and conductivity change across an MIEC membrane in the POM reactor, where *L* represents the thickness of membrane

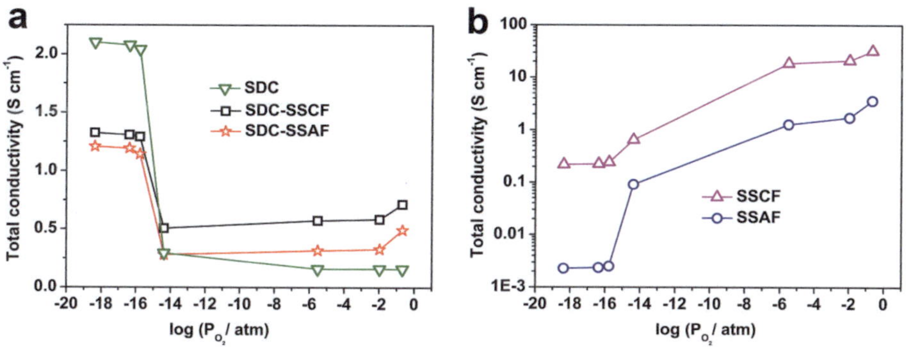

Fig. 9.12 Dependence of the total conductivities on oxygen partial pressure for different samples at 950 °C (Ref. [68])

wt.% $Sm_{0.6}Sr_{0.4}Cr_{0.3}Fe_{0.7}O_{3-\delta}$ (SDC–SSCF) was designed to prove the above idea [68]. The conductivity of this new dual-phase membrane was higher than SDC–SSAF under various oxygen partial pressures, especially under the oxygen partial pressure range of $10^{-2}-10^{-15}$ atm, as shown in Fig. 9.12. Its oxygen permeation flux reached 7.6 mL $cm^{-2}min^{-1}$ in the POM reactor at 950 °C, and the flux was about 1.8 times that of SDC–SSAF under the same conditions, although the oxygen permeation flux of SDC–SSCF was lower than that of SDC–SSAF under an oxygen partial pressure gradient of 0.21 atm/0.005 atm. This result indicates that the improvement of electronic conductivity in the intermediate–low oxygen partial pressure range is effective to improve the oxygen permeation flux of dual-phase membranes for POM reaction. In addition, over a period of 220 h, >95 % methane conversion and >98 % CO selectivity were retained at 950 °C. The SEM and EDS results confirm the good stability of SDC–SSCF as a membrane reactor for syngas production. And the performance of the SDC–SSCF membrane could be further improved by reducing its thickness, for example, by fabricating asymmetric membranes via tape casting or other techniques.

Table 9.2 lists the performance of dual-phase membranes reported in literatures for the partial oxidation of methane.

9.3.3 Activation of the POM Reaction in MIEC Membrane Reactors

Activation processes were reported during the initial stage of the partial oxidation of methane to syngas in membrane reactors as indicated by a suddenly significant increase in oxygen permeation flux and CO selectivity, as shown in Fig. 9.13. At the beginning of the membrane reaction, the Ni-based catalyst is gradually reduced from its oxidation state to Ni^0 nanoparticles dispersed on a γ-Al_2O_3 support

Table 9.2 Performance of dual-phase membranes for the partial oxidation of methane reaction

Membrane materials	Membrane configurations	Thickness (mm)	Temperature (°C)	J_{O2} (mL cm^{-2} min^{-1})	Catalysts	Conversion and selectivity	Ref.
60 vol% YSZ–40 vol% $La_{0.8}Sr_{0.2}MnO_{3-\delta}$	Hollow fiber	0.075	950	~3	30 vol% NiO/70 vol% YSZ	~36 %, ~92 %	[69]
60 wt% $Ce_{0.8}Gd_{0.2}O_{2-\delta}$–40 wt% $Gd_{0.2}Sr_{0.8}FeO_{3-\delta}$	Disk	0.5	950	2~5	LiLaNiO/γ-Al_2O_3	> 98 %, > 98 %	[70]
75 wt% $Ce_{0.85}Sm_{0.15}O_{2-\delta}$–25 wt% $Sm_{0.6}Sr_{0.4}FeO_{3-\delta}$	Disk	0.6	950	4.2	LiLaNiO/γ-Al_2O_3	> 98 %, > 98 %	[71]
75 wt% $Ce_{0.85}Sm_{0.15}O_{2-\delta}$–25 wt% $Sm_{0.6}Sr_{0.4}Al_{0.3}Fe_{0.7}O_{3-\delta}$	Disk	0.5	950	4.3	LiLaNiO/γ-Al_2O_3	> 98 %, > 98 %	[59]
$Zr_{0.86}Y_{0.14}O_{1.92}$–$La_{0.8}Sr_{0.2}Cr_{0.5}Fe_{0.5}O_{3-\delta}$	Hollow fiber	0.08	950	7.9	33 wt% Ru/LSCF	~91 %, ~90 %	[72, 73]
60 wt% $Ce_{0.9}Pr_{0.1}O_{2-\delta}$–40 wt% $Pr_{0.6}Sr_{0.4}FeO_{3-\delta}$	Disk	0.6	950	4.4	Ni-based catalysts	> 99 %, > 97 %	[74]
Ag/$Ce_{0.9}Gd_{0.1}O_{2-x}$	Disk	1	750	0.18	Nickel foam	21 %, 92 %	[75]
$Ce_{0.8}Sm_{0.2}O_{2-\delta}$/$La_{0.8}Sr_{0.2}CrO_{3-\delta}$	Tube	0.78	950	0.42	$SrTiO_3$	17 %, 74 %	[76]
75 wt% $Ce_{0.85}Sm_{0.15}O_{1.925}$-25 wt% $Sm_{0.6}Sr_{0.4}Cr_{0.3}Fe_{0.7}O_{3-\delta}$	Disk	0.5	950	7.6	LiLaNiO/γ-Al_2O_3	>95 %, >97 %	[68]

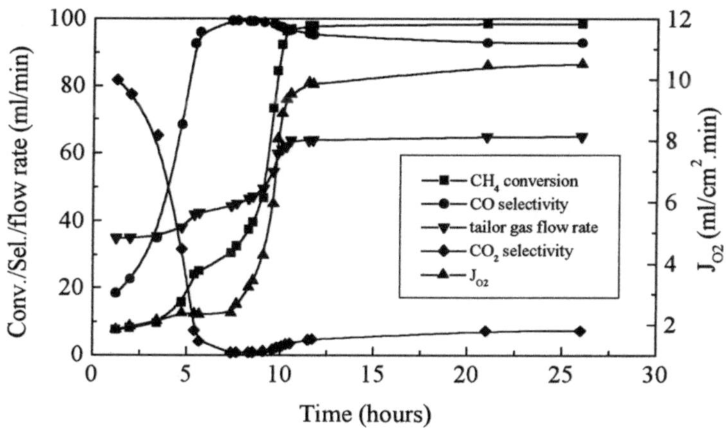

Fig. 9.13 Time dependence of methane conversion, CO selectivity, CO_2 selectivity, tail gas flow rate, and oxygen permeation flux at the catalyst activation stage at 850 °C (Reproduced from Ref. [10] with permission of Elsevier)

[10]. However, Ni^0 is difficult to form due to the low reaction activity of methane, and the main products are CO_2 and water at this stage; thus, the oxygen permeation flux, methane conversion, and CO selectivity are low and increase slowly. However, with the increase of Ni^0, the activity of the catalyst gradually increases. As long as enough Ni^0 is produced, the methane conversion and CO selectivity quickly increase and produce large amounts of CO and hydrogen that in turn speed up the activation/reduction of the catalyst. Finally, the self-catalysis process results in a sudden increase in oxygen permeation flux, methane conversion, and CO selectivity. Different activation processes were found on different MIEC membranes. When different membrane reactors were filled with the same catalyst and operated at the same temperature, different activation periods were observed. Tong et al. found that the unsteady-state period of the activation process of the BZCF membrane reactor for methane conversion was only 2 h at 850 °C, which is much shorter than that of the BSCF membrane reactor [14]. Li et al. reported a 6-h period activation process for the $BaCe_{0.1}Co_{0.4}Fe_{0.5}O_{3-\delta}$ membrane reactor at 875 °C, and the activation process could be shortened to 35 min by increasing the operation temperature to 950 °C [16]. Zhu et al. almost did not observe the activation process (less than 20 min) on the $BaCe_{0.15}Fe_{0.85}O_{3-\delta}$ membrane reactor at 850 °C [33].

The large difference in the activation times of different MIEC membrane reactors may result from their different stabilities under CO_2 atmosphere. At the initial stage of the membrane reaction, the main product at the methane side is CO_2, which can react with membrane materials and destroy the perovskite structure. Frequently, a dense layer of carbonate is formed on the membrane surface. However, the partial pressure of CO_2 gradually decreases with the increase of Ni-based catalyst activity, and as a result, a porous layer is formed through the decomposition of the carbonate. Thus, it is clear that a membrane with higher stability under CO_2 atmosphere or operated at higher temperature undergoes a shorter activation process for the partial oxidation of methane to syngas. With the doping of high valence cations and decrease of cobalt content in the B site, the stability of the MIEC membranes under CO_2 atmosphere can be gradually improved, and consequently the activation period gradually shortened from 20 h to tens of minutes. The SDC–SSAF dual-phase membrane could retain its crystalline structure and oxygen permeation flux even under a pure CO_2 atmosphere [77], so no activation period was observed when the membrane was employed for methane conversion. Similarly, the $(Pr_{0.9}La_{0.1})_2(Ni_{0.74}Cu_{0.21}Ga_{0.05})O_{4+\delta}$(PLNCG) membrane reactor, which was found to be stable under CO_2 atmosphere, immediately reached its steady state for methane conversion to syngas at 900 °C [78, 79].

It can be inferred from the above discussion that the formation of a porous layer on the membrane surface at the methane side may terminate as long as the membrane reaction system reaches its steady state. The thickness of the porous layer may not change remarkably with the prolongation of the on-stream. However, no experiment that verifies this deduction was reported up to now. For CO_2-tolerant dual-phase membranes, no carbonate is formed on the membrane surface; thus, no porous layer is formed after the long-term on-stream. Of course, the decomposition of the crystalline structure by reduction of H_2, CO, and CH_4 is also an important source for the formation of a porous layer, but it is not the primary cause of the appearance of the activation period. Furthermore, no activation period was reported in the literature when methane was replaced by hydrogen as the feeding gas to the cobalt-based perovskite membranes. As discussed above, the activation period may not appear if the membrane is stable under CO_2 atmosphere. Therefore, the step of catalyst activation (reduction of a Ni-based catalyst to Ni^0) has a weak influence on the activation period of the MIEC membrane reactors for methane conversion.

The oxygen permeation flux through the MIEC membrane reactors for methane conversion does not increase with the decrease of membrane thickness for many membranes, even though they are thicker than the characteristic thickness determined under an oxygen partial pressure gradient of air/inert gas. The characteristic thickness of a BSCF membrane is around 0.5 mm for oxygen permeation, and the oxygen permeation flux of the membrane increases almost linearly with the decrease of the thickness, when it is thicker than 1.0 mm. For example, at 900 °C the oxygen permeation flux of the BSCF tubular membrane increases from 1.53 mL $cm^{-2}min^{-1}$ at a thickness of 1.70 mm to 2.20 mL $cm^{-2}min^{-1}$ at a thickness of 1.18 mm. However, when the membranes were employed for methane conversion to syngas, they both exhibited the same oxygen permeation flux [38]. Wang and coworkers found that the oxygen permeation flux during the methane conversion to syngas for the $BaCo_{0.7}Fe_{0.2}Ta_{0.1}O_{3-\delta}$ membrane remained almost constant as the thickness was reduced from 1.6 to 0.8 mm, as shown in Fig. 9.14 [18]. The membrane reactors operated under similar conditions afforded similar methane conversion and CO selectivity. In their subsequent work on the $BaCo_{0.7}Fe_{0.2}Ta_{0.1}O_{3-\delta}$ hollow fiber membrane reactor, although the membrane thickness was decreased to about 0.2 mm, the oxygen permeation flux was still similar to the

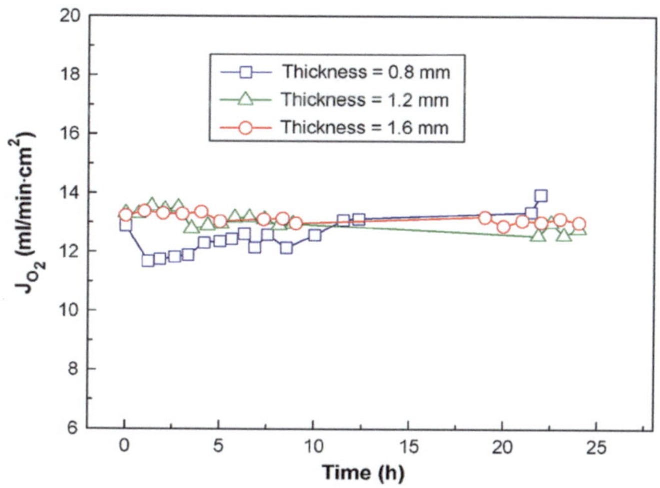

Fig. 9.14 Dependence of oxygen permeation flux on time for the $BaCo_{0.7}Fe_{0.2}Ta_{0.1}O_{3-\delta}$ membrane with different thicknesses at 900 °C. Experimental conditions: air = 200 ml/min, 43 % He diluted methane = 50 ml/min, and membrane thicknesses = 0.8 mm, 1.2 mm, and 1.6 mm (Reproduced from Ref. [18] with permission of Elsevier)

above disk-type membranes [80]. These results indicate that the limiting step of the oxygen permeation is the surface exchange under the membrane reactor conditions.

However, for cobalt-free membrane reactors, their oxygen permeation fluxes increase with the reduction of membrane thickness. Zhu et al. found that at 850 °C the oxygen permeation flux of the $BaCe_{0.15}Fe_{0.85}O_{3-\delta}$ membrane reactor increases from 3.0 mL cm^{-2}min^{-1} at a thickness of 1.5 mm to 4.2 mL cm^{-2}min^{-1} at a thickness of 1.0 mm [33]. Hong et al. reported that at 900 °C the oxygen permeation flux of $La_{0.4}Ba_{0.6}Zn_{0.2}Fe_{0.8}O_{3-\delta}$ reached 3.8 mL cm^{-2}min^{-1} at a thickness of 1.0 mm and it was improved to 8.0 mL cm^{-2}min^{-1} at a thickness of 0.5 mm [34]. Takahashi et al. developed a $La_{0.6}Sr_{0.4}Ti_{0.3}Fe_{0.7}O_{3-\delta}$ membrane for the conversion of methane to syngas and found that the oxygen permeation flux through the membrane increased linearly in proportion to the reciprocal of membrane thickness as the thickness was decreased from 0.5 mm (~2 mL cm^{-2}min^{-1} at 900 °C) to 0.1 mm (~10 mL cm^{-2}min^{-1} at 900 °C) [81]. All these data show that the oxygen permeation through the cobalt-free membrane is controlled by bulk diffusion not by surface exchange. Therefore, the question arises as to why the oxygen permeation through the cobalt-based membranes is controlled by surface exchange, while that through cobalt-free membranes is controlled by bulk diffusion as they are all employed for the methane conversion to syngas reaction?

A common phenomenon occurring on cobalt-based membranes is the formation of a porous layer with thickness up to 100 microns after a period of methane conversion to syngas. Accordingly, the formation of the porous layer is the result of forming–decomposing of carbonates and the decomposition of perovskite structure by reductive gases. Usually, the porous layer is made of carbonates, simple oxides, metal cobalt, hydroxides, etc.; obviously these compounds have low catalytic activation toward either oxygen evolution or partial oxidation reactions. Therefore, exchange resistance on the gas–solid interface is large. Although porous layers are also found on cobalt-free membranes, their thicknesses are lower than several microns. Sometimes, for very stable membranes, such as the SDC–SSAF dual-phase membrane, no porous layer is observed on the membrane surface [59]. Thus, a lower interfacial exchange resistance occurs on the cobalt-free membranes. From the above analysis, it can be expected that the oxygen permeation flux of the cobalt-free membranes can be significantly improved by fabricating them into asymmetric structures with a thin dense layer for the conversion of methane to syngas, and it may reach higher oxygen permeation fluxes than those achieved through cobalt-based perovskite membranes for methane conversion. In view of this, the stable cobalt-free membranes are promising for the construction of membrane reactors for practical applications of methane conversion.

9.3.4 Mechanic Stability

It is interesting to note that some membranes more stable than BSCF and BCFN, such as $La_{0.6}Sr_{0.4}Co_{0.2}Fe_{0.8}O_{3-\delta}$ and $La_{0.8}Sr_{0.2}Co_{0.1}Fe_{0.8}Cr_{0.1}O_{3-\delta}$, cannot retain their structure during the on-stream methane conversion to syngas. Jin et al. reported that in most of their syngas production experiments using tubular $La_{0.6}Sr_{0.4}Co_{0.2}Fe_{0.8}O_{3-\delta}$ membrane reactors, the membrane broke apart after that methane was introduced in the tube side (air flowing through the shell side) at high temperatures for about 3–7 h [36]. Bouwmeester reported that when the $La_{0.8}Sr_{0.2}Co_{0.1}Fe_{0.8}Cr_{0.1}O_{3-\delta}$ membrane was used as membrane reactor for the conversion of methane to syngas, it cracked after on-stream for 350 h at 900 °C [82].

When an MIEC membrane is embedded in a reactor for methane conversion to syngas, one of its sides is exposed to air and another to a strongly reductive atmosphere (oxygen partial pressure down to 10^{-21} atm). The reduction-induced expansion at the syngas side and oxidation-induced contraction at the air side are inevitable. Thus, tensile and compressive stresses build up at the air and syngas side, respectively. Hendriksen et al. reported a detailed investigation on the mechanic stability of MIEC membranes for syngas production [83]. Figure 9.15 shows the calculated maximum tensile stress in a syngas membrane composed of $La_{0.6}Sr_{0.4}Co_{0.2}Fe_{0.8}O_{3-\delta}$ as a function of membrane thickness under two different assumptions of electrode kinetics. The calculation is based on the conditions that the oxygen partial pressures at the syngas and air side are 10^{-19} and 0.2 atm, respectively, and the operation temperature is 800 °C. As shown in Fig. 9.15, the stress value rises with the increase of membrane thickness. This result reveals that a greater oxygen partial pressure difference could build across a thicker membrane and consequently lead to the significant difference in local lattice expansion and

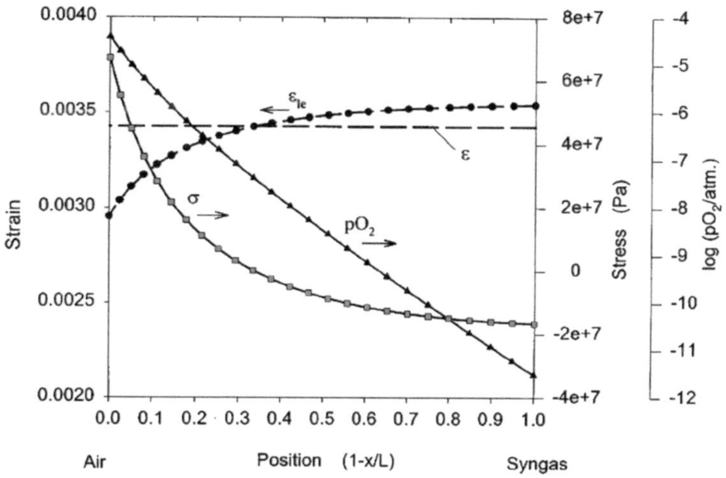

Fig. 9.15 Calculated p_{O_2}-, strain, and stress profiles across a $La_{0.6}Sr_{0.4}Co_{0.2}Fe_{0.8}O_{3-\delta}$ membrane exposed to a p_{O_2} of 10^{-19} atm at the syngas side and a p_{O_2} of 0.2 atm on the air side ($T = 800^{\circ}C$). The following values of the parameters were assumed: $r_{an} = 0.8\ cm^2$, $r_{cat} = 0.4\ cm^2$, $E = 100$ GPa, and $\nu = 0.30$ (Reproduced from Ref. [83] with permission of Elsevier)

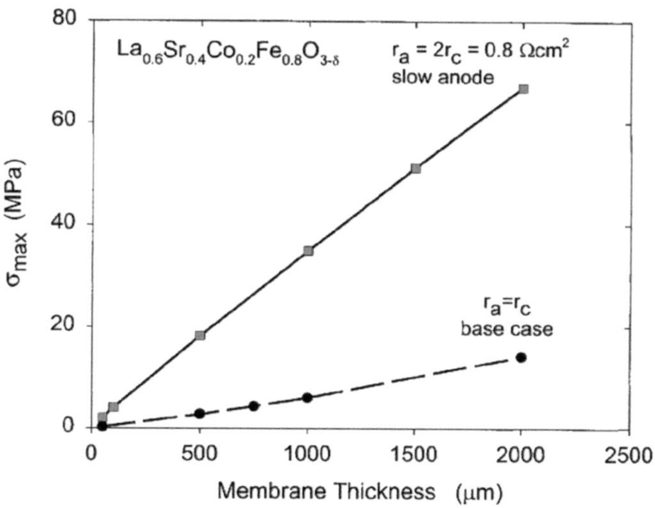

Fig. 9.16 Calculated maximum tensile stress in a syngas membrane of $La_{0.6}Sr_{0.4}Co_{0.2}Fe_{0.8}O_{3-\delta}$ under "base case" conditions as a function of membrane thickness under two different assumptions of the electrode kinetics (Reproduced from Ref. [83] with permission of Elsevier)

mechanical stresses. If the interfacial oxygen exchange kinetics at the air side are faster than those at the syngas side, the maximum tensile stresses are much higher than in the case of the two sides having equal exchange resistances. As shown in Fig. 9.16, the maximum tensile stress is located close to the air side surface, and it is proportional to the difference between overall expansion and local lattice expansion [83]. The improvement of the interfacial oxygen exchange kinetics could enlarge the maximum tensile stress, since the oxygen partial pressure difference between the two surfaces would be raised, consequently leading to the increase of difference in local lattice expansion. However, the tensile stress can be gradually released through plastic deformation if the membrane materials have high creep rate.

It is interesting to note that MIEC membranes can protect themselves through phase decomposition to decrease the interfacial oxygen exchange kinetics and through plastic deformation to release the tensile stress. BSCF is a typical example of a membrane that protects itself from breakage by forming porous layers with slow interfacial oxygen exchange kinetics on both sides and releasing the tensile stress through plastic deformation. However, if a membrane is stable and difficult to be decomposed under the membrane reaction conditions, and if the membrane material has low creep rate, then the great tensile stress could break the membrane. Obviously, $La_{0.6}Sr_{0.4}Co_{0.2}Fe_{0.8}O_{3-\delta}$ and $La_{0.8}Sr_{0.2}Co_{0.1}Fe_{0.8}Cr_{0.1}O_{3-\delta}$ are typical examples of this situation. It can be inferred from the above discussion that from the point of view of membrane configuration, the tubular shape is preferred with air flowing to the tube side and methane flowing to the shell side. Ba-based MIEC perovskite materials usually have higher creep rate than the La-based counterparts. Thus, although La-based membranes have higher structural stability than Ba-based membranes, they are still prone to breaking under syngas production conditions if their mechanical strength cannot sustain the tensile stress.

9.3.5 *MIEC Membrane Reactors for Fuel Conversion to Syngas or Hydrogen*

The conversion of hydrocarbons to hydrogen for onboard mobile devices was thought to be an attractive alternative way of supplying hydrogen for proton exchange membrane fuel cells (PEMFC). Mixed reforming of hydrocarbons and alcohols to hydrogen in MIEC membrane reactors attracted some attention in recent years. Zhu et al. explored the production of hydrogen in a BSCF MIEC membrane reactor by steam and partial oxidation of heptane [84]. A 100 % n-heptane conversion and 95–97 % hydrogen selectivity were achieved at 850 °C. The membrane reactor was steadily operated for more than 100 h, and the catalyst LiLaNiO/γ-Al_2O_3 exhibited a good activity and a favorable stability during the on-stream. In a subsequent simulation study, gasoline was used to replace heptane in a 1.4-mm-thick BSCF membrane reactor, leading to 100 % conversion, 90 % CO selectivity, 95 % H_2 selectivity, and 8.0 mL $cm^{-2}min^{-1}$ oxygen permeation flux during a 500-h period on-stream at 850 °C [85]. Coke oven gas (COG) is a by-product of the coke making industry, containing ~25 % methane, 45–60 % hydrogen, 5–9 % CO, and minor impurities such as CO_2, hydrocarbons, nitrogen, and oxygen. In China, only about 20 % of the produced COG is simply utilized as fuel, while a large amount of

COG is directly burnt at the end of an opened chimney and then directly discharged into the atmosphere, causing serious air pollution and energy dissipation. Ding and coworkers proposed the conversion of COG into hydrogen in MIEC membrane reactors. The conversion of COG to syngas was conducted in a disk-type 1.0-mm-thick $BaCo_{0.7}Fe_{0.2}Nb_{0.1}O_{3-\delta}$ membrane reactor with a NiO/MgO solid-solution catalyst [86]. During a steadily on-stream for more than 100 h at 875 °C, 95 % CH_4 conversion, 80 % H_2 selectivity, 106 % CO selectivity, and a 16.3 mL $cm^{-2}min^{-1}$ oxygen permeation flux were achieved.

9.4 Selective Oxidation of Hydrocarbons to Value-Added Products

9.4.1 Oxidation Coupling of Methane to Ethane and Ethylene

Except for the indirect route, i.e., via syngas as intermediate product, the direct conversion of methane to value-added products is technologically attractive due to its great economic benefits. Among the numerous direct conversion methods of methane, such as the selective oxidation to methanol and formaldehyde, oxygen-free aromatization to aromatic hydrocarbons, and oxygen-free coupling to ethylene, the selective oxidation coupling of methane (OCM) to ethane and ethylene (C_2) is deemed as a promising route to the direct utilization of methane for the production of important chemicals. However, the selective oxidation of methane affords products that have higher reactivity than methane; therefore the deep oxidation to CO_2 and CO is inevitable. As a result, the selectivity of the reaction toward the target products is low. The OCM route to ethane and ethylene was invented by Keller and Bhasin in the 1980s [87] and has received significant attention from both industrial and academic communities. Most of the research on this reaction process was conducted using a regular fixed-bed reactor by co-feeding oxygen and methane. Since C_2 products are much more reactive with oxygen than methane, it is difficult to improve their selectivity to a commercially target value, i.e., >85 %, at the methane conversion of >35 %. In other words, a C_2 yield of >30 % should be achieved to realize certain economic benefits for the commercialization of the OCM process. However, during the past 30 years, the highest yield achieved was only 25 % using regular fixed-bed reactors. Therefore, new reactors were proposed as potential ways to solve the problem through changing the reaction kinetics of the OCM reaction.

MIEC membrane reactors can selectively supply oxygen to the reaction system and maintain a low oxygen partial pressure; thus, deep oxidation reactions can be partially inhibited, leading to potentially improved C_2 selectivity and yields. Thus, it is evident that the oxygen input is controllable in MIEC membrane reactors. Lin and coworkers found that fluorite-type membranes showed both high oxygen permeability and catalytic activity toward the OCM reaction [88]. In a disk-type $Bi_{1.5}Y_{0.5}O_3$ membrane reactor, the C_2 selectivity was 30 % higher than that in traditional co-feeding reactors. The maximum C_2 selectivity and yield reached >90 % and 17 % at 850 °C and 900 °C, respectively. The C_2 production rate could be effectively improved by enlarging the membrane surface. Subsequently, another fluorite-type $Bi_{1.5}Y_{0.3}Sm_{0.2}O_3$ tubular membrane for the OCM reaction was prepared. Under optimal conditions, a methane conversion of 35 %, C_2 yield of 22 %, and C_2 selectivity of 62 % were obtained in the tubular membrane reactor [89]. Wang et al. found that as the BSCF tubular membrane reactor was filled with a 20 %La–10 %Sr/CaO catalyst, the methane conversion, C_2 selectivity, and C_2 yield were all greatly improved compared with the membrane reactor in the absence of catalyst [90]. Oliver et al. investigated the OCM reaction in BSCF membrane reactors with the membrane surface modified by three different catalysts and found that the membrane coated with a LaSr/CaO catalyst gave the highest C_2 yield up to 18.4 % at 950 °C [91]. Tan et al. conducted the OCM reaction in their LSCF hollow fiber membrane reactor without filling any catalyst [92]. The LSCF powder was found to have no catalytic activity toward the OCM reaction, but it did show high selectivity up to 71.9 % in the LSCF hollow fiber membrane reactor. If the lumen side of the fiber membrane was piled with an $SrTi_{0.9}Li_{0.1}O_3$ catalyst, the methane conversion and oxygen permeation flux were remarkably increased, together with a decrease of C_2 selectivity. At optimal temperatures, a maximum C_2 yield close to 21 % could be achieved in the packed hollow fiber membrane reactor.

Although many efforts were made to develop MIEC membrane reactors for the OCM reaction, the methane conversion, C_2 selectivity, and C_2 yield are still lower than the values obtained in a fixed-bed reactor with optimal catalysts. To further enhance the C_2 yield in MIEC membrane reactors, efforts should focus on the reaction kinetics of the OCM process and membrane surface modification to match the oxygen permeation and OCM reaction kinetics.

9.4.2 Oxidation Dehydrogenation of Light Alkanes

The conversion of light alkanes to olefins is becoming profitable in view of the depletion of petroleum resources in the future. Stream cracking of ethane to ethylene and catalytic dehydrogenation of propane and butanes to propene and butenes, respectively, have been commercialized for more than 20 years. However, huge energy consumption and coke deposit induced catalyst deactivation during these processes are difficult issues to solve. Selective oxidative dehydrogenation is another potential economical route for the conversion of light alkanes to olefins that avoids an external energy input into the reactor and coke deposit induced catalyst deactivation. One of challenges of this process is how to achieve high olefin selectivity, because the olefin products are more reactive than the fed alkanes, i.e., alkanes are easier to be deeply oxidized to CO_x by O_2. One solution to inhibit deep oxidation reactions is to use lattice oxygen (O^{2-}) according to a Mars–van Krevelen type mechanism other than gaseous oxygen (O_2) as oxidant in a periodic shift reactor. High olefin selectivity >90 % can be easily obtained using this type of reactor.

An MIEC membrane can convert gaseous oxygen into lattice oxygen, but the lattice oxygen is going to reunite to the gaseous oxygen as it arrives at the other side of the MIEC membrane surface. If the catalytic oxidation reaction kinetics are fast enough, i.e., the alkane molecules react faster with the lattice oxygen than the reunion of lattice to gaseous oxygen, the MIEC membrane could continuously supply lattice oxygen to react with alkanes. Thus, it can be expected that high olefin selectivity can be achieved in MIEC membrane reactors. A BSCF disk-type membrane reactor without filling any catalyst was employed for the selective oxidation of ethane to ethylene [93]. During a 100 h on-stream at 800 °C, 80 % ethylene selectivity and a per pass ethylene yield of 67 % were obtained. However, 53.7 % ethylene selectivity was achieved under the same conditions in a fixed-bed reactor by co-feeding oxygen and ethane. When the BSCF membrane surface was modified by Pd or V/MgO catalyst particles for the oxidation reaction, much higher ethylene yields up to 76 % were obtained at 777 °C [94]. Considering the strong thermal effect of the oxidation reaction, single run conversions are usually controlled to be lower than 50 % to avoid temperature runaway. As shown in Fig. 9.17, if the ethane conversion was maintained lower than 50 %, high ethylene selectivity even up to 95 % could be achieved. The above results verified the concept on the MIEC membrane continuous supplying lattice oxygen and revealed the feasibility for practical applications.

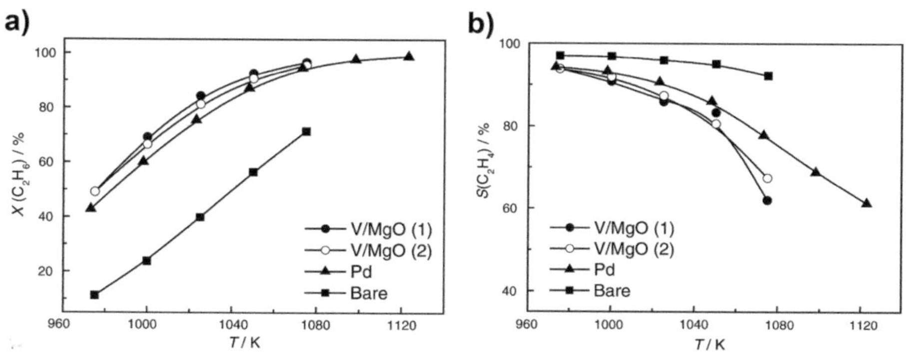

Fig. 9.17 Ethane conversion (**a**) and ethylene selectivity (**b**) as a function of temperature for a bare (■), Pd nanocluster modified (▲), and V/MgO micron grain modified (●, ○) membrane. In the case of V/MgO, the experiment was carried a second time to evaluate the reproducibility, denoting V/MgO (1) and V/MgO (2), respectively. The air side compartment was fed with 50 ml min^{-1} at total pressure of 1.2 bar; the reaction side was supplied with 37 ml $min^{-1}C_2H_6$ in He at C_2H_6 partial pressure of 0.25 bar (Reproduced from Ref. [94] with permission of Elsevier)

9.4.3 Other Reactions

In the industry, platinum–rhodium alloy gauzes (5–10 % Rh) are used as catalysts for ammonia oxidation to nitric acid at high temperature with a NO yield of 94–96 %. N_2 and N_2O are the main by-products of this important industrial process. This process has been well optimized in the past years; however, it is still necessary to develop new economical and environmentally friendly processes to avoid precious metal loss during the reaction and reduce N_2O emissions. MIEC perovskite oxides are good catalysts for ammonia oxidation to NO with high NO and low N_2O selectivity. Of course, the employment of MIEC membrane reactors was suggested in the ammonia oxidation reaction. Perez–Ramirez et al. used a lanthanum ferrite-based perovskite membrane reactor for the ammonia oxidation reaction and achieved high NO selectivity (98 %) and no N_2O formation [95]. Therefore, the perovskite membrane reactor for ammonia oxidation provides a new path for the formation of NO due to its high efficiency and minimal environmental pollution. Particularly, the membrane reactor integrates the separation and reaction process in a single reactive separation unit.

Methane dehydroaromatization has attracted considerable attention from both academia and industry because of its potential economic benefits. An MIEC membrane reactor was employed for the reaction aiming at shifting the reaction to the right by using the permeated oxygen to consume the resultant hydrogen. The production of steam by hydrogen combustion suppresses coke formation and extends the lifetime of the catalytic system. After 1000 min on-stream, the methane conversion and aromatic selectivity were stable at 10 and 30 %, respectively, for the membrane reactor; however, under the same reaction conditions, the two performances were 7 % and 6 %, respectively, for the traditional fixed-bed reactor [96].

9.5 Selective Removal of Oxygen from the Reaction System: Water Splitting for Hydrogen Production

Water is considered to be an ideal source of hydrogen because it is clean, abundant, and renewable. The ΔG for the water-splitting reaction reaches zero at temperatures as high as 4057 °C under a pressure of 1 bar. The water-splitting reaction is a typical equilibrium limited reaction, and its conversion under the equilibrium state is only 0.1 % even at high temperatures up to 1600 °C. Although this reaction is conceptually simple, efficient hydrogen production by water splitting is difficult to be realized due to the low equilibrium constant (i.e., $K_p \approx 2 \times 10^{-8}$ at 950 °C). However, meaningful amounts of hydrogen can be generated at moderate temperatures if the equilibrium is shifted toward water splitting by using an oxygen transport membrane to remove oxygen from the reaction system.

In the MIEC membrane reactors introduced above in Sects. 9.3 and 9.4, the role of the MIEC membrane is to control the input of oxygen. Of course, the MIEC

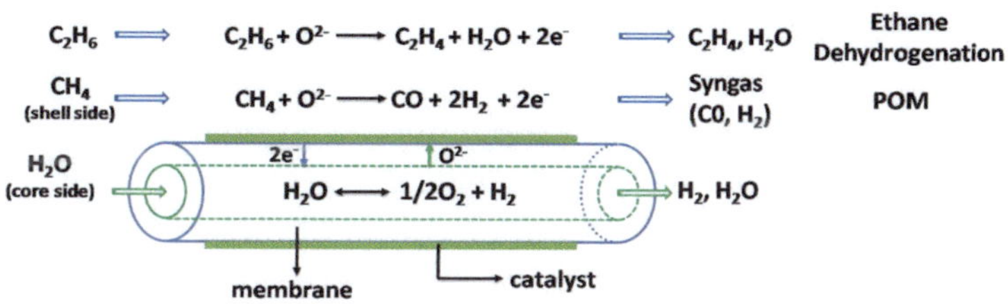

Fig. 9.18 Schematic diagram of water splitting coupled with other catalytic oxidation reactions in MIEC membrane reactors

membrane can also be used to remove oxygen from a reaction system and shift the reaction equilibrium to the right. Following the concept of process intensification, researches were conducted on coupling water splitting with other catalytic oxidation reactions in MIEC membrane reactors. The acquirement of other important chemicals makes the coupling in membrane reactors more meaningful in terms of energy saving and process intensification. Up to now, the couplings between the water-splitting reaction and the partial oxidation of methane, ethane dehydrogenation, and oxidative steam reforming of ethanol have been investigated. Figure 9.18 schematically shows the concept of water splitting coupled with other catalytic oxidation reactions in MIEC membrane reactors. One side of the membrane (defined as the hydrogen-production side) is exposed to a steam atmosphere where water splits into hydrogen and oxygen at high temperature, while the other side (defined as the oxygen-permeation side) is swept by reductive gases (such as methane, ethane, etc.) that react with the oxygen permeated from the hydrogen-production side to produce targeted chemicals and reduce the oxygen partial pressure. As a result, an oxygen chemical potential gradient is established across the membrane to drive oxygen ions that are continuously transported from the hydrogen-production side to the oxygen-permeation side. Theoretically, 100 % pure hydrogen can be obtained from the hydrogen-production side of the membrane after cooling and drying. As methane is fed to the shell side, the permeated oxygen from the core side to the shell side is consumed by the reaction $CH_4 + 1/2O_2 \rightarrow CO + 2H_2$ [3]. The value of this process is not only to produce pure hydrogen but also synthesis gas that can be used to generate a wide variety of valuable hydrocarbons and oxygenates. The idea of using methane as reductive gas on the oxygen-permeation side was proposed by Balachandran et al. in their study on a GDC–Ni dual-phase membrane for hydrogen production via water splitting [97], although the products in the oxygen-permeation side were not detected. Wang and Caro et al. systematically studied the coupling reaction in a $BaZr_xCo_yFe_zO_{3-\delta}$ $(x + y + z = 1)$ hollow fiber membrane reactor [3].

The assets of the coupling effects are listed as follows:

1. Pure H_2 devoid of CO produced through fossil fuels in one step
2. High value-added chemicals produced with high selectivity
3. Reaction–reaction coupling allowing high process intensification
4. No safety risk during the operation of the membrane reactor for oxidation reactions

9.5.1 Membrane Materials

As shown in Chaps. 6 and 7, although many types of materials could be fabricated into membranes for oxygen permeation up to now, few of them are suitable to be used as membrane reactors for hydrogen production through water splitting due to the harsh environment containing steam and reducing gases. In an early report, Naito and Arashi conducted a study on hydrogen production through water splitting in an MIEC ZrO_2 - TiO_2 - Y_2O_3 membrane reactor. A hydrogen production rate of about $0.6\ mL\ cm^{-2}min^{-1}$ was achieved at high temperatures up to 1683 °C [98]. The hydrogen production rate is mainly limited by the low electronic conductivity of the membrane material. Recently, to lower the operation temperature and improve the hydrogen production rate, some new membranes have been used to produce hydrogen from water splitting. These membrane materials can be classified into two groups: (1) perovskite-related membranes and (2) dual-phase membranes made of metal–fluorite or perovskite–fluorite composite materials.

9.5.1.1 Perovskite-Related Membranes

Balachandran et al. employed $La_{0.7}Sr_{0.3}Cu_{0.2}Fe_{0.8}O_{3-\delta}$ (LSCF7328) perovskite membranes in MIEC membrane reactors for hydrogen production from water splitting [99, 100]. A LSCF7328 asymmetric membrane was fabricated with a 22-μm-thick dense layer and a porous substrate by using the same material to prevent the mismatch of thermal expansion or contraction during either sintering or measurement. A hydrogen production rate of 4.7 $mL\ cm^{-2}min^{-1}$ was obtained in the asymmetric membrane reactor at 900 °C, with 49 vol% water vapor fed to the hydrogen-production side and 99.5 % CO fed to the oxygen-permeation side. Besides, other perovskite materials, such as $BaZr_xCo_yFe_zO_{3-\delta}(x + y + z = 1)$ [3] and $SrCo_{0.4}Fe_{0.5}Zr_{0.1}O_{3-\delta}$ [101], were also explored for hydrogen production through water splitting.

There is a need to develop cobalt (or copper)-free perovskite materials since perovskite materials containing either copper or cobalt do not possess enough stability under a reducing atmosphere. The $BaFe_{0.9}Zr_{0.1}O_{3-\delta}$ (BFZ) material was investigated by Balachandran et al. [102] as a potential Co (or Cu)-free perovskite material for hydrogen production through water splitting. A 1.56-mm-thick BFZ disk-type membrane gave a hydrogen production rate of $3.2\ mL\ cm^{-2}min^{-1}$ at 900 °C with 49 vol% water vapor fed to the hydrogen-production side and 99.5 % CO fed to the oxygen-permeation side. In order to achieve a high hydrogen production, tubular membranes were fabricated to get a larger membrane area than that of disk-type membranes. A value up to 22 $mL\ min^{-1}$ was achieved on a BFZ tubular membrane with dimensions OD = 9.6 mm, ID = 7.5 mm, and L = 57.4 mm [102]. However, the area-specific hydrogen production rates of the disks were more than double those of the tubes. It was thought that the temperature and average oxygen partial pressure gradient across the tubular membrane were lower than those for disks under nominally identical external conditions, resulting in a lower hydrogen production rate for the tubular membrane than it was expected at 900 °C.

Owing to its high chemical stability and mechanical integrity, a perovskite-related intergrowth oxide composed of $SrFeCo_{0.5}O_x$ was used in an MIEC membrane reactor for hydrogen production from water splitting [103]. This material was reported to have high ambipolar conductivity and stability when it was used to construct a membrane reactor for the partial oxidation of methane to syngas. A ~20-μm-thick $SrFeCo_{0.5}O_x$ thin film on a porous substrate was fabricated by a colloidal casting method. The largest hydrogen production rate of 17.4 mL $cm^{-2}min^{-1}$ was obtained at 900 °C on the $SrFeCo_{0.5}O_x$ asymmetric membrane. However, a phase transition limited the application of SFC as an MIEC membrane at temperatures lower than 825 °C [103].

9.5.1.2 Dual-Phase Membranes

To increase the electronic conductivity of solid oxygen ionic conducting oxides, metals were used as the electron-conducting phase to construct dual-phase membranes. Metallic Ni has good electronic conductivity and excellent catalytic activity toward the water-splitting reaction. Ceria-based solid oxides show high oxygen ionic conductivities and high stability under low oxygen partial pressure atmosphere. Taking all these factors into consideration, a fluorite–metal dual-phase membrane made of Ni and gadolina-doped ceria (GDC–Ni, Ni 40 vol% in the dual-phase membrane) was prepared [97]. A hydrogen production rate of 4.0 mL $cm^{-2}min^{-1}$ was achieved on a 130-μm -thick GDC–Ni disk-type dual-phase membrane at the temperature of 900 °C. This measurement was accomplished by flowing humidified nitrogen (water partial pressure was 0.49 atm) over the hydrogen-production side of the disk and 80 % H_2/balance He over the oxygen-permeation side. However, metallic Ni easily migrates to the membrane surface under a high temperature steam, which leads to a significant change in the microstructure of the dual-phase membrane and affects the stability of hydrogen production.

In view of the commercialization of membrane technology, stability is more important than permeability. In order to enhance the stability of MIEC membranes, fluorite–perovskite dual-phase membranes were developed. A fluorite–perovskite dual-phase membrane containing 60 vol% $Gd_{0.2}Ce_{0.8}O_{2-\delta}$ (GDC) and 40 vol% $Gd_{0.08}Sr_{0.88}Ti_{0.95}Al_{0.05}O_{3\pm\delta}$ (GSTA) was developed for hydrogen production from water splitting [104]. For this membrane, GDC acts as an oxygen ionic conductor, while GSTA acts as an n-type electronic conductor with negligible oxygen ionic conductivity under operating conditions. Asymmetric GDC–GSTA membranes with different dense layer thicknesses were prepared by the spin-coating method. An asymmetric GDC–GSTA membrane with a dense layer of 25 μm in thickness gave a hydrogen production rate of 9.67 mL $cm^{-2}min^{-1}$ at 900 °C. The hydrogen production rate could be further enhanced by surface modification.

9.5.2 Stability Under High Oxygen Partial Pressure Gradient

Unlike the traditional MIEC membrane reactors shown in Sect. 9.3 of this chapter that operate under an oxidizing environment on one side and a reducing one on the other side, for the hydrogen production process from water splitting, both sides of the MIEC membranes are under reductive environments. Therefore, most MIEC membranes are unstable under the harsh operation conditions for hydrogen production from water splitting. Up to now, there was no report about the long-term stability of hydrogen production from water splitting in an MIEC membrane reactor. It was stated that a 22-μm-thick LSCF7328 asymmetric membrane was stable during a ~80-h testing for hydrogen production by measuring the hydrogen production rate at the beginning and end of the test [105]. Rafael et al. studied the stability of a $La_{0.6}Sr_{0.4}Co_{0.2}Fe_{0.8}O_{3-\delta}$ microtubular membrane for hydrogen production by flowing methane on the oxygen-permeation side [106]. Traces of methane were detected in the outlet of the hydrogen-production side, indicating a gas leakage through microcracks in the membrane. The amount of methane leaked began to increase after 120 h and abruptly increased after 260 h. This result reveals the failure of the membrane caused by axial stress due to the temperature gradient along the length of the membrane and chemical stress due to oxidation and reduction of the perovskite membrane. Besides, the high water partial pressure on the hydrogen-production side damaged the surface structure of the membrane. Compared with fresh surfaces, the surface on the oxygen-permeation side was decorated with small crystallites after the long-term operation, and the spot energy dispersive X-ray spectroscopy analysis of the crystallites demonstrated that they were enriched with cobalt. On the other hand, SEM analysis of the hydrogen-production side surface showed the appearance of needlelike crystals that were also observed in the cross section. These phenomena indicated that both surfaces of the membrane were seriously damaged by the atmosphere of hydrogen produced from water splitting and methane conversion, respectively.

However, unlike the MIEC membrane reactor for partial oxidation of methane to syngas, as both sides of MIEC membranes used to produce hydrogen are exposed to a reductive atmosphere, it is evident that the oxygen partial pressure difference induced tensile stress will be smaller for the latter than for the former. Thus, it is desirable to find stable membranes for hydrogen production because only the stability under reductive atmosphere is required, while the membrane stability under both oxidative and reductive atmosphere is simultaneously required for partial oxidation of methane to syngas.

9.5.3 *Production of Ammonia and Liquid Fuel Synthesis Gas in One Membrane Reactor*

Ammonia synthesized via the Haber–Bosch process is the second most produced chemical in the world and is used to produce fertilizers to increase food production. Natural gas is usually used as feedstock to produce ammonia synthesis gas (ASG, $H_2/N_2 = 3$). This process is accompanied by large atmospheric CO_2 emissions, i.e., ~400 million metric tons annually. In a typical ASG production process, six main steps are required to convert cleaned natural gas to ASG, i.e., steam reforming, air reforming, water–gas shifting at high and low temperatures, CO_2 separation, and methanation. This long route technology for ASG production requires expensive capital investment and consumes significant energy. In the whole ammonia synthesis process, the ASG production contributes to ~84 % of energy consumption [107]; therefore great innovation in ASG production technology is imperative to reduce energy consumption in the ammonia synthesis industry. On the other hand, natural gas is a typical feedstock to produce liquid fuel synthesis gas (LFSG, $H_2/CO = 2$) in three main steps, i.e., steam reforming, air separation, and mixed reforming. LFSG production contributes to most of the costs and energy consumption in the entire gas-to-liquid process. A large portion of energy is consumed by air separation for pure oxygen production.

Recently, we presented a membrane reactor with an MIEC membrane and catalyst to coproduce ASG and LFSG from water, air, and methane in one step [108]. Figure 9.19 shows the concept of this membrane reactor. Air and water with

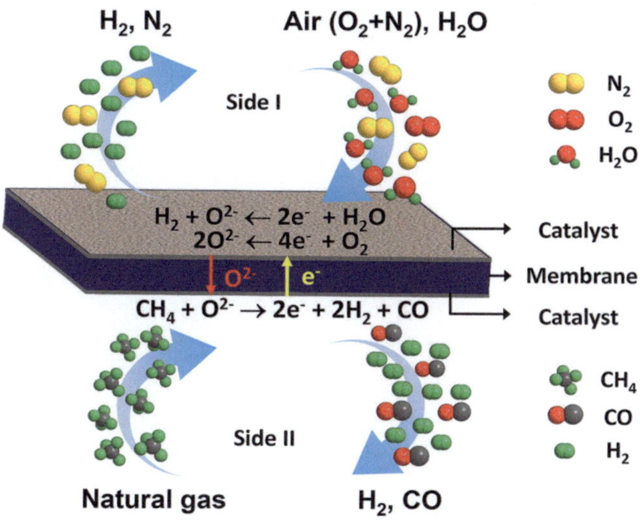

Fig. 9.19 Concept of the one-step production of the two synthesis gases for ammonia ($H_2/N_2 = 3:1$) and liquid fuel ($H_2/CO = 2:1$) from water, air, and natural gas in a membrane reactor (Reproduced from Ref. [108] with permission of Wiley)

an appropriate ratio are fed to one side (side I) of the membrane, while methane is fed to the other side (side II). At high temperatures, oxygen from air and water on side I permeates through the MIEC membrane via diffusion of oxygen ions to side II, where the ions react with methane to form LFSG; meanwhile electrons migrate from side II to side I to keep the electric neutrality of the whole process. The LFSG has extremely low oxygen partial pressure (e.g., 10^{-21} atm) on side II, which drives the permeation of oxygen from side I to side II. In this process, the amount of oxygen in the air is limited in the feed; thus, oxygen molecules can completely permeate to side II. Thermodynamic calculations show that the equilibrium conversion of the water-splitting reaction can be as high as 96 % at 900 °C if oxygen is rapidly removed by the POM reaction via the MIEC membrane. Therefore, the resultant gases on side I with a suitable H_2/N_2 ratio are ready for ammonia synthesis after drying; simultaneously, the produced LFSG on side II with a suitable H_2/CO ratio is ready for liquid fuel synthesis. The benefits of this membrane reactor are summarized in Fig. 9.20. The most important advantage is to shorten the six-step process (steam reforming, air reforming, water–gas shifting at high and low temperatures, CO_2 separation, and methanation) for ASG production and three-step process (steam reforming, oxygen separation, and mixed reforming) for LFSG production to a one-step process. The highly integrated membrane reactor allows much higher energy efficiency (~63 % energy saving) than the traditional complex processes. The concept was realized in a $Ba_{0.98}Ce_{0.05}Fe_{0.95}O_{3-\delta}$ (BCF) membrane reactor, and the operation conditions, i.e., methane concentration, methane flow rate, temperature, and steam flow rate, were investigated in detail. Figure 9.21 shows the time-dependent performance of the catalytic membrane reactor at 900 °C

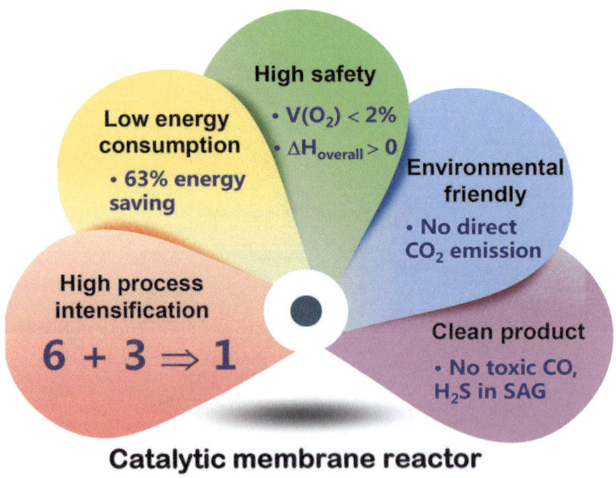

Fig. 9.20 Merits of the membrane reactor for the production of ASG and LFSG compared with industrial processes. "6" represents a six-step process for the industrial production of ASG; "3" represents a three-step process for the industrial production of LFSG; "1" represents the one-step process for the coproduction of ASG and LFSG in the membrane reactor (Reproduced from Ref. [108] with permission of Wiley)

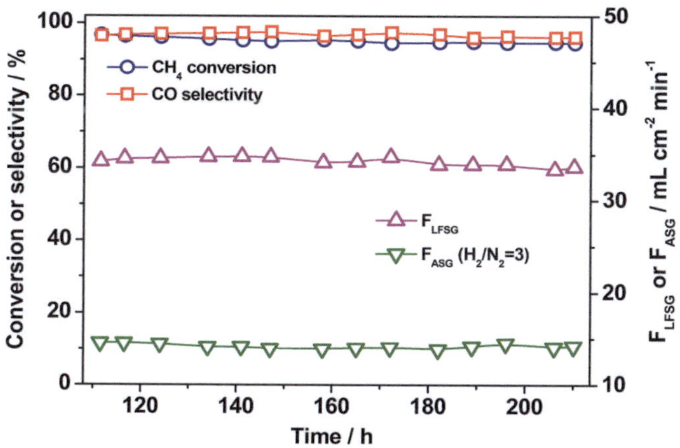

Fig. 9.21 Stable operation of the membrane reactor for the production of ASG and LFSG at 900 °C. H_2O steam: 200 mL min^{-1}; CH_4 : 9.0 mL min^{-1} (Reproduced from Ref. [108] with permission of Wiley)

for the two syngases production. CH_4 conversion and CO selectivity were high (94–98 %), and the ASG ($H_2/N_2 = 3$) and LFSG ($H_2/CO = 2$) production rates remained at ~14 and ~34 mL $cm^{-2}min^{-1}$, respectively. These results indicate that the membrane reactor can be steadily operated to high efficiently coproduce ASG and LFSG. For a factory with a production capability of 0.3 million tons of ammonia and one million tons of methanol per year, it is enough to construct a membrane reactor with the membrane area of 10,000 m^2, which corresponds to a membrane reactor volume of just 100 m^3. Hence, it is evident that the catalytic membrane reactor delivers a new concept on process intensification and the success of this new concept may trigger innovation in traditional natural gas chemical industries.

References

1. Gryaznov VM (1963) Mechanism of the catalytic redistribution of hydrogen in unsaturated cyclic hydrocarbons. Russ Chem Rev 32:188–200
2. Wang HH, Cong Y, Yang WS (2002) High selectivity of oxidative dehydrogenation of ethane to ethylene in an oxygen permeable membrane reactor. Chem Commun 14:1468–1469
3. Jiang HQ, Wang HH, Werth S, Schiestel T, Caro J (2008) Simultaneous production of hydrogen and synthesis gas by combining water splitting with partial oxidation of methane in a hollow-fiber membrane reactor. Angew Chem Int Ed 47:9341–9344
4. Bredesen R, Sogge J (1996) Paper presented at the United Nations Applications of Innovative Membrane Technology in Chemical Industry. Chem/Sem 21/R.12. Cetaro, Calabria, Italy, 1–4 May 1996
5. Balachandran U, Kleefisch MS, Kobylinski TP, Morissette SL, Pei S, Poeppel RB, Udovich CA (1994) Mixed crystalline metal oxide contains at least strontium, cobalt, iron and opt

lanthanide metal – has non-perovskite structure for use as dense ceramic membrane to transport oxygen in partial oxidation reactor. Patent: WO9424065-A
6. Balachandran U, Dusek JT, Sweeney SM, Poeppel RB, Mieville RL, Maiya PS, KleefischMS PS, Kobylinski TP, Udovich CA, Bose AC (1995) Methane to syngas via ceramic membrane. Am Ceram Soc Bull 74:71–75
7. Balachandran U, Dusek JT, Maiya PS, Ma B, Mieville RL, Kleefisch MS, Udovich CA (1997) Ceramic membrane reactor for converting methane to syngas. Catal Today 36:265–272
8. Pie S, Kleefisch MS, Kobylinski TP, Faber J, Udovich CA, Zhang-McCoy V, Dabrowski B, Balachandran U, Mieville RL, Peoppel RB (1995) Failure mechanisms of ceramic membrane reactors in partial oxidation of methane to synthesis gas. Catal Lett 30:201–212
9. Tsai CY, Dixon AG, Moser WR, Ma YH (1997) Dense perovskite membrane reactor for partial oxidation of methane to syngas. AIChE J 43:2741–2748
10. Shao ZP, Dong H, Xiong GX, Cong Y, Yang WS (2001) Performance of a mixed-conducting ceramic membrane reactor with high oxygen permeability for methane conversion. J Membr Sci 183:181–192
11. Wu ZT, Jin WQ, Xu NP (2006) Oxygen permeability and stability of Al_2O_3-doped $SrCo_{0.8}Fe_{0.2}O_{3-\delta}$ mixed conducting oxides. J Membr Sci 279:320–327
12. Kharton VV, Sobyanin VA, Belyaev VD, Semin GL, Veniaminov SA, Tsipis EV, Yaremchenko AA, Valente AA, Marozau IP, Frade JR, Rocha J (2004) Methane oxidation on the surface of mixed-conducting $La_{0.3}Sr_{0.7}Co_{0.8}Ga_{0.2}O_{3-\delta}$. Catal Commun 5:311–316
13. Bayraktar D, Clemens F, Diethelm S, Graule T, Van Herle J, Holtappels P (2006) Production and properties of substituted $LaFeO_3$-perovskite tubular membranes for partial oxidation of methane to syngas. J Eur Ceram Soc 27:2455–2461
14. Tong JH, Yang WS, Cai R, Zhu BC, Lin LW (2002) Novel and ideal zirconium-based dense membrane reactors for partial oxidation of methane to syngas. Catal Lett 78:129–137
15. Wang HH, Tablet C, Schiestel T, Werth S, Caro J (2006) Partial oxidation of methane to syngas in a perovskite hollow fiber membrane reactor. Catal Commun 7:907–912
16. Li QM, Zhu XF, He YF, Yang WS (2010) Partial oxidation of methane in $BaCe_{0.1}Co_{0.4}Fe_{0.5}O_{3-\delta}$ membrane reactor. Catal Today 149:185–190
17. Makoto H, Kazunari D, Michhikazu H, Takashi T (2006) $Ba_{1.0}Co_{0.7}Fe_{0.2}Nb_{0.1}O_{3-\delta}$ Dense ceramic as an oxygen permeable membrane for partial oxidation of methane to synthesis gas. Chem Lett 35:1326–1327
18. Luo HX, Wei YY, Jiang HQ, Yuan WH, Lv YX, Caro J, Wang HH (2010) Performance of a ceramic membrane reactor with high oxygen flux Ta-containing perovskite for the partial oxidation of methane to syngas. J Membr Sci 350:154–160
19. Yang L, Tan L, Gu XH, Jin WQ, Zhang LX, Xu NP (2003) A new series of Sr(Co, Fe, Zr) $O_{3-\delta}$ perovskite-type membrane materials for oxygen permeation. Ind Eng Chem Res 42:2299–2305
20. Dong XL, Liu ZK, He YJ, Jin WQ, Xu NP (2009) $SrAl_2O_4$-improved $SrCo_{0.8}Fe_{0.2}O_{3-\delta}$ mixed-conducting membrane for effective production of hydrogen from methane. J Membr Sci 331:109–116
21. Ikeguchi M, Mimura T, Sekine Y, Kikuchi E, Matsukata M (2005) Reaction and oxygen permeation studies in $Sm_{0.4}Ba_{0.6}Co_{0.2}Fe_{0.8}O_{3-\delta}$ membrane reactor for partial oxidation of methane to syngas. Appl Catal A Gen 290:212–220
22. Diethelm S, SfeirJ CF, Van Herle J, Favrat D (2004) Planar and tubular perovskite-type membrane reactors for the partial oxidation of methane to syngas. J Solid State Electrochem 8:611–617
23. Ishihara T, Yamada T, Arikawa H, Nishiguchi H, Takita Y (2000) Mixed electronic–oxide ionic conductivity and oxygen permeating property of Fe-, Co- or Ni-doped $LaGaO_3$ perovskite oxide. Solid State Ionics 135:631–636
24. Ishihara T, Tsuruta Y, Todaka T, Nishiguchi H, Takita Y (2002) Fe doped $LaGaO_3$perovskite oxide as an oxygen separating membrane for CH_4 partial oxidation. Solid State Ionics 152:709–714

25. Viveta A, Geffroy PM, Chartier T, Del Gallo P, Richet N (2011) $La_{(1-x)}Sr_xFe_{(1-y)}Ga_yO_{3-\delta}$ perovskite membrane: oxygen semi-permeation, thermal expansion coefficient and chemical stability under reducing conditions. J Membr Sci 372:373–379
26. Ritchie JT, Richardson JT, Luss D (2001) Ceramic membrane reactor for synthesis gas production. AIChE J 47:2092–2101
27. Yamaji K, Negishi H, Horita T, Sakai N, Yokokawa H (2000) Vaporization process of Ga from doped $LaGaO_3$ electrolytes in reducing atmospheres. Solid State Ionics 135:389–396
28. Rakshit SK, Parida SC, Singh Z, Prasad R, Venugopal V (2004) Thermodynamic properties of ternary oxides in the system Ba-Fe-O using solid-state electrochemical cells with oxide and fluoride ion conducting electrolytes. J Solid State Chem 177:1146–1156
29. Zhu XF, Wang HH, Cong Y, Yang WS (2006) Structure and oxygen permeation of cerium light doped $BaFeO_{3-\delta}$ ceramic membranes. Solid State Ionics 117:2917–2921
30. Watanabe K, Takauchi D, Yuasa M, Kida T, Shimanoe K, Teraoka Y, Yamazoe N (2009) Oxygen permeation properties of Co-free perovskite-type oxide membranes based on $BaFe_{1-y}Zr_yO_3$. J Electrochem Soc 156:E81–E85
31. Liu XT, Zhao HL, Yang JY, Li Y, Chen T, Lu XG, Ding WZ, Li FS (2011) Lattice characteristics, structure stability and oxygen permeability of $BaFe_{1-x}Y_xO_{3-\delta}$ ceramic membranes. J Membr Sci 383:235–240
32. Watenabe K, Yuasa M, Kida T, Teraoka Y, Yamazoe N, Shimanoe K (2010) High-performance oxygen-permeable membranes with an asymmetric structure using $Ba_{0.95}La_{0.05}FeO_{3-\delta}$perovskite-type oxide. Adv Mater 22:2367–2370
33. Zhu XF, Wang HH, Cong Y, Yang WS (2006) Partial oxidation of methane to syngas in $BaCe_{0.15}Fe_{0.85}O_{3-\delta}$ membrane reactors. Catal Lett 111:179–185
34. Gong ZL, Hong L (2011) Integration of air separation and partial oxidation of methane in the $La_{0.4}Ba_{0.6}Fe_{0.8}Zn_{0.2}O_{3-\delta}$membrane reactor. J Membr Sci 380:81–86
35. Kharton VV, Shaula AL, Snijkers FMM, Cooymans JFC, Luyten JJ, Yaremchenko AA, Valente AA, Tsipis EV, Frade JR, Marques FMB, Rocha J (2005) Processing, stability and oxygen permeability of Sr(Fe, Al)O_3-based ceramic membranes. J Membr Sci 252:215–225
36. Jin WQ, Li SG, Huang P, Xu NP, Shi J, Lin YS (2000) Tubular lanthanum cobaltite perovskite-type membrane reactors for partial oxidation of methane to syngas. J Membr Sci 166:13–22
37. Hamakawa S, Hayakawa T, Suzuki K, Murata K, Takehira K, Yoshino S, Nakamura J, Uchijima T (2000) Methane conversion into synthesis gas using an electrochemical membrane reactor. Solid State Ionics 136–137:761–766
38. Wang HH, Cong Y, Yang WS (2003) Investigation on the partial oxidation of methane to syngas in a tubular $Ba_{0.5}Sr_{0.5}Co_{0.8}Fe_{0.2}O_{3-\delta}$ membrane reactor. Catal Today 82:157–166
39. Kharton VV, Yaremchenko AA, Valente AA, Sobyanin VA, Belyaev VD, Semin GL, Veniaminov SA, Tsipis EV, Shaula AL, Frade JR, Rocha J (2005) Methane oxidation over Fe-, Co-, Ni- and V-containing mixed conductors. Solid State Ionics 176:781–791
40. Lu H, Tong JH, Cong Y, Yang WS (2005) Partial oxidation of methane in $Ba_{0.5}Sr_{0.5}Co_{0.8}Fe_{0.2}O_{3-\delta}$ membrane reactor at high pressures. Catal Today 104:154–159
41. Hoang DL, Chan SH (2006) Effect of reactor dimensions on the performance of an O_2 pump integrated partial oxidation reformer – a modelling approach. Int J Hydrog Energy 31:1–12
42. Delbos C, Lebain G, Richet N, Bertail C (2010) Performances of tubular mixed conducting $La_{0.8}Sr_{0.2}Fe_{0.7}Ga_{0.3}O_{3-\delta}$ membrane reactor for under pressure methane conversion to syngas. Catal Today 156:146–152
43. Zhang YW, Liu J, Ding WZ, Lu XG (2011) Performance of an oxygen-permeable membrane reactor for partial oxidation of methane in coke oven gas to syngas. Fuel 90:324–330
44. Babakhani EG, Towfighi J, Taheri Z, Pour AN, Zekordi M, Taheri A (2012) Partial oxidation of methane in $Ba_{0.5}Sr_{0.5}Co_{0.8}Fe_{0.1}Ni_{0.1}O_{3-\delta}$ceramic membrane reactor. J Nat Gas Chem 21:519–525
45. Araki S, Yamamoto H, Hoshi Y, Lu JF, Hakuta Y, Hayashi H, Ohashi T, Sato K, Nishioka M, Inoue T, Hikazudani S, Hamakawa S (2012) Synthesis of $Ca_{0.8}Sr_{0.2}Ti_{0.7}Fe_{0.3}O_{3-\delta}$ thin film

membranes and its application to the partial oxidation of methane. Solid State Ionics 221:43–49
46. Song SD, Zhang P, Han MF, Singhal SC (2012) Oxygen permeation and partial oxidation of methane reaction in $Ba_{0.9}Co_{0.7}Fe_{0.2}Nb_{0.1}O_{3-\delta}$oxygen permeation membrane. J Membr Sci 415–416:654–662
47. Zhang YW, Su K, Zeng FL, Ding WZ, Lu XG (2013) A novel tubular oxygen-permeable membrane reactor for partial oxidation of CH_4 in coke oven gas to syngas. Int J Hydrog Energy 38:8783–8789
48. Kathiraser Y, Kawi S (2013) $La_{0.6}Sr_{0.4}Co_{0.8}Ga_{0.2}O_{3-\delta}$(LSCG) hollow fiber membrane reactor: partial oxidation of methane at medium temperature. AIChE J 59:3874–3885
49. Liao Q, Chen Y, Wei YY, Zhou LY, Wang HH (2014) Performance of U-shaped $BaCo_{0.7}Fe_{0.2}Ta_{0.1}O_{3-\delta}$ hollow-fiber membranes reactor with high oxygen permeation for methane conversion. Chem Eng J 237:146–152
50. Zhu JW, Guo SB, Liu GP, Liu ZK, Zhang ZC, Jin WQ (2015) A robust mixed-conducting multichannel hollow fiber membrane reactor. AIChE J 61:2592–2599
51. Doorn VRHE, Bouwmeester HJM, Burggraaf AJ (1998) Kinetic decomposition of $La_{0.3}Sr_{0.7}CoO_{3-\delta}$ perovskite membranes during oxygen permeation. Solid State Ionics 111:263–272
52. Palcut M, Wiik K, Grande T (2007) Cation self-diffusion in $LaCoO_3$ and La_2CoO_4 studied by diffusion couple experiments. J Phys Chem B 111:2299–2308
53. Smith JB, Norby T (2006) Cation self-diffusion in $LaFeO_3$ measured by the solid state reaction method. Solid State Ionics 177:639–646
54. Palcut M, Christensen JS, Wiik K, Grande T (2008) Impurity diffusion of ^{141}Pr in $LaMnO_3$, $LaCoO_3$ and $LaFeO_3$ materials. Phys Chem Chem Phys 10:6544–6552
55. Kishimoto H, Sakai N, Horita T, Yamaji K, Brito ME, Yokokawa H (2007) Cation transport behavior in SOFC cathode materials of $La_{0.8}Sr_{0.2}CoO_3$ and $La_{0.8}Sr_{0.2}FeO_3$ with perovskite structure. Solid State Ionics 178:1317–1325
56. Harvey SP, De Souza RA, Martin M (2012) Diffusion of La and Mn in $Ba_{0.5}Sr_{0.5}Co_{0.8}Fe_{0.2}O_{3-\delta}$ polycrystalline ceramics. Energy Environ Sci 5:5803–5813
57. Kubicek M, Rupp GM, Huber S, Penn A, Opitz AK, Bernardi J, Stöger-Pollach M, Huttera H, Fleig J (2014) Cation diffusion in $La_{0.6}Sr_{0.4}CoO_{3-\delta}$ below 800 °C and its relevance for Sr segregation. Phys Chem Chem Phys 16:2715–2726
58. Chen CS, Feng SJ, Ran S, Zhu DC, Liu W, Bouwmeester HJM (2003) Conversion of methane to syngas by a membrane-based oxidation-reforming process. Angew Chem Int Ed 42:5196–5198
59. Zhu XF, Li QM, Cong Y, Yang WS (2008) Syngas generation in a membrane reactor with a highly stable ceramic composite membrane. Catal Commun 10:309–312
60. Tsipis EV, Patrakeev MV, Kharton VV, Yaremchenko AA, Mather GC, Shaula AL, Leonidov IA, Kozhevnikov VL, Frade JR (2005) Transport properties and thermal expansion of Ti-substituted $La_{1-x}Sr_xFeO_{3-\delta}$ ($x = 0.5$–0.7). Solid State Sci 7:355–365
61. Patrakeev MV, Bahteeva JA, Mitberg EB, Leonidov IA, Kozhevnikov VL, Poeppelmeier KR (2003) Electron/hole and ion transport in $La_{1-x}Sr_xFeO_{3-\delta}$. J Solid State Chem 172:219–231
62. Patrakeeva MV, Leonidov IA, Kozhevnikov VL, Kharton VV (2004) Ion–electron transport in strontium ferrites: relationships with structural features and stability. Solid State Sci 6:907–913
63. Kharton VV, Yaremchenko AA, Shaula AL, Viskup AP, Marques FMB, Frade JR, Naumovich EN, Casanova JR, Marozau IP (2004) Oxygen permeability and thermal expansion of ferrite-based mixed conducting ceramics. Defect Diffus Forum 226–228:141–159
64. Kozhevnikov VL, Leonidov IA, Bahteeva JA, Patrakeev MV, Mitberg EB, Poeppelmeier KR (2004) Disordering and mixed conductivity in the solid solution $LaSr_2Fe_{3-y}Cr_yO_{8+\delta}$. Chem Mater 16:5014–5020
65. Fowler DE, Haag JM, Boland C, Bierschenk DM, Barnett SA, Poeppelmeier KR (2014) Stable, low polarization resistance solid oxide fuel cell anodes: $La_{1-x}Sr_xCr_{1-x}Fe_xO_{3-\delta}$ ($x=$ 0.2–0.67). Chem Mater 26:3113–3120

66. Haag JM, Barnett SA, Richardson JW, Poeppelmeier KR (2010) Structural and chemical evolution of the SOFC anode $La_{0.30}Sr_{0.70}Fe_{0.70}Cr_{0.30}O_{3-\delta}$ upon reduction and oxidation: an in situ neutron diffraction study. Chem Mater 22:3283–3289
67. Yoo J, Kim S, Choi H, Rhim Y, Lim J, Lee S, Jacobson AJ (2011) Measurement of electrical conductivity of $La_{0.2}Sr_{0.8}Cr_{0.2}Fe_{0.8}O_{3-\delta}$ using gas-tight electrochemical cells. J Electroceram 26:56–62
68. Cai LL, Li WP, Cao ZW, Zhu XF, Yang WS (2016) Improving oxygen permeation of MIEC membrane reactor by enhancing the electronic conductivity under intermediate-low oxygen partial pressures. J Membr Sci 520: 607–615
69. Wu ZT, Wang B, Li K (2010) A novel dual-layer ceramic hollow fibre membrane reactor for methane conversion. J Membr Sci 352:63–70
70. Zhu XF, Yang WS (2008) Composite membrane based on ionic conductor and mixed conductor for oxygen permeation. AIChE J 54:665–672
71. Zhu XF, Li QM, He YF, Cong Y, Yang WS (2010) Oxygen permeation and partial oxidation of methane in dual-phase membrane reactors. J Membr Sci 360:454–460
72. Liu JJ, Zhang SQ, Wang WD, Gao JF, Liu W, Chen CS (2012) Partial oxidation of methane in a $Zr_{0.84}Y_{0.16}O_{1.92}$-$La_{0.8}Sr_{0.2}Cr_{0.5}Fe_{0.5}O_{3-\delta}$ hollow fiber membrane reactor targeting solid oxide fuel cell applications. J Power Sources 217:287–290
73. Liu JJ, Liu T, Wang WD, Gao JF, Chen CS (2012) $Zr_{0.84}Y_{0.16}O_{1.92}$-$La_{0.8}Sr_{0.2}Cr_{0.5}Fe_{0.5}O_{3-\delta}$ dual-phase composite hollow fiber membrane targeting chemical reactor applications. J Membr Sci 389:435–440
74. Luo HX, Jiang HQ, Klande T, Cao ZW, Liang FY, Wang HH, Caro J (2012) Novel cobalt-free, noble metal-free oxygen-permeable $40Pr_{0.6}Sr_{0.4}FeO_{3-\delta}$-$60Ce_{0.9}Pr_{0.1}O_{2-\delta}$ dual-phase membrane. Chem Mater 24:2148–2154
75. Ruiz-Trejo E, Boldrin P, Medley-Hallam JL, Darr J, Atkinson A, Brandon NP (2015) Partial oxidation of methane using silver/gadolinia-doped ceria composite membranes. Chem Eng Sci 127:269–275
76. Tian TF, Wang WD, Zhan MC, Chen CS (2010) Catalytic partial oxidation of methane over $SrTiO_3$ with oxygen-permeable membrane reactor. Catal Commun 11:624–628
77. Zhu XF, Liu Y, Cong Y, Yang WS (2013) $Ce_{0.85}Sm_{0.15}O_{1.925}$– $Sm_{0.6}Sr_{0.4}Al_{0.3}Fe_{0.7}O_3$ dual-phase membranes: one-pot synthesis and stability in a CO_2 atmosphere. Solid State Ionics 253:57–63
78. Wei YY, Liao Q, Li Z, Wang HH, Feldhoff A, Caro J (2014) Partial oxidation of methane in hollow-fiber membrane reactors based on alkaline-earth metal-free CO_2-tolerant oxide. AIChE J 60:3587–3595
79. Tang J, Wei YY, ZhouLY LZ, Wang HH (2012) Oxygen permeation through a CO2-tolerant mixed conducting oxide $(Pr_{0.9}La_{0.1})_2(Ni_{0.74}Cu_{0.21}Ga_{0.05})O_{4+\delta}$. AIChE J 58:2473–2478
80. Liao Q, Chen Y, Wei YY, Zhou LY, Wang HH (2014) Performance of U-shaped $BaCo_{0.7}Fe_{0.2}Ta_{0.1}O_{3-\delta}$ hollow-fiber membranes reactor with high oxygen permeation for methane conversion. Chem Eng J 237:146–152
81. Takahashi Y, Kawahara A, Suzuki T, Hirano M, Shin W (2010) Perovskite membrane of $La_{1-x}Sr_xTi_{1-y}Fe_yO_{3-\delta}$ for partial oxidation of methane to syngas. Solid State Ionics 181:300–305
82. Bouwmeester HJM (2003) Dense ceramic membranes for methane conversion. Catal Today 82:141–150
83. Hendriksen PV, Larsen PH, Mogensen M, Poulsen FW, Wiik K (2000) Prospects and problems of dense oxygen permeable membranes. Catal Today 56:283–295
84. Zhu WL, Han W, Xiong GX, Yang WS (2005) Mixed reforming of heptane to syngas in the $Ba_{0.5}Sr_{0.5}Co_{0.8}Fe_{0.2}O_3$ membrane reactor. Catal Today 104:149–153
85. Zhu WL, Han W, Xiong GX, Yang WS (2006) Mixed reforming of simulated gasoline to hydrogen in a BSCF membrane reactor. Catal Today 118:39–43
86. Yang ZB, Ding WZ, Zhang YY, Lu XG, Zhang YW, Shen PJ (2010) Catalytic partial oxidation of coke oven gas to syngas in an oxygen permeation membrane reactor combined with NiO/MgO catalyst. Int J Hydrog Energy 35:6239–6247

87. Keller GE, Bhasin MM (1982) Synthesis of ethylene via oxidative coupling of methane: I. Determination of active catalysts. J Catal 73:9–19
88. Zeng Y, Lin YS (1997) Oxidative coupling of methane on oxygen-semipermeable yttria-doped bismuth oxide ceramics in a reducing atmosphere. Ind Eng Chem Res 36:277–283
89. Akin FT, Lin YS (2002) Controlled oxidative coupling of methane by ionic conducting ceramic membrane. Catal Lett 78:239–242
90. Wang HH, Cong Y, Yang WS (2005) Oxidative coupling of methane in $Ba_{0.5}Sr_{0.5}Co_{0.8}Fe_{0.2}O_{3-\delta}$ tubular membrane reactors. Catal Today 104:160–167
91. Olivier L, Haag S, Mirodatos C, Veen ACV (2009) Oxidative coupling of methane using catalyst modified dense perovskitemembrane reactors. Catal Today 142:34–41
92. Tan XY, Pang ZB, Gu Z, Liu SM (2007) Catalytic perovskite hollow fibre membrane reactors for methane oxidative coupling. J Membr Sci 302:109–114
93. Wang HH, Cong Y, Yang WS (2002) High selectivity of oxidative dehydrogenation of ethane to ethylene in an oxygen permeable membrane reactor. Chem Commun 14:1468–1469
94. Rebeilleau-Dassonneville M, Rosini S, Veen VAC, Farrusseng D, Mirodatos C (2005) Oxidative activation of ethane on catalytic modified dense ionic oxygen conducting membranes. Catal Today 104:131–137
95. Pérez-Ramírez J, Vigeland B (2005) Perovskite membranes in ammonia oxidation: towards process intensification in nitric acid manufacture. Angew Chem Int Ed 44:1112–1115
96. Cao ZW, Jiang HQ, Luo HX, Baumann S, Meulenberg WA, Assmann J, Mleczko L, Liu Y, Caro J (2013) Natural gas to fuels and chemicals: improved methane aromatization in an oxygen-permeable membrane reactor. Angew Chem Int Ed 52:13794–13797
97. Balachandran U, Lee TH, Wang S, Dorris SE (2004) Use of mixed conducting membranes to produce hydrogen by water dissociation. Int J Hydrog Energy 29:291–296
98. Naito H, Arashi H (1995) Hydrogen production from direct water splitting at high temperature using a ZrO_2-TiO_2-Y_2O_3 membrane. Solid State Ionics 79:366–370
99. Park CY, Lee TH, Dorris SE, Balachandran U (2008) $La_{0.7}Sr_{0.3}Cu_{0.2}Fe_{0.8}O_{3-\delta}$ as oxygen transport membrane for producing hydrogen via water splitting. ECS Trans 13:393–403
100. Park CY, Lee TH, Dorris SE, Balachandran U (2010) Hydrogen production from fossil and renewable sources using an oxygen transport membrane. Int J Hydrog Energy 35:4103–4110
101. Zhu N, Dong XL, Liu ZK, Zhang GR, Jin WQ, Xu NP (2012) Toward highly-effective and sustainable hydrogen production: bio-ethanol oxidative steam reforming coupled with water splitting in a thin tubular membrane reactor. Chem Commun 48:7137–7139
102. Park CY, Lee TH, Dorris SE, Balachandran U (2013) A cobalt-free oxygen transport membrane, $BaFe_{0.9}Zr_{0.1}O_{3-\delta}$, and its application for producing hydrogen. Int J Hydrog Energy 38:6450–6459
103. Balachandran U, Dusek JT, Mieville RL, Poeppel RB, Kleefisch MS, Pei S, Kobylinski TP, Udovich CA, Bose AC (1995) Dense ceramic membranes for partial oxidation of methane to syngas. Appl Catal A Gen 133:19–29
104. Wang HB, Gopalan S, Pal UB (2011) Hydrogen generation and separation using $Gd_{0.2}Ce_{0.8}O_{1.9-\delta}$-$Gd_{0.08}Sr_{0.88}Ti_{0.95}Al_{0.05}O_{3\pm\delta}$ mixed ionic and electronic conducting membranes. Electrochim Acta 56:6989–6996
105. Park CY, Lee TH, Dorris SE, Lu Y, Balachandran U (2011) Oxygen permeation and coal-gas-assisted hydrogen production using oxygen transport membranes. Int J Hydrog Energy 36:9345–9354
106. Franca RV, Thursfield A, Metcalfe IS (2012) $La_{0.6}Sr_{0.4}Co_{0.2}Fe_{0.8}O_{3-\delta}$ microtubular membranes for hydrogen production from water splitting. J Membr Sci 389:173–181
107. Michalsky R, Parman BJ, Amanor-Boadu V, Pfromm PH (2012) Solar thermochemical production of ammonia from water, air and sunlight: thermodynamic and economic analyses. Energy 42:251–260
108. Li WP, Zhu XF, Chen SG, Yang WS (2016) Integration of nine steps into one membrane reactor to produce synthesis gases for ammonia and liquid fuel. Angew Chem Int Ed 55:8566–8570

Chapter 10
Progress on the Commercialization of MIEC Membrane Technology

Abstract Except for pure oxygen production, MIEC membranes can be used in membrane reactors for natural gas conversion. However, pure oxygen production is the process with the greatest potential for commercialization of the MIEC membrane technology in the near future. Air Products & Chemicals Inc. (APCI) is the pioneer on the commercialization of MIEC membrane technology for oxygen production, leading a team that includes several companies, institutes, and universities. In this chapter, the great progresses achieved by the APCI team are introduced, including the development history, membrane module design, membrane sealing, and module performance. Additionally, the progresses achieved by Aachen University are also briefly introduced.

Keywords Commercialization of MIEC membrane • Membrane modules • Module sealing • Module performance • Pure oxygen production

10.1 Air Separation for Pure Oxygen Production

Pure oxygen separated from air is an important raw material for many industrial processes, such as steel, glass, alumina, and ammonia production, for which oxygen plants are usually built with high investment costs based on the cryogenic distillation technology, as shown in Fig. 10.1. Cryogenics is currently the most developed technology for large-scale oxygen production. It comprises air compressing, cooling, expanding, and further cooling until the air is liquefied, and then the liquid air is separated in a distillation tower with a column tray number larger than 100 based on the difference in boiling point between oxygen and nitrogen. After more than 100 years' development, the overall thermodynamic efficiency of the modern-day cryogenic air separation unit has reached its theoretical limit; few technical breakthroughs could lead to a significant reduction of the cost of oxygen. Its production capability in a typical cryogenic plant is 2000 t per day (TPD). High-purity oxygen >99 % is produced accompanied with the production of nitrogen and argon; however, the energy consumption increases sharply if the purity is enhanced from 97 to 99 %, as shown in Fig. 10.2 [1]. Other frequently mentioned technologies for oxygen production are pressure swing adsorption (PSA) and vacuum swing adsorption (VSA) based on the adsorption difference between oxygen and nitrogen

X. Zhu, W. Yang, *Mixed Conducting Ceramic Membranes*, Green Chemistry and Sustainable Technology, DOI 10.1007/978-3-662-53534-9_10

Fig. 10.1 A typical air separation unit based on the cryogenic distillation technology

Fig. 10.2 Specific energy consumption (E) of a cryogenic air distillation process with respect to oxygen purity (y) (Reproduced from Belaissaoui et al. [1] with permission of Elsevier)

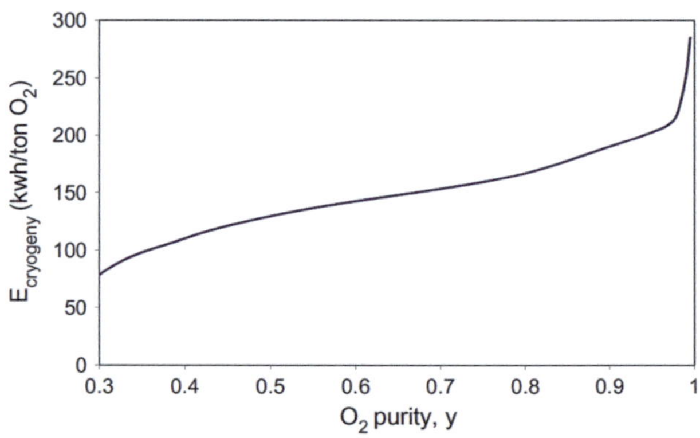

on adsorbents. Argon has an adsorption property similar to oxygen molecules; thus the oxygen purity is typically in the range of 90–95 %. Small- to medium-scale plants with a production capability lower than 100 TPD were built since PSA and VSA need less equipment investment as compared to the same scale cryogenic oxygen separation plant. The adsorption separation technologies are adequate for applications in the fields in which high-purity oxygen is not needed, such as oxygen-rich combustion, wastewater treatment, and paper manufacturing. However, PSA and VSA technologies are still considered as too energy intensive to be accepted for oxygen production in large scale. The membrane separation technique has attracted growing interest during the last decades as an alternative approach. Polymeric oxygen separation membranes, which can produce oxygen with a purity of 25–40 % in small scale (<25TPD), are reported to be potentially attractive for oxygen-rich air production. However, owing to their limited oxygen/nitrogen separation factor and low permeability, the commercially available polymeric membranes cannot economically compete in the production of oxygen with a high purity in large scale with cryogenics, PSA, and VSA. All of the above referred technologies have drawbacks that prevented cost-effective high-purity oxygen production and resulted in extremely high cost for many potential applications, such as the oxyfuel process for power generation and CO_2 capture.

MIEC membranes are regarded as a revolutionary new air separation technology to efficiently produce oxygen with purity greater than 99 % due to their infinite separation factor toward oxygen and high permeability. This technology has the potential to lower the cost of oxygen production by +30 % as compared to the cryogenic distillation technology. The scale of MIEC membranes can be small, medium, and even large depending on the application requirements. Additionally, MIEC membranes can be combined with other industrial processes acting as membrane reactors for process intensification and energy savings. Recently, a research and development alliance led by Air Products & Chemicals Inc. (APCI) has achieved great progress in the commercialization of the MIEC membrane technology for pure oxygen production. The following sections focus on the introduction of the APCI technology, while progresses achieved in other institutes are also briefly introduced.

10.2 APCI's Technology

10.2.1 Brief Overview of the Development of MIEC Membrane Technology

In 1989, APCI started to develop the MIEC membrane technology for the purpose of commercialization of the low-cost oxygen separation process [2]. It was only 4 years after the first report by Teraoka and coworkers on MIEC membranes for oxygen separation, and at that time, only two papers had been published [3]. APCI evaluated the economy of the MIEC membrane separation process, which involved compressing the feed air, pulling a vacuum on the oxygen product, or a combination of the two, and concluded that the MIEC membrane technology could result in a cost saving of at least 30 % over current PSA, VSA, or cryogenic distillation technologies. Therefore, it was believed that this new oxygen separation technology could arouse a revolution in the oxygen production industry, allowing additional applications of pure oxygen and thus boosting the oxygen market if high-purity oxygen could be produced at lower costs. In the initial stage (1989–1993), APCI sponsored research activities and achieved significant technological advances in powder preparation of membrane compositions, dense membrane fabrication, and thin-film-membrane deposition [2]. Furthermore, the deep understanding of the oxygen transport mechanisms through MIEC membranes facilitated APCI to design proprietary membrane structures and modules. Then APCI gained financial support from the Advanced Technology Program for a 3-year project to advance the MIEC membrane technology. During this stage (1994–1997), APCI contracted Ceramatec, Inc., to aid the development of a laboratory prototype unit. Although a functional laboratory prototype unit was not successfully built up at this stage, a critical technique for the unit, i.e., ceramic–steel seals, was developed. Additionally, many approaches developed for the design, construction, and testing of a laboratory prototype unit during this project encouraged their additional R&D support that continues today. The progress in the early stage persuaded the Department of Energy (DOE) to fund the R&D of MIEC membrane technology and the overall goal of the project to create an oxygen separation system that could achieve a 30 % cost reduction compared to the cryogenic distillation-based technology. APCI assembled a team including Ceramatec, Concepts/NREC, Eltron Research, McDermott Technologies (SOFCo EFS Holdings), Pennsylvania State University, Siemens Westinghouse, Texaco Gasification (GE Energy Gasification), and the University of Pennsylvania to advance the technology and the knowledge gained in the project supported by the Advanced Technology Program. The team drew up a multiphase plan on the commercialization of the MIEC membrane technology for pure oxygen production:

- Phase 1: Technical feasibility (0.1 TPD O_2)
- Phase 2: Prototype testing (1–5 TPD O_2)
- Phase 3: Intermediate-scale testing (100 TPD O_2)
- Phase 4: Reaction-driven membranes
- Phase 5: Ceramic module fabrication to supply 2000-TPD O_2 facility
- Planning next phase: Energy scale development (2000 TPD O_2)

Phase I was completed in 2001. During this phase, the activities focused on the optimization of MIEC membrane materials and the design, construction, and operation of membrane modules with oxygen production capability of 0.1 TPD. This material is represented by the formula $Ln_xA_{1-x}Co_{1-y-z}Fe_yCu_zO_{3-\delta}$, where Ln and A are selected from the lanthanide block and the alkaline earth elements, respectively. It was thought that these materials would provide a favorable balance of oxygen flux, resistance to degradation at high oxygen partial pressures, favorable sintering properties, and thermal expansion matching with other separation parts. APCI focused on the development of the MIEC membrane modules. Throughout this stage, planar ceramic wafers were designed and fabricated as building blocks for the low-cost high-performance MIEC membrane modules [4, 5]. The success of the membrane modules allowed APCI and its partners to build tonnage quantity oxygen plants and optimize the process to decrease the energy consumption. The MIEC membrane technology was combined with advanced power generation systems, as shown in Fig. 10.3 [6–8]. The inputs are high-pressure air, heat, and fuel, while the outputs are pure oxygen, power, and steam. Phase II was completed in 2006. During this phase, the activities focused on a 5-TPD subscale engineering prototype, as shown in Fig. 10.4 [6–8]. Six modules are installed in a pressure

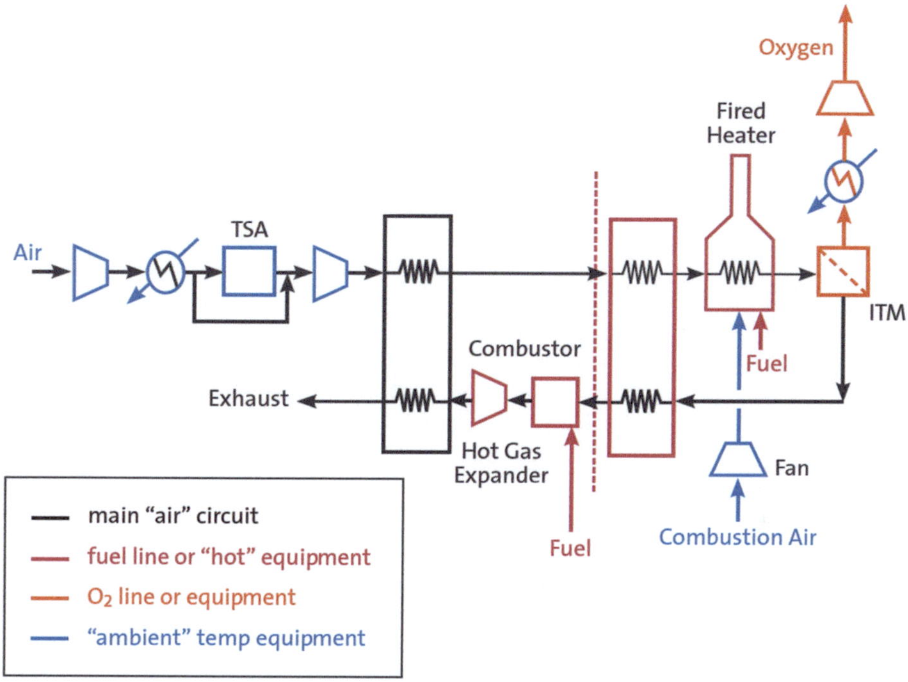

Fig. 10.3 MIEC membrane integrated with a gas turbine-based power cycle (Reproduced from Anderson et al. [8] with permission of Elsevier)

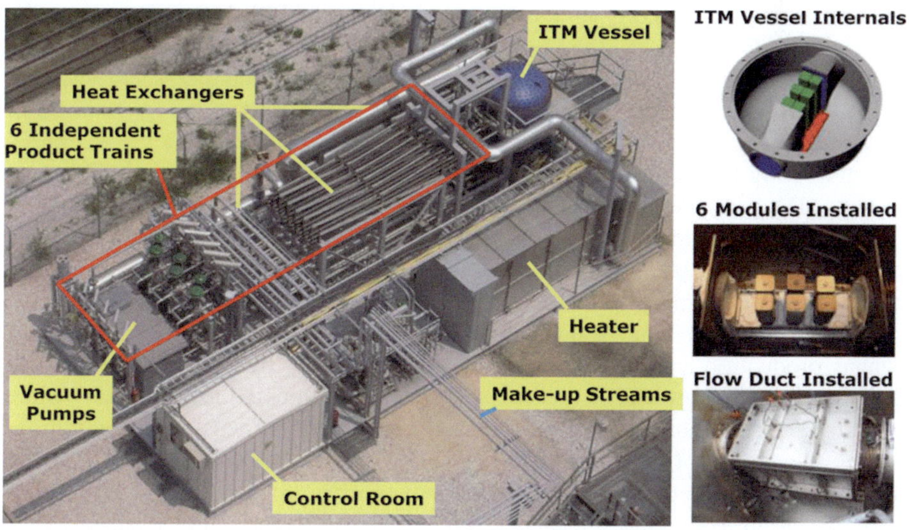

Fig. 10.4 Subscale engineering prototype capable of testing up to six 144-wafer ITM oxygen modules (Reproduced from Bose et al. [7] with permission of Springer)

Fig. 10.5 Intermediate-scale test unit capable of producing up to 100 TPD of oxygen (Reproduced from Anderson et al. [8] with permission of Elsevier)

vessel, and every module has an oxygen production capability of half tons. Six independent product trains were constructed for each module. The 5-TPD subscale engineering prototype steadily runs over 1000 days with pressure and temperature cycling. An oxygen purity of >99 % was acquired with a stable flux at target values. The success of the prototype stimulated APCI and its partners to build an engineering test facility with an oxygen production capability of 100–150 TPD and test the full size MIEC modules in the pressure vessel under a variety of operating conditions, which were the main tasks of phase III. The intermediate-scale test unit (ISTU) was in its start-up phase for operation in 2014, as shown in Fig. 10.5 [6]. In phase V, the main tasks will focus on designing, building, and operating a ceramic manufacturing facility to supply a 2000-TPD MIEC membrane test unit. The goal in phase IV is to develop MIEC membrane reactors for syngas production. APCI regards this technology as a revolutionary technology, which integrates oxygen production and syngas generation in a single unit. This process shows reduced capital cost in the syngas generation process and overall syngas/H_2/CO_2 production process by 30 % and 20 %, respectively. Additionally, the membrane technology for syngas production exhibits improved thermal efficiency by 3 % compared to the traditional technology. APCI and its partners have completed the laboratory-scale materials and membrane development and are currently focusing on building a process development unit with a syngas or H_2 production capability of 27 Nm^3/h. Their next step is to develop a subscale engineering prototype with a capability of 280 Nm^3/h in 2016 and an early commercial-scale unit with a capability of 10,000 Nm^3/h in 2018 [6].

10.2.2 *MIEC Membrane Module Design and Fabrication*

As mentioned in Chap. 1, the frequently reported geometric configurations of MIEC membranes are planar, tubular, and hollow fibroid. APCI and its partners evaluated both planar and tubular membranes for oxygen separation and chose the planar configuration to construct their MIEC membrane modules. The planar configuration has high area/volume ratio compared with the tubular shape, which can be as high as 400 m^2/m^3 for plates but lower than 50 m^2/m^3 for tubes. The permeation flux of an MIEC membrane is inversely proportional to the thickness of the dense layer if bulk diffusion dominates. Therefore, it is indispensable to develop techniques for the preparation of MIEC membranes that have a thin dense layer for selective oxygen permeation and a thick porous layer to support the dense layer. The planar design is amenable to standard tape-casting techniques for ceramic plate production in large scale, especially for ceramic plates with multilayers. For the tubular configuration, a large-scale continuous production is available using the extrusion method; however the time-consuming characteristic and low repeatability are observed if the dip-coating method is used to prepare a tube membrane with a thin dense layer.

Figure 10.6 shows the planar membrane structure invented by the APCI researchers [4–7]. The planar membrane has a thin dense layer with a thickness of 30–100 μm and a porous layer with a thickness of ~500 μm. Two membranes are assembled together with the support layer of one membrane facing to the support layer of another membrane and the dense slotted backbones located in the middle. The porous support layer and the dense slotted backbones have the same chemical composition of the dense layer to avoid the cracks induced by the difference in thermal expansion coefficients of different layers. The composite plate is called membrane wafer. The four sides of a membrane wafer are sealed with the same

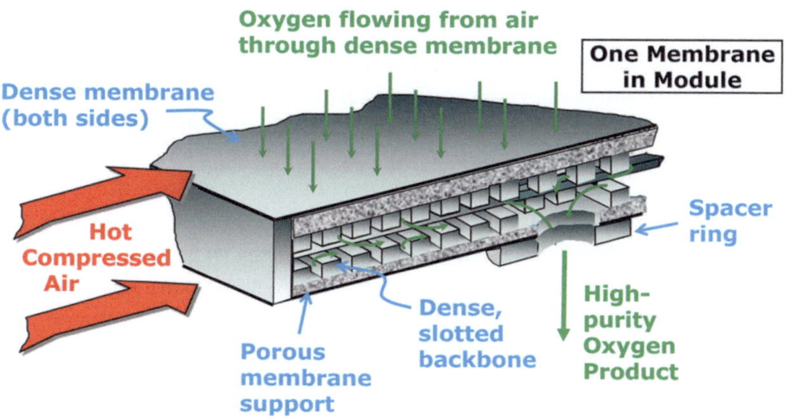

Fig. 10.6 MIEC membrane module architecture—planar wafer stack design (Reproduced from Bose et al. [7] with permission of Springer)

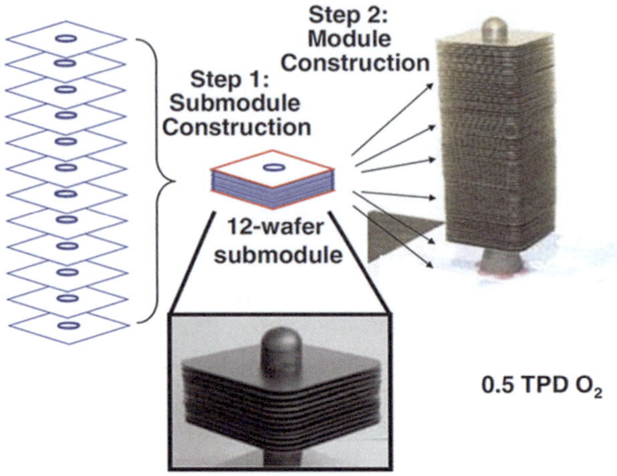

Fig. 10.7 Construction steps of the commercial modules (Reproduced from Bose et al. [7] with permission of Springer)

material as the membrane by using a patented technique, which will be introduced in the following section. There is a hole located at the center of each wafer, as shown in Fig. 10.7. The wafers are joined together to form a 12-wafer submodule with a ceramic spacer ring located at the center to keep a consistent gap between adjacent wafers. High-pressure air is fed to the outside module, and the permeated oxygen is withdrawn under vacuum via the tube formed by the connected wafers and ceramic spacer rings. The 12-wafer submodule has been successfully tested for oxygen permeation and has an oxygen production capability of 0.1 TPD. Larger commercial-scale modules with production capabilities of 0.5 or 1.0 TPD are fabricated by jointing the submodules. A conical tube at the bottom and an end cap at the top are fitted to the commercial-scale module. The conical tube has two functions: one is to provide structural stability to the module, and the other is to facilitate the connection between the module and an internally insulated metal pipe. In view of the commercial concept of MIEC membranes for oxygen production, a vessel is equipped with an array of multi-wafer modules which are arranged in a common flow duct and connected through a series of manifolds to an oxygen header below, as shown in Figs. 10.8 and 10.9 [6–8]. Each module in the vessel has an oxygen production capability of approximately 1 TPD. This type of design fitted the MIEC membrane separation unit with an oxygen production capability of 100–1000s TPD. The commissioned 100-TPD intermediate-scale test unit that is under construction adopts this design, as shown in Fig. 10.5.

A standard ceramic fabrication technology is used to prepare the MIEC membranes. The traditional solid-state reaction method was used to prepare the ceramic powders. After the powder is milled to an appropriate particle size, it is mixed with solvents (ethanol and toluene), dispersants, binders, and plasticizers to produce a stable slurry with a well-defined viscosity. The slurry is then casted onto a PET carrier using the tape-casting technique (see Chap. 4) to form green ceramic tapes after drying at a proper temperature. Green ceramic tapes with various thicknesses are prepared by adjusting the gap of the doctor blade. The green ceramic tape is then featured by laser

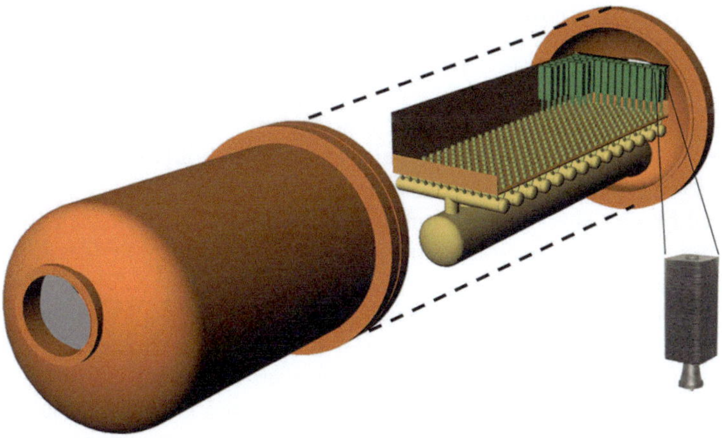

Fig. 10.8 Commercial-scale vessel concept (Reproduced from Bose et al. [7] with permission of Springer)

Fig. 10.9 MIEC membrane modules installed in the flow duct of the intermediate-scale test unit (Reproduced from Anderson et al. [8] with permission of Elsevier)

cutters or mechanical punches into the dimensions and architecture required for the individual layers of the membrane components. The correct combination of layers is laminated together and then sintered. Then, the sintered wafers are jointed with ceramic spacer rings to form a module during another thermal treatment. Currently, the ceramic fabrication scale-up is in commissioning in Tooele near Salt Lake City for the 2000-TPD MIEC membrane test unit.

10.2.3 Module Sealing

In order to prepare the MIEC membrane module shown in Fig. 10.6, the key technology required is the ceramic–ceramic sealing. The four sides of a wafer and the space between a wafer and a ceramic spacer ring should be sealed by a proper sealant with similar mechanical, chemical, and expansion characteristics as the MIEC membrane. The sealing should be robust so to remain leak-free at high temperature and in a high-pressure atmosphere. There are three different methods to seal the membrane module, i.e., glass or glass–ceramic sealing, metallic braze sealing, and all-ceramic sealing. The first two types are well known and widely used in the ceramic industry, for instance, in solid oxide fuel cells and sensors. Nowadays, most of the research works focus on the glass, glass–ceramic, and braze sealing, and much experience has been acquired from industrial processes. Therefore, the first idea for sealing the MIEC membrane modules would be using the first two types of sealing. However, MIEC membranes have high thermal expansion coefficients at high temperatures (typically larger than $20 \times 10^{-6}\ K^{-1}$) and special expansion curves due to the chemical expansion under a certain temperature and oxygen partial pressure. Therefore, it is very difficult to develop a sealant that has the same thermal expansion characteristics as those of MIEC membranes.

Glass and glass–ceramics based on silica, boron oxides, and phosphates are the most widely used sealing materials in a variety of ceramic assemblies, as a good sealing is easily achieved at the softening temperature of glass. Additionally, the stress induced by thermal or chemical expansion can be released due to the flexibility of glass at the temperature above its softening point. However, chemical reactions between glass and MIEC membranes are inevitable at high temperature, and the produced new compounds most likely have different thermal and chemical properties from those of the MIEC membrane material and glass. Although the volatilization of silicon, boron, and phosphorus species at high temperature is extremely slow, they could nonetheless be deposited on the MIEC membrane surfaces and lead to the degradation of the oxygen permeation flux during a long operating period. The braze sealing materials which can tolerate high-temperature air are precious metals, such as Pt, Au, and Ag. However, the sealing containing Pt or Au is too expensive and thus would impair the overall economics of the MIEC membrane technology, while the less expensive Ag cannot sustain a long-period sealing due to its volatilization at elevated temperature.

APCI invented an all-ceramic sealing technology to seal their MIEC membranes with a number of benefits as compared with the glass or braze sealing [7]. It was called "transient liquid phase (TLP) method." The ceramic sealing has the same or similar chemical composition as the MIEC membrane. To seal parts, a TLP material is applied to the sealing interface, and then the assembly is heated to a temperature that allows the formation of a liquid phase. The liquid phase fills the gaps at the interface and diffuses into the parent materials being joined due to capillary forces. Subsequently, the liquid reacts with the surrounding materials and produces a new compound that has a higher melting point than the liquid phase. Thus, after the completion of the reaction, the interface and pores are sealed by the newly produced ceramic oxide, which has similar or compatible chemical composition to the MIEC membrane. For example, the $(La_{0.75}Ca_{0.25})_{1.01}FeO_3$ perovskite membrane can be sealed by itself through the TLP method. The sealing material to seal the membrane consists of two parts: one is 20 wt% CaO–80 wt% Fe_2O_3, which has a melting point of approximately 1205 °C, and the other is 84.2 wt% La_2O_3–15.8 wt% Fe_2O_3, which has a melting point higher than the sintering temperature of the $(La_{0.75}Ca_{0.25})_{1.01}FeO_3$ perovskite membrane. The weight ratio of the two parts is 1:2.07. After the sealing material is applied to the interface to be sealed, the assembly is heated to 1250 °C to melt the CaO–Fe_2O_3 composite. Then, the assembly is maintained at this temperature for a certain period in order to allow the reaction of the liquid phase CaO–Fe_2O_3 compound with the solid phase La_2O_3–Fe_2O_3 and the interfaces to be sealed. Finally, a solid phase with chemical composition of $(La_{0.75}Ca_{0.25})_{1.01}FeO_3$ is formed. The TLP method has many advantages as compared with the other methods using glass, glass–ceramics, and brazes to seal the MIEC membranes. The sealing material has the same or similar chemical composition as the MIEC membrane, thus the sealing has the same chemical and thermal properties as the membrane material, and it is not necessary to be concerned with reactivity or chemical incompatibility between the sealing material and the membrane. Additionally, the membrane can be operated at a higher temperature than the sealing temperature, while that is impossible for the glass, glass–ceramic, and braze sealing. This invention is one of the key enabling technologies for implementing planar ceramic membranes.

10.2.4 Module Performance

APCI tested a membrane module containing 12 wafers in a technology demonstration unit for 15,000 h. This module was operated at feed flow rates, and process conditions matched those of the target commercial operation. High-pressure air up to 1.7 MPa with a flow rate of 60 Nm^3/h was fed to the demonstration unit, and a vacuum pump was installed to withdraw the oxygen permeated through the membrane module. Temperature and pressure cycles, as well as changes in pressure, were deliberately and not deliberately carried out to evaluate the module performance during the long-term operation. Figure 10.10 shows the time-dependent performance at constant temperature and pressure [8]. The purity of oxygen remained >99 % during the 15,000 h on-stream, but the oxygen permeation flux decreased by about 10 % after 10,000 h running. The degradation in permeation flux was attributed to the macroscopic wafer displacement, as shown in Fig. 10.11 a and b, which was induced by the change of microstructure of the wafers due to the ceramic creep effect under high pressure [8]. The result of creep can be observed in

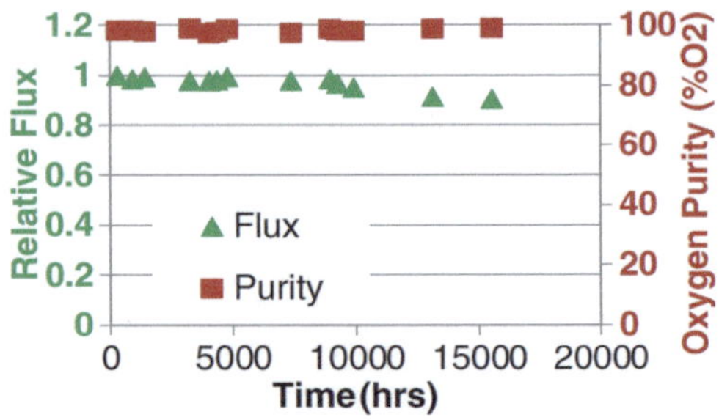

Fig. 10.10 Long-term testing of the 12-wafer module in the technology demonstration unit (Reproduced from Anderson et al. [8] with permission of Elsevier)

(a)

(b)

(c)

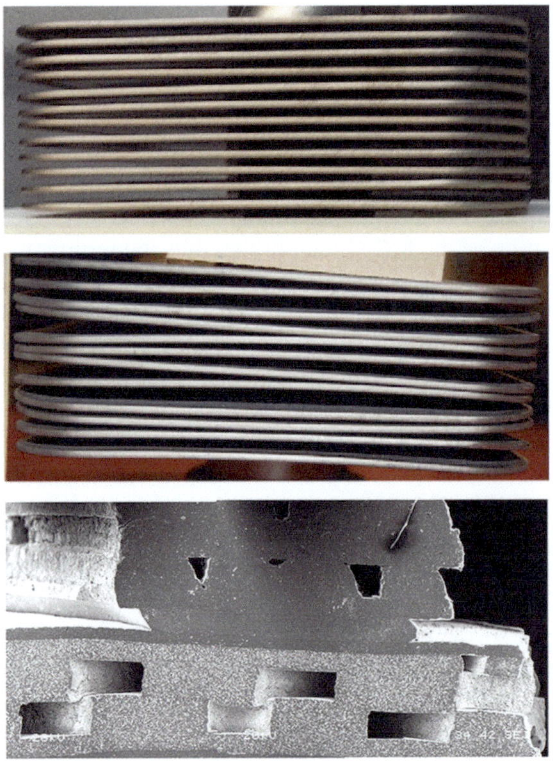

Fig. 10.11 Before testing (**a**) and after testing (**b**) pictures of the 12-wafer module tested for over 15,000 h, showing evidence of compressive creep. (**c**) Example of deformation and decreased cross-sectional area of the channels due to creep, as seen in the channels at the bottom middle of the image (Reproduced from Anderson et al. [8] with permission of Elsevier)

the SEM picture shown in Fig. 10.11c, which shows the cross-sectional area of the internal flow channels of the membrane wafers at the central manifold decreased after the long term on-stream. The deformation of the channel structure increases the concentration polarization and thus degrades the oxygen permeation.

A subscale engineering prototype was operated for over 1100 total days, which also demonstrated the target commercial flux at full size modules (0.5 TPD). High oxygen purity up to 99.9 % was achieved. The planned temperature and pressure cycles, as shown in Fig. 10.12, successfully demonstrated that the subscale

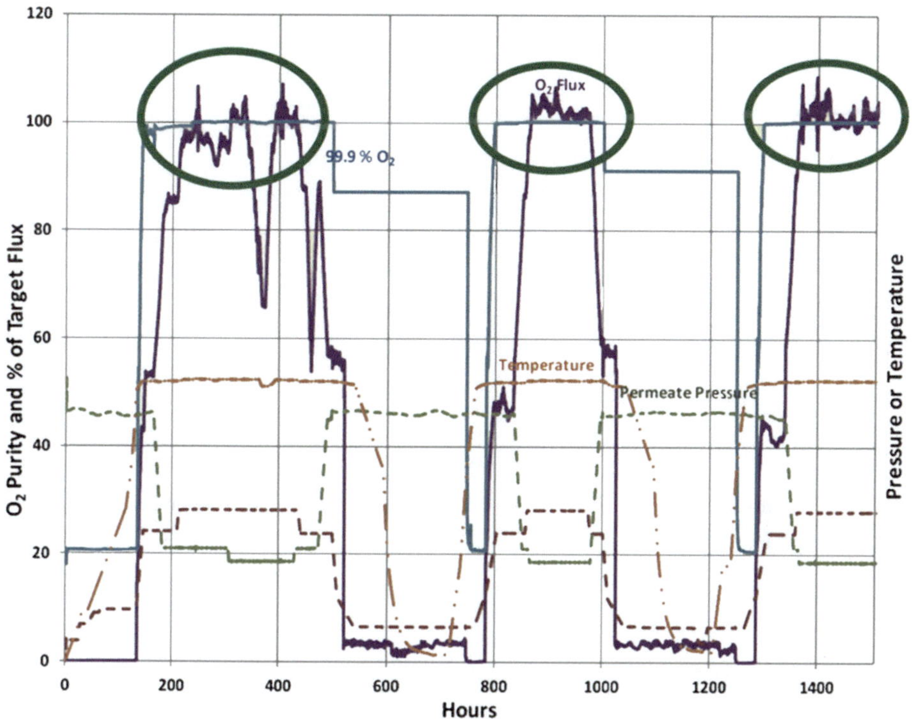

Fig. 10.12 Oxygen purity and flux of a full size MIEC membrane module through temperature and pressure cycles (Reproduced from Anderson et al. [8] with permission of Elsevier)

engineering prototype had repeatable performance during temperature and pressure cycling tests [6]. Two 1-TPD modules were tested and showed a 0.7–1-TPD oxygen capacity over 22 days on-stream [6]. An intermediate-scale test unit containing two rows of modules (28 modules) was operated for 300 h at a production capacity of 16 TPD (equal to 467 Nm^3/h). In the first row, there are five modules with the air side coated by porous layers for oxygen activation and nine modules without porous layers; in the second row, the two numbers are six and eight, respectively. Figure 10.13 shows the comparison of the fluxes of modules with and without air side porous layers and fluxes of modules in the first and second rows [8]. Modules with porous layers exhibited about 9 % higher oxygen permeation flux than those without porous layers, and the permeation flux of the first row was slightly higher than the second row because the first row extracted some of the oxygen in air before reaching the second row. The oxygen production rate of the unit and oxygen purity was steadily maintained with no indication of decay during this demonstration. It is the largest air separation unit that has been successfully run based on the MIEC membrane technology up to now.

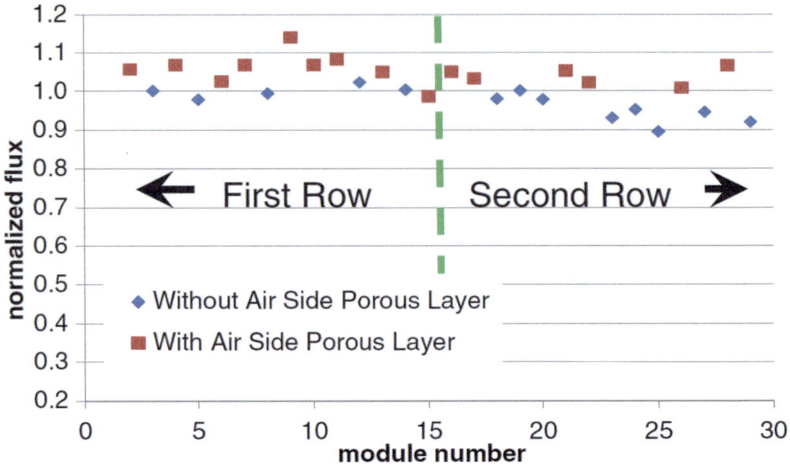

Fig. 10.13 Comparison of the fluxes of modules with and without air side coated with porous layers and fluxes of modules in the first and second rows (Reproduced from Anderson et al. [8] with permission of Elsevier)

10.3 Aachen University's Technology

The target of Aachen University in the context of the development of the MIEC membrane technology is to combine it with the oxycoal technology for CO_2 capture and power generation [9, 10]. They adopted the three-end rather than four-end operation. To fulfill the four-end operation, in which flue gas (containing CO_2, H_2O, and SO_2) is used as the sweep gas to carry the permeated oxygen to a boiler, a membrane material which can tolerate the flue gas needed to be developed. However, no MIEC membrane material with high oxygen permeability can maintain a stable oxygen permeation flux under the sweep of the flue gas. For the three-end operation, i.e., air separation for pure oxygen production, the membrane was fed with high-pressure air (15–20 bar), and the permeated oxygen was withdrawn by a vacuum pump. $Ba_{0.5}Sr_{0.5}Co_{0.8}Fe_{0.2}O_{3-\delta}$ (BSCF) first developped by Yang's group in Dalian was chosen as the MIEC membrane material due to its extremely high oxygen permeability among the perovskite-type MIEC membranes. The sintering temperature of the BSCF ceramics was optimized based on the testing results of fracture bending stress and porosity. The highest strength and lowest porosity were achieved at sintering temperatures of 1075–1100 °C for the cold-isostatic pressing samples. The one-end closed tube was chosen as the membrane geometric shape based on many potential advantages, such as high packing density, easier to join than honeycombs or hollow fibers, easy to change in case of a defect, several manufacturing alternatives, and possible proof test before installation. The BSCF green tubes were prepared by cold-isostatic pressing at 175 MPa and then sintered at the optimal temperature for 5 h. The tube is 500 mm in length, 15 mm in

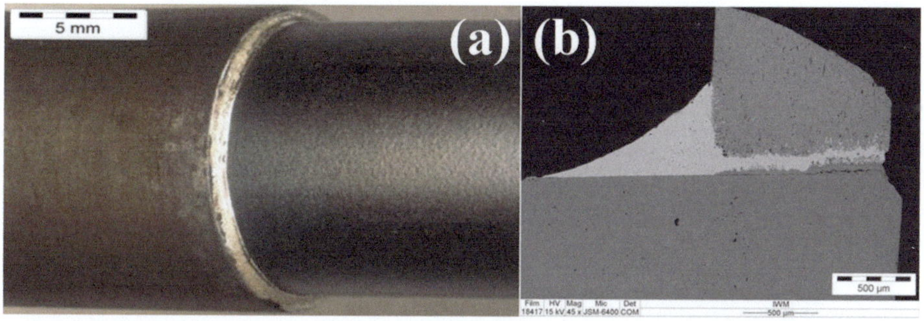

Fig. 10.14 Picture of the AgCuO reactive air brazing joint with a BSCF tube, (**a**) digital photo and (**b**) SEM picture of the cross section (Reproduced from Pfaff et al. [10] with permission of Wiley)

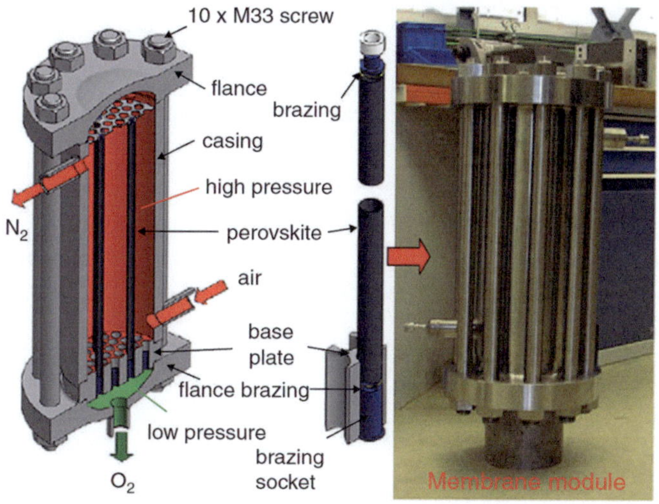

Fig. 10.15 120-kW oxycoal pilot plant developed by RWTH Aachen University; membrane module: outer heating, 1-m^2 membrane area, 42 perovskite tubes with the length of 500 mm and diameter of 15 mm, 150-kg HT steel, $T_{max} = 850$ °C, $P_{max} = 20$ bar (Reproduced from Dong and Jin [11] with permission of Elsevier)

diameter, and 0.8 mm in wall thickness. Ag doped with 4 % CuO was used to seal the gap between the BSCF tube and the X15CrNiSi25-20 alloy, as shown in Fig. 10.14. The joint stress in the ceramic was high after cooling (5 K min^{-1}) from the brazing temperature (955 °C) to room temperature, but it decreases fast from the initial 81.5 to 2.8 MPa after holding the assembly at 850 °C for 10 h. A membrane module containing 42 tubes with a total membrane area of 1 m^2 was built, as shown in Fig. 10.15. This module can supply oxygen 15 L min^{-1} or 0.031 TPD. The membranes demonstrated an oxygen permeation flux of 1.46 NmL cm^{-2} $min^{-}1$ under an oxygen pressure gradient of $P_{O_2,feed}/P_{O_2,perm.} = 5$. The building of

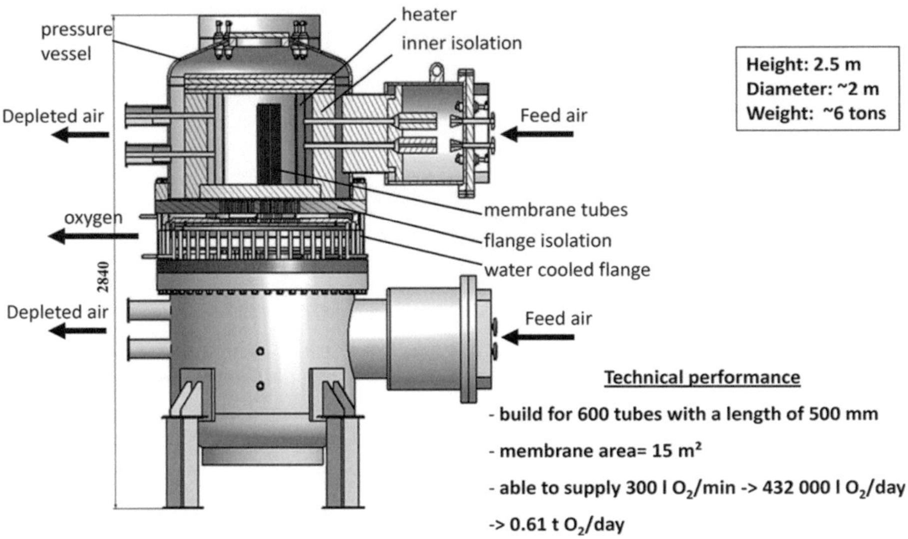

Fig. 10.16 Schematic diagram of the pilot-scale module containing 600 BSCF tube for oxygen separation (Reproduced from Pfaff et al. [10] with permission of Wiley)

a larger pilot module, as shown in Fig. 10.16, containing 600 tubes (corresponding to a membrane area of 15 m^2) started in August 2011. This pilot module could have an oxygen production capacity of 300 L min^{-1} or 0.61 TPD.

References

1. Belaissaoui B, Le Moullec Y, Hagi H, Favre E (2014) Energy efficiency of oxygen enriched air production technologies: Cryogeny vs membranes. Sep Purif Technol 125:142–150
2. Air Products and Chemicals Inc (1993) Project title: energy-efficient oxygen production using novel ion-transport membranes
3. Teraoka Y, Zhang HM, Furukawa N, Yamazoe N (1985) Oxygen permeation through perovskite-type oxides. Chem Lett 14:1743–1746
4. Armstrong PA, Bennett DL, Ted Foster EP, Stein VE (2002) Ceramic membrane development for oxygen supply to gasification applications. In: Proceedings of the gasification technologies conference, San Francisco, 27–30 Oct 2002
5. Armstrong PA, Bennett DL, Foster EP, Stein VE (2004) ITM Oxygen for gasification. In: Proceedings of the gasification technologies conference, Washington DC, 3–6 Oct 2004
6. White V, Repasky JM, Carolan MF, Armstrong PA, Stein VE, Foster EP (2014) ITM oxygen technology for gasification applications. In: Proceedings of the 12th European gasification conference, Rotterdam, 10–13 Mar 2014
7. Bose AC, Stiegel GJ, Armstrong PA, Halper BJ, Foster EP (2009) Progress in ion transport membranes for gas separation applications. In: Bose AC (ed) Inorganic membranes for energy and environmental applications. Springer, New York, pp 3–25

8. Anderson LL, Armstrong PA, Broekhuis RR, Carolan MF, Chen J, Hutcheon MD, Lewinsohn CA, Miller CF, Repasky JM, Taylor DM, Woods CM (2016) Advances in ion transport membrane technology for oxygen and syngas production. Solid State Ionics 288:331–337
9. Kaletsch A, Pfaff EM, Broeckmann C, Modifell M, Nauels N (2011) Pilot module for oxygen separation with BSCF membranes. In: Proceedings of the 2nd ICEPE, Frankfurt am Main, 20–22 Jun 2011
10. Pfaff EM, Kaletsch A, Broeckmann C (2012) Design of a mixed ionic/electronic conducting oxygen transport membrane pilot module. Chem Eng Technol 35:455–463
11. Dong XL, Jin WQ (2012) Mixed conducting ceramic membranes for high efficiency power generation with CO_2 capture. Curr Opin Chem Eng 1:163–170

MIX
Papier aus verantwortungsvollen Quellen
Paper from responsible sources
FSC® C105338

If you have any concerns about our products,
you can contact us on
ProductSafety@springernature.com

In case Publisher is established outside the EU,
the EU authorized representative is:
Springer Nature Customer Service Center GmbH
Europaplatz 3, 69115 Heidelberg, Germany

Printed by Libri Plureos GmbH
in Hamburg, Germany